College Trigonometry

Version $\lfloor \pi \rfloor$
Corrected Edition

by

Carl Stitz, Ph.D.
Lakeland Community College

Jeff Zeager, Ph.D.
Lorain County Community College

July 4, 2013

Acknowledgements

While the cover of this textbook lists only two names, the book as it stands today would simply not exist if not for the tireless work and dedication of several people. First and foremost, we wish to thank our families for their patience and support during the creative process. We would also like to thank our students - the sole inspiration for the work. Among our colleagues, we wish to thank Rich Basich, Bill Previts, and Irina Lomonosov, who not only were early adopters of the textbook, but also contributed materials to the project. Special thanks go to Katie Cimperman, Terry Dykstra, Frank LeMay, and Rich Hagen who provided valuable feedback from the classroom. Thanks also to David Stumpf, Ivana Gorgievska, Jorge Gerszonowicz, Kathryn Arocho, Heather Bubnick, and Florin Muscutariu for their unwaivering support (and sometimes defense) of the book. From outside the classroom, we wish to thank Don Anthan and Ken White, who designed the electric circuit applications used in the text, as well as Drs. Wendy Marley and Marcia Ballinger for the Lorain CCC enrollment data used in the text. The authors are also indebted to the good folks at our schools' bookstores, Gwen Sevtis (Lakeland CC) and Chris Callahan (Lorain CCC), for working with us to get printed copies to the students as inexpensively as possible. We would also like to thank Lakeland folks Jeri Dickinson, Mary Ann Blakeley, Jessica Novak, and Corrie Bergeron for their enthusiasm and promotion of the project. The administrations at both schools have also been very supportive of the project, so from Lakeland, we wish to thank Dr. Morris W. Beverage, Jr., President, Dr. Fred Law, Provost, Deans Don Anthan and Dr. Steve Oluic, and the Board of Trustees. From Lorain County Community College, we wish to thank Dr. Roy A. Church, Dr. Karen Wells, and the Board of Trustees. From the Ohio Board of Regents, we wish to thank former Chancellor Eric Fingerhut, Darlene McCoy, Associate Vice Chancellor of Affordability and Efficiency, and Kelly Bernard. From OhioLINK, we wish to thank Steve Acker, John Magill, and Stacy Brannan. We also wish to thank the good folks at WebAssign, most notably Chris Hall, COO, and Joel Hollenbeck (former VP of Sales.) Last, but certainly not least, we wish to thank all the folks who have contacted us over the interwebs, most notably Dimitri Moonen and Joel Wordsworth, who gave us great feedback, and Antonio Olivares who helped debug the source code.

10 Foundations of Trigonometry **693**

10.1 Angles and their Measure . . . 693

10.1.1 Applications of Radian Measure: Circular Motion . . . 706

10.1.2 Exercises . . . 709

10.1.3 Answers . . . 712

10.2 The Unit Circle: Cosine and Sine . . . 717

10.2.1 Beyond the Unit Circle . . . 730

10.2.2 Exercises . . . 736

10.2.3 Answers . . . 740

10.3 The Six Circular Functions and Fundamental Identities . . . 744

10.3.1 Beyond the Unit Circle . . . 752

10.3.2 Exercises . . . 759

10.3.3 Answers . . . 766

10.4 Trigonometric Identities . . . 770

10.4.1 Exercises . . . 782

10.4.2 Answers . . . 787

10.5 Graphs of the Trigonometric Functions . . . 790

10.5.1 Graphs of the Cosine and Sine Functions . . . 790

10.5.2 Graphs of the Secant and Cosecant Functions . . . 800

10.5.3 Graphs of the Tangent and Cotangent Functions . . . 804

10.5.4 Exercises . . . 809

10.5.5 Answers . . . 811

10.6 The Inverse Trigonometric Functions . . . 819

10.6.1 Inverses of Secant and Cosecant: Trigonometry Friendly Approach . . . 827

10.6.2 Inverses of Secant and Cosecant: Calculus Friendly Approach . . . 830

10.6.3 Calculators and the Inverse Circular Functions. . . . 833

10.6.4 Solving Equations Using the Inverse Trigonometric Functions. . . . 838

10.6.5 Exercises . . . 841

10.6.6 Answers . . . 849

10.7 Trigonometric Equations and Inequalities . . . 857

10.7.1 Exercises . . . 874

10.7.2 Answers . . . 877

11 Applications of Trigonometry **881**

11.1 Applications of Sinusoids . . . 881

11.1.1 Harmonic Motion . . . 885

11.1.2 Exercises . . . 891

11.1.3 Answers . . . 894

11.2 The Law of Sines . . . 896

11.2.1 Exercises . . . 904

11.2.2 Answers . . . 908

11.3 The Law of Cosines . . . 910

11.3.1 Exercises . . . 916
11.3.2 Answers . . . 918
11.4 Polar Coordinates . . . 919
11.4.1 Exercises . . . 930
11.4.2 Answers . . . 932
11.5 Graphs of Polar Equations . . . 938
11.5.1 Exercises . . . 958
11.5.2 Answers . . . 963
11.6 Hooked on Conics Again . . . 973
11.6.1 Rotation of Axes . . . 973
11.6.2 The Polar Form of Conics . . . 981
11.6.3 Exercises . . . 986
11.6.4 Answers . . . 987
11.7 Polar Form of Complex Numbers . . . 991
11.7.1 Exercises . . . 1004
11.7.2 Answers . . . 1007
11.8 Vectors . . . 1012
11.8.1 Exercises . . . 1027
11.8.2 Answers . . . 1031
11.9 The Dot Product and Projection . . . 1034
11.9.1 Exercises . . . 1043
11.9.2 Answers . . . 1045
11.10 Parametric Equations . . . 1048
11.10.1 Exercises . . . 1059
11.10.2 Answers . . . 1063

Index **1069**

Preface

Thank you for your interest in our book, but more importantly, thank you for taking the time to read the Preface. I always read the Prefaces of the textbooks which I use in my classes because I believe it is in the Preface where I begin to understand the authors - who they are, what their motivation for writing the book was, and what they hope the reader will get out of reading the text. Pedagogical issues such as content organization and how professors and students should best use a book can usually be gleaned out of its Table of Contents, but the reasons behind the choices authors make should be shared in the Preface. Also, I feel that the Preface of a textbook should demonstrate the authors' love of their discipline and passion for teaching, so that I come away believing that they really want to help students and not just make money. Thus, I thank my fellow Preface-readers again for giving me the opportunity to share with you the need and vision which guided the creation of this book and passion which both Carl and I hold for Mathematics and the teaching of it.

Carl and I are natives of Northeast Ohio. We met in graduate school at Kent State University in 1997. I finished my Ph.D in Pure Mathematics in August 1998 and started teaching at Lorain County Community College in Elyria, Ohio just two days after graduation. Carl earned his Ph.D in Pure Mathematics in August 2000 and started teaching at Lakeland Community College in Kirtland, Ohio that same month. Our schools are fairly similar in size and mission and each serves a similar population of students. The students range in age from about 16 (Ohio has a Post-Secondary Enrollment Option program which allows high school students to take college courses for free while still in high school.) to over 65. Many of the "non-traditional" students are returning to school in order to change careers. A majority of the students at both schools receive some sort of financial aid, be it scholarships from the schools' foundations, state-funded grants or federal financial aid like student loans, and many of them have lives busied by family and job demands. Some will be taking their Associate degrees and entering (or re-entering) the workforce while others will be continuing on to a four-year college or university. Despite their many differences, our students share one common attribute: they do not want to spend $200 on a College Algebra book.

The challenge of reducing the cost of textbooks is one that many states, including Ohio, are taking quite seriously. Indeed, state-level leaders have started to work with faculty from several of the colleges and universities in Ohio and with the major publishers as well. That process will take considerable time so Carl and I came up with a plan of our own. We decided that the best way to help our students right now was to write our own College Algebra book and give it away electronically for free. We were granted sabbaticals from our respective institutions for the Spring semester of 2009 and actually began writing the textbook on December 16, 2008. Using an open-source text editor called TexNicCenter and an open-source distribution of LaTeX called MikTex 2.7, Carl and I wrote and edited all of the text, exercises and answers and created all of the graphs (using Metapost within LaTeX) for Version $0.\overline{9}$ in about eight months. (We choose to create a text in only black and white to keep printing costs to a minimum for those students who prefer a printed edition. This somewhat Spartan page layout stands in sharp relief to the explosion of colors found in most other College Algebra texts, but neither Carl nor I believe the four-color print adds anything of value.) I used the book in three sections of College Algebra at Lorain County Community College in the Fall of 2009 and Carl's colleague, Dr. Bill Previts, taught a section of College Algebra at Lakeland with the book that semester as well. Students had the option of downloading the book as a .pdf file from our website www.stitz-zeager.com or buying a low-cost printed version from our colleges' respective bookstores. (By giving this book away for free electronically, we end the cycle of new editions appearing every 18 months to curtail the used book market.) During Thanksgiving break in November 2009, many additional exercises written by Dr. Previts were added and the typographical errors found by our students and others were corrected. On December 10, 2009, Version $\sqrt{2}$ was released. The book remains free for download at our website and by using Lulu.com as an on-demand printing service, our bookstores are now able to provide a printed edition for just under $19. Neither Carl nor I have, or will ever, receive any royalties from the printed editions. As a contribution back to the open-source community, all of the LaTeX files used to compile the book are available for free under a Creative Commons License on our website as well. That way, anyone who would like to rearrange or edit the content for their classes can do so as long as it remains free.

The only disadvantage to not working for a publisher is that we don't have a paid editorial staff. What we have instead, beyond ourselves, is friends, colleagues and unknown people in the open-source community who alert us to errors they find as they read the textbook. What we gain in not having to report to a publisher so dramatically outweighs the lack of the paid staff that we have turned down every offer to publish our book. (As of the writing of this Preface, we've had three offers.) By maintaining this book by ourselves, Carl and I retain all creative control and keep the book our own. We control the organization, depth and rigor of the content which means we can resist the pressure to diminish the rigor and homogenize the content so as to appeal to a mass market. A casual glance through the Table of Contents of most of the major publishers' College Algebra books reveals nearly isomorphic content in both order and depth. Our Table of Contents shows a different approach, one that might be labeled "Functions First." To truly use The Rule of Four, that is, in order to discuss each new concept algebraically, graphically, numerically and verbally, it seems completely obvious to us that one would need to introduce functions first. (Take a moment and compare our ordering to the classic "equations first, then the Cartesian Plane and THEN functions" approach seen in most of the major players.) We then introduce a class of functions and discuss the equations, inequalities (with a heavy emphasis on sign diagrams) and applications which involve functions in that class. The material is presented at a level that definitely prepares a student for Calculus while giving them relevant Mathematics which can be used in other classes as well. Graphing calculators are used sparingly and only as a tool to enhance the Mathematics, not to replace it. The answers to nearly all of the computational homework exercises are given in the text and we have gone to great lengths to write some very thought provoking discussion questions whose answers are not given. One will notice that our exercise sets are much shorter than the traditional sets of nearly 100 "drill and kill" questions which build skill devoid of understanding. Our experience has been that students can do about 15-20 homework exercises a night so we very carefully chose smaller sets of questions which cover all of the necessary skills and get the students thinking more deeply about the Mathematics involved.

Critics of the Open Educational Resource movement might quip that "open-source is where bad content goes to die," to which I say this: take a serious look at what we offer our students. Look through a few sections to see if what we've written is bad content in your opinion. I see this open-source book not as something which is "free and worth every penny", but rather, as a high quality alternative to the business as usual of the textbook industry and I hope that you agree. If you have any comments, questions or concerns please feel free to contact me at jeff@stitz-zeager.com or Carl at carl@stitz-zeager.com.

Jeff Zeager
Lorain County Community College
January 25, 2010

Chapter 10

Foundations of Trigonometry

10.1 Angles and their Measure

This section begins our study of Trigonometry and to get started, we recall some basic definitions from Geometry. A **ray** is usually described as a 'half-line' and can be thought of as a line segment in which one of the two endpoints is pushed off infinitely distant from the other, as pictured below. The point from which the ray originates is called the **initial point** of the ray.

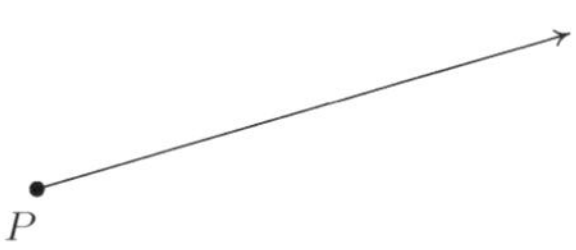

A ray with initial point P.

When two rays share a common initial point they form an **angle** and the common initial point is called the **vertex** of the angle. Two examples of what are commonly thought of as angles are

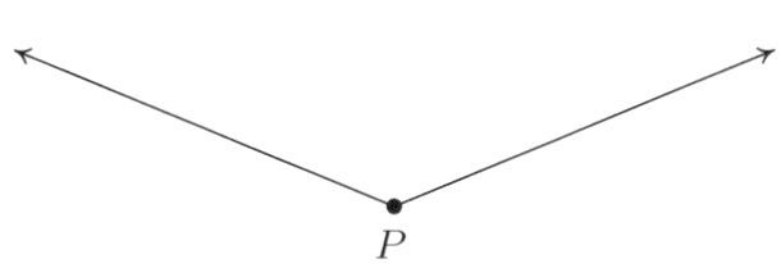

An angle with vertex P.

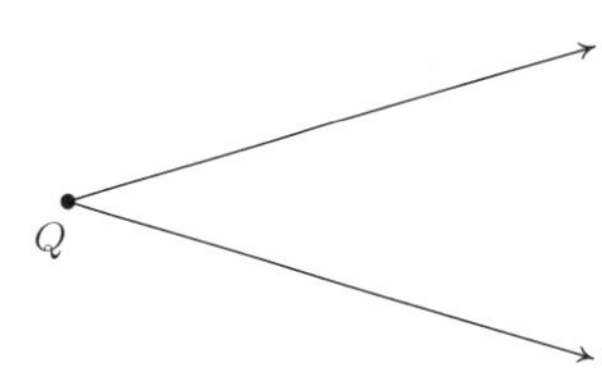

An angle with vertex Q.

However, the two figures below also depict angles - albeit these are, in some sense, extreme cases. In the first case, the two rays are directly opposite each other forming what is known as a **straight angle**; in the second, the rays are identical so the 'angle' is indistinguishable from the ray itself.

A straight angle.

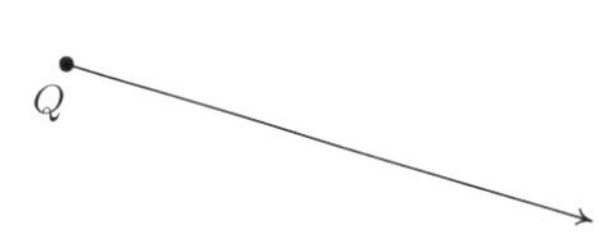

The **measure of an angle** is a number which indicates the amount of rotation that separates the rays of the angle. There is one immediate problem with this, as pictured below.

Which amount of rotation are we attempting to quantify? What we have just discovered is that we have at least two angles described by this diagram.[1] Clearly these two angles have different measures because one appears to represent a larger rotation than the other, so we must label them differently. In this book, we use lower case Greek letters such as α (alpha), β (beta), γ (gamma) and θ (theta) to label angles. So, for instance, we have

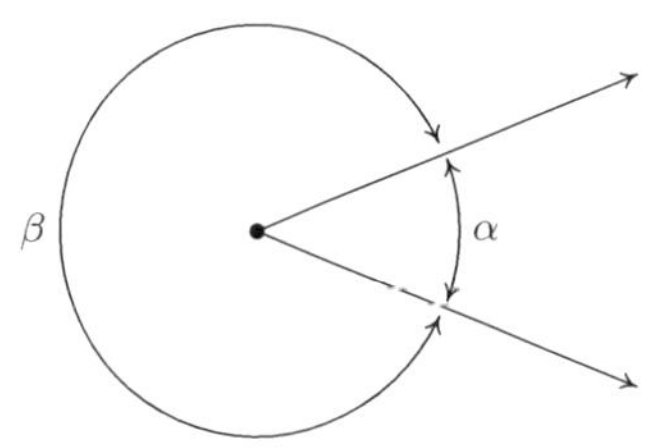

One commonly used system to measure angles is **degree measure**. Quantities measured in degrees are denoted by the familiar '$^\circ$' symbol. One complete revolution as shown below is 360°, and parts of a revolution are measured proportionately.[2] Thus half of a revolution (a straight angle) measures $\frac{1}{2}(360^\circ) = 180^\circ$, a quarter of a revolution (a **right angle**) measures $\frac{1}{4}(360^\circ) = 90^\circ$ and so on.

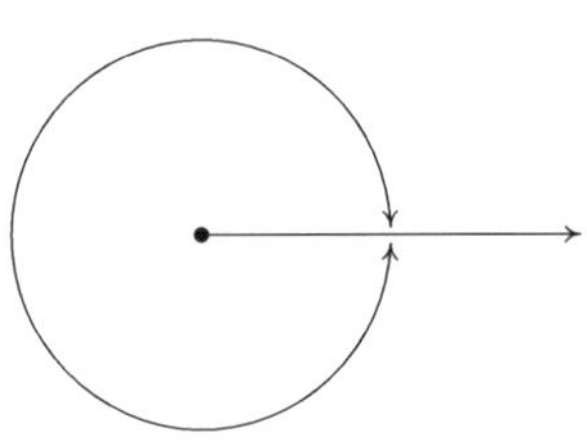
One revolution $\leftrightarrow$ 360°

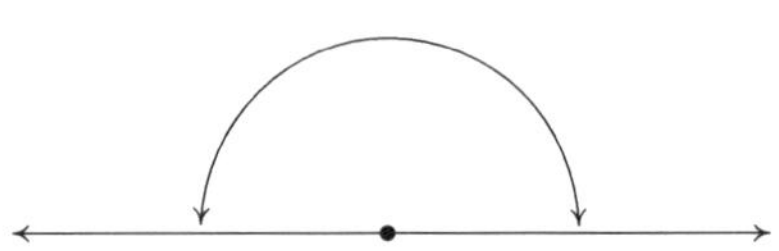
180°

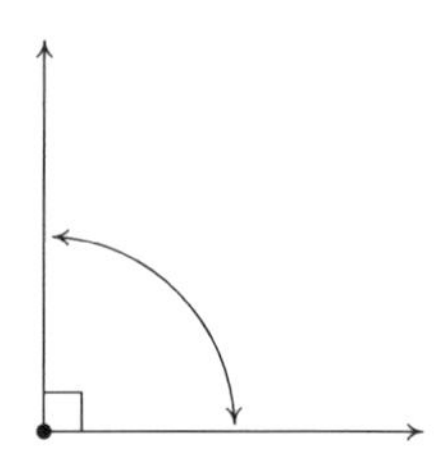
90°

Note that in the above figure, we have used the small square '□' to denote a right angle, as is commonplace in Geometry. Recall that if an angle measures strictly between 0° and 90° it is called an **acute angle** and if it measures strictly between 90° and 180° it is called an **obtuse angle**. It is important to note that, theoretically, we can know the measure of any angle as long as we

[1]The phrase 'at least' will be justified in short order.
[2]The choice of '360' is most often attributed to the Babylonians.

know the proportion it represents of entire revolution.[3] For instance, the measure of an angle which represents a rotation of $\frac{2}{3}$ of a revolution would measure $\frac{2}{3}(360^\circ) = 240^\circ$, the measure of an angle which constitutes only $\frac{1}{12}$ of a revolution measures $\frac{1}{12}(360^\circ) = 30^\circ$ and an angle which indicates no rotation at all is measured as 0°.

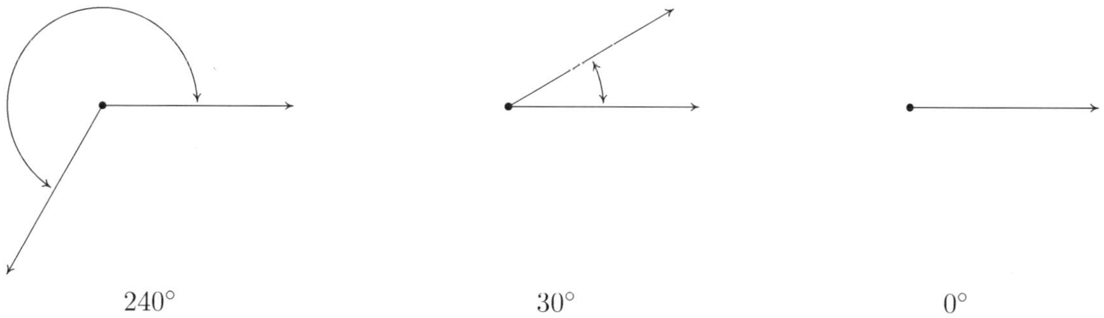

Using our definition of degree measure, we have that 1° represents the measure of an angle which constitutes $\frac{1}{360}$ of a revolution. Even though it may be hard to draw, it is nonetheless not difficult to imagine an angle with measure smaller than 1°. There are two ways to subdivide degrees. The first, and most familiar, is **decimal degrees**. For example, an angle with a measure of 30.5° would represent a rotation halfway between 30° and 31°, or equivalently, $\frac{30.5}{360} = \frac{61}{720}$ of a full rotation. This can be taken to the limit using Calculus so that measures like $\sqrt{2}^\circ$ make sense.[4] The second way to divide degrees is the **Degree - Minute - Second** (**DMS**) system. In this system, one degree is divided equally into sixty minutes, and in turn, each minute is divided equally into sixty seconds.[5] In symbols, we write $1^\circ = 60'$ and $1' = 60''$, from which it follows that $1^\circ = 3600''$. To convert a measure of 42.125° to the DMS system, we start by noting that $42.125^\circ = 42^\circ + 0.125^\circ$. Converting the partial amount of degrees to minutes, we find $0.125^\circ\left(\frac{60'}{1^\circ}\right) = 7.5' = 7' + 0.5'$. Converting the partial amount of minutes to seconds gives $0.5'\left(\frac{60''}{1'}\right) = 30''$. Putting it all together yields

$$\begin{aligned} 42.125^\circ &= 42^\circ + 0.125^\circ \\ &= 42^\circ + 7.5' \\ &= 42^\circ + 7' + 0.5' \\ &= 42^\circ + 7' + 30'' \\ &= 42^\circ 7' 30'' \end{aligned}$$

On the other hand, to convert $117^\circ 15' 45''$ to decimal degrees, we first compute $15'\left(\frac{1^\circ}{60'}\right) = \frac{1}{4}^\circ$ and $45''\left(\frac{1^\circ}{3600''}\right) = \frac{1}{80}^\circ$. Then we find

[3] This is how a protractor is graded.

[4] Awesome math pun aside, this is the same idea behind defining irrational exponents in Section 6.1.

[5] Does this kind of system seem familiar?

$$\begin{aligned} 117°15'45'' &= 117° + 15' + 45'' \\ &= 117° + \tfrac{1}{4}° + \tfrac{1}{80}° \\ &= \tfrac{9381}{80}° \\ &= 117.2625° \end{aligned}$$

Recall that two acute angles are called **complementary angles** if their measures add to $90°$. Two angles, either a pair of right angles or one acute angle and one obtuse angle, are called **supplementary angles** if their measures add to $180°$. In the diagram below, the angles α and β are supplementary angles while the pair γ and θ are complementary angles.

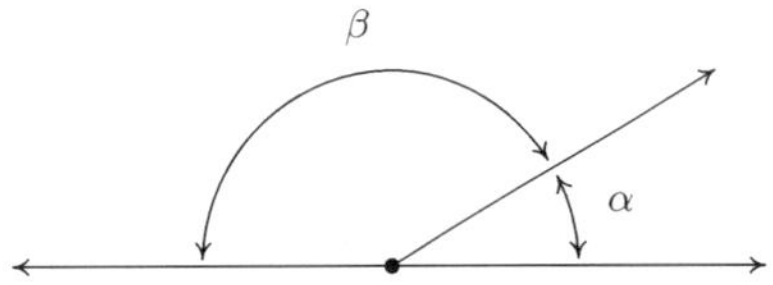

Supplementary Angles

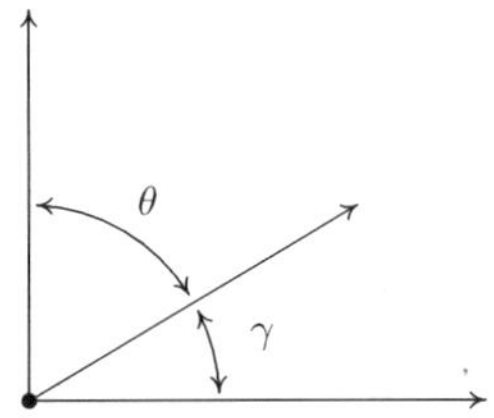

Complementary Angles

In practice, the distinction between the angle itself and its measure is blurred so that the sentence 'α is an angle measuring $42°$' is often abbreviated as '$\alpha = 42°$.' It is now time for an example.

Example 10.1.1. Let $\alpha = 111.371°$ and $\beta = 37°28'17''$.

1. Convert α to the DMS system. Round your answer to the nearest second.
2. Convert β to decimal degrees. Round your answer to the nearest thousandth of a degree.
3. Sketch α and β.
4. Find a supplementary angle for α.
5. Find a complementary angle for β.

Solution.

1. To convert α to the DMS system, we start with $111.371° = 111° + 0.371°$. Next we convert $0.371°\left(\frac{60'}{1°}\right) = 22.26'$. Writing $22.26' = 22' + 0.26'$, we convert $0.26'\left(\frac{60''}{1'}\right) = 15.6''$. Hence,

$$\begin{aligned} 111.371° &= 111° + 0.371° \\ &= 111° + 22.26' \\ &= 111° + 22' + 0.26' \\ &= 111° + 22' + 15.6'' \\ &= 111°22'15.6'' \end{aligned}$$

Rounding to seconds, we obtain $\alpha \approx 111°22'16''$.

2. To convert β to decimal degrees, we convert $28' \left(\frac{1^\circ}{60'}\right) = \frac{7}{15}^\circ$ and $17'' \left(\frac{1^\circ}{3600''}\right) = \frac{17}{3600}^\circ$. Putting it all together, we have

$$\begin{aligned} 37^\circ 28' 17'' &= 37^\circ + 28' + 17'' \\ &= 37^\circ + \frac{7}{15}^\circ + \frac{17}{3600}^\circ \\ &= \frac{134897}{3600}^\circ \\ &\approx 37.471^\circ \end{aligned}$$

3. To sketch α, we first note that $90^\circ < \alpha < 180^\circ$. If we divide this range in half, we get $90^\circ < \alpha < 135^\circ$, and once more, we have $90^\circ < \alpha < 112.5^\circ$. This gives us a pretty good estimate for α, as shown below.[6] Proceeding similarly for β, we find $0^\circ < \beta < 90^\circ$, then $0^\circ < \beta < 45^\circ$, $22.5^\circ < \beta < 45^\circ$, and lastly, $33.75^\circ < \beta < 45^\circ$.

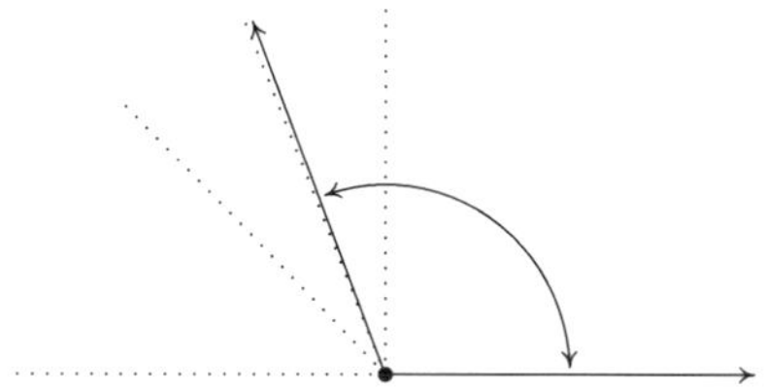

Angle α

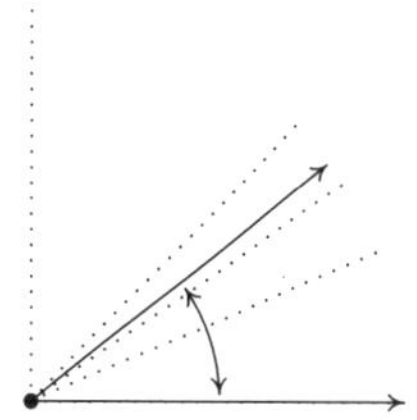

Angle β

4. To find a supplementary angle for α, we seek an angle θ so that $\alpha + \theta = 180^\circ$. We get $\theta = 180^\circ - \alpha = 180^\circ - 111.371^\circ = 68.629^\circ$.

5. To find a complementary angle for β, we seek an angle γ so that $\beta + \gamma = 90^\circ$. We get $\gamma = 90^\circ - \beta = 90^\circ - 37^\circ 28' 17''$. While we could reach for the calculator to obtain an approximate answer, we choose instead to do a bit of sexagesimal[7] arithmetic. We first rewrite $90^\circ = 90^\circ 0' 0'' = 89^\circ 60' 0'' = 89^\circ 59' 60''$. In essence, we are 'borrowing' $1^\circ = 60'$ from the degree place, and then borrowing $1' = 60''$ from the minutes place.[8] This yields, $\gamma = 90^\circ - 37^\circ 28' 17'' = 89^\circ 59' 60'' - 37^\circ 28' 17'' = 52^\circ 31' 43''$. □

Up to this point, we have discussed only angles which measure between 0° and 360°, inclusive. Ultimately, we want to use the arsenal of Algebra which we have stockpiled in Chapters 1 through 9 to not only solve geometric problems involving angles, but also to extend their applicability to other real-world phenomena. A first step in this direction is to extend our notion of 'angle' from merely measuring an extent of rotation to quantities which can be associated with real numbers. To that end, we introduce the concept of an **oriented angle**. As its name suggests, in an oriented

[6] If this process seems hauntingly familiar, it should. Compare this method to the Bisection Method introduced in Section 3.3.

[7] Like 'latus rectum,' this is also a real math term.

[8] This is the exact same kind of 'borrowing' you used to do in Elementary School when trying to find 300 − 125. Back then, you were working in a base ten system; here, it is base sixty.

angle, the direction of the rotation is important. We imagine the angle being swept out starting from an **initial side** and ending at a **terminal side**, as shown below. When the rotation is counter-clockwise[9] from initial side to terminal side, we say that the angle is **positive**; when the rotation is clockwise, we say that the angle is **negative**.

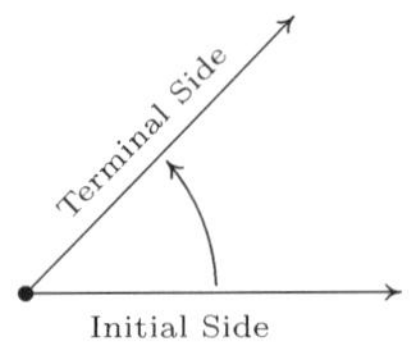

A positive angle, 45°

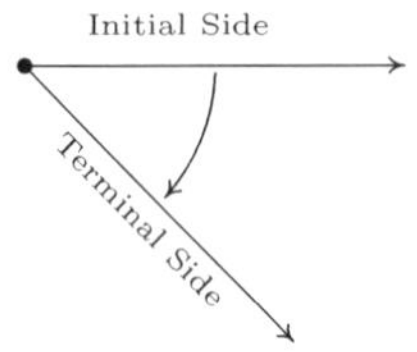

A negative angle, −45°

At this point, we also extend our allowable rotations to include angles which encompass more than one revolution. For example, to sketch an angle with measure 450° we start with an initial side, rotate counter-clockwise one complete revolution (to take care of the 'first' 360°) then continue with an additional 90° counter-clockwise rotation, as seen below.

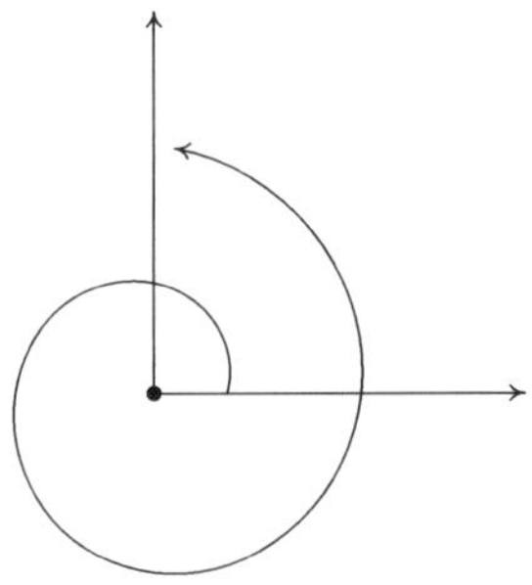
450°

To further connect angles with the Algebra which has come before, we shall often overlay an angle diagram on the coordinate plane. An angle is said to be in **standard position** if its vertex is the origin and its initial side coincides with the positive x-axis. Angles in standard position are classified according to where their terminal side lies. For instance, an angle in standard position whose terminal side lies in Quadrant I is called a 'Quadrant I angle'. If the terminal side of an angle lies on one of the coordinate axes, it is called a **quadrantal angle**. Two angles in standard position are called **coterminal** if they share the same terminal side.[10] In the figure below, $\alpha = 120°$ and $\beta = -240°$ are two coterminal Quadrant II angles drawn in standard position. Note that $\alpha = \beta + 360°$, or equivalently, $\beta = \alpha - 360°$. We leave it as an exercise to the reader to verify that coterminal angles always differ by a multiple of 360°.[11] More precisely, if α and β are coterminal angles, then $\beta = \alpha + 360° \cdot k$ where k is an integer.[12]

[9]'widdershins'
[10]Note that by being in standard position they automatically share the same initial side which is the positive x-axis.
[11]It is worth noting that all of the pathologies of Analytic Trigonometry result from this innocuous fact.
[12]Recall that this means $k = 0, \pm 1, \pm 2, \ldots$.

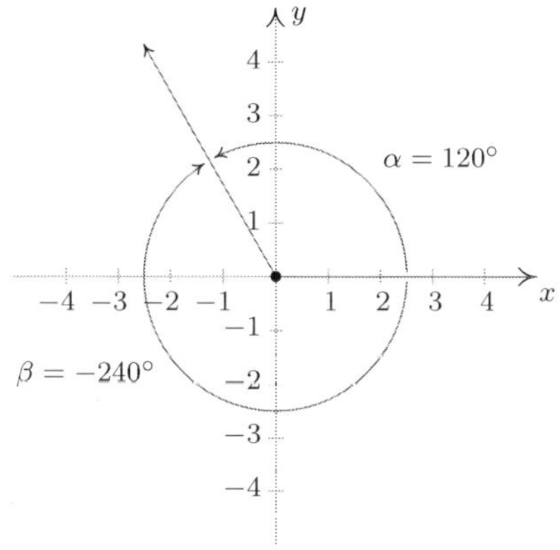

Two coterminal angles, $\alpha = 120^\circ$ and $\beta = -240^\circ$, in standard position.

Example 10.1.2. Graph each of the (oriented) angles below in standard position and classify them according to where their terminal side lies. Find three coterminal angles, at least one of which is positive and one of which is negative.

1. $\alpha = 60^\circ$
2. $\beta = -225^\circ$
3. $\gamma = 540^\circ$
4. $\phi = -750^\circ$

Solution.

1. To graph $\alpha = 60^\circ$, we draw an angle with its initial side on the positive x-axis and rotate counter-clockwise $\frac{60^\circ}{360^\circ} = \frac{1}{6}$ of a revolution. We see that α is a Quadrant I angle. To find angles which are coterminal, we look for angles θ of the form $\theta = \alpha + 360^\circ \cdot k$, for some integer k. When $k = 1$, we get $\theta = 60^\circ + 360^\circ = 420^\circ$. Substituting $k = -1$ gives $\theta = 60^\circ - 360^\circ = -300^\circ$. Finally, if we let $k = 2$, we get $\theta = 60^\circ + 720^\circ = 780^\circ$.

2. Since $\beta = -225^\circ$ is negative, we start at the positive x-axis and rotate *clockwise* $\frac{225^\circ}{360^\circ} = \frac{5}{8}$ of a revolution. We see that β is a Quadrant II angle. To find coterminal angles, we proceed as before and compute $\theta = -225^\circ + 360^\circ \cdot k$ for integer values of k. We find 135°, -585° and 495° are all coterminal with -225°.

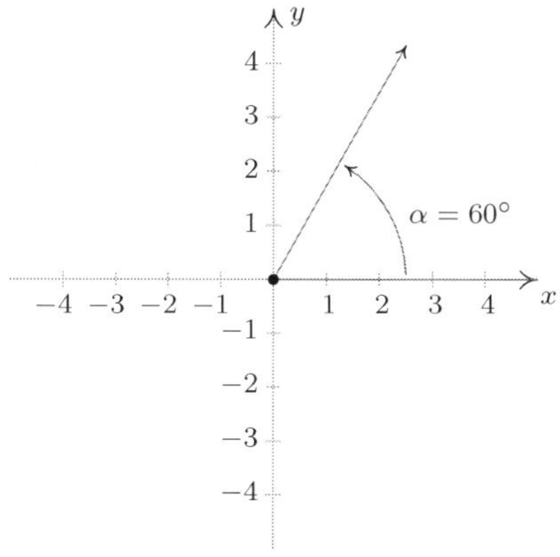

$\alpha = 60^\circ$ in standard position.

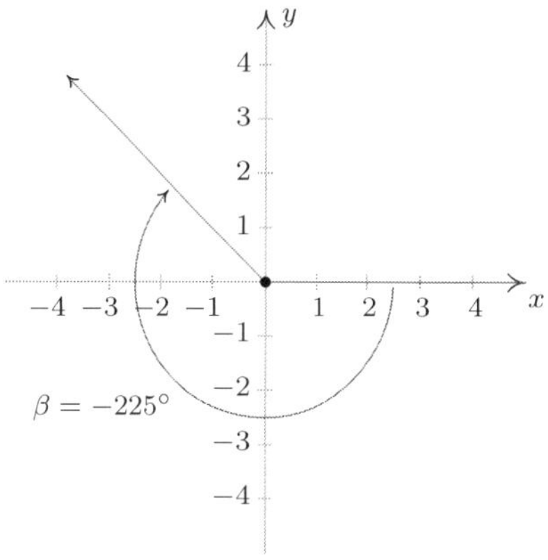

$\beta = -225^\circ$ in standard position.

3. Since $\gamma = 540^\circ$ is positive, we rotate counter-clockwise from the positive x-axis. One full revolution accounts for 360°, with 180°, or $\frac{1}{2}$ of a revolution remaining. Since the terminal side of γ lies on the negative x-axis, γ is a quadrantal angle. All angles coterminal with γ are of the form $\theta = 540^\circ + 360^\circ \cdot k$, where k is an integer. Working through the arithmetic, we find three such angles: 180°, -180° and 900°.

4. The Greek letter ϕ is pronounced 'fee' or 'fie' and since ϕ is negative, we begin our rotation clockwise from the positive x-axis. Two full revolutions account for 720°, with just 30° or $\frac{1}{12}$ of a revolution to go. We find that ϕ is a Quadrant IV angle. To find coterminal angles, we compute $\theta = -750^\circ + 360^\circ \cdot k$ for a few integers k and obtain -390°, -30° and 330°.

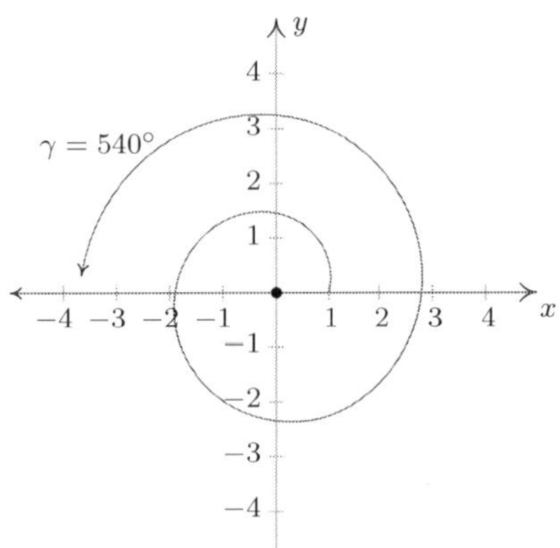

$\gamma = 540^\circ$ in standard position.

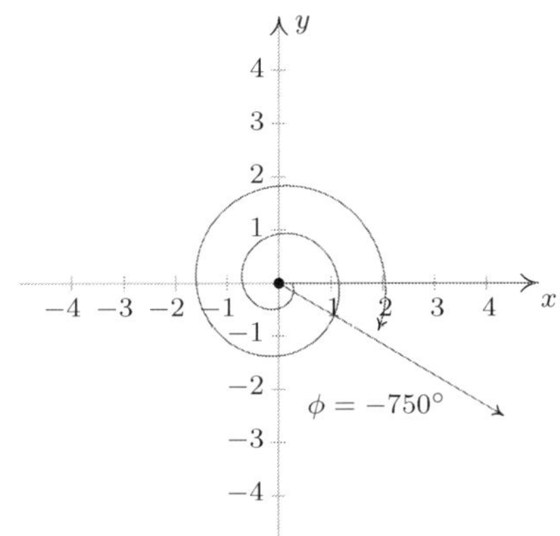

$\phi = -750^\circ$ in standard position.

□

Note that since there are infinitely many integers, any given angle has infinitely many coterminal angles, and the reader is encouraged to plot the few sets of coterminal angles found in Example 10.1.2 to see this. We are now just one step away from completely marrying angles with the real numbers and the rest of Algebra. To that end, we recall this definition from Geometry.

Definition 10.1. The real number π is defined to be the ratio of a circle's circumference to its diameter. In symbols, given a circle of circumference C and diameter d,

$$\pi = \frac{C}{d}$$

While Definition 10.1 is quite possibly the 'standard' definition of π, the authors would be remiss if we didn't mention that buried in this definition is actually a theorem. As the reader is probably aware, the number π is a mathematical constant - that is, it doesn't matter *which* circle is selected, the ratio of its circumference to its diameter will have the same value as any other circle. While this is indeed true, it is far from obvious and leads to a counterintuitive scenario which is explored in the Exercises. Since the diameter of a circle is twice its radius, we can quickly rearrange the equation in Definition 10.1 to get a formula more useful for our purposes, namely: $2\pi = \frac{C}{r}$

This tells us that for any circle, the ratio of its circumference to its radius is also always constant; in this case the constant is 2π. Suppose now we take a *portion* of the circle, so instead of comparing the entire circumference C to the radius, we compare some arc measuring s units in length to the radius, as depicted below. Let θ be the **central angle** subtended by this arc, that is, an angle whose vertex is the center of the circle and whose determining rays pass through the endpoints of the arc. Using proportionality arguments, it stands to reason that the ratio $\frac{s}{r}$ should also be a constant among all circles, and it is this ratio which defines the **radian measure** of an angle.

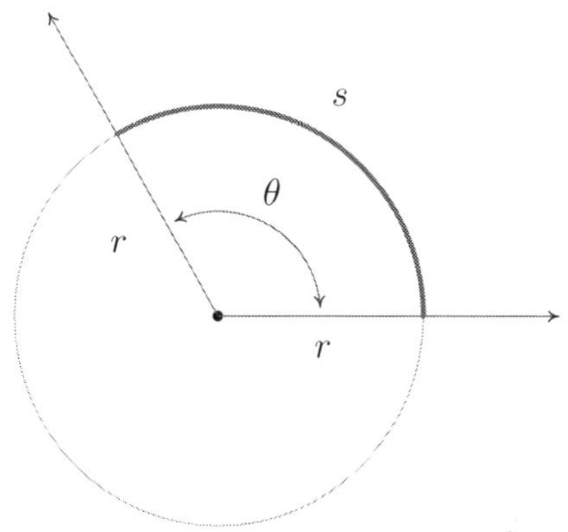

The radian measure of θ is $\frac{s}{r}$.

To get a better feel for radian measure, we note that an angle with radian measure 1 means the corresponding arc length s equals the radius of the circle r, hence $s = r$. When the radian measure is 2, we have $s = 2r$; when the radian measure is 3, $s = 3r$, and so forth. Thus the radian measure of an angle θ tells us how many 'radius lengths' we need to sweep out along the circle to subtend the angle θ.

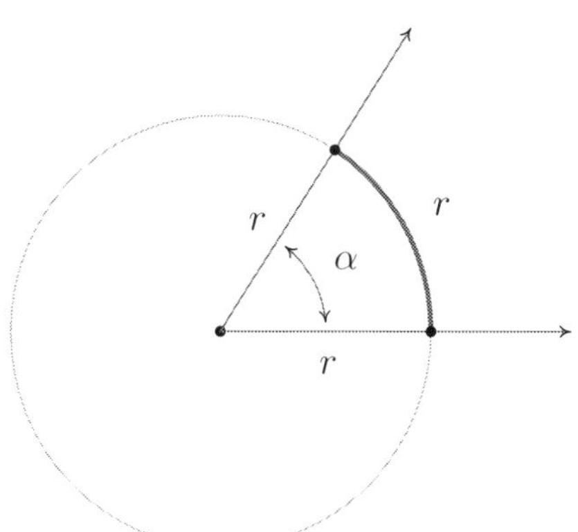

α has radian measure 1

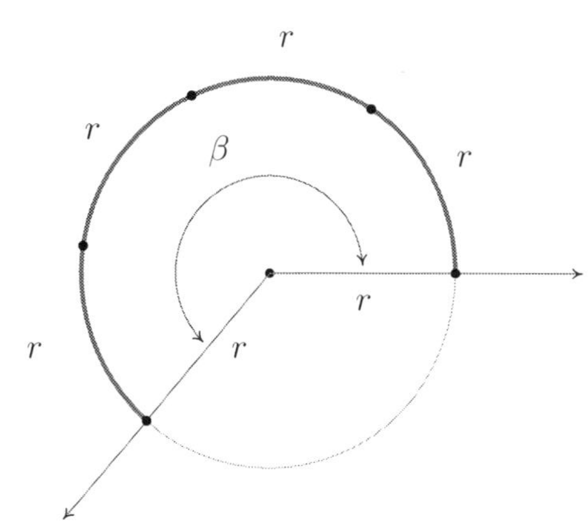

β has radian measure 4

Since one revolution sweeps out the entire circumference $2\pi r$, one revolution has radian measure $\frac{2\pi r}{r} = 2\pi$. From this we can find the radian measure of other central angles using proportions, just like we did with degrees. For instance, half of a revolution has radian measure $\frac{1}{2}(2\pi) = \pi$, a quarter revolution has radian measure $\frac{1}{4}(2\pi) = \frac{\pi}{2}$, and so forth. Note that, by definition, the radian measure of an angle is a length divided by another length so that these measurements are actually dimensionless and are considered 'pure' numbers. For this reason, we do not use any symbols to denote radian measure, but we use the word 'radians' to denote these dimensionless units as needed. For instance, we say one revolution measures '2π radians,' half of a revolution measures 'π radians,' and so forth.

As with degree measure, the distinction between the angle itself and its measure is often blurred in practice, so when we write '$\theta = \frac{\pi}{2}$', we mean θ is an angle which measures $\frac{\pi}{2}$ radians.[13] We extend radian measure to oriented angles, just as we did with degrees beforehand, so that a positive measure indicates counter-clockwise rotation and a negative measure indicates clockwise rotation.[14] Much like before, two positive angles α and β are supplementary if $\alpha + \beta = \pi$ and complementary if $\alpha + \beta = \frac{\pi}{2}$. Finally, we leave it to the reader to show that when using radian measure, two angles α and β are coterminal if and only if $\beta = \alpha + 2\pi k$ for some integer k.

Example 10.1.3. Graph each of the (oriented) angles below in standard position and classify them according to where their terminal side lies. Find three coterminal angles, at least one of which is positive and one of which is negative.

1. $\alpha = \dfrac{\pi}{6}$
2. $\beta = -\dfrac{4\pi}{3}$
3. $\gamma = \dfrac{9\pi}{4}$
4. $\phi = -\dfrac{5\pi}{2}$

Solution.

1. The angle $\alpha = \frac{\pi}{6}$ is positive, so we draw an angle with its initial side on the positive x-axis and rotate counter-clockwise $\frac{(\pi/6)}{2\pi} = \frac{1}{12}$ of a revolution. Thus α is a Quadrant I angle. Coterminal angles θ are of the form $\theta = \alpha + 2\pi \cdot k$, for some integer k. To make the arithmetic a bit easier, we note that $2\pi = \frac{12\pi}{6}$, thus when $k = 1$, we get $\theta = \frac{\pi}{6} + \frac{12\pi}{6} = \frac{13\pi}{6}$. Substituting $k = -1$ gives $\theta = \frac{\pi}{6} - \frac{12\pi}{6} = -\frac{11\pi}{6}$ and when we let $k = 2$, we get $\theta = \frac{\pi}{6} + \frac{24\pi}{6} = \frac{25\pi}{6}$.

2. Since $\beta = -\frac{4\pi}{3}$ is negative, we start at the positive x-axis and rotate clockwise $\frac{(4\pi/3)}{2\pi} = \frac{2}{3}$ of a revolution. We find β to be a Quadrant II angle. To find coterminal angles, we proceed as before using $2\pi = \frac{6\pi}{3}$, and compute $\theta = -\frac{4\pi}{3} + \frac{6\pi}{3} \cdot k$ for integer values of k. We obtain $\frac{2\pi}{3}$, $-\frac{10\pi}{3}$ and $\frac{8\pi}{3}$ as coterminal angles.

[13] The authors are well aware that we are now identifying radians with real numbers. We will justify this shortly.

[14] This, in turn, endows the subtended arcs with an orientation as well. We address this in short order.

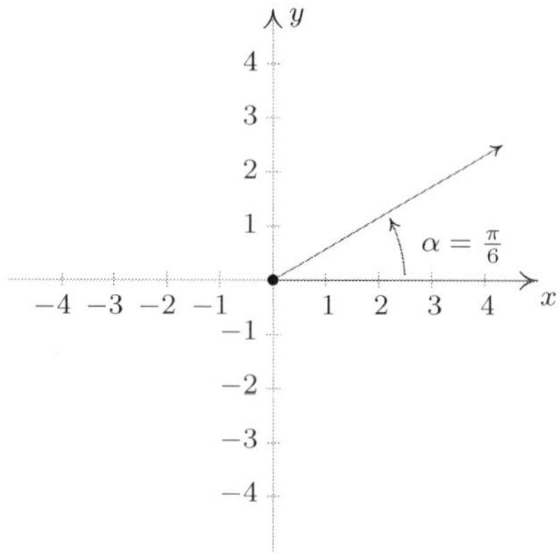

$\alpha = \frac{\pi}{6}$ in standard position.

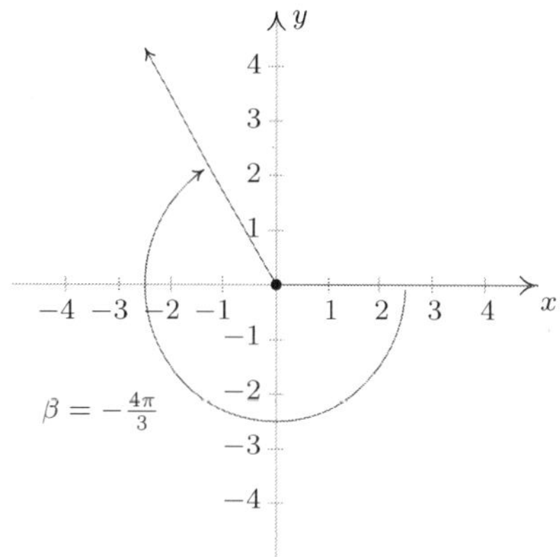

$\beta = -\frac{4\pi}{3}$ in standard position.

3. Since $\gamma = \frac{9\pi}{4}$ is positive, we rotate counter-clockwise from the positive x-axis. One full revolution accounts for $2\pi = \frac{8\pi}{4}$ of the radian measure with $\frac{\pi}{4}$ or $\frac{1}{8}$ of a revolution remaining. We have γ as a Quadrant I angle. All angles coterminal with γ are of the form $\theta = \frac{9\pi}{4} + \frac{8\pi}{4} \cdot k$, where k is an integer. Working through the arithmetic, we find: $\frac{\pi}{4}$, $-\frac{7\pi}{4}$ and $\frac{17\pi}{4}$.

4. To graph $\phi = -\frac{5\pi}{2}$, we begin our rotation clockwise from the positive x-axis. As $2\pi = \frac{4\pi}{2}$, after one full revolution clockwise, we have $\frac{\pi}{2}$ or $\frac{1}{4}$ of a revolution remaining. Since the terminal side of ϕ lies on the negative y-axis, ϕ is a quadrantal angle. To find coterminal angles, we compute $\theta = -\frac{5\pi}{2} + \frac{4\pi}{2} \cdot k$ for a few integers k and obtain $-\frac{\pi}{2}$, $\frac{3\pi}{2}$ and $\frac{7\pi}{2}$.

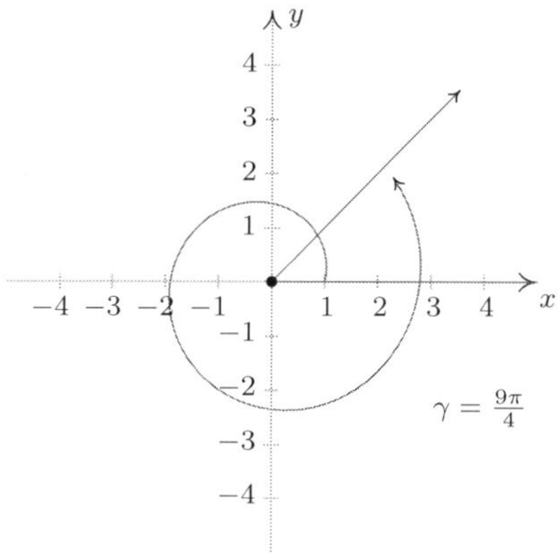

$\gamma = \frac{9\pi}{4}$ in standard position.

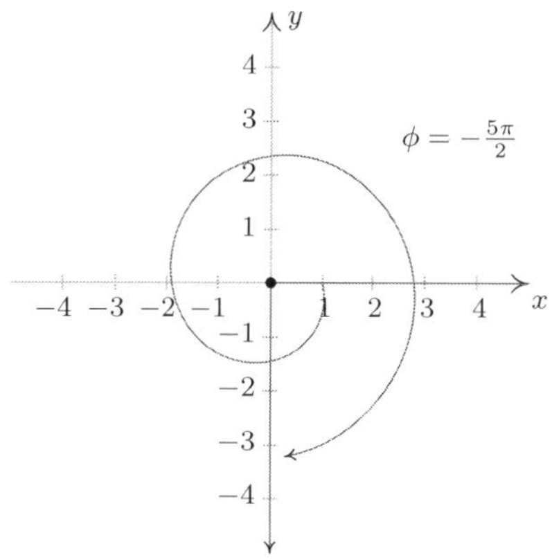

$\phi = -\frac{5\pi}{2}$ in standard position.

□

It is worth mentioning that we could have plotted the angles in Example 10.1.3 by first converting them to degree measure and following the procedure set forth in Example 10.1.2. While converting back and forth from degrees and radians is certainly a good skill to have, it is best that you learn to 'think in radians' as well as you can 'think in degrees'. The authors would, however, be derelict in our duties if we ignored the basic conversion between these systems altogether. Since one revolution counter-clockwise measures 360° and the same angle measures 2π radians, we can use the proportion $\frac{2\pi \text{ radians}}{360^\circ}$, or its reduced equivalent, $\frac{\pi \text{ radians}}{180^\circ}$, as the conversion factor between the two systems. For example, to convert 60° to radians we find $60^\circ \left(\frac{\pi \text{ radians}}{180^\circ}\right) = \frac{\pi}{3}$ radians, or simply $\frac{\pi}{3}$. To convert from radian measure back to degrees, we multiply by the ratio $\frac{180^\circ}{\pi \text{ radian}}$. For example, $-\frac{5\pi}{6}$ radians is equal to $\left(-\frac{5\pi}{6} \text{ radians}\right)\left(\frac{180^\circ}{\pi \text{ radians}}\right) = -150^\circ$.[15] Of particular interest is the fact that an angle which measures 1 in radian measure is equal to $\frac{180^\circ}{\pi} \approx 57.2958^\circ$.

We summarize these conversions below.

Equation 10.1. Degree - Radian Conversion:

- To convert degree measure to radian measure, multiply by $\dfrac{\pi \text{ radians}}{180^\circ}$
- To convert radian measure to degree measure, multiply by $\dfrac{180^\circ}{\pi \text{ radians}}$

In light of Example 10.1.3 and Equation 10.1, the reader may well wonder what the allure of radian measure is. The numbers involved are, admittedly, much more complicated than degree measure. The answer lies in how easily angles in radian measure can be identified with real numbers. Consider the Unit Circle, $x^2+y^2=1$, as drawn below, the angle θ in standard position and the corresponding arc measuring s units in length. By definition, and the fact that the Unit Circle has radius 1, the radian measure of θ is $\dfrac{s}{r} = \dfrac{s}{1} = s$ so that, once again blurring the distinction between an angle and its measure, we have $\theta = s$. In order to identify real numbers with oriented angles, we make good use of this fact by essentially 'wrapping' the real number line around the Unit Circle and associating to each real number t an *oriented* arc on the Unit Circle with initial point $(1,0)$.

Viewing the vertical line $x=1$ as another real number line demarcated like the y-axis, given a real number $t>0$, we 'wrap' the (vertical) interval $[0,t]$ around the Unit Circle in a counter-clockwise fashion. The resulting arc has a length of t units and therefore the corresponding angle has radian measure equal to t. If $t<0$, we wrap the interval $[t,0]$ *clockwise* around the Unit Circle. Since we have defined clockwise rotation as having negative radian measure, the angle determined by this arc has radian measure equal to t. If $t=0$, we are at the point $(1,0)$ on the x-axis which corresponds to an angle with radian measure 0. In this way, we identify each real number t with the corresponding angle with radian measure t.

[15]Note that the negative sign indicates clockwise rotation in both systems, and so it is carried along accordingly.

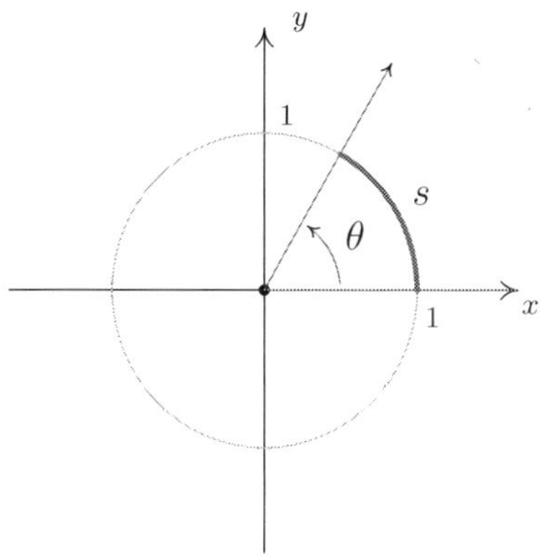

On the Unit Circle, $\theta = s$.

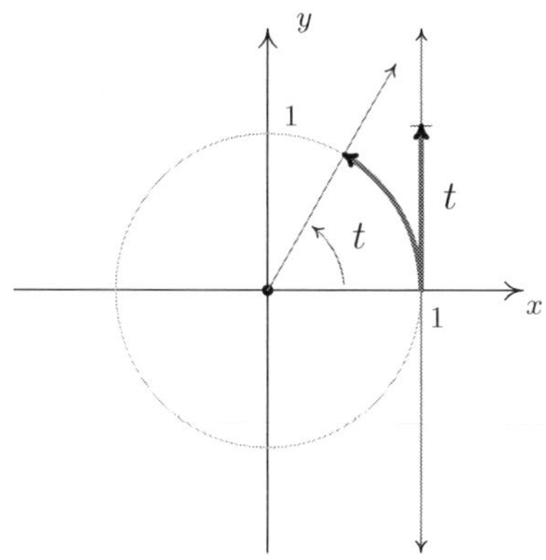

Identifying $t > 0$ with an angle.

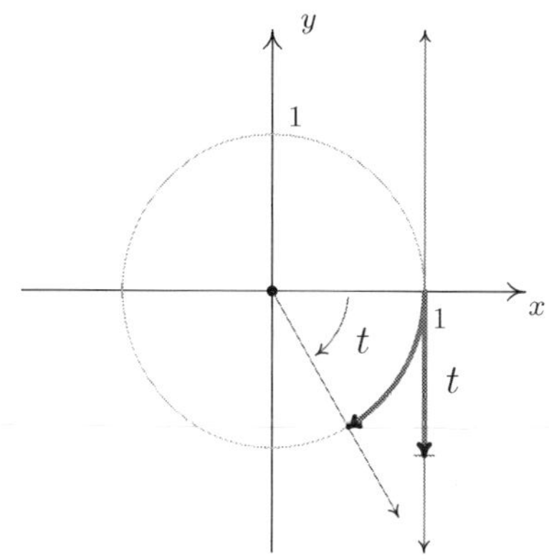

Identifying $t < 0$ with an angle.

Example 10.1.4. Sketch the oriented arc on the Unit Circle corresponding to each of the following real numbers.

1. $t = \dfrac{3\pi}{4}$
2. $t = -2\pi$
3. $t = -2$
4. $t = 117$

Solution.

1. The arc associated with $t = \frac{3\pi}{4}$ is the arc on the Unit Circle which subtends the angle $\frac{3\pi}{4}$ in radian measure. Since $\frac{3\pi}{4}$ is $\frac{3}{8}$ of a revolution, we have an arc which begins at the point $(1, 0)$ proceeds counter-clockwise up to midway through Quadrant II.

2. Since one revolution is 2π radians, and $t = -2\pi$ is negative, we graph the arc which begins at $(1, 0)$ and proceeds *clockwise* for one full revolution.

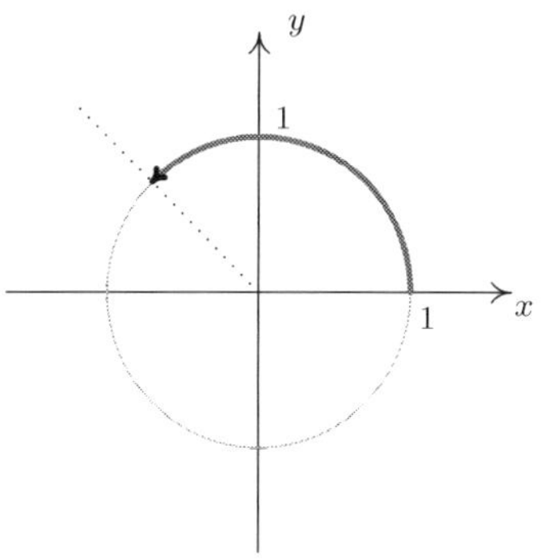

$t = \frac{3\pi}{4}$

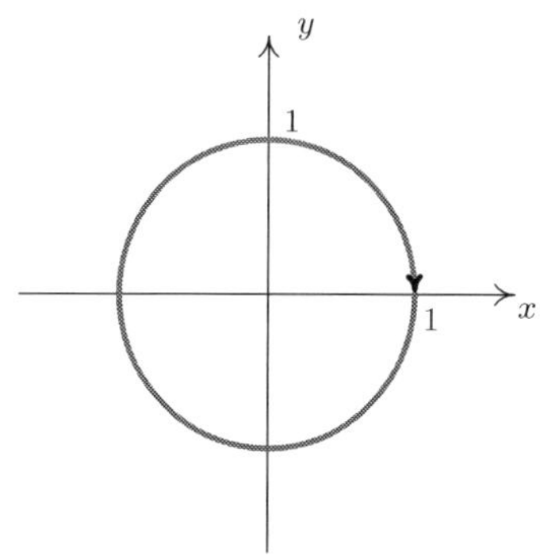

$t = -2\pi$

3. Like $t = -2\pi$, $t = -2$ is negative, so we begin our arc at $(1, 0)$ and proceed clockwise around the unit circle. Since $\pi \approx 3.14$ and $\frac{\pi}{2} \approx 1.57$, we find that rotating 2 radians clockwise from the point $(1, 0)$ lands us in Quadrant III. To more accurately place the endpoint, we proceed as we did in Example 10.1.1, successively halving the angle measure until we find $\frac{5\pi}{8} \approx 1.96$ which tells us our arc extends just a bit beyond the quarter mark into Quadrant III.

4. Since 117 is positive, the arc corresponding to $t = 117$ begins at $(1, 0)$ and proceeds counter-clockwise. As 117 is much greater than 2π, we wrap around the Unit Circle several times before finally reaching our endpoint. We approximate $\frac{117}{2\pi}$ as 18.62 which tells us we complete 18 revolutions counter-clockwise with 0.62, or just shy of $\frac{5}{8}$ of a revolution to spare. In other words, the terminal side of the angle which measures 117 radians in standard position is just short of being midway through Quadrant III.

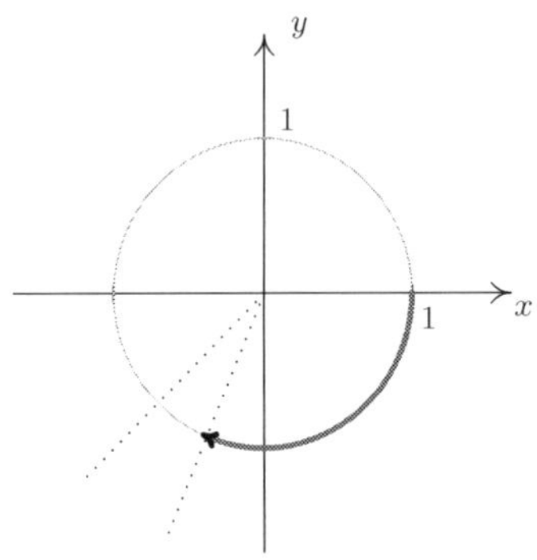

$t = -2$

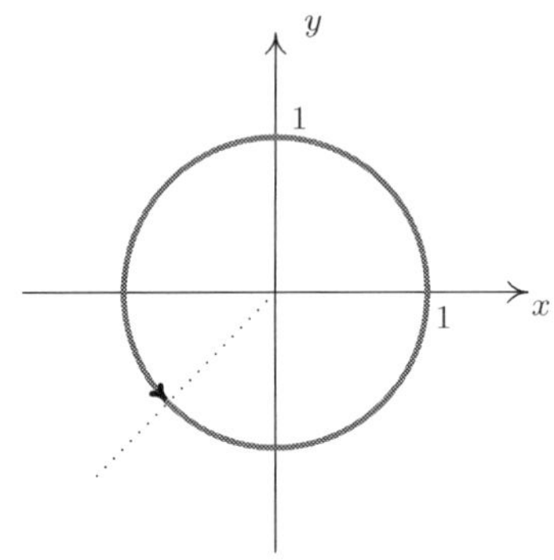

$t = 117$

□

10.1.1 Applications of Radian Measure: Circular Motion

Now that we have paired angles with real numbers via radian measure, a whole world of applications awaits us. Our first excursion into this realm comes by way of circular motion. Suppose an object is moving as pictured below along a circular path of radius r from the point P to the point Q in an amount of time t.

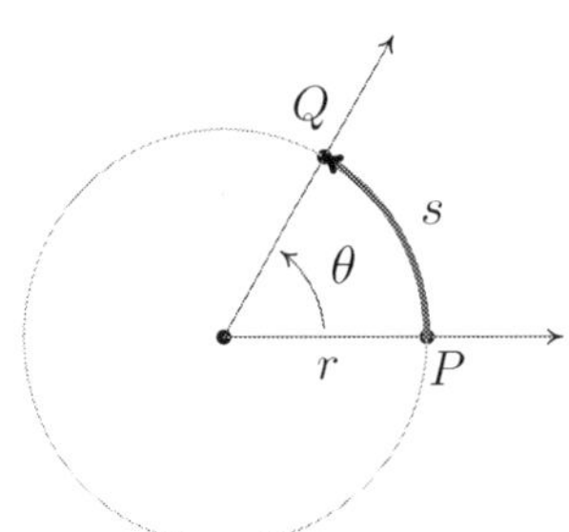

Here s represents a *displacement* so that $s > 0$ means the object is traveling in a counter-clockwise direction and $s < 0$ indicates movement in a clockwise direction. Note that with this convention the formula we used to define radian measure, namely $\theta = \frac{s}{r}$, still holds since a negative value of s incurred from a clockwise displacement matches the negative we assign to θ for a clockwise rotation. In Physics, the **average velocity** of the object, denoted $\overline{v}$ and read as 'v-bar', is defined as the average rate of change of the position of the object with respect to time.[16] As a result, we

[16]See Definition 2.3 in Section 2.1 for a review of this concept.

have $\overline{v} = \frac{\text{displacement}}{\text{time}} = \dfrac{s}{t}$. The quantity $\overline{v}$ has units of $\frac{\text{length}}{\text{time}}$ and conveys two ideas: the direction in which the object is moving and how fast the position of the object is changing. The contribution of direction in the quantity $\overline{v}$ is either to make it positive (in the case of counter-clockwise motion) or negative (in the case of clockwise motion), so that the quantity $|\overline{v}|$ quantifies how fast the object is moving - it is the **speed** of the object. Measuring θ in radians we have $\theta = \dfrac{s}{r}$ thus $s = r\theta$ and

$$\overline{v} = \frac{s}{t} = \frac{r\theta}{t} = r \cdot \frac{\theta}{t}$$

The quantity $\dfrac{\theta}{t}$ is called the **average angular velocity** of the object. It is denoted by $\overline{\omega}$ and is read 'omega-bar'. The quantity $\overline{\omega}$ is the average rate of change of the angle θ with respect to time and thus has units $\frac{\text{radians}}{\text{time}}$. If $\overline{\omega}$ is constant throughout the duration of the motion, then it can be shown[17] that the average velocities involved, namely $\overline{v}$ and $\overline{\omega}$, are the same as their instantaneous counterparts, v and ω, respectively. In this case, v is simply called the 'velocity' of the object and is the instantaneous rate of change of the position of the object with respect to time.[18] Similarly, ω is called the 'angular velocity' and is the instantaneous rate of change of the angle with respect to time.

If the path of the object were 'uncurled' from a circle to form a line segment, then the velocity of the object on that line segment would be the same as the velocity on the circle. For this reason, the quantity v is often called the *linear* velocity of the object in order to distinguish it from the *angular* velocity, ω. Putting together the ideas of the previous paragraph, we get the following.

Equation 10.2. Velocity for Circular Motion: For an object moving on a circular path of radius r with constant angular velocity ω, the (linear) velocity of the object is given by $v = r\omega$.

We need to talk about units here. The units of v are $\frac{\text{length}}{\text{time}}$, the units of r are length only, and the units of ω are $\frac{\text{radians}}{\text{time}}$. Thus the left hand side of the equation $v = r\omega$ has units $\frac{\text{length}}{\text{time}}$, whereas the right hand side has units length $\cdot \frac{\text{radians}}{\text{time}} = \frac{\text{length}\cdot\text{radians}}{\text{time}}$. The supposed contradiction in units is resolved by remembering that radians are a dimensionless quantity and angles in radian measure are identified with real numbers so that the units $\frac{\text{length}\cdot\text{radians}}{\text{time}}$ reduce to the units $\frac{\text{length}}{\text{time}}$. We are long overdue for an example.

Example 10.1.5. Assuming that the surface of the Earth is a sphere, any point on the Earth can be thought of as an object traveling on a circle which completes one revolution in (approximately) 24 hours. The path traced out by the point during this 24 hour period is the Latitude of that point. Lakeland Community College is at 41.628° north latitude, and it can be shown[19] that the radius of the earth at this Latitude is approximately 2960 miles. Find the linear velocity, in miles per hour, of Lakeland Community College as the world turns.

Solution. To use the formula $v = r\omega$, we first need to compute the angular velocity ω. The earth makes one revolution in 24 hours, and one revolution is 2π radians, so $\omega = \frac{2\pi \text{ radians}}{24 \text{ hours}} = \frac{\pi}{12 \text{ hours}}$,

[17] You guessed it, using Calculus ...

[18] See the discussion on Page 161 for more details on the idea of an 'instantaneous' rate of change.

[19] We will discuss how we arrived at this approximation in Example 10.2.6.

where, once again, we are using the fact that radians are real numbers and are dimensionless. (For simplicity's sake, we are also assuming that we are viewing the rotation of the earth as counter-clockwise so $\omega > 0$.) Hence, the linear velocity is

$$v = 2960\,\text{miles} \cdot \frac{\pi}{12\,\text{hours}} \approx 775\,\frac{\text{miles}}{\text{hour}}$$

□

It is worth noting that the quantity $\frac{1\,\text{revolution}}{24\,\text{hours}}$ in Example 10.1.5 is called the **ordinary frequency** of the motion and is usually denoted by the variable f. The ordinary frequency is a measure of how often an object makes a complete cycle of the motion. The fact that $\omega = 2\pi f$ suggests that ω is also a frequency. Indeed, it is called the **angular frequency** of the motion. On a related note, the quantity $T = \dfrac{1}{f}$ is called the **period** of the motion and is the amount of time it takes for the object to complete one cycle of the motion. In the scenario of Example 10.1.5, the period of the motion is 24 hours, or one day.

The concepts of frequency and period help frame the equation $v = r\omega$ in a new light. That is, if ω is fixed, points which are farther from the center of rotation need to travel faster to maintain the same angular frequency since they have farther to travel to make one revolution in one period's time. The distance of the object to the center of rotation is the radius of the circle, r, and is the 'magnification factor' which relates ω and v. We will have more to say about frequencies and periods in Section 11.1. While we have exhaustively discussed velocities associated with circular motion, we have yet to discuss a more natural question: if an object is moving on a circular path of radius r with a fixed angular velocity (frequency) ω, what is the position of the object at time t? The answer to this question is the very heart of Trigonometry and is answered in the next section.

10.1.2 EXERCISES

In Exercises 1 - 4, convert the angles into the DMS system. Round each of your answers to the nearest second.

1. $63.75°$
2. $200.325°$
3. $-317.06°$
4. $179.999°$

In Exercises 5 - 8, convert the angles into decimal degrees. Round each of your answers to three decimal places.

5. $125°50'$
6. $-32°10'12''$
7. $502°35'$
8. $237°58'43''$

In Exercises 9 - 28, graph the oriented angle in standard position. Classify each angle according to where its terminal side lies and then give two coterminal angles, one of which is positive and the other negative.

9. $330°$
10. $-135°$
11. $120°$
12. $405°$
13. $-270°$
14. $\frac{5\pi}{6}$
15. $-\frac{11\pi}{3}$
16. $\frac{5\pi}{4}$
17. $\frac{3\pi}{4}$
18. $-\frac{\pi}{3}$
19. $\frac{7\pi}{2}$
20. $\frac{\pi}{4}$
21. $-\frac{\pi}{2}$
22. $\frac{7\pi}{6}$
23. $-\frac{5\pi}{3}$
24. 3π
25. -2π
26. $-\frac{\pi}{4}$
27. $\frac{15\pi}{4}$
28. $-\frac{13\pi}{6}$

In Exercises 29 - 36, convert the angle from degree measure into radian measure, giving the exact value in terms of π.

29. $0°$
30. $240°$
31. $135°$
32. $-270°$
33. $-315°$
34. $150°$
35. $45°$
36. $-225°$

In Exercises 37 - 44, convert the angle from radian measure into degree measure.

37. π
38. $-\frac{2\pi}{3}$
39. $\frac{7\pi}{6}$
40. $\frac{11\pi}{6}$
41. $\frac{\pi}{3}$
42. $\frac{5\pi}{3}$
43. $-\frac{\pi}{6}$
44. $\frac{\pi}{2}$

In Exercises 45 - 49, sketch the oriented arc on the Unit Circle which corresponds to the given real number.

45. $t = \frac{5\pi}{6}$
46. $t = -\pi$
47. $t = 6$
48. $t = -2$
49. $t = 12$

50. A yo-yo which is 2.25 inches in diameter spins at a rate of 4500 revolutions per minute. How fast is the edge of the yo-yo spinning in miles per hour? Round your answer to two decimal places.

51. How many revolutions per minute would the yo-yo in exercise 50 have to complete if the edge of the yo-yo is to be spinning at a rate of 42 miles per hour? Round your answer to two decimal places.

52. In the yo-yo trick 'Around the World,' the performer throws the yo-yo so it sweeps out a vertical circle whose radius is the yo-yo string. If the yo-yo string is 28 inches long and the yo-yo takes 3 seconds to complete one revolution of the circle, compute the speed of the yo-yo in miles per hour. Round your answer to two decimal places.

53. A computer hard drive contains a circular disk with diameter 2.5 inches and spins at a rate of 7200 RPM (revolutions per minute). Find the linear speed of a point on the edge of the disk in miles per hour.

54. A rock got stuck in the tread of my tire and when I was driving 70 miles per hour, the rock came loose and hit the inside of the wheel well of the car. How fast, in miles per hour, was the rock traveling when it came out of the tread? (The tire has a diameter of 23 inches.)

55. The Giant Wheel at Cedar Point is a circle with diameter 128 feet which sits on an 8 foot tall platform making its overall height is 136 feet. (Remember this from Exercise 17 in Section 7.2?) It completes two revolutions in 2 minutes and 7 seconds.[20] Assuming the riders are at the edge of the circle, how fast are they traveling in miles per hour?

56. Consider the circle of radius r pictured below with central angle θ, measured in radians, and subtended arc of length s. Prove that the area of the shaded sector is $A = \frac{1}{2}r^2\theta$.

    (Hint: Use the proportion $\frac{A}{\text{area of the circle}} = \frac{s}{\text{circumference of the circle}}$.)

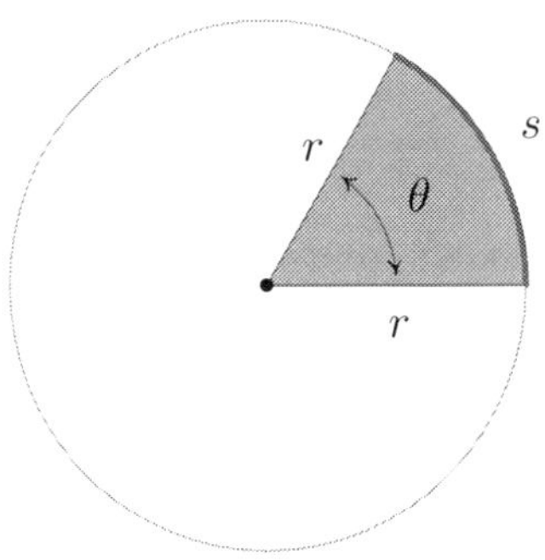

[20]Source: Cedar Point's webpage.

In Exercises 57 - 62, use the result of Exercise 56 to compute the areas of the circular sectors with the given central angles and radii.

57. $\theta = \frac{\pi}{6}, \; r = 12$

58. $\theta = \frac{5\pi}{4}, \; r = 100$

59. $\theta = 330^{\circ}, \; r = 9.3$

60. $\theta = \pi, \; r = 1$

61. $\theta = 240^{\circ}, \; r = 5$

62. $\theta = 1^{\circ}, \; r = 117$

63. Imagine a rope tied around the Earth at the equator. Show that you need to add only 2π feet of length to the rope in order to lift it one foot above the ground around the entire equator. (You do NOT need to know the radius of the Earth to show this.)

64. With the help of your classmates, look for a proof that π is indeed a constant.

10.1.3 Answers

1. $63°45'$
2. $200°19'30''$
3. $-317°3'36''$
4. $179°59'56''$
5. $125.833°$
6. $-32.17°$
7. $502.583°$
8. $237.979°$
9. $330°$ is a Quadrant IV angle coterminal with $690°$ and $-30°$

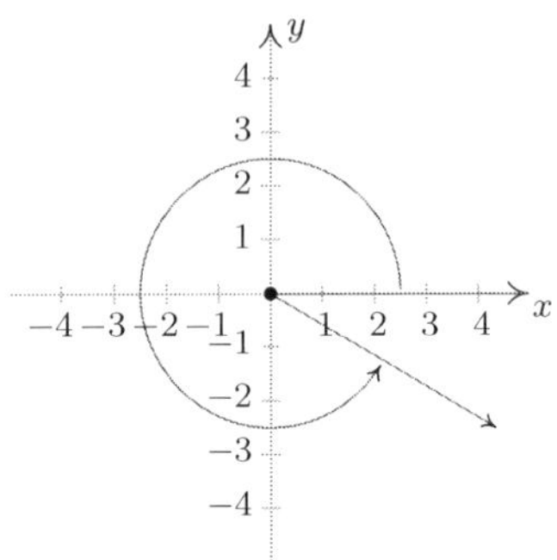

10. $-135°$ is a Quadrant III angle coterminal with $225°$ and $-495°$

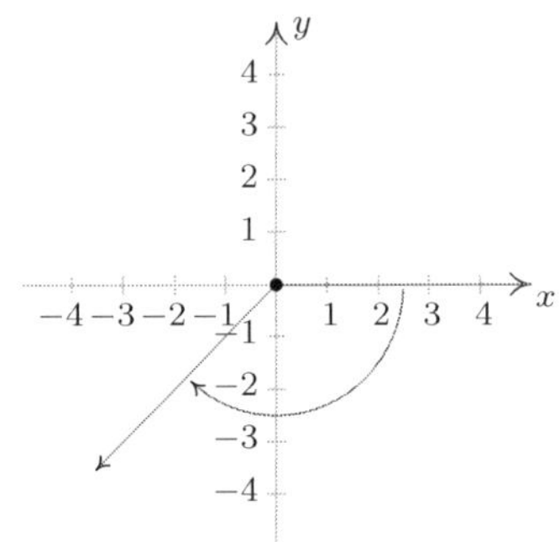

11. $120°$ is a Quadrant II angle coterminal with $480°$ and $-240°$

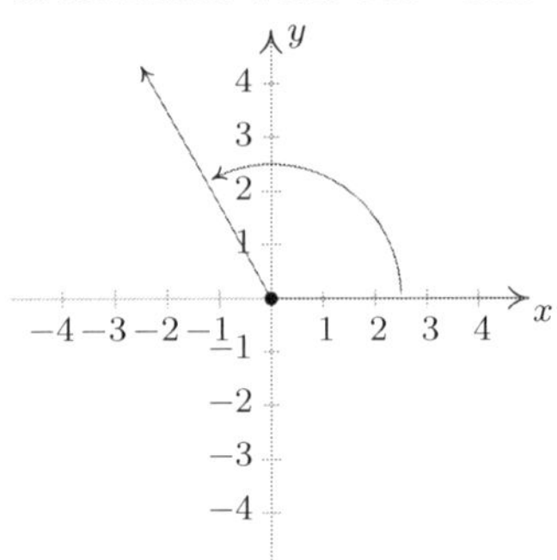

12. $405°$ is a Quadrant I angle coterminal with $45°$ and $-315°$

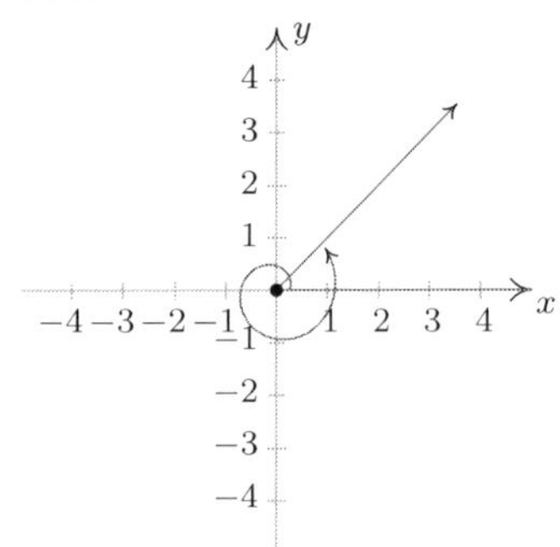

13. $-270°$ lies on the positive y-axis coterminal with $90°$ and $-630°$

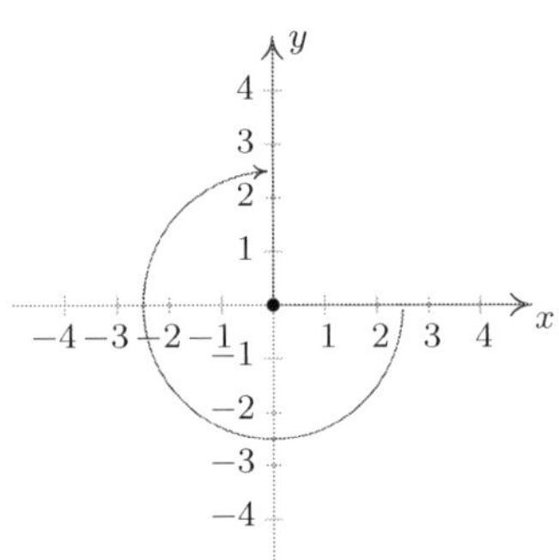

14. $\frac{5\pi}{6}$ is a Quadrant II angle coterminal with $\frac{17\pi}{6}$ and $-\frac{7\pi}{6}$

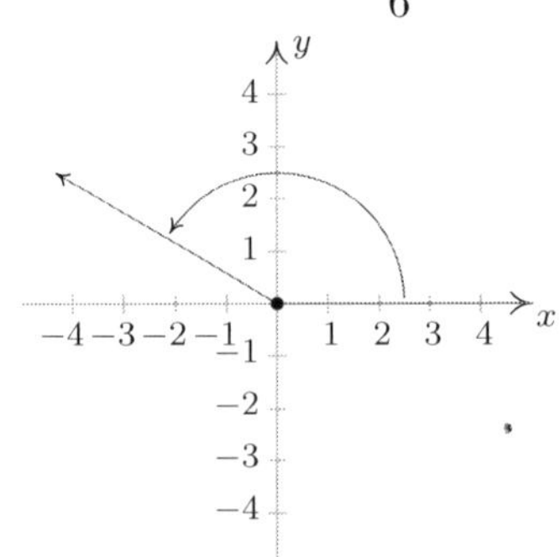

15. $-\frac{11\pi}{3}$ is a Quadrant I angle
coterminal with $\frac{\pi}{3}$ and $-\frac{5\pi}{3}$

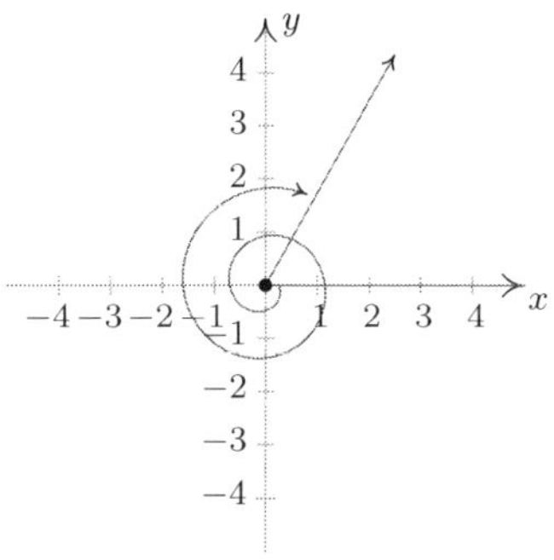

16. $\frac{5\pi}{4}$ is a Quadrant III angle
coterminal with $\frac{13\pi}{4}$ and $-\frac{3\pi}{4}$

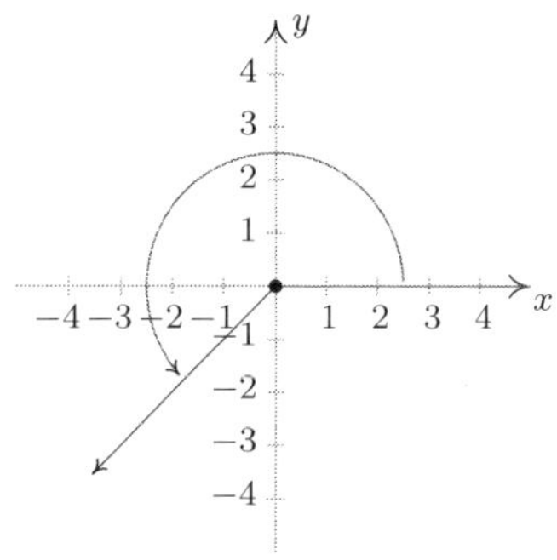

17. $\frac{3\pi}{4}$ is a Quadrant II angle
coterminal with $\frac{11\pi}{4}$ and $-\frac{5\pi}{4}$

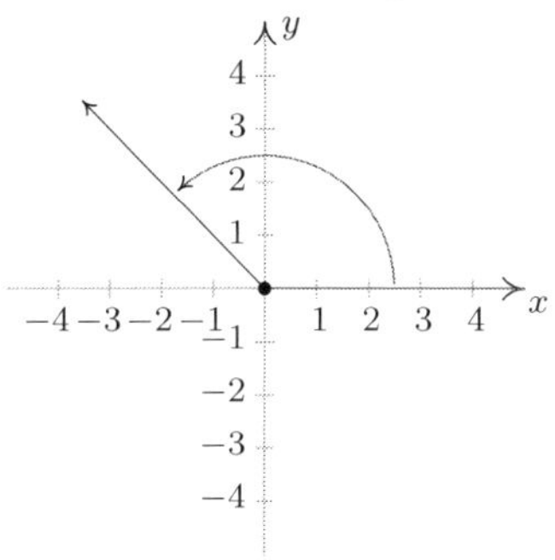

18. $-\frac{\pi}{3}$ is a Quadrant IV angle
coterminal with $\frac{5\pi}{3}$ and $-\frac{7\pi}{3}$

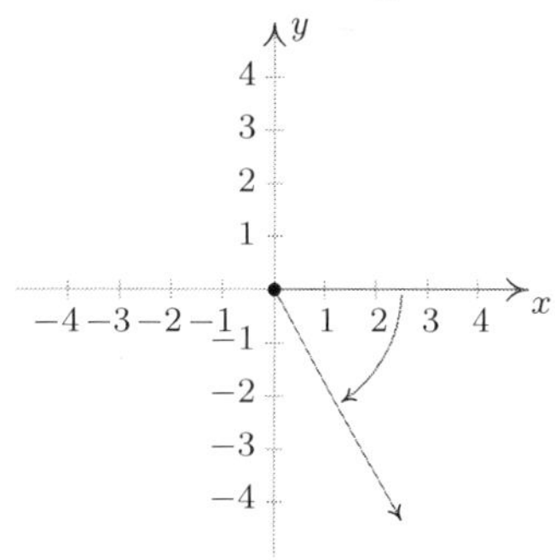

19. $\frac{7\pi}{2}$ lies on the negative y-axis
coterminal with $\frac{3\pi}{2}$ and $-\frac{\pi}{2}$

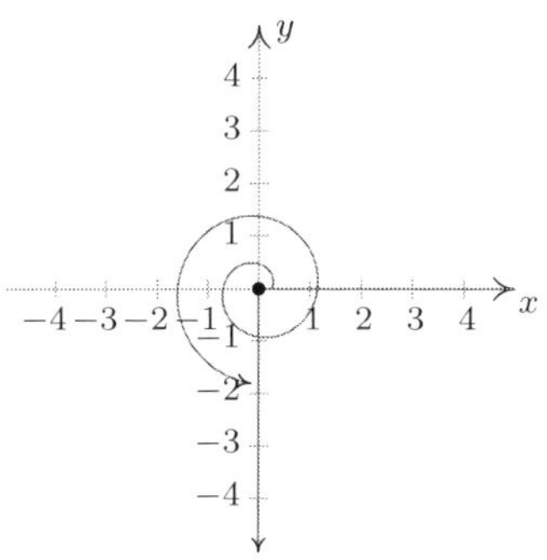

20. $\frac{\pi}{4}$ is a Quadrant I angle
coterminal with $\frac{9\pi}{4}$ and $-\frac{7\pi}{4}$

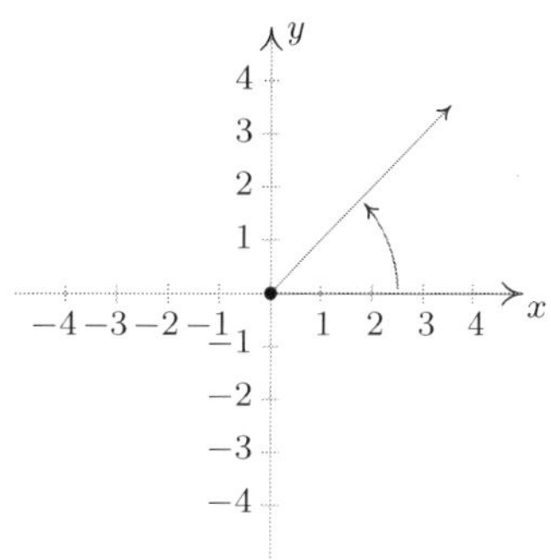

21. $-\frac{\pi}{2}$ lies on the negative y-axis

coterminal with $\frac{3\pi}{2}$ and $-\frac{5\pi}{2}$

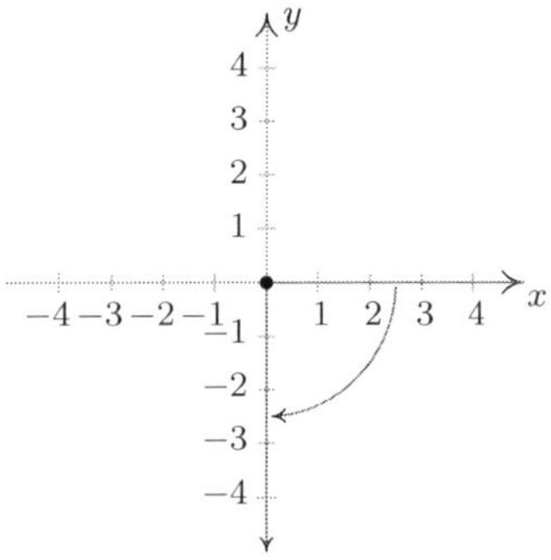

22. $\frac{7\pi}{6}$ is a Quadrant III angle

coterminal with $\frac{19\pi}{6}$ and $-\frac{5\pi}{6}$

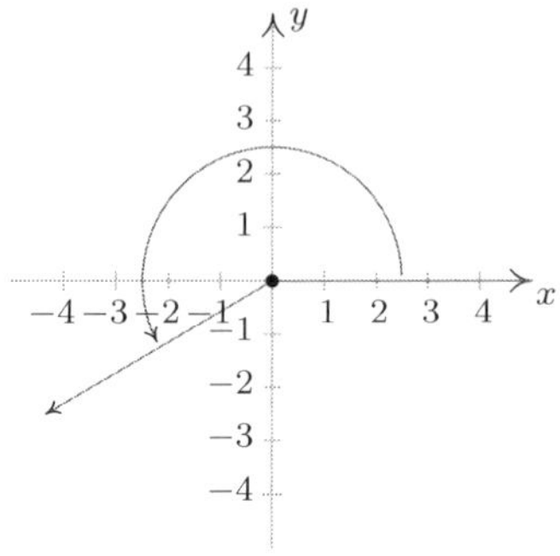

23. $-\frac{5\pi}{3}$ is a Quadrant I angle

coterminal with $\frac{\pi}{3}$ and $-\frac{11\pi}{3}$

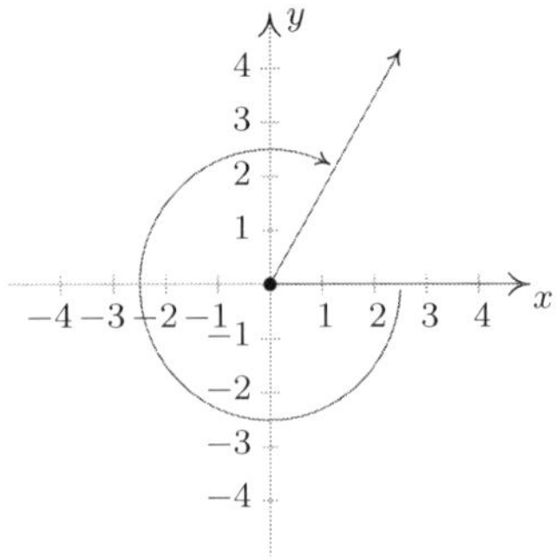

24. 3π lies on the negative x-axis

coterminal with π and $-\pi$

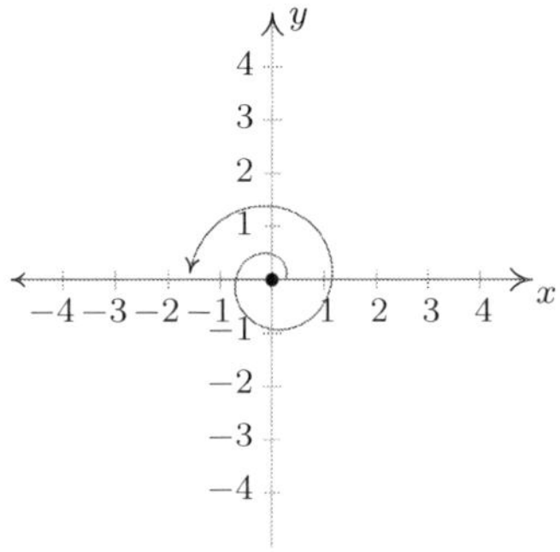

25. -2π lies on the positive x-axis

coterminal with 2π and -4π

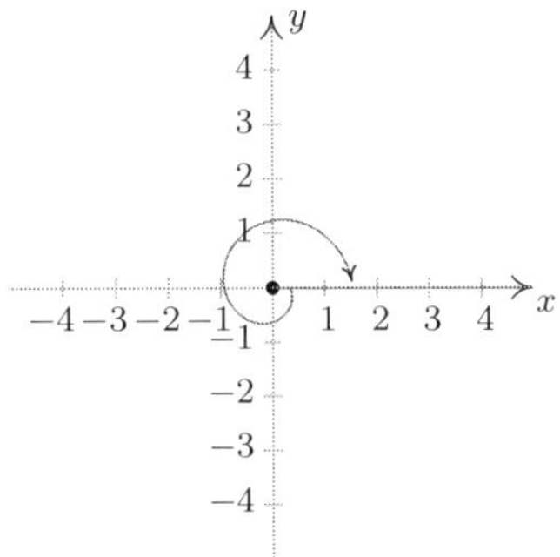

26. $-\frac{\pi}{4}$ is a Quadrant IV angle

coterminal with $\frac{7\pi}{4}$ and $-\frac{9\pi}{4}$

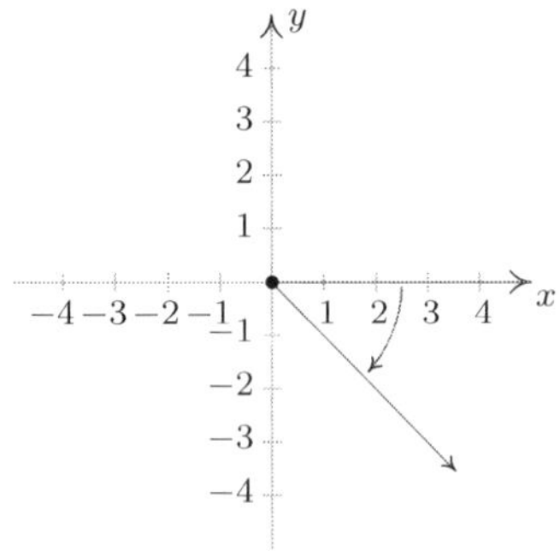

27. $\frac{15\pi}{4}$ is a Quadrant IV angle coterminal with $\frac{7\pi}{4}$ and $-\frac{\pi}{4}$

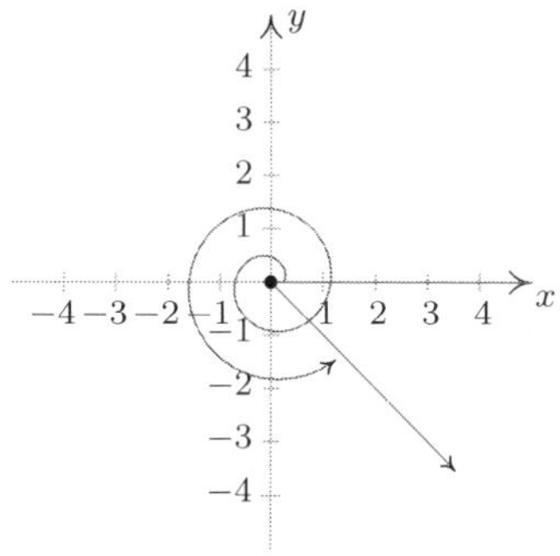

28. $-\frac{13\pi}{6}$ is a Quadrant IV angle coterminal with $\frac{11\pi}{6}$ and $-\frac{\pi}{6}$

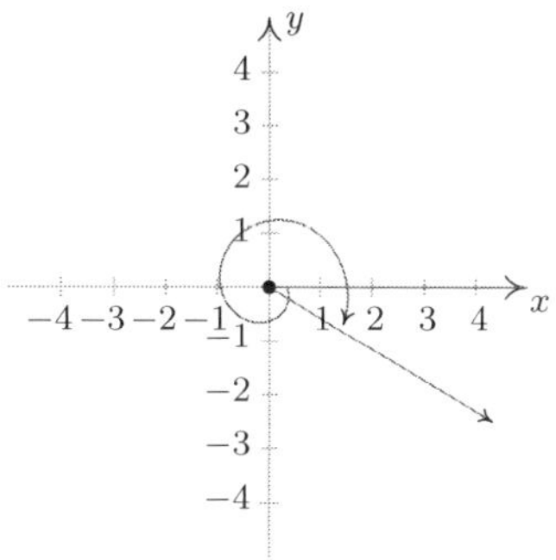

29. 0

30. $\frac{4\pi}{3}$

31. $\frac{3\pi}{4}$

32. $-\frac{3\pi}{2}$

33. $-\frac{7\pi}{4}$

34. $\frac{5\pi}{6}$

35. $\frac{\pi}{4}$

36. $-\frac{5\pi}{4}$

37. 180°

38. -120°

39. 210°

40. 330°

41. 60°

42. 300°

43. -30°

44. 90°

45. $t = \frac{5\pi}{6}$

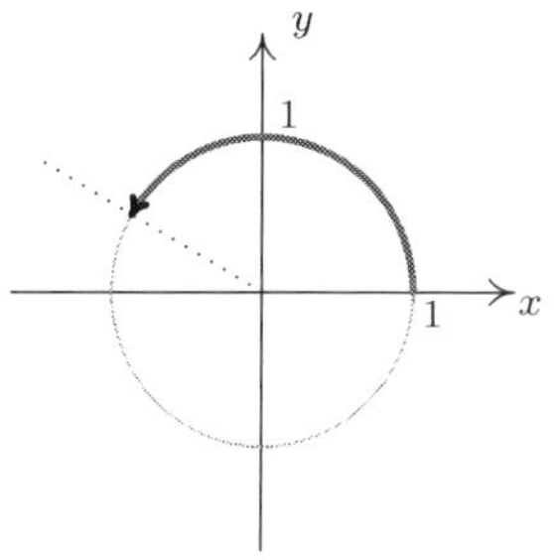

46. $t = -\pi$

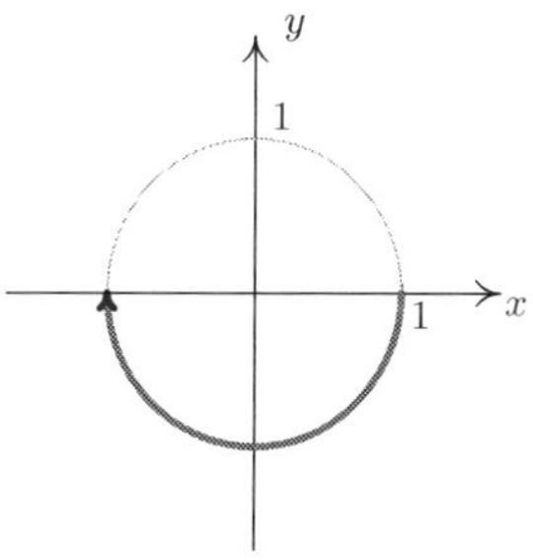

47. $t = 6$

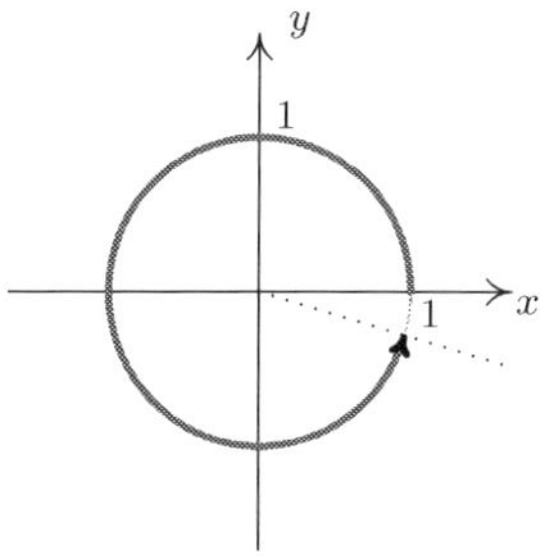

48. $t = -2$

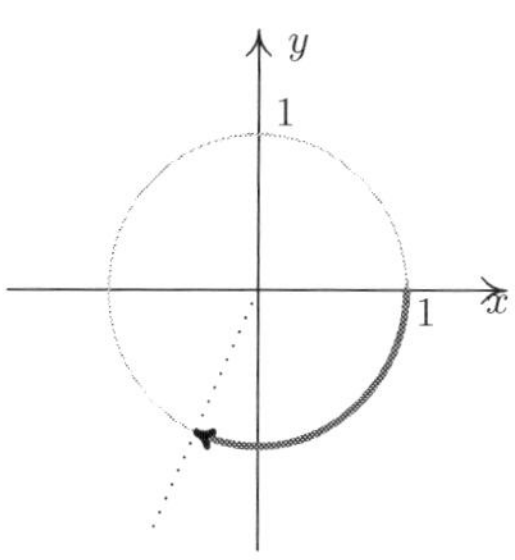

49. $t = 12$ (between 1 and 2 revolutions)

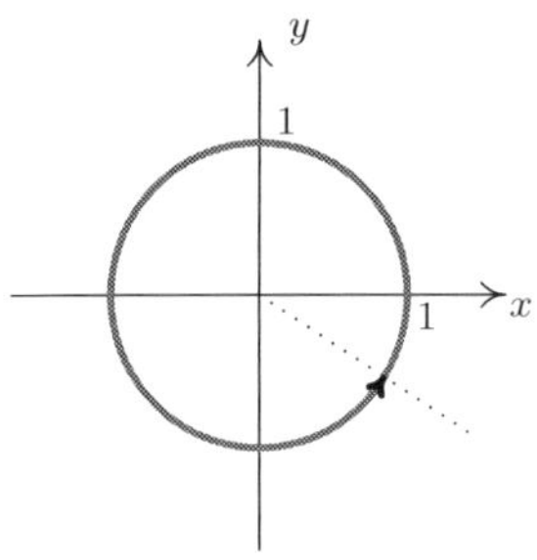

50. About 30.12 miles per hour

51. About 6274.52 revolutions per minute

52. About 3.33 miles per hour

53. About 53.55 miles per hour

54. 70 miles per hour

55. About 4.32 miles per hour

57. 12π square units

58. 6250π square units

59. $79.2825\pi \approx 249.07$ square units

60. $\frac{\pi}{2}$ square units

61. $\frac{50\pi}{3}$ square units

62. $38.025\pi \approx 119.46$ square units

10.2 THE UNIT CIRCLE: COSINE AND SINE

In Section 10.1.1, we introduced circular motion and derived a formula which describes the linear velocity of an object moving on a circular path at a constant angular velocity. One of the goals of this section is describe the *position* of such an object. To that end, consider an angle θ in standard position and let P denote the point where the terminal side of θ intersects the Unit Circle. By associating the point P with the angle θ, we are assigning a *position* on the Unit Circle to the angle θ. The x-coordinate of P is called the **cosine** of θ, written $\cos(\theta)$, while the y-coordinate of P is called the **sine** of θ, written $\sin(\theta)$.[1] The reader is encouraged to verify that these rules used to match an angle with its cosine and sine do, in fact, satisfy the definition of a function. That is, for each angle θ, there is only one associated value of $\cos(\theta)$ and only one associated value of $\sin(\theta)$.

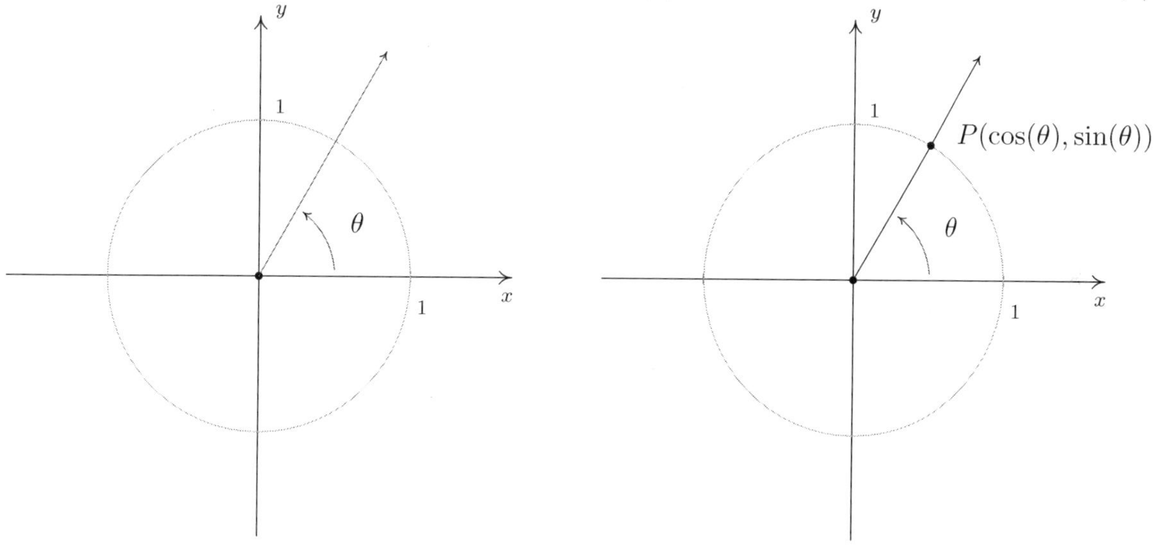

Example 10.2.1. Find the cosine and sine of the following angles.

1. $\theta = 270^\circ$
2. $\theta = -\pi$
3. $\theta = 45^\circ$
4. $\theta = \frac{\pi}{6}$
5. $\theta = 60^\circ$

Solution.

1. To find $\cos(270^\circ)$ and $\sin(270^\circ)$, we plot the angle $\theta = 270^\circ$ in standard position and find the point on the terminal side of θ which lies on the Unit Circle. Since 270° represents $\frac{3}{4}$ of a counter-clockwise revolution, the terminal side of θ lies along the negative y-axis. Hence, the point we seek is $(0, -1)$ so that $\cos(270^\circ) = 0$ and $\sin(270^\circ) = -1$.

2. The angle $\theta = -\pi$ represents one half of a clockwise revolution so its terminal side lies on the negative x-axis. The point on the Unit Circle that lies on the negative x-axis is $(-1, 0)$ which means $\cos(-\pi) = -1$ and $\sin(-\pi) = 0$.

[1] The etymology of the name 'sine' is quite colorful, and the interested reader is invited to research it; the 'co' in 'cosine' is explained in Section 10.4.

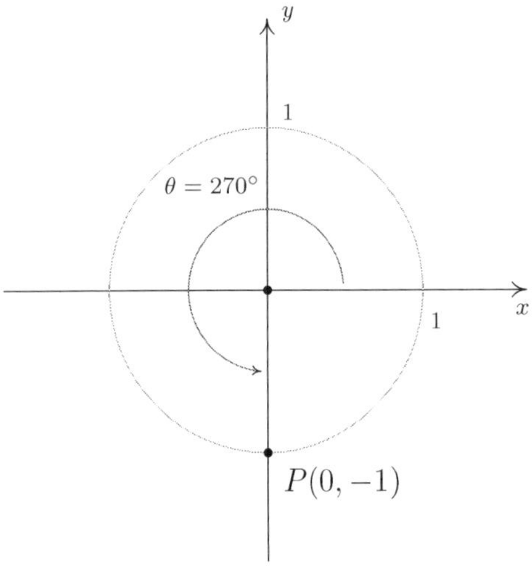

Finding $\cos(270°)$ and $\sin(270°)$

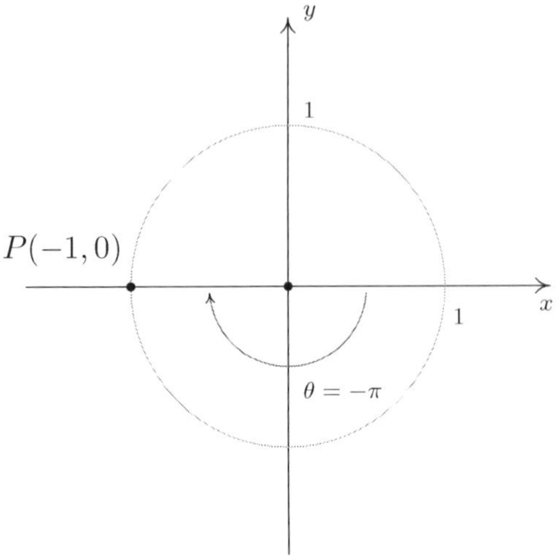

Finding $\cos(-\pi)$ and $\sin(-\pi)$

3. When we sketch $\theta = 45°$ in standard position, we see that its terminal does not lie along any of the coordinate axes which makes our job of finding the cosine and sine values a bit more difficult. Let $P(x, y)$ denote the point on the terminal side of θ which lies on the Unit Circle. By definition, $x = \cos(45°)$ and $y = \sin(45°)$. If we drop a perpendicular line segment from P to the x-axis, we obtain a $45° - 45° - 90°$ right triangle whose legs have lengths x and y units. From Geometry,[2] we get $y = x$. Since $P(x, y)$ lies on the Unit Circle, we have $x^2 + y^2 = 1$. Substituting $y = x$ into this equation yields $2x^2 = 1$, or $x = \pm\sqrt{\frac{1}{2}} = \pm\frac{\sqrt{2}}{2}$. Since $P(x, y)$ lies in the first quadrant, $x > 0$, so $x = \cos(45°) = \frac{\sqrt{2}}{2}$ and with $y = x$ we have $y = \sin(45°) = \frac{\sqrt{2}}{2}$.

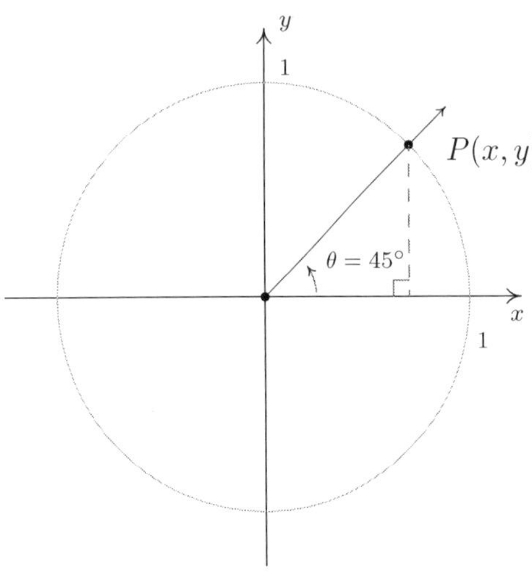

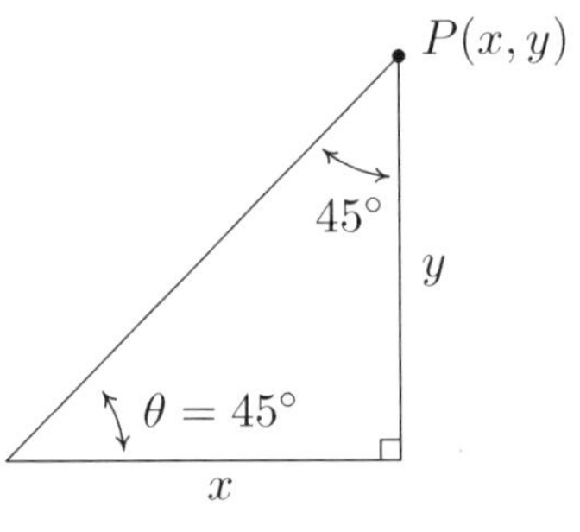

[2]Can you show this?

4. As before, the terminal side of $\theta = \frac{\pi}{6}$ does not lie on any of the coordinate axes, so we proceed using a triangle approach. Letting $P(x, y)$ denote the point on the terminal side of θ which lies on the Unit Circle, we drop a perpendicular line segment from P to the x-axis to form a $30° - 60° - 90°$ right triangle. After a bit of Geometry[3] we find $y = \frac{1}{2}$ so $\sin\left(\frac{\pi}{6}\right) = \frac{1}{2}$. Since $P(x, y)$ lies on the Unit Circle, we substitute $y = \frac{1}{2}$ into $x^2 + y^2 = 1$ to get $x^2 = \frac{3}{4}$, or $x = \pm\frac{\sqrt{3}}{2}$. Here, $x > 0$ so $x = \cos\left(\frac{\pi}{6}\right) = \frac{\sqrt{3}}{2}$.

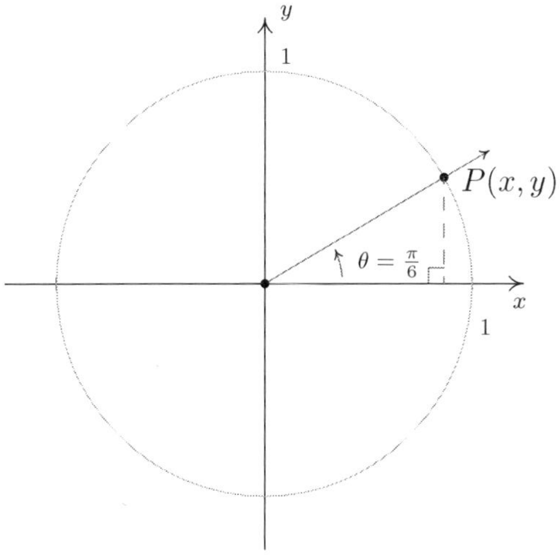

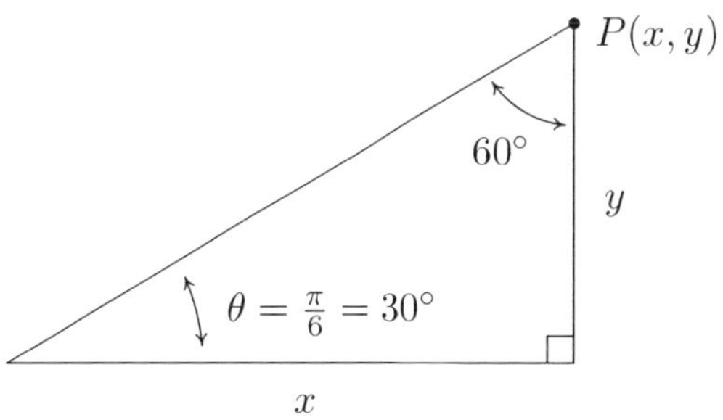

5. Plotting $\theta = 60°$ in standard position, we find it is not a quadrantal angle and set about using a triangle approach. Once again, we get a $30° - 60° - 90°$ right triangle and, after the usual computations, find $x = \cos(60°) = \frac{1}{2}$ and $y = \sin(60°) = \frac{\sqrt{3}}{2}$.

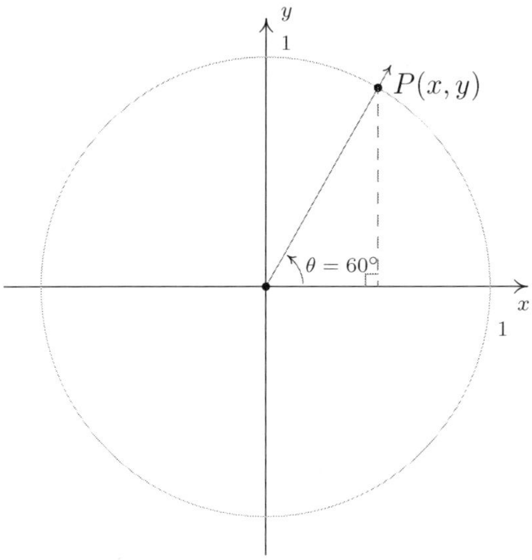

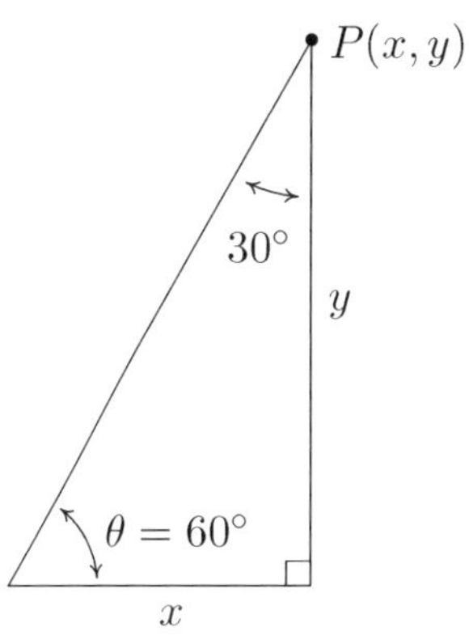

□

[3] Again, can you show this?

In Example 10.2.1, it was quite easy to find the cosine and sine of the quadrantal angles, but for non-quadrantal angles, the task was much more involved. In these latter cases, we made good use of the fact that the point $P(x, y) = (\cos(\theta), \sin(\theta))$ lies on the Unit Circle, $x^2 + y^2 = 1$. If we substitute $x = \cos(\theta)$ and $y = \sin(\theta)$ into $x^2 + y^2 = 1$, we get $(\cos(\theta))^2 + (\sin(\theta))^2 = 1$. An unfortunate[4] convention, which the authors are compelled to perpetuate, is to write $(\cos(\theta))^2$ as $\cos^2(\theta)$ and $(\sin(\theta))^2$ as $\sin^2(\theta)$. Rewriting the identity using this convention results in the following theorem, which is without a doubt one of the most important results in Trigonometry.

Theorem 10.1. The Pythagorean Identity: For any angle θ, $\cos^2(\theta) + \sin^2(\theta) = 1$.

The moniker 'Pythagorean' brings to mind the Pythagorean Theorem, from which both the Distance Formula and the equation for a circle are ultimately derived.[5] The word 'Identity' reminds us that, regardless of the angle θ, the equation in Theorem 10.1 is always true. If one of $\cos(\theta)$ or $\sin(\theta)$ is known, Theorem 10.1 can be used to determine the other, up to a $(\pm)$ sign. If, in addition, we know where the terminal side of θ lies when in standard position, then we can remove the ambiguity of the $(\pm)$ and completely determine the missing value as the next example illustrates.

Example 10.2.2. Using the given information about θ, find the indicated value.

1. If θ is a Quadrant II angle with $\sin(\theta) = \frac{3}{5}$, find $\cos(\theta)$.
2. If $\pi < \theta < \frac{3\pi}{2}$ with $\cos(\theta) = -\frac{\sqrt{5}}{5}$, find $\sin(\theta)$.
3. If $\sin(\theta) = 1$, find $\cos(\theta)$.

Solution.

1. When we substitute $\sin(\theta) = \frac{3}{5}$ into The Pythagorean Identity, $\cos^2(\theta) + \sin^2(\theta) = 1$, we obtain $\cos^2(\theta) + \frac{9}{25} = 1$. Solving, we find $\cos(\theta) = \pm\frac{4}{5}$. Since θ is a Quadrant II angle, its terminal side, when plotted in standard position, lies in Quadrant II. Since the x-coordinates are negative in Quadrant II, $\cos(\theta)$ is too. Hence, $\cos(\theta) = -\frac{4}{5}$.
2. Substituting $\cos(\theta) = -\frac{\sqrt{5}}{5}$ into $\cos^2(\theta) + \sin^2(\theta) = 1$ gives $\sin(\theta) = \pm\frac{2}{\sqrt{5}} = \pm\frac{2\sqrt{5}}{5}$. Since we are given that $\pi < \theta < \frac{3\pi}{2}$, we know θ is a Quadrant III angle. Hence both its sine and cosine are negative and we conclude $\sin(\theta) = -\frac{2\sqrt{5}}{5}$.
3. When we substitute $\sin(\theta) = 1$ into $\cos^2(\theta) + \sin^2(\theta) = 1$, we find $\cos(\theta) = 0$. □

Another tool which helps immensely in determining cosines and sines of angles is the symmetry inherent in the Unit Circle. Suppose, for instance, we wish to know the cosine and sine of $\theta = \frac{5\pi}{6}$. We plot θ in standard position below and, as usual, let $P(x, y)$ denote the point on the terminal side of θ which lies on the Unit Circle. Note that the terminal side of θ lies $\frac{\pi}{6}$ radians short of one half revolution. In Example 10.2.1, we determined that $\cos\left(\frac{\pi}{6}\right) = \frac{\sqrt{3}}{2}$ and $\sin\left(\frac{\pi}{6}\right) = \frac{1}{2}$. This means

[4]This is unfortunate from a 'function notation' perspective. See Section 10.6.

[5]See Sections 1.1 and 7.2 for details.

that the point on the terminal side of the angle $\frac{\pi}{6}$, when plotted in standard position, is $\left(\frac{\sqrt{3}}{2}, \frac{1}{2}\right)$. From the figure below, it is clear that the point $P(x, y)$ we seek can be obtained by reflecting that point about the y-axis. Hence, $\cos\left(\frac{5\pi}{6}\right) = -\frac{\sqrt{3}}{2}$ and $\sin\left(\frac{5\pi}{6}\right) = \frac{1}{2}$.

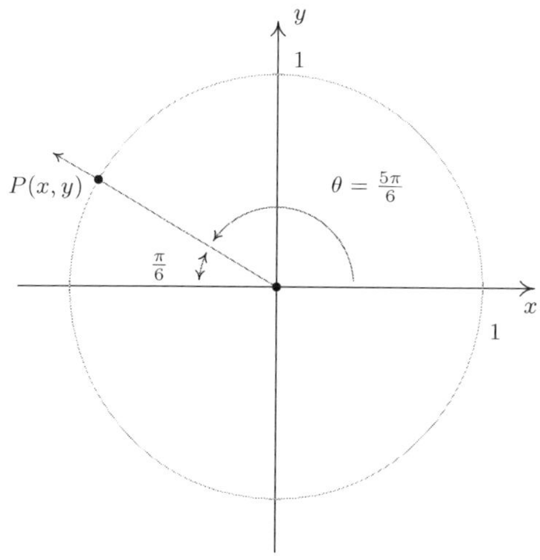

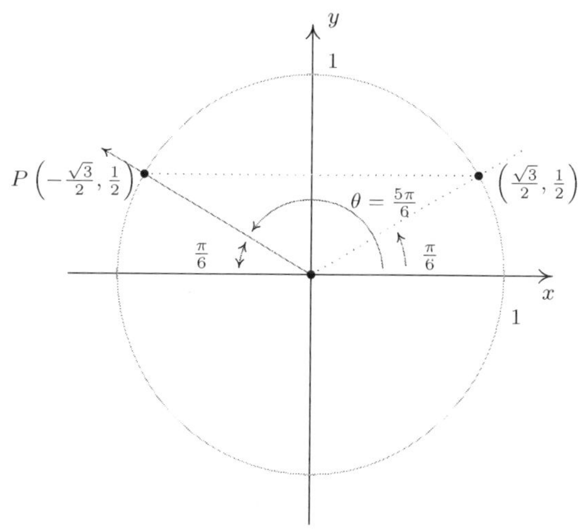

In the above scenario, the angle $\frac{\pi}{6}$ is called the **reference angle** for the angle $\frac{5\pi}{6}$. In general, for a non-quadrantal angle θ, the reference angle for θ (usually denoted α) is the *acute* angle made between the terminal side of θ and the x-axis. If θ is a Quadrant I or IV angle, α is the angle between the terminal side of θ and the *positive* x-axis; if θ is a Quadrant II or III angle, α is the angle between the terminal side of θ and the *negative* x-axis. If we let P denote the point $(\cos(\theta), \sin(\theta))$, then P lies on the Unit Circle. Since the Unit Circle possesses symmetry with respect to the x-axis, y-axis and origin, regardless of where the terminal side of θ lies, there is a point Q symmetric with P which determines θ's reference angle, α as seen below.

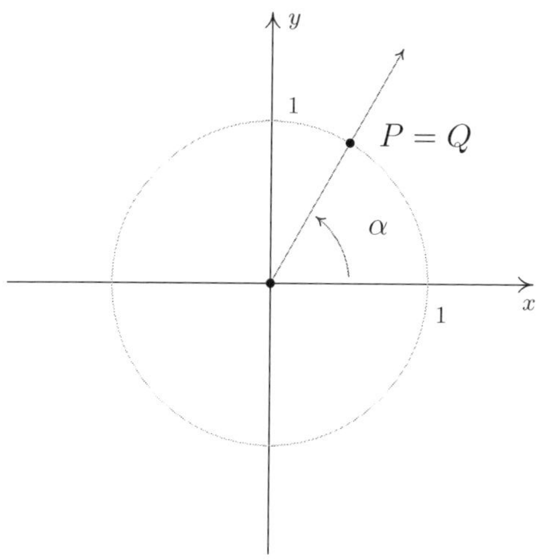

Reference angle α for a Quadrant I angle

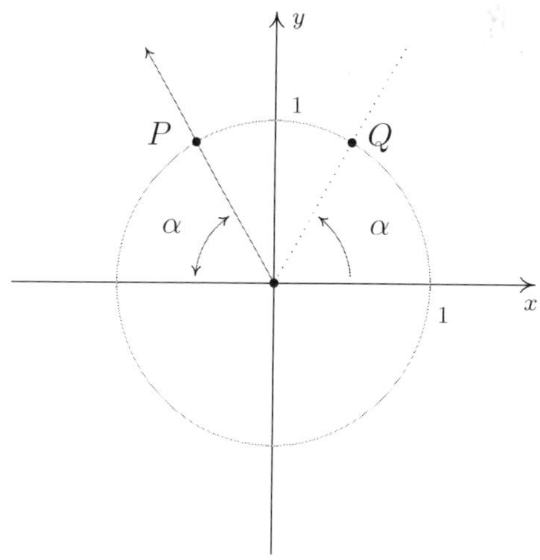

Reference angle α for a Quadrant II angle

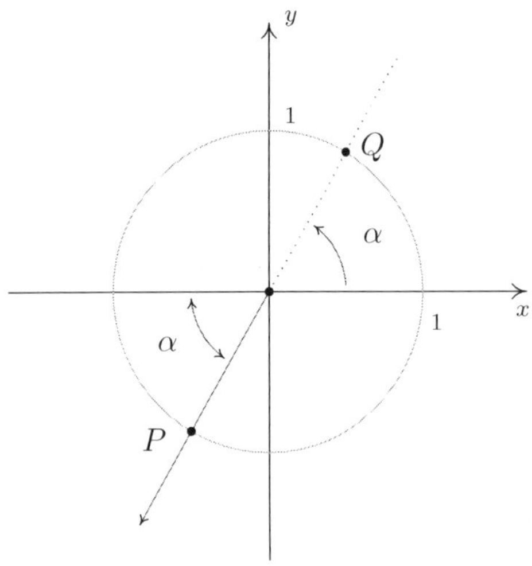

Reference angle α for a Quadrant III angle

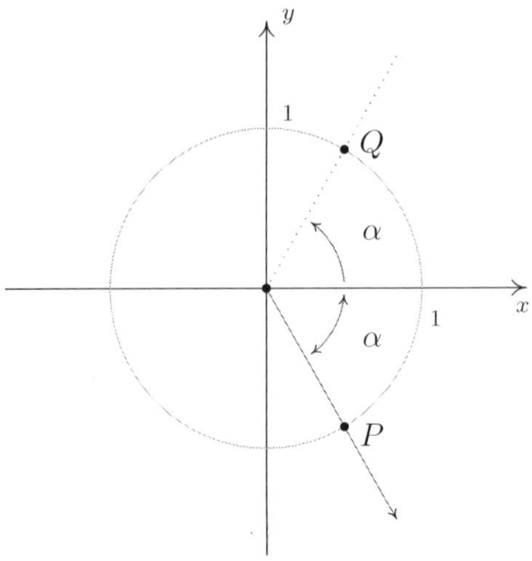

Reference angle α for a Quadrant IV angle

We have just outlined the proof of the following theorem.

Theorem 10.2. Reference Angle Theorem. Suppose α is the reference angle for θ. Then $\cos(\theta) = \pm\cos(\alpha)$ and $\sin(\theta) = \pm\sin(\alpha)$, where the choice of the $(\pm)$ depends on the quadrant in which the terminal side of θ lies.

In light of Theorem 10.2, it pays to know the cosine and sine values for certain common angles. In the table below, we summarize the values which we consider essential and must be memorized.

Cosine and Sine Values of Common Angles

θ(degrees)	θ(radians)	$\cos(\theta)$	$\sin(\theta)$
0°	0	1	0
30°	$\frac{\pi}{6}$	$\frac{\sqrt{3}}{2}$	$\frac{1}{2}$
45°	$\frac{\pi}{4}$	$\frac{\sqrt{2}}{2}$	$\frac{\sqrt{2}}{2}$
60°	$\frac{\pi}{3}$	$\frac{1}{2}$	$\frac{\sqrt{3}}{2}$
90°	$\frac{\pi}{2}$	0	1

Example 10.2.3. Find the cosine and sine of the following angles.

1. $\theta = 225^\circ$
2. $\theta = \frac{11\pi}{6}$
3. $\theta = -\frac{5\pi}{4}$
4. $\theta = \frac{7\pi}{3}$

Solution.

1. We begin by plotting $\theta = 225^\circ$ in standard position and find its terminal side overshoots the negative x-axis to land in Quadrant III. Hence, we obtain θ's reference angle α by subtracting: $\alpha = \theta - 180^\circ = 225^\circ - 180^\circ = 45^\circ$. Since θ is a Quadrant III angle, both $\cos(\theta) < 0$ and

$\sin(\theta) < 0$. The Reference Angle Theorem yields: $\cos(225^\circ) = -\cos(45^\circ) = -\frac{\sqrt{2}}{2}$ and $\sin(225^\circ) = -\sin(45^\circ) = -\frac{\sqrt{2}}{2}$.

2. The terminal side of $\theta = \frac{11\pi}{6}$, when plotted in standard position, lies in Quadrant IV, just shy of the positive x-axis. To find θ's reference angle α, we subtract: $\alpha = 2\pi - \theta = 2\pi - \frac{11\pi}{6} = \frac{\pi}{6}$. Since θ is a Quadrant IV angle, $\cos(\theta) > 0$ and $\sin(\theta) < 0$, so the Reference Angle Theorem gives: $\cos\left(\frac{11\pi}{6}\right) = \cos\left(\frac{\pi}{6}\right) = \frac{\sqrt{3}}{2}$ and $\sin\left(\frac{11\pi}{6}\right) = -\sin\left(\frac{\pi}{6}\right) = -\frac{1}{2}$.

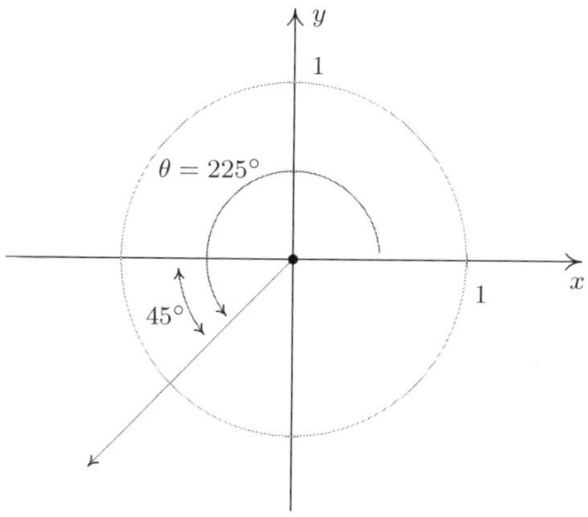

Finding $\cos(225^\circ)$ and $\sin(225^\circ)$

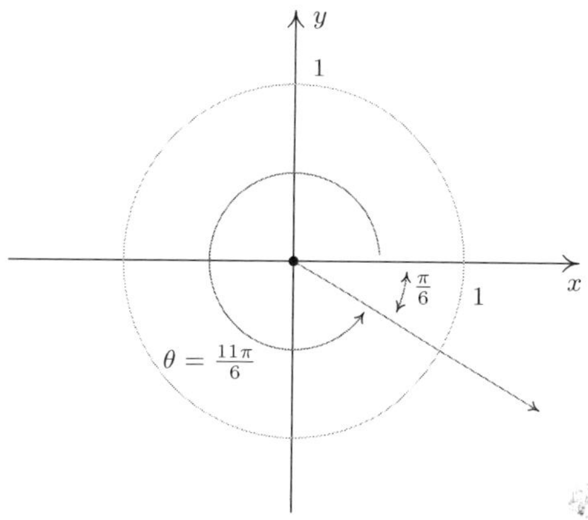

Finding $\cos\left(\frac{11\pi}{6}\right)$ and $\sin\left(\frac{11\pi}{6}\right)$

3. To plot $\theta = -\frac{5\pi}{4}$, we rotate *clockwise* an angle of $\frac{5\pi}{4}$ from the positive x-axis. The terminal side of θ, therefore, lies in Quadrant II making an angle of $\alpha = \frac{5\pi}{4} - \pi = \frac{\pi}{4}$ radians with respect to the negative x-axis. Since θ is a Quadrant II angle, the Reference Angle Theorem gives: $\cos\left(-\frac{5\pi}{4}\right) = -\cos\left(\frac{\pi}{4}\right) = -\frac{\sqrt{2}}{2}$ and $\sin\left(-\frac{5\pi}{4}\right) = \sin\left(\frac{\pi}{4}\right) = \frac{\sqrt{2}}{2}$.

4. Since the angle $\theta = \frac{7\pi}{3}$ measures more than $2\pi = \frac{6\pi}{3}$, we find the terminal side of θ by rotating one full revolution followed by an additional $\alpha = \frac{7\pi}{3} - 2\pi = \frac{\pi}{3}$ radians. Since θ and α are coterminal, $\cos\left(\frac{7\pi}{3}\right) = \cos\left(\frac{\pi}{3}\right) = \frac{1}{2}$ and $\sin\left(\frac{7\pi}{3}\right) = \sin\left(\frac{\pi}{3}\right) = \frac{\sqrt{3}}{2}$.

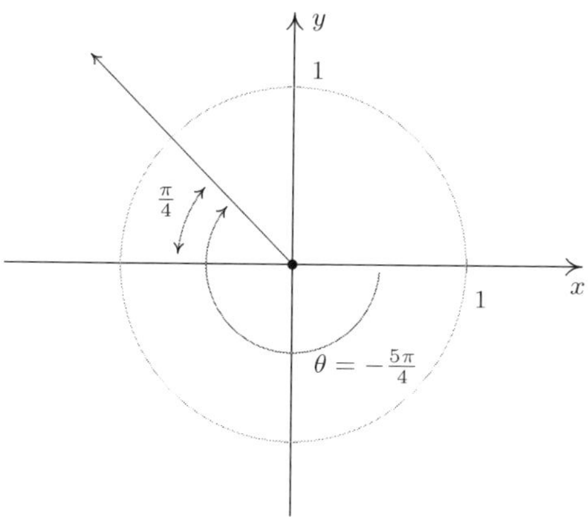

Finding $\cos\left(-\frac{5\pi}{4}\right)$ and $\sin\left(-\frac{5\pi}{4}\right)$

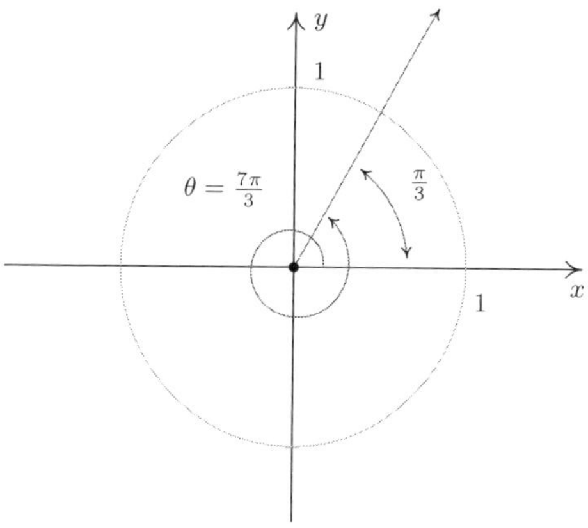

Finding $\cos\left(\frac{7\pi}{3}\right)$ and $\sin\left(\frac{7\pi}{3}\right)$ □

The reader may have noticed that when expressed in radian measure, the reference angle for a non-quadrantal angle is easy to spot. Reduced fraction multiples of π with a denominator of 6 have $\frac{\pi}{6}$ as a reference angle, those with a denominator of 4 have $\frac{\pi}{4}$ as their reference angle, and those with a denominator of 3 have $\frac{\pi}{3}$ as their reference angle.[6] The Reference Angle Theorem in conjunction with the table of cosine and sine values on Page 722 can be used to generate the following figure, which the authors feel should be committed to memory.

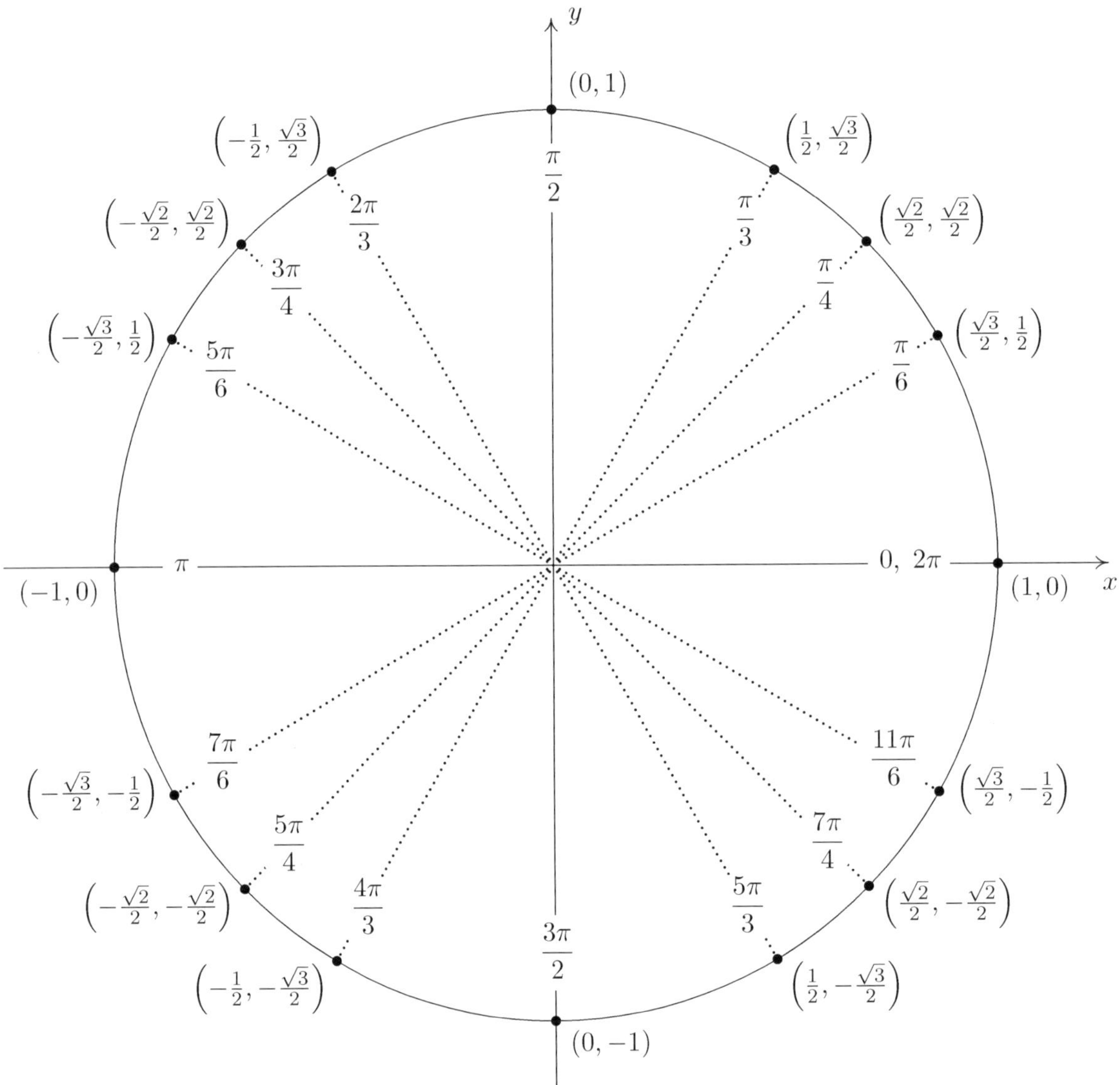

Important Points on the Unit Circle

[6]For once, we have something convenient about using radian measure in contrast to the abstract theoretical nonsense about using them as a 'natural' way to match oriented angles with real numbers!

The next example summarizes all of the important ideas discussed thus far in the section.

Example 10.2.4. Suppose α is an acute angle with $\cos(\alpha) = \frac{5}{13}$.

1. Find $\sin(\alpha)$ and use this to plot α in standard position.

2. Find the sine and cosine of the following angles:

   (a) $\theta = \pi + \alpha$ (b) $\theta = 2\pi - \alpha$ (c) $\theta = 3\pi - \alpha$ (d) $\theta = \frac{\pi}{2} + \alpha$

Solution.

1. Proceeding as in Example 10.2.2, we substitute $\cos(\alpha) = \frac{5}{13}$ into $\cos^2(\alpha) + \sin^2(\alpha) = 1$ and find $\sin(\alpha) = \pm\frac{12}{13}$. Since α is an acute (and therefore Quadrant I) angle, $\sin(\alpha)$ is positive. Hence, $\sin(\alpha) = \frac{12}{13}$. To plot α in standard position, we begin our rotation on the positive x-axis to the ray which contains the point $(\cos(\alpha), \sin(\alpha)) = \left(\frac{5}{13}, \frac{12}{13}\right)$.

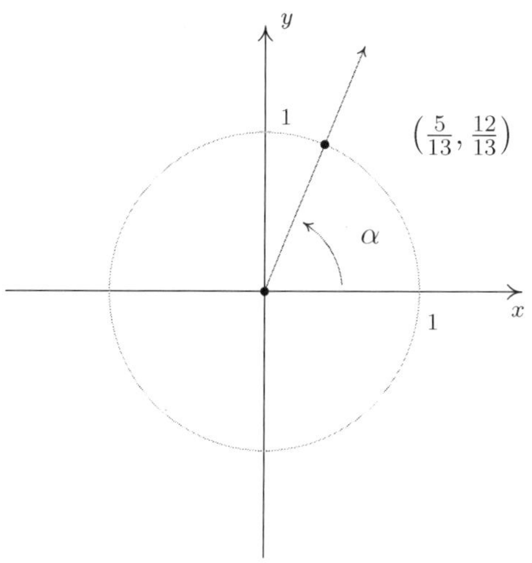

Sketching α

2. (a) To find the cosine and sine of $\theta = \pi + \alpha$, we first plot θ in standard position. We can imagine the sum of the angles $\pi+\alpha$ as a sequence of two rotations: a rotation of π radians followed by a rotation of α radians.[7] We see that α is the reference angle for θ, so by The Reference Angle Theorem, $\cos(\theta) = \pm\cos(\alpha) = \pm\frac{5}{13}$ and $\sin(\theta) = \pm\sin(\alpha) = \pm\frac{12}{13}$. Since the terminal side of θ falls in Quadrant III, both $\cos(\theta)$ and $\sin(\theta)$ are negative, hence, $\cos(\theta) = -\frac{5}{13}$ and $\sin(\theta) = -\frac{12}{13}$.

[7] Since $\pi + \alpha = \alpha + \pi$, θ may be plotted by reversing the order of rotations given here. You should do this.

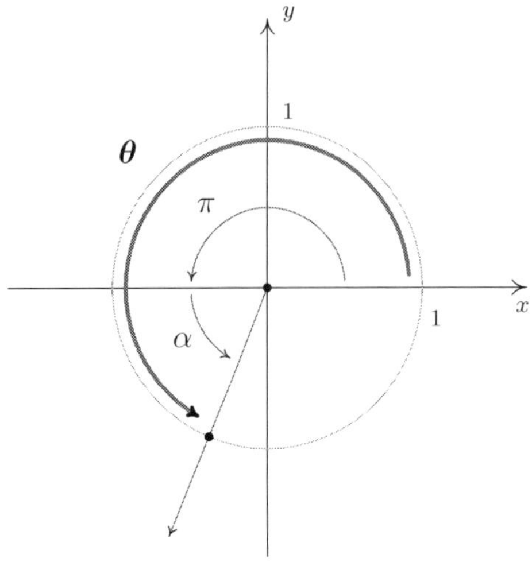

Visualizing $\boldsymbol{\theta = \pi + \alpha}$

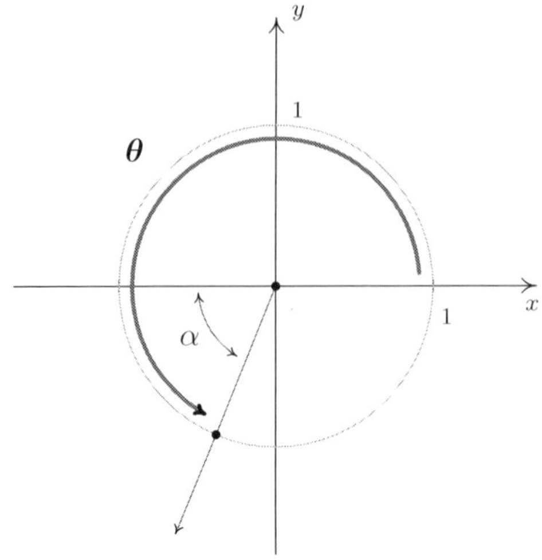

$\boldsymbol{\theta}$ has reference angle α

(b) Rewriting $\theta = 2\pi - \alpha$ as $\theta = 2\pi + (-\alpha)$, we can plot θ by visualizing one complete revolution counter-clockwise followed by a *clockwise* revolution, or 'backing up,' of α radians. We see that α is θ's reference angle, and since θ is a Quadrant IV angle, the Reference Angle Theorem gives: $\cos(\theta) = \frac{5}{13}$ and $\sin(\theta) = -\frac{12}{13}$.

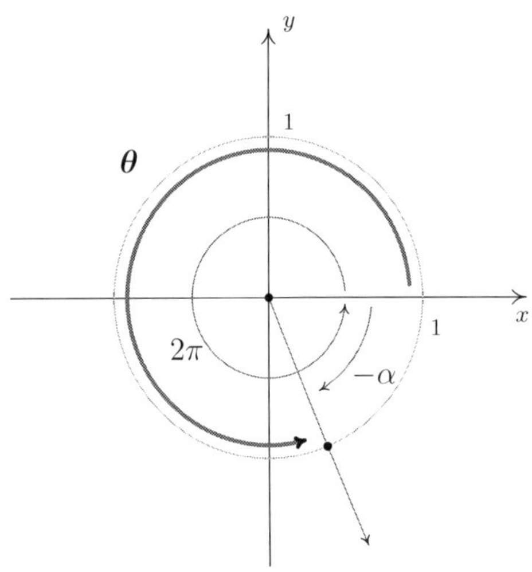

Visualizing $\boldsymbol{\theta = 2\pi - \alpha}$

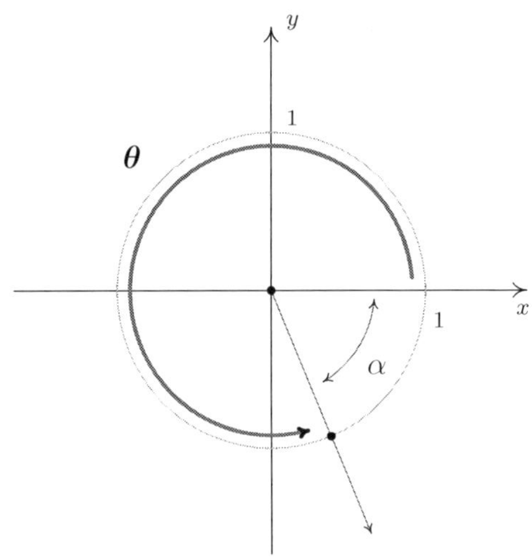

$\boldsymbol{\theta}$ has reference angle α

(c) Taking a cue from the previous problem, we rewrite $\theta = 3\pi - \alpha$ as $\theta = 3\pi + (-\alpha)$. The angle 3π represents one and a half revolutions counter-clockwise, so that when we 'back up' α radians, we end up in Quadrant II. Using the Reference Angle Theorem, we get $\cos(\theta) = -\frac{5}{13}$ and $\sin(\theta) = \frac{12}{13}$.

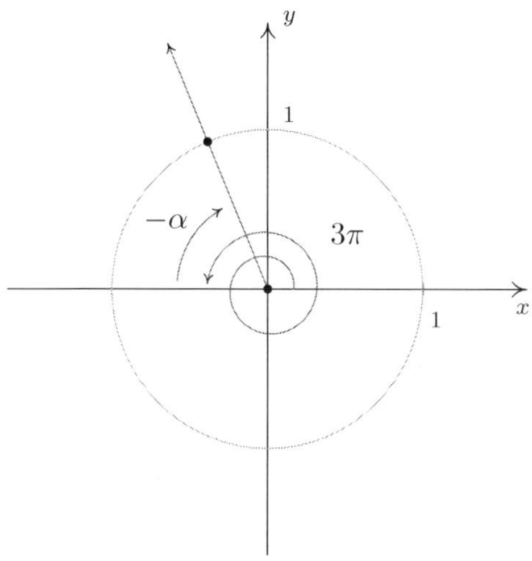

Visualizing $3\pi - \alpha$

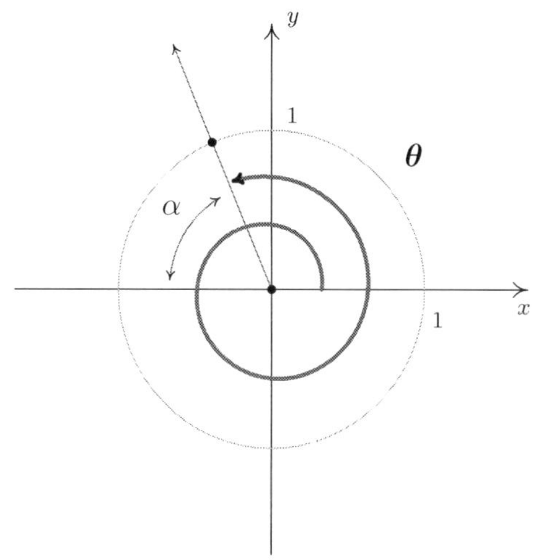

$\boldsymbol{\theta}$ has reference angle α

(d) To plot $\theta = \frac{\pi}{2} + \alpha$, we first rotate $\frac{\pi}{2}$ radians and follow up with α radians. The reference angle here is *not* α, so The Reference Angle Theorem is not immediately applicable. (It's important that you see why this is the case. Take a moment to think about this before reading on.) Let $Q(x, y)$ be the point on the terminal side of θ which lies on the Unit Circle so that $x = \cos(\theta)$ and $y = \sin(\theta)$. Once we graph α in standard position, we use the fact that equal angles subtend equal chords to show that the dotted lines in the figure below are equal. Hence, $x = \cos(\theta) = -\frac{12}{13}$. Similarly, we find $y = \sin(\theta) = \frac{5}{13}$.

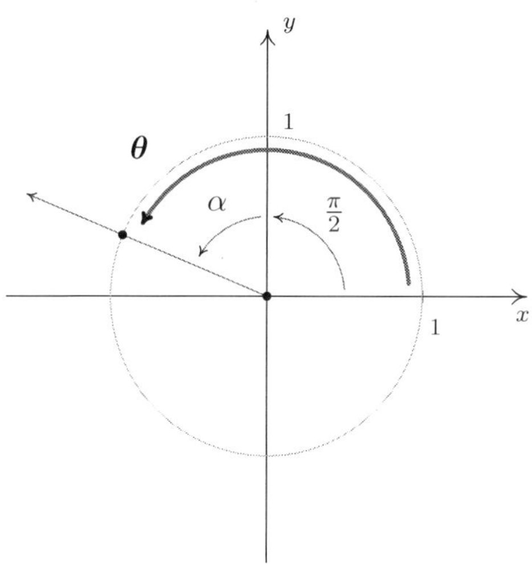

Visualizing $\boldsymbol{\theta = \frac{\pi}{2} + \alpha}$

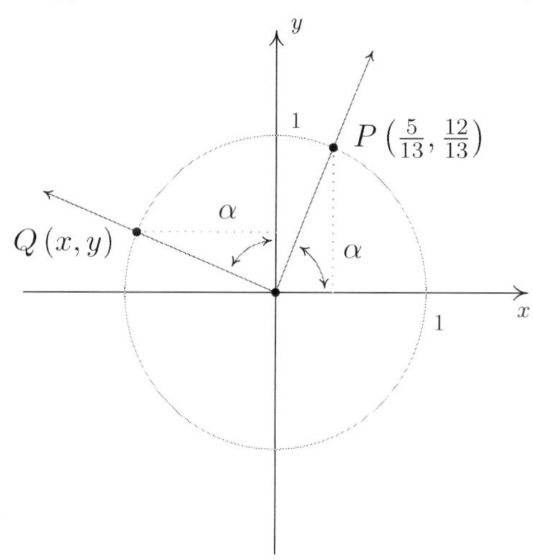

Using symmetry to determine $Q(x, y)$

□

Our next example asks us to solve some very basic trigonometric equations.[8]

Example 10.2.5. Find all of the angles which satisfy the given equation.

1. $\cos(\theta) = \dfrac{1}{2}$
2. $\sin(\theta) = -\dfrac{1}{2}$
3. $\cos(\theta) = 0$.

Solution. Since there is no context in the problem to indicate whether to use degrees or radians, we will default to using radian measure in our answers to each of these problems. This choice will be justified later in the text when we study what is known as Analytic Trigonometry. In those sections to come, radian measure will be the *only* appropriate angle measure so it is worth the time to become "fluent in radians" now.

1. If $\cos(\theta) = \frac{1}{2}$, then the terminal side of θ, when plotted in standard position, intersects the Unit Circle at $x = \frac{1}{2}$. This means θ is a Quadrant I or IV angle with reference angle $\frac{\pi}{3}$.

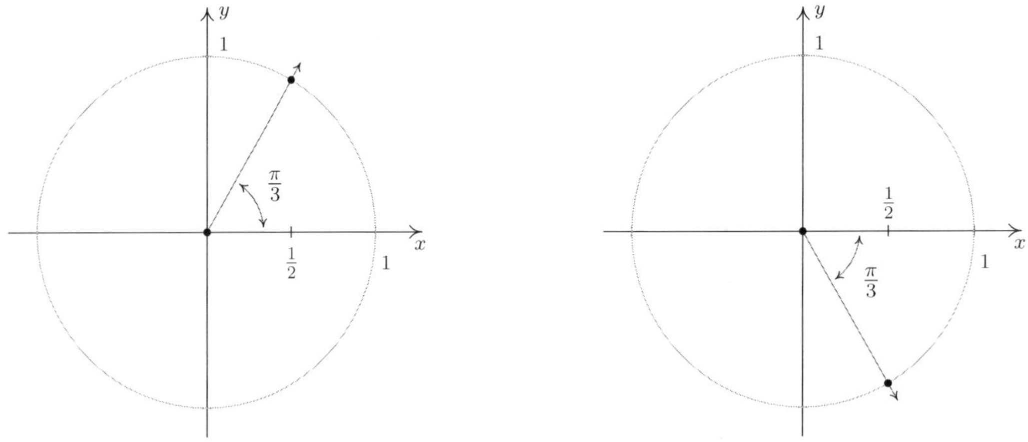

One solution in Quadrant I is $\theta = \frac{\pi}{3}$, and since all other Quadrant I solutions must be coterminal with $\frac{\pi}{3}$, we find $\theta = \frac{\pi}{3} + 2\pi k$ for integers k.[9] Proceeding similarly for the Quadrant IV case, we find the solution to $\cos(\theta) = \frac{1}{2}$ here is $\frac{5\pi}{3}$, so our answer in this Quadrant is $\theta = \frac{5\pi}{3} + 2\pi k$ for integers k.

2. If $\sin(\theta) = -\frac{1}{2}$, then when θ is plotted in standard position, its terminal side intersects the Unit Circle at $y = -\frac{1}{2}$. From this, we determine θ is a Quadrant III or Quadrant IV angle with reference angle $\frac{\pi}{6}$.

[8] We will study trigonometric equations more formally in Section 10.7. Enjoy these relatively straightforward exercises while they last!

[9] Recall in Section 10.1, two angles in radian measure are coterminal if and only if they differ by an integer multiple of 2π. Hence to describe all angles coterminal with a given angle, we add $2\pi k$ for integers $k = 0, \pm 1, \pm 2, \ldots$.

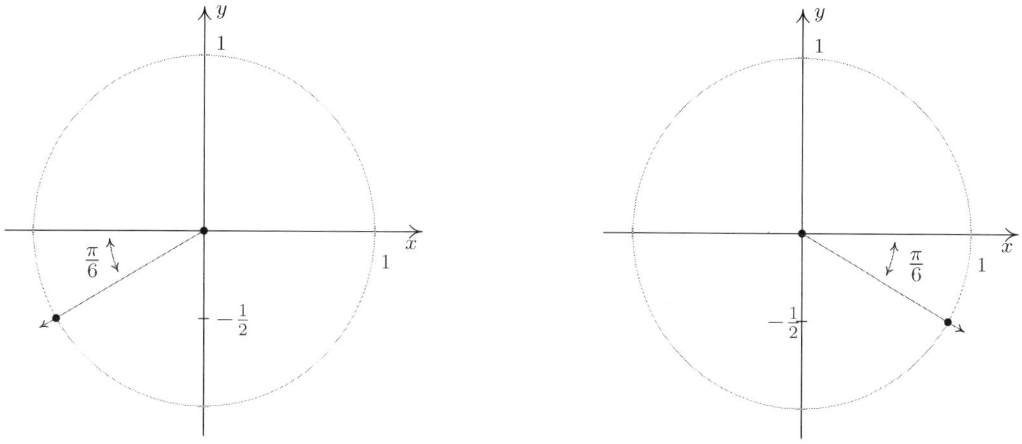

In Quadrant III, one solution is $\frac{7\pi}{6}$, so we capture all Quadrant III solutions by adding integer multiples of 2π: $\theta = \frac{7\pi}{6} + 2\pi k$. In Quadrant IV, one solution is $\frac{11\pi}{6}$ so all the solutions here are of the form $\theta = \frac{11\pi}{6} + 2\pi k$ for integers k.

3. The angles with $\cos(\theta) = 0$ are quadrantal angles whose terminal sides, when plotted in standard position, lie along the y-axis.

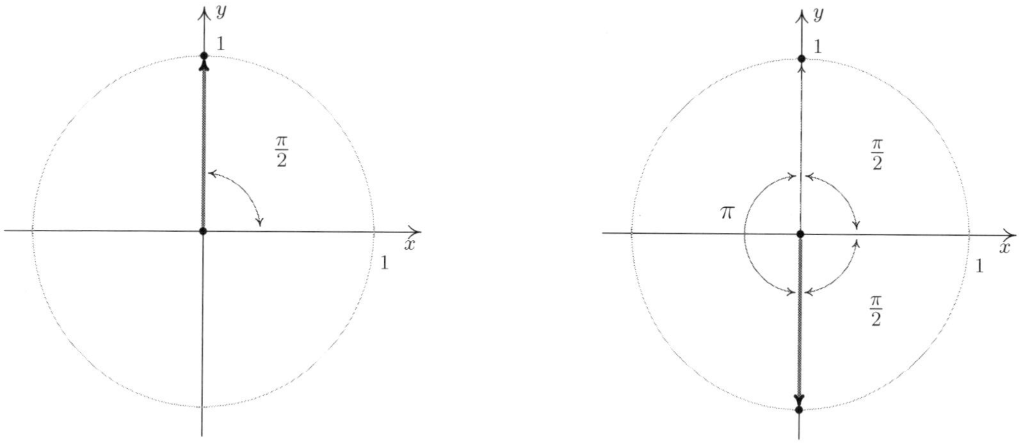

While, technically speaking, $\frac{\pi}{2}$ isn't a reference angle we can nonetheless use it to find our answers. If we follow the procedure set forth in the previous examples, we find $\theta = \frac{\pi}{2} + 2\pi k$ and $\theta = \frac{3\pi}{2} + 2\pi k$ for integers, k. While this solution is correct, it can be shortened to $\theta = \frac{\pi}{2} + \pi k$ for integers k. (Can you see why this works from the diagram?) □

One of the key items to take from Example 10.2.5 is that, in general, solutions to trigonometric equations consist of infinitely many answers. To get a feel for these answers, the reader is encouraged to follow our mantra from Chapter 9 - that is, 'When in doubt, write it out!' This is especially important when checking answers to the exercises. For example, another Quadrant IV solution to $\sin(\theta) = -\frac{1}{2}$ is $\theta = -\frac{\pi}{6}$. Hence, the family of Quadrant IV answers to number 2 above could just have easily been written $\theta = -\frac{\pi}{6} + 2\pi k$ for integers k. While on the surface, this family may look different than the stated solution of $\theta = \frac{11\pi}{6} + 2\pi k$ for integers k, we leave it to the reader to show they represent the same list of angles.

10.2.1 Beyond the Unit Circle

We began the section with a quest to describe the position of a particle experiencing circular motion. In defining the cosine and sine functions, we assigned to each angle a position on the *Unit* Circle. In this subsection, we broaden our scope to include circles of radius r centered at the origin. Consider for the moment the *acute* angle θ drawn below in standard position. Let $Q(x, y)$ be the point on the terminal side of θ which lies on the circle $x^2 + y^2 = r^2$, and let $P(x', y')$ be the point on the terminal side of θ which lies on the Unit Circle. Now consider dropping perpendiculars from P and Q to create two right triangles, ΔOPA and ΔOQB. These triangles are similar,[10] thus it follows that $\frac{x}{x'} = \frac{r}{1} = r$, so $x = rx'$ and, similarly, we find $y = ry'$. Since, by definition, $x' = \cos(\theta)$ and $y' = \sin(\theta)$, we get the coordinates of Q to be $x = r\cos(\theta)$ and $y = r\sin(\theta)$. By reflecting these points through the x-axis, y-axis and origin, we obtain the result for all non-quadrantal angles θ, and we leave it to the reader to verify these formulas hold for the quadrantal angles.

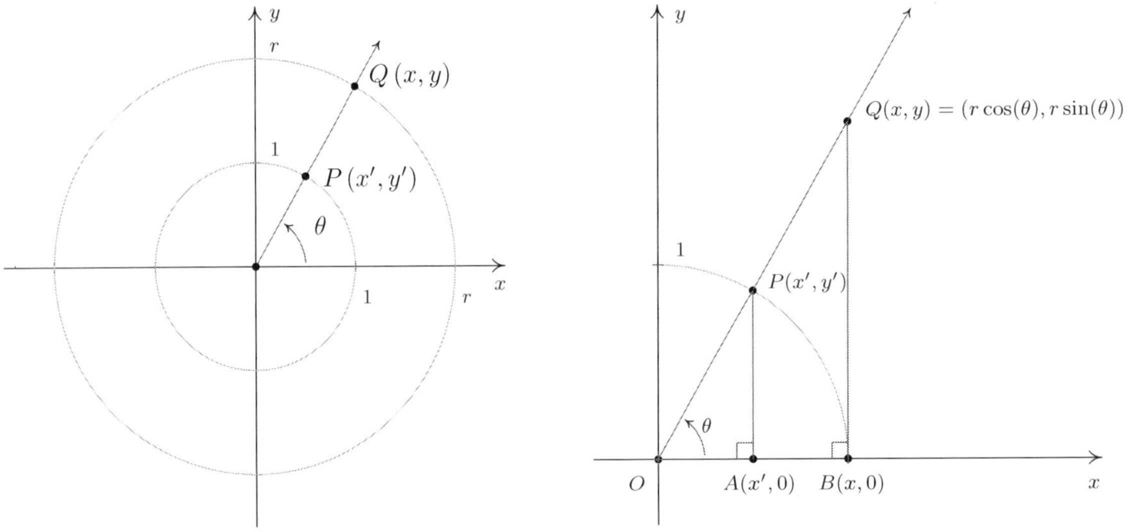

Not only can we describe the coordinates of Q in terms of $\cos(\theta)$ and $\sin(\theta)$ but since the radius of the circle is $r = \sqrt{x^2 + y^2}$, we can also express $\cos(\theta)$ and $\sin(\theta)$ in terms of the coordinates of Q. These results are summarized in the following theorem.

Theorem 10.3. If $Q(x, y)$ is the point on the terminal side of an angle θ, plotted in standard position, which lies on the circle $x^2 + y^2 = r^2$ then $x = r\cos(\theta)$ and $y = r\sin(\theta)$. Moreover,

$$\cos(\theta) = \frac{x}{r} = \frac{x}{\sqrt{x^2 + y^2}} \quad \text{and} \quad \sin(\theta) = \frac{y}{r} = \frac{y}{\sqrt{x^2 + y^2}}$$

[10] Do you remember why?

Note that in the case of the Unit Circle we have $r = \sqrt{x^2 + y^2} = 1$, so Theorem 10.3 reduces to our definitions of $\cos(\theta)$ and $\sin(\theta)$.

Example 10.2.6.

1. Suppose that the terminal side of an angle θ, when plotted in standard position, contains the point $Q(4, -2)$. Find $\sin(\theta)$ and $\cos(\theta)$.

2. In Example 10.1.5 in Section 10.1, we approximated the radius of the earth at 41.628° north latitude to be 2960 miles. Justify this approximation if the radius of the Earth at the Equator is approximately 3960 miles.

Solution.

1. Using Theorem 10.3 with $x = 4$ and $y = -2$, we find $r = \sqrt{(4)^2 + (-2)^2} = \sqrt{20} = 2\sqrt{5}$ so that $\cos(\theta) = \frac{x}{r} = \frac{4}{2\sqrt{5}} = \frac{2\sqrt{5}}{5}$ and $\sin(\theta) = \frac{y}{r} = \frac{-2}{2\sqrt{5}} = -\frac{\sqrt{5}}{5}$.

2. Assuming the Earth is a sphere, a cross-section through the poles produces a circle of radius 3960 miles. Viewing the Equator as the x-axis, the value we seek is the x-coordinate of the point $Q(x, y)$ indicated in the figure below.

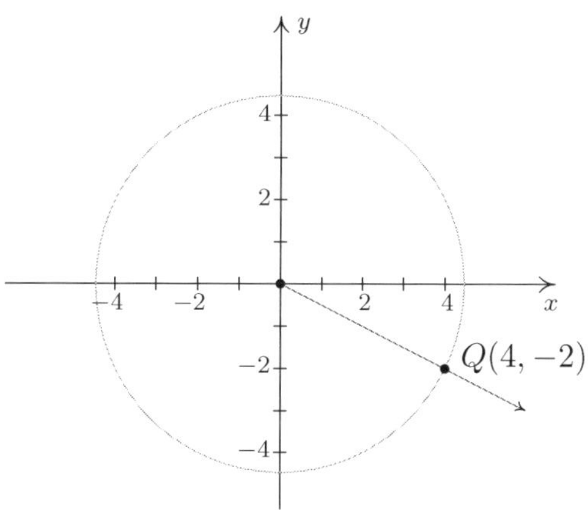

The terminal side of θ contains $Q(4, -2)$

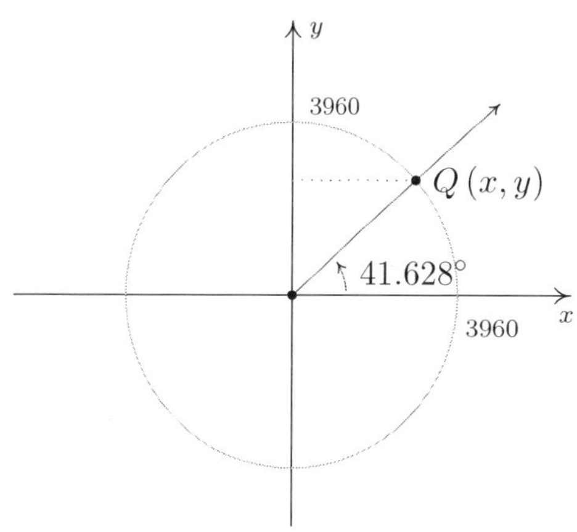

A point on the Earth at 41.628°N

Using Theorem 10.3, we get $x = 3960\cos(41.628^\circ)$. Using a calculator in 'degree' mode, we find $3960\cos(41.628^\circ) \approx 2960$. Hence, the radius of the Earth at North Latitude 41.628° is approximately 2960 miles.

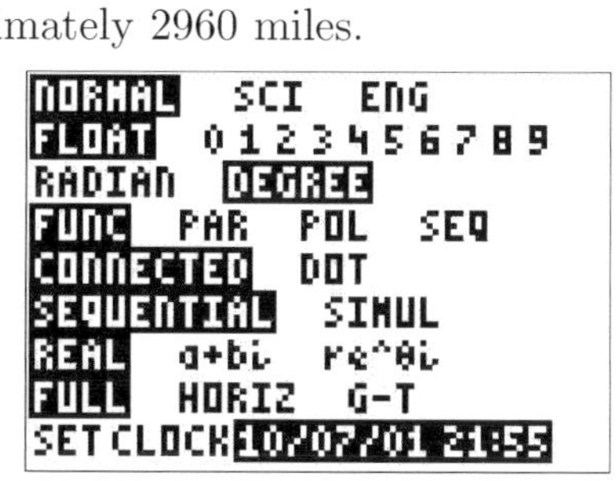

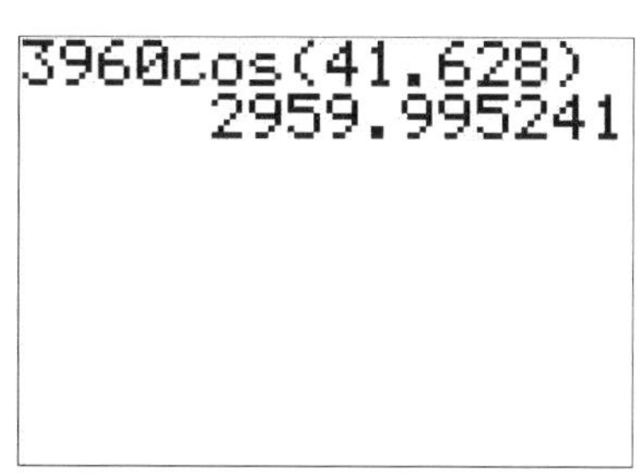

□

Theorem 10.3 gives us what we need to describe the position of an object traveling in a circular path of radius r with constant angular velocity ω. Suppose that at time t, the object has swept out an angle measuring θ radians. If we assume that the object is at the point $(r, 0)$ when $t = 0$, the angle θ is in standard position. By definition, $\omega = \frac{\theta}{t}$ which we rewrite as $\theta = \omega t$. According to Theorem 10.3, the location of the object $Q(x, y)$ on the circle is found using the equations $x = r\cos(\theta) = r\cos(\omega t)$ and $y = r\sin(\theta) = r\sin(\omega t)$. Hence, at time t, the object is at the point $(r\cos(\omega t), r\sin(\omega t))$. We have just argued the following.

Equation 10.3. Suppose an object is traveling in a circular path of radius r centered at the origin with constant angular velocity ω. If $t = 0$ corresponds to the point $(r, 0)$, then the x and y coordinates of the object are functions of t and are given by $x = r\cos(\omega t)$ and $y = r\sin(\omega t)$. Here, $\omega > 0$ indicates a counter-clockwise direction and $\omega < 0$ indicates a clockwise direction.

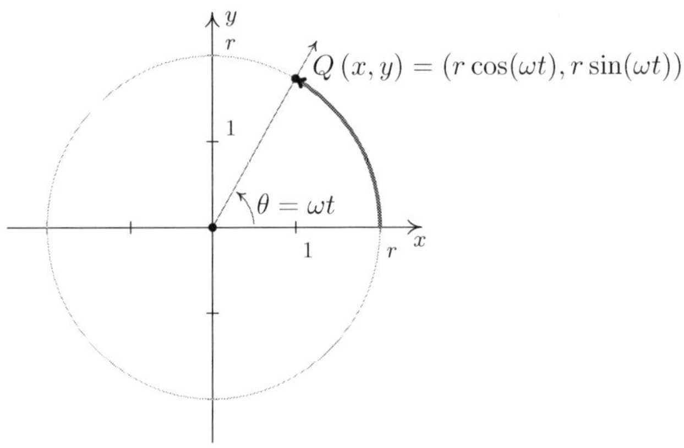

Equations for Circular Motion

Example 10.2.7. Suppose we are in the situation of Example 10.1.5. Find the equations of motion of Lakeland Community College as the earth rotates.

Solution. From Example 10.1.5, we take $r = 2960$ miles and and $\omega = \frac{\pi}{12 \text{ hours}}$. Hence, the equations of motion are $x = r\cos(\omega t) = 2960\cos\left(\frac{\pi}{12}t\right)$ and $y = r\sin(\omega t) = 2960\sin\left(\frac{\pi}{12}t\right)$, where x and y are measured in miles and t is measured in hours. □

In addition to circular motion, Theorem 10.3 is also the key to developing what is usually called 'right triangle' trigonometry.[11] As we shall see in the sections to come, many applications in trigonometry involve finding the measures of the angles in, and lengths of the sides of, right triangles. Indeed, we made good use of some properties of right triangles to find the exact values of the cosine and sine of many of the angles in Example 10.2.1, so the following development shouldn't be that much of a surprise. Consider the generic right triangle below with corresponding acute angle θ. The side with length a is called the side of the triangle *adjacent* to θ; the side with length b is called the side of the triangle *opposite* θ; and the remaining side of length c (the side opposite the

[11]You may have been exposed to this in High School.

right angle) is called the hypotenuse. We now imagine drawing this triangle in Quadrant I so that the angle θ is in standard position with the adjacent side to θ lying along the positive x-axis.

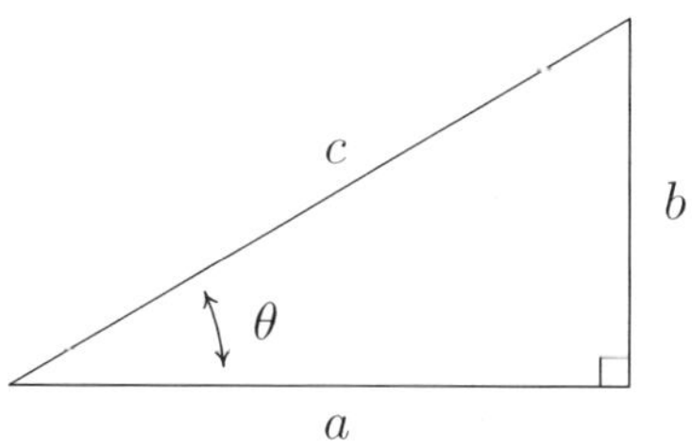

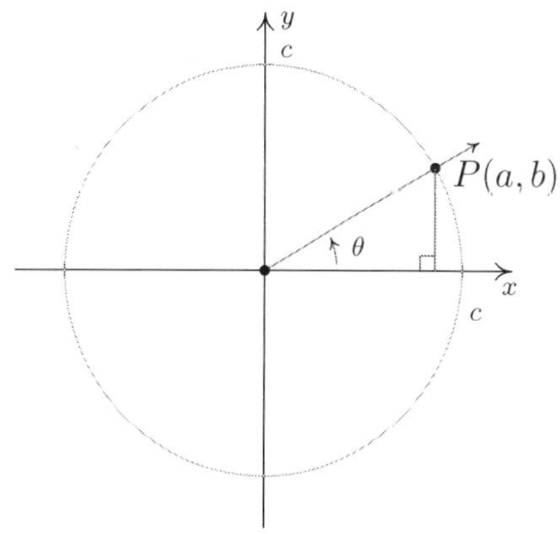

According to the Pythagorean Theorem, $a^2 + b^2 = c^2$, so that the point $P(a, b)$ lies on a circle of radius c. Theorem 10.3 tells us that $\cos(\theta) = \frac{a}{c}$ and $\sin(\theta) = \frac{b}{c}$, so we have determined the cosine and sine of θ in terms of the lengths of the sides of the right triangle. Thus we have the following theorem.

Theorem 10.4. Suppose θ is an acute angle residing in a right triangle. If the length of the side adjacent to θ is a, the length of the side opposite θ is b, and the length of the hypotenuse is c, then $\cos(\theta) = \dfrac{a}{c}$ and $\sin(\theta) = \dfrac{b}{c}$.

Example 10.2.8. Find the measure of the missing angle and the lengths of the missing sides of:

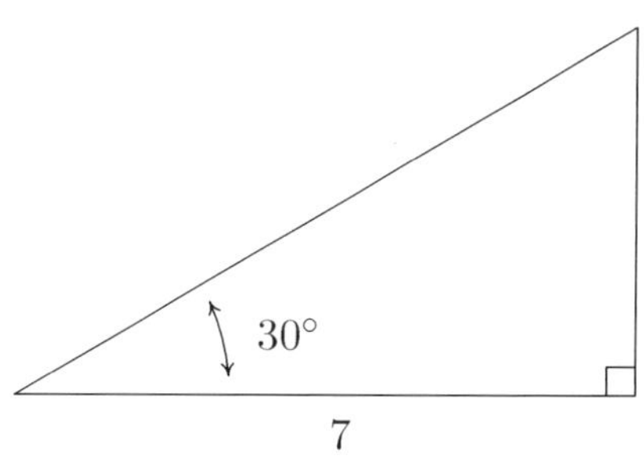

Solution. The first and easiest task is to find the measure of the missing angle. Since the sum of angles of a triangle is 180°, we know that the missing angle has measure $180^\circ - 30^\circ - 90^\circ = 60^\circ$. We now proceed to find the lengths of the remaining two sides of the triangle. Let c denote the length of the hypotenuse of the triangle. By Theorem 10.4, we have $\cos(30^\circ) = \frac{7}{c}$, or $c = \frac{7}{\cos(30^\circ)}$. Since $\cos(30^\circ) = \frac{\sqrt{3}}{2}$, we have, after the usual fraction gymnastics, $c = \frac{14\sqrt{3}}{3}$. At this point, we have two ways to proceed to find the length of the side opposite the 30° angle, which we'll denote b. We know the length of the adjacent side is 7 and the length of the hypotenuse is $\frac{14\sqrt{3}}{3}$, so we could use the Pythagorean Theorem to find the missing side and solve $(7)^2 + b^2 = \left(\frac{14\sqrt{3}}{3}\right)^2$ for b. Alternatively, we could use Theorem 10.4, namely that $\sin(30^\circ) = \frac{b}{c}$. Choosing the latter, we find $b = c\sin(30^\circ) = \frac{14\sqrt{3}}{3} \cdot \frac{1}{2} = \frac{7\sqrt{3}}{3}$. The triangle with all of its data is recorded below.

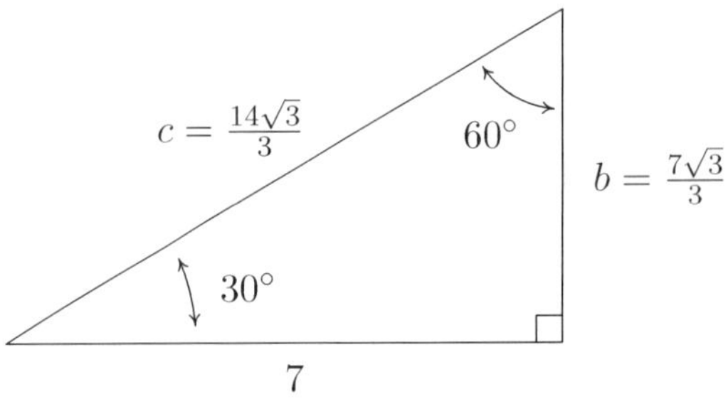

□

We close this section by noting that we can easily extend the functions cosine and sine to real numbers by identifying a real number t with the angle $\theta = t$ radians. Using this identification, we define $\cos(t) = \cos(\theta)$ and $\sin(t) = \sin(\theta)$. In practice this means expressions like $\cos(\pi)$ and $\sin(2)$ can be found by regarding the inputs as angles in radian measure or real numbers; the choice is the reader's. If we trace the identification of real numbers t with angles θ in radian measure to its roots on page 704, we can spell out this correspondence more precisely. For each real number t, we associate an oriented arc t units in length with initial point $(1, 0)$ and endpoint $P(\cos(t), \sin(t))$.

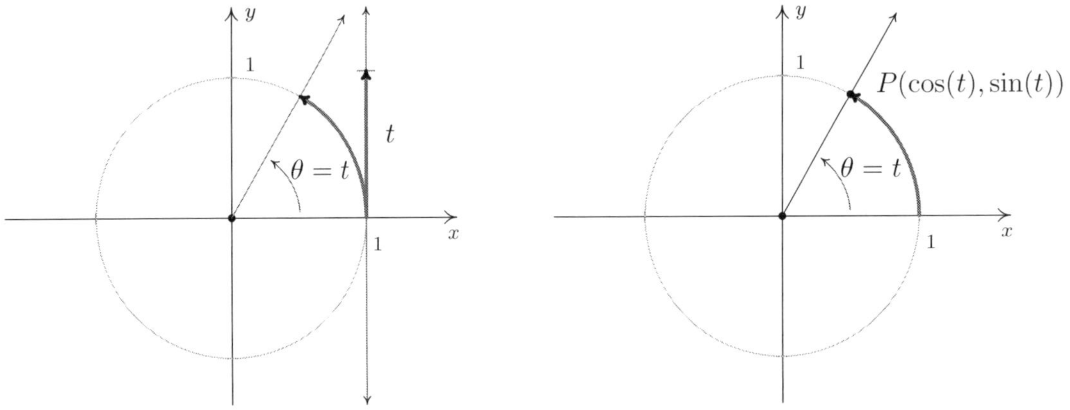

In the same way we studied polynomial, rational, exponential, and logarithmic functions, we will study the trigonometric functions $f(t) = \cos(t)$ and $g(t) = \sin(t)$. The first order of business is to find the domains and ranges of these functions. Whether we think of identifying the real number t with the angle $\theta = t$ radians, or think of wrapping an oriented arc around the Unit Circle to find coordinates on the Unit Circle, it should be clear that both the cosine and sine functions are defined for all real numbers t. In other words, the domain of $f(t) = \cos(t)$ and of $g(t) = \sin(t)$ is $(-\infty, \infty)$. Since $\cos(t)$ and $\sin(t)$ represent x- and y-coordinates, respectively, of points on the Unit Circle, they both take on all of the values between -1 an 1, inclusive. In other words, the range of $f(t) = \cos(t)$ and of $g(t) = \sin(t)$ is the interval $[-1, 1]$. To summarize:

Theorem 10.5. Domain and Range of the Cosine and Sine Functions:

- The function $f(t) = \cos(t)$
  - has domain $(-\infty, \infty)$
  - has range $[-1, 1]$
- The function $g(t) = \sin(t)$
  - has domain $(-\infty, \infty)$
  - has range $[-1, 1]$

Suppose, as in the Exercises, we are asked to solve an equation such as $\sin(t) = -\frac{1}{2}$. As we have already mentioned, the distinction between t as a real number and as an angle $\theta = t$ radians is often blurred. Indeed, we solve $\sin(t) = -\frac{1}{2}$ in the exact same manner[12] as we did in Example 10.2.5 number 2. Our solution is only cosmetically different in that the variable used is t rather than θ: $t = \frac{7\pi}{6} + 2\pi k$ or $t = \frac{11\pi}{6} + 2\pi k$ for integers, k. We will study the cosine and sine functions in greater detail in Section 10.5. Until then, keep in mind that any properties of cosine and sine developed in the following sections which regard them as functions of *angles* in *radian* measure apply equally well if the inputs are regarded as *real numbers*.

[12]Well, to be pedantic, we would be technically using 'reference numbers' or 'reference arcs' instead of 'reference angles' – but the idea is the same.

10.2.2 EXERCISES

In Exercises 1 - 20, find the exact value of the cosine and sine of the given angle.

1. $\theta = 0$
2. $\theta = \dfrac{\pi}{4}$
3. $\theta = \dfrac{\pi}{3}$
4. $\theta = \dfrac{\pi}{2}$
5. $\theta = \dfrac{2\pi}{3}$
6. $\theta = \dfrac{3\pi}{4}$
7. $\theta = \pi$
8. $\theta = \dfrac{7\pi}{6}$
9. $\theta = \dfrac{5\pi}{4}$
10. $\theta = \dfrac{4\pi}{3}$
11. $\theta = \dfrac{3\pi}{2}$
12. $\theta = \dfrac{5\pi}{3}$
13. $\theta = \dfrac{7\pi}{4}$
14. $\theta = \dfrac{23\pi}{6}$
15. $\theta = -\dfrac{13\pi}{2}$
16. $\theta = -\dfrac{43\pi}{6}$
17. $\theta = -\dfrac{3\pi}{4}$
18. $\theta = -\dfrac{\pi}{6}$
19. $\theta = \dfrac{10\pi}{3}$
20. $\theta = 117\pi$

In Exercises 21 - 30, use the results developed throughout the section to find the requested value.

21. If $\sin(\theta) = -\dfrac{7}{25}$ with θ in Quadrant IV, what is $\cos(\theta)$?
22. If $\cos(\theta) = \dfrac{4}{9}$ with θ in Quadrant I, what is $\sin(\theta)$?
23. If $\sin(\theta) = \dfrac{5}{13}$ with θ in Quadrant II, what is $\cos(\theta)$?
24. If $\cos(\theta) = -\dfrac{2}{11}$ with θ in Quadrant III, what is $\sin(\theta)$?
25. If $\sin(\theta) = -\dfrac{2}{3}$ with θ in Quadrant III, what is $\cos(\theta)$?
26. If $\cos(\theta) = \dfrac{28}{53}$ with θ in Quadrant IV, what is $\sin(\theta)$?
27. If $\sin(\theta) = \dfrac{2\sqrt{5}}{5}$ and $\dfrac{\pi}{2} < \theta < \pi$, what is $\cos(\theta)$?
28. If $\cos(\theta) = \dfrac{\sqrt{10}}{10}$ and $2\pi < \theta < \dfrac{5\pi}{2}$, what is $\sin(\theta)$?
29. If $\sin(\theta) = -0.42$ and $\pi < \theta < \dfrac{3\pi}{2}$, what is $\cos(\theta)$?
30. If $\cos(\theta) = -0.98$ and $\dfrac{\pi}{2} < \theta < \pi$, what is $\sin(\theta)$?

In Exercises 31 - 39, find all of the angles which satisfy the given equation.

31. $\sin(\theta) = \dfrac{1}{2}$

32. $\cos(\theta) = -\dfrac{\sqrt{3}}{2}$

33. $\sin(\theta) = 0$

34. $\cos(\theta) = \dfrac{\sqrt{2}}{2}$

35. $\sin(\theta) = \dfrac{\sqrt{3}}{2}$

36. $\cos(\theta) = -1$

37. $\sin(\theta) = -1$

38. $\cos(\theta) = \dfrac{\sqrt{3}}{2}$

39. $\cos(\theta) = -1.001$

In Exercises 40 - 48, solve the equation for t. (See the comments following Theorem 10.5.)

40. $\cos(t) = 0$

41. $\sin(t) = -\dfrac{\sqrt{2}}{2}$

42. $\cos(t) = 3$

43. $\sin(t) = -\dfrac{1}{2}$

44. $\cos(t) = \dfrac{1}{2}$

45. $\sin(t) = -2$

46. $\cos(t) = 1$

47. $\sin(t) = 1$

48. $\cos(t) = -\dfrac{\sqrt{2}}{2}$

In Exercises 49 - 54, use your calculator to approximate the given value to three decimal places. Make sure your calculator is in the proper angle measurement mode!

49. $\sin(78.95^\circ)$

50. $\cos(-2.01)$

51. $\sin(392.994)$

52. $\cos(207^\circ)$

53. $\sin(\pi^\circ)$

54. $\cos(e)$

In Exercises 55 - 58, find the measurement of the missing angle and the lengths of the missing sides. (See Example 10.2.8)

55. Find θ, b, and c.

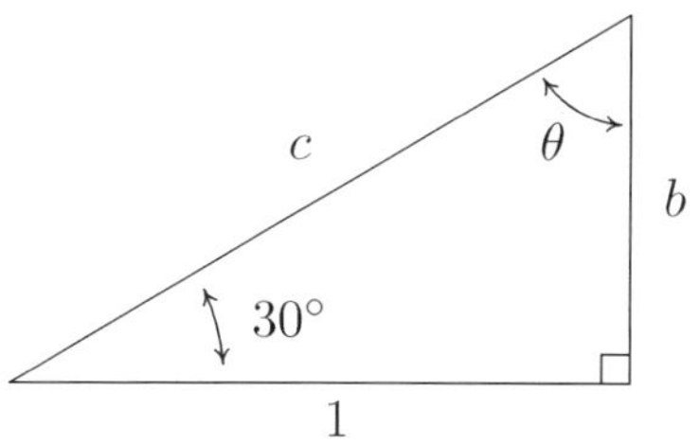

56. Find θ, a, and c.

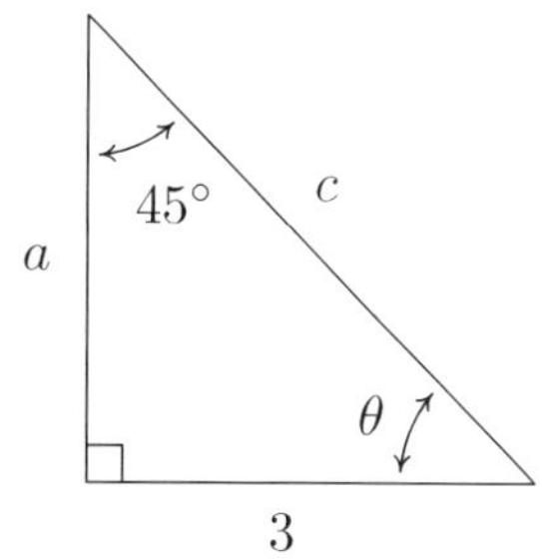

57. Find α, a, and b.

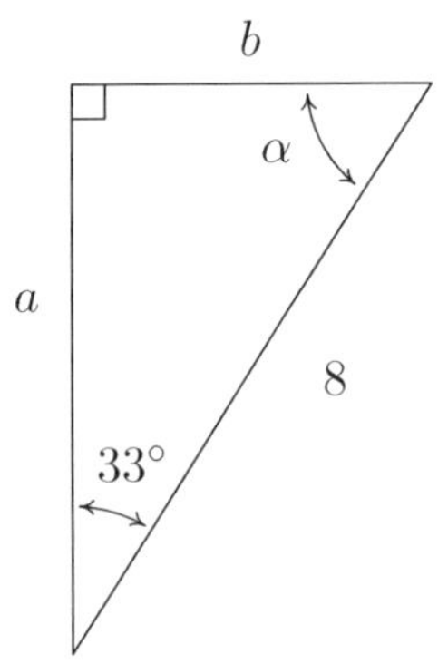

58. Find β, a, and c.

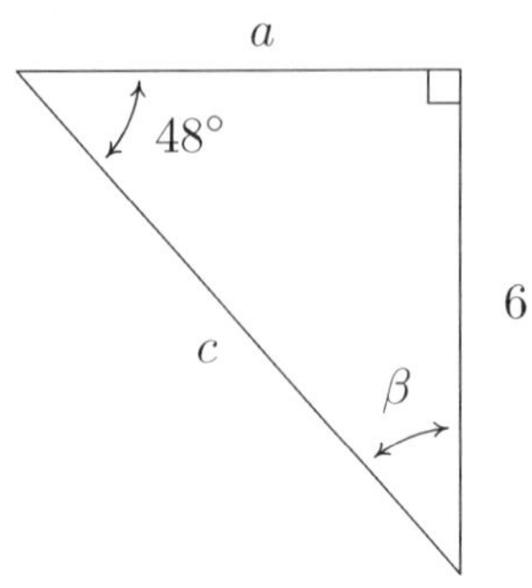

In Exercises 59 - 64, assume that θ is an acute angle in a right triangle and use Theorem 10.4 to find the requested side.

59. If $\theta = 12^\circ$ and the side adjacent to θ has length 4, how long is the hypotenuse?

60. If $\theta = 78.123^\circ$ and the hypotenuse has length 5280, how long is the side adjacent to θ?

61. If $\theta = 59^\circ$ and the side opposite θ has length 117.42, how long is the hypotenuse?

62. If $\theta = 5^\circ$ and the hypotenuse has length 10, how long is the side opposite θ?

63. If $\theta = 5^\circ$ and the hypotenuse has length 10, how long is the side adjacent to θ?

64. If $\theta = 37.5^\circ$ and the side opposite θ has length 306, how long is the side adjacent to θ?

In Exercises 65 - 68, let θ be the angle in standard position whose terminal side contains the given point then compute $\cos(\theta)$ and $\sin(\theta)$.

65. $P(-7, 24)$

66. $Q(3, 4)$

67. $R(5, -9)$

68. $T(-2, -11)$

In Exercises 69 - 72, find the equations of motion for the given scenario. Assume that the center of the motion is the origin, the motion is counter-clockwise and that $t = 0$ corresponds to a position along the positive x-axis. (See Equation 10.3 and Example 10.1.5.)

69. A point on the edge of the spinning yo-yo in Exercise 50 from Section 10.1.

    Recall: The diameter of the yo-yo is 2.25 inches and it spins at 4500 revolutions per minute.

70. The yo-yo in exercise 52 from Section 10.1.

    Recall: The radius of the circle is 28 inches and it completes one revolution in 3 seconds.

71. A point on the edge of the hard drive in Exercise 53 from Section 10.1.

    Recall: The diameter of the hard disk is 2.5 inches and it spins at 7200 revolutions per minute.

72. A passenger on the Big Wheel in Exercise 55 from Section 10.1.

    Recall: The diameter is 128 feet and completes 2 revolutions in 2 minutes, 7 seconds.

73. Consider the numbers: 0, 1, 2, 3, 4. Take the square root of each of these numbers, then divide each by 2. The resulting numbers should look hauntingly familiar. (See the values in the table on 722.)

74. Let α and β be the two acute angles of a right triangle. (Thus α and β are complementary angles.) Show that $\sin(\alpha) = \cos(\beta)$ and $\sin(\beta) = \cos(\alpha)$. The fact that co-functions of complementary angles are equal in this case is not an accident and a more general result will be given in Section 10.4.

75. In the scenario of Equation 10.3, we assumed that at $t = 0$, the object was at the point $(r, 0)$. If this is not the case, we can adjust the equations of motion by introducing a 'time delay.' If $t_0 > 0$ is the first time the object passes through the point $(r, 0)$, show, with the help of your classmates, the equations of motion are $x = r\cos(\omega(t - t_0))$ and $y = r\sin(\omega(t - t_0))$.

10.2.3 Answers

1. $\cos(0) = 1, \ \sin(0) = 0$

2. $\cos\left(\frac{\pi}{4}\right) = \frac{\sqrt{2}}{2}, \ \sin\left(\frac{\pi}{4}\right) = \frac{\sqrt{2}}{2}$

3. $\cos\left(\frac{\pi}{3}\right) = \frac{1}{2}, \ \sin\left(\frac{\pi}{3}\right) = \frac{\sqrt{3}}{2}$

4. $\cos\left(\frac{\pi}{2}\right) = 0, \ \sin\left(\frac{\pi}{2}\right) = 1$

5. $\cos\left(\frac{2\pi}{3}\right) = -\frac{1}{2}, \ \sin\left(\frac{2\pi}{3}\right) = \frac{\sqrt{3}}{2}$

6. $\cos\left(\frac{3\pi}{4}\right) = -\frac{\sqrt{2}}{2}, \ \sin\left(\frac{3\pi}{4}\right) = \frac{\sqrt{2}}{2}$

7. $\cos(\pi) = -1, \ \sin(\pi) = 0$

8. $\cos\left(\frac{7\pi}{6}\right) = -\frac{\sqrt{3}}{2}, \ \sin\left(\frac{7\pi}{6}\right) = -\frac{1}{2}$

9. $\cos\left(\frac{5\pi}{4}\right) = -\frac{\sqrt{2}}{2}, \ \sin\left(\frac{5\pi}{4}\right) = -\frac{\sqrt{2}}{2}$

10. $\cos\left(\frac{4\pi}{3}\right) = -\frac{1}{2}, \ \sin\left(\frac{4\pi}{3}\right) = -\frac{\sqrt{3}}{2}$

11. $\cos\left(\frac{3\pi}{2}\right) = 0, \ \sin\left(\frac{3\pi}{2}\right) = -1$

12. $\cos\left(\frac{5\pi}{3}\right) = \frac{1}{2}, \ \sin\left(\frac{5\pi}{3}\right) = -\frac{\sqrt{3}}{2}$

13. $\cos\left(\frac{7\pi}{4}\right) = \frac{\sqrt{2}}{2}, \ \sin\left(\frac{7\pi}{4}\right) = -\frac{\sqrt{2}}{2}$

14. $\cos\left(\frac{23\pi}{6}\right) = \frac{\sqrt{3}}{2}, \ \sin\left(\frac{23\pi}{6}\right) = -\frac{1}{2}$

15. $\cos\left(-\frac{13\pi}{2}\right) = 0, \ \sin\left(-\frac{13\pi}{2}\right) = -1$

16. $\cos\left(-\frac{43\pi}{6}\right) = -\frac{\sqrt{3}}{2}, \ \sin\left(-\frac{43\pi}{6}\right) = \frac{1}{2}$

17. $\cos\left(-\frac{3\pi}{4}\right) = -\frac{\sqrt{2}}{2}, \ \sin\left(-\frac{3\pi}{4}\right) = -\frac{\sqrt{2}}{2}$

18. $\cos\left(-\frac{\pi}{6}\right) = \frac{\sqrt{3}}{2}, \ \sin\left(-\frac{\pi}{6}\right) = -\frac{1}{2}$

19. $\cos\left(\frac{10\pi}{3}\right) = -\frac{1}{2}, \ \sin\left(\frac{10\pi}{3}\right) = -\frac{\sqrt{3}}{2}$

20. $\cos(117\pi) = -1, \ \sin(117\pi) = 0$

21. If $\sin(\theta) = -\frac{7}{25}$ with θ in Quadrant IV, then $\cos(\theta) = \frac{24}{25}$.

22. If $\cos(\theta) = \frac{4}{9}$ with θ in Quadrant I, then $\sin(\theta) = \frac{\sqrt{65}}{9}$.

23. If $\sin(\theta) = \frac{5}{13}$ with θ in Quadrant II, then $\cos(\theta) = -\frac{12}{13}$.

24. If $\cos(\theta) = -\frac{2}{11}$ with θ in Quadrant III, then $\sin(\theta) = -\frac{\sqrt{117}}{11}$.

25. If $\sin(\theta) = -\frac{2}{3}$ with θ in Quadrant III, then $\cos(\theta) = -\frac{\sqrt{5}}{3}$.

26. If $\cos(\theta) = \frac{28}{53}$ with θ in Quadrant IV, then $\sin(\theta) = -\frac{45}{53}$.

27. If $\sin(\theta) = \frac{2\sqrt{5}}{5}$ and $\frac{\pi}{2} < \theta < \pi$, then $\cos(\theta) = -\frac{\sqrt{5}}{5}$.

28. If $\cos(\theta) = \frac{\sqrt{10}}{10}$ and $2\pi < \theta < \frac{5\pi}{2}$, then $\sin(\theta) = \frac{3\sqrt{10}}{10}$.

29. If $\sin(\theta) = -0.42$ and $\pi < \theta < \frac{3\pi}{2}$, then $\cos(\theta) = -\sqrt{0.8236} \approx -0.9075$.

30. If $\cos(\theta) = -0.98$ and $\frac{\pi}{2} < \theta < \pi$, then $\sin(\theta) = \sqrt{0.0396} \approx 0.1990$.

31. $\sin(\theta) = \frac{1}{2}$ when $\theta = \frac{\pi}{6} + 2\pi k$ or $\theta = \frac{5\pi}{6} + 2\pi k$ for any integer k.

32. $\cos(\theta) = -\frac{\sqrt{3}}{2}$ when $\theta = \frac{5\pi}{6} + 2\pi k$ or $\theta = \frac{7\pi}{6} + 2\pi k$ for any integer k.

33. $\sin(\theta) = 0$ when $\theta = \pi k$ for any integer k.

34. $\cos(\theta) = \frac{\sqrt{2}}{2}$ when $\theta = \frac{\pi}{4} + 2\pi k$ or $\theta = \frac{7\pi}{4} + 2\pi k$ for any integer k.

35. $\sin(\theta) = \frac{\sqrt{3}}{2}$ when $\theta = \frac{\pi}{3} + 2\pi k$ or $\theta = \frac{2\pi}{3} + 2\pi k$ for any integer k.

36. $\cos(\theta) = -1$ when $\theta = (2k+1)\pi$ for any integer k.

37. $\sin(\theta) = -1$ when $\theta = \frac{3\pi}{2} + 2\pi k$ for any integer k.

38. $\cos(\theta) = \frac{\sqrt{3}}{2}$ when $\theta = \frac{\pi}{6} + 2\pi k$ or $\theta = \frac{11\pi}{6} + 2\pi k$ for any integer k.

39. $\cos(\theta) = -1.001$ never happens

40. $\cos(t) = 0$ when $t = \frac{\pi}{2} + \pi k$ for any integer k.

41. $\sin(t) = -\frac{\sqrt{2}}{2}$ when $t = \frac{5\pi}{4} + 2\pi k$ or $t = \frac{7\pi}{4} + 2\pi k$ for any integer k.

42. $\cos(t) = 3$ never happens.

43. $\sin(t) = -\frac{1}{2}$ when $t = \frac{7\pi}{6} + 2\pi k$ or $t = \frac{11\pi}{6} + 2\pi k$ for any integer k.

44. $\cos(t) = \frac{1}{2}$ when $t = \frac{\pi}{3} + 2\pi k$ or $t = \frac{5\pi}{3} + 2\pi k$ for any integer k.

45. $\sin(t) = -2$ never happens

46. $\cos(t) = 1$ when $t = 2\pi k$ for any integer k.

47. $\sin(t) = 1$ when $t = \frac{\pi}{2} + 2\pi k$ for any integer k.

48. $\cos(t) = -\frac{\sqrt{2}}{2}$ when $t = \frac{3\pi}{4} + 2\pi k$ or $t = \frac{5\pi}{4} + 2\pi k$ for any integer k.

49. $\sin(78.95^\circ) \approx 0.981$

50. $\cos(-2.01) \approx -0.425$

51. $\sin(392.994) \approx -0.291$

52. $\cos(207^\circ) \approx -0.891$

53. $\sin(\pi^\circ) \approx 0.055$

54. $\cos(e) \approx -0.912$

55. $\theta = 60^\circ$, $b = \frac{\sqrt{3}}{3}$, $c = \frac{2\sqrt{3}}{3}$

56. $\theta = 45^\circ$, $a = 3$, $c = 3\sqrt{2}$

57. $\alpha = 57^\circ$, $a = 8\cos(33^\circ) \approx 6.709$, $b = 8\sin(33^\circ) \approx 4.357$

58. $\beta = 42^\circ$, $c = \frac{6}{\sin(48^\circ)} \approx 8.074$, $a = \sqrt{c^2 - 6^2} \approx 5.402$

59. The hypotenuse has length $\frac{4}{\cos(12^\circ)} \approx 4.089$.

60. The side adjacent to θ has length $5280\cos(78.123^\circ) \approx 1086.68$.

61. The hypotenuse has length $\frac{117.42}{\sin(59^\circ)} \approx 136.99$.

62. The side opposite θ has length $10\sin(5^\circ) \approx 0.872$.

63. The side adjacent to θ has length $10\cos(5^\circ) \approx 9.962$.

64. The hypotenuse has length $c = \frac{306}{\sin(37.5^\circ)} \approx 502.660$, so the side adjacent to θ has length $\sqrt{c^2 - 306^2} \approx 398.797$.

65. $\cos(\theta) = -\frac{7}{25}$, $\sin(\theta) = \frac{24}{25}$

66. $\cos(\theta) = \frac{3}{5}$, $\sin(\theta) = \frac{4}{5}$

67. $\cos(\theta) = \frac{5\sqrt{106}}{106}$, $\sin(\theta) = -\frac{9\sqrt{106}}{106}$

68. $\cos(\theta) = -\frac{2\sqrt{5}}{25}$, $\sin(\theta) = -\frac{11\sqrt{5}}{25}$

69. $r = 1.125$ inches, $\omega = 9000\pi \, \frac{\text{radians}}{\text{minute}}$, $x = 1.125\cos(9000\pi\, t)$, $y = 1.125\sin(9000\pi\, t)$. Here x and y are measured in inches and t is measured in minutes.

70. $r = 28$ inches, $\omega = \frac{2\pi}{3} \frac{\text{radians}}{\text{second}}$, $x = 28\cos\left(\frac{2\pi}{3}t\right)$, $y = 28\sin\left(\frac{2\pi}{3}t\right)$. Here x and y are measured in inches and t is measured in seconds.

71. $r = 1.25$ inches, $\omega = 14400\pi \frac{\text{radians}}{\text{minute}}$, $x = 1.25\cos(14400\pi\, t)$, $y = 1.25\sin(14400\pi\, t)$. Here x and y are measured in inches and t is measured in minutes.

72. $r = 64$ feet, $\omega = \frac{4\pi}{127} \frac{\text{radians}}{\text{second}}$, $x = 64\cos\left(\frac{4\pi}{127}t\right)$, $y = 64\sin\left(\frac{4\pi}{127}t\right)$. Here x and y are measured in feet and t is measured in seconds

10.3 The Six Circular Functions and Fundamental Identities

In section 10.2, we defined $\cos(\theta)$ and $\sin(\theta)$ for angles θ using the coordinate values of points on the Unit Circle. As such, these functions earn the moniker **circular functions**.[1] It turns out that cosine and sine are just two of the six commonly used circular functions which we define below.

Definition 10.2. The Circular Functions: Suppose θ is an angle plotted in standard position and $P(x, y)$ is the point on the terminal side of θ which lies on the Unit Circle.

- The **cosine** of θ, denoted $\cos(\theta)$, is defined by $\cos(\theta) = x$.
- The **sine** of θ, denoted $\sin(\theta)$, is defined by $\sin(\theta) = y$.
- The **secant** of θ, denoted $\sec(\theta)$, is defined by $\sec(\theta) = \dfrac{1}{x}$, provided $x \neq 0$.
- The **cosecant** of θ, denoted $\csc(\theta)$, is defined by $\csc(\theta) = \dfrac{1}{y}$, provided $y \neq 0$.
- The **tangent** of θ, denoted $\tan(\theta)$, is defined by $\tan(\theta) = \dfrac{y}{x}$, provided $x \neq 0$.
- The **cotangent** of θ, denoted $\cot(\theta)$, is defined by $\cot(\theta) = \dfrac{x}{y}$, provided $y \neq 0$.

While we left the history of the name 'sine' as an interesting research project in Section 10.2, the names 'tangent' and 'secant' can be explained using the diagram below. Consider the acute angle θ below in standard position. Let $P(x, y)$ denote, as usual, the point on the terminal side of θ which lies on the Unit Circle and let $Q(1, y')$ denote the point on the terminal side of θ which lies on the vertical line $x = 1$.

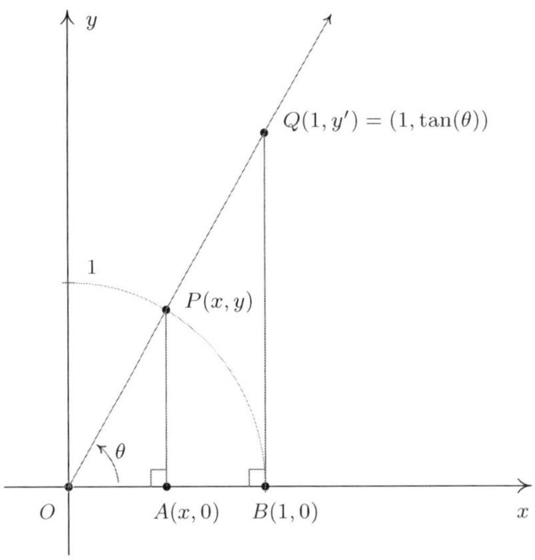

[1] In Theorem 10.4 we also showed cosine and sine to be functions of an angle residing in a right triangle so we could just as easily call them *trigonometric* functions. In later sections, you will find that we do indeed use the phrase 'trigonometric function' interchangeably with the term 'circular function'.

The word 'tangent' comes from the Latin meaning 'to touch,' and for this reason, the line $x = 1$ is called a *tangent* line to the Unit Circle since it intersects, or 'touches', the circle at only one point, namely $(1, 0)$. Dropping perpendiculars from P and Q creates a pair of similar triangles ΔOPA and ΔOQB. Thus $\frac{y'}{y} = \frac{1}{x}$ which gives $y' = \frac{y}{x} = \tan(\theta)$, where this last equality comes from applying Definition 10.2. We have just shown that for acute angles θ, $\tan(\theta)$ is the y-coordinate of the point on the terminal side of θ which lies on the line $x = 1$ which is *tangent* to the Unit Circle. Now the word 'secant' means 'to cut', so a secant line is any line that 'cuts through' a circle at two points.[2] The line containing the terminal side of θ is a secant line since it intersects the Unit Circle in Quadrants I and III. With the point P lying on the Unit Circle, the length of the hypotenuse of ΔOPA is 1. If we let h denote the length of the hypotenuse of ΔOQB, we have from similar triangles that $\frac{h}{1} = \frac{1}{x}$, or $h = \frac{1}{x} = \sec(\theta)$. Hence for an acute angle θ, $\sec(\theta)$ is the length of the line segment which lies on the secant line determined by the terminal side of θ and 'cuts off' the tangent line $x = 1$. Not only do these observations help explain the names of these functions, they serve as the basis for a fundamental inequality needed for Calculus which we'll explore in the Exercises.

Of the six circular functions, only cosine and sine are defined for all angles. Since $\cos(\theta) = x$ and $\sin(\theta) = y$ in Definition 10.2, it is customary to rephrase the remaining four circular functions in terms of cosine and sine. The following theorem is a result of simply replacing x with $\cos(\theta)$ and y with $\sin(\theta)$ in Definition 10.2.

Theorem 10.6. Reciprocal and Quotient Identities:

- $\sec(\theta) = \dfrac{1}{\cos(\theta)}$, provided $\cos(\theta) \neq 0$; if $\cos(\theta) = 0$, $\sec(\theta)$ is undefined.
- $\csc(\theta) = \dfrac{1}{\sin(\theta)}$, provided $\sin(\theta) \neq 0$; if $\sin(\theta) = 0$, $\csc(\theta)$ is undefined.
- $\tan(\theta) = \dfrac{\sin(\theta)}{\cos(\theta)}$, provided $\cos(\theta) \neq 0$; if $\cos(\theta) = 0$, $\tan(\theta)$ is undefined.
- $\cot(\theta) = \dfrac{\cos(\theta)}{\sin(\theta)}$, provided $\sin(\theta) \neq 0$; if $\sin(\theta) = 0$, $\cot(\theta)$ is undefined.

It is high time for an example.

Example 10.3.1. Find the indicated value, if it exists.

1. $\sec(60^\circ)$
2. $\csc\left(\frac{7\pi}{4}\right)$
3. $\cot(3)$
4. $\tan(\theta)$, where θ is any angle coterminal with $\frac{3\pi}{2}$.
5. $\cos(\theta)$, where $\csc(\theta) = -\sqrt{5}$ and θ is a Quadrant IV angle.
6. $\sin(\theta)$, where $\tan(\theta) = 3$ and $\pi < \theta < \frac{3\pi}{2}$.

[2]Compare this with the definition given in Section 2.1.

Solution.

1. According to Theorem 10.6, $\sec(60^\circ) = \frac{1}{\cos(60^\circ)}$. Hence, $\sec(60^\circ) = \frac{1}{(1/2)} = 2$.

2. Since $\sin\left(\frac{7\pi}{4}\right) = -\frac{\sqrt{2}}{2}$, $\csc\left(\frac{7\pi}{4}\right) = \frac{1}{\sin\left(\frac{7\pi}{4}\right)} = \frac{1}{-\sqrt{2}/2} = -\frac{2}{\sqrt{2}} = -\sqrt{2}$.

3. Since $\theta = 3$ radians is not one of the 'common angles' from Section 10.2, we resort to the calculator for a decimal approximation. Ensuring that the calculator is in radian mode, we find $\cot(3) = \frac{\cos(3)}{\sin(3)} \approx -7.015$.

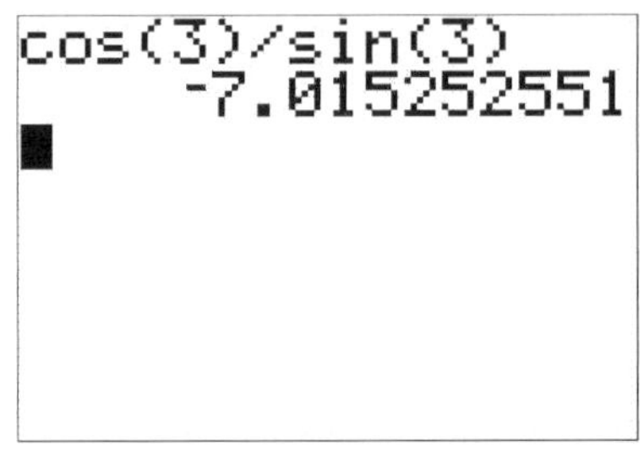

4. If θ is coterminal with $\frac{3\pi}{2}$, then $\cos(\theta) = \cos\left(\frac{3\pi}{2}\right) = 0$ and $\sin(\theta) = \sin\left(\frac{3\pi}{2}\right) = -1$. Attempting to compute $\tan(\theta) = \frac{\sin(\theta)}{\cos(\theta)}$ results in $\frac{-1}{0}$, so $\tan(\theta)$ is undefined.

5. We are given that $\csc(\theta) = \frac{1}{\sin(\theta)} = -\sqrt{5}$ so $\sin(\theta) = -\frac{1}{\sqrt{5}} = -\frac{\sqrt{5}}{5}$. As we saw in Section 10.2, we can use the Pythagorean Identity, $\cos^2(\theta) + \sin^2(\theta) = 1$, to find $\cos(\theta)$ by knowing $\sin(\theta)$. Substituting, we get $\cos^2(\theta) + \left(-\frac{\sqrt{5}}{5}\right)^2 = 1$, which gives $\cos^2(\theta) = \frac{4}{5}$, or $\cos(\theta) = \pm\frac{2\sqrt{5}}{5}$. Since θ is a Quadrant IV angle, $\cos(\theta) > 0$, so $\cos(\theta) = \frac{2\sqrt{5}}{5}$.

6. If $\tan(\theta) = 3$, then $\frac{\sin(\theta)}{\cos(\theta)} = 3$. Be careful - this does **NOT** mean we can take $\sin(\theta) = 3$ and $\cos(\theta) = 1$. Instead, from $\frac{\sin(\theta)}{\cos(\theta)} = 3$ we get: $\sin(\theta) = 3\cos(\theta)$. To relate $\cos(\theta)$ and $\sin(\theta)$, we once again employ the Pythagorean Identity, $\cos^2(\theta) + \sin^2(\theta) = 1$. Solving $\sin(\theta) = 3\cos(\theta)$ for $\cos(\theta)$, we find $\cos(\theta) = \frac{1}{3}\sin(\theta)$. Substituting this into the Pythagorean Identity, we find $\sin^2(\theta) + \left(\frac{1}{3}\sin(\theta)\right)^2 = 1$. Solving, we get $\sin^2(\theta) = \frac{9}{10}$ so $\sin(\theta) = \pm\frac{3\sqrt{10}}{10}$. Since $\pi < \theta < \frac{3\pi}{2}$, θ is a Quadrant III angle. This means $\sin(\theta) < 0$, so our final answer is $\sin(\theta) = -\frac{3\sqrt{10}}{10}$. □

While the Reciprocal and Quotient Identities presented in Theorem 10.6 allow us to always reduce problems involving secant, cosecant, tangent and cotangent to problems involving cosine and sine, it is not always convenient to do so.[3] It is worth taking the time to memorize the tangent and cotangent values of the common angles summarized below.

[3]As we shall see shortly, when solving equations involving secant and cosecant, we usually convert back to cosines and sines. However, when solving for tangent or cotangent, we usually stick with what we're dealt.

Tangent and Cotangent Values of Common Angles

θ(degrees)	θ(radians)	$\tan(\theta)$	$\cot(\theta)$
0°	0	0	undefined
30°	$\frac{\pi}{6}$	$\frac{\sqrt{3}}{3}$	$\sqrt{3}$
45°	$\frac{\pi}{4}$	1	1
60°	$\frac{\pi}{3}$	$\sqrt{3}$	$\frac{\sqrt{3}}{3}$
90°	$\frac{\pi}{2}$	undefined	0

Coupling Theorem 10.6 with the Reference Angle Theorem, Theorem 10.2, we get the following.

Theorem 10.7. Generalized Reference Angle Theorem. The values of the circular functions of an angle, if they exist, are the same, up to a sign, of the corresponding circular functions of its reference angle. More specifically, if α is the reference angle for θ, then: $\cos(\theta) = \pm\cos(\alpha)$, $\sin(\theta) = \pm\sin(\alpha)$, $\sec(\theta) = \pm\sec(\alpha)$, $\csc(\theta) = \pm\csc(\alpha)$, $\tan(\theta) = \pm\tan(\alpha)$ and $\cot(\theta) = \pm\cot(\alpha)$. The choice of the $(\pm)$ depends on the quadrant in which the terminal side of θ lies.

We put Theorem 10.7 to good use in the following example.

Example 10.3.2. Find all angles which satisfy the given equation.

1. $\sec(\theta) = 2$
2. $\tan(\theta) = \sqrt{3}$
3. $\cot(\theta) = -1$.

Solution.

1. To solve $\sec(\theta) = 2$, we convert to cosines and get $\frac{1}{\cos(\theta)} = 2$ or $\cos(\theta) = \frac{1}{2}$. This is the exact same equation we solved in Example 10.2.5, number 1, so we know the answer is: $\theta = \frac{\pi}{3} + 2\pi k$ or $\theta = \frac{5\pi}{3} + 2\pi k$ for integers k.

2. From the table of common values, we see $\tan\left(\frac{\pi}{3}\right) = \sqrt{3}$. According to Theorem 10.7, we know the solutions to $\tan(\theta) = \sqrt{3}$ must, therefore, have a reference angle of $\frac{\pi}{3}$. Our next task is to determine in which quadrants the solutions to this equation lie. Since tangent is defined as the ratio $\frac{y}{x}$ of points (x, y) on the Unit Circle with $x \neq 0$, tangent is positive when x and y have the same sign (i.e., when they are both positive or both negative.) This happens in Quadrants I and III. In Quadrant I, we get the solutions: $\theta = \frac{\pi}{3} + 2\pi k$ for integers k, and for Quadrant III, we get $\theta = \frac{4\pi}{3} + 2\pi k$ for integers k. While these descriptions of the solutions are correct, they can be combined into one list as $\theta = \frac{\pi}{3} + \pi k$ for integers k. The latter form of the solution is best understood looking at the geometry of the situation in the diagram below.[4]

[4]See Example 10.2.5 number 3 in Section 10.2 for another example of this kind of simplification of the solution.

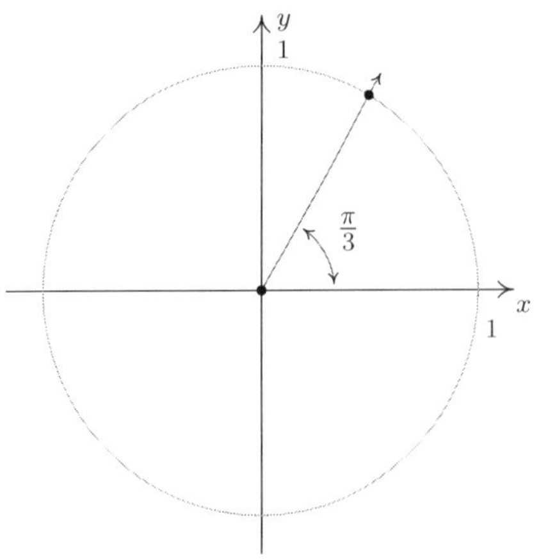

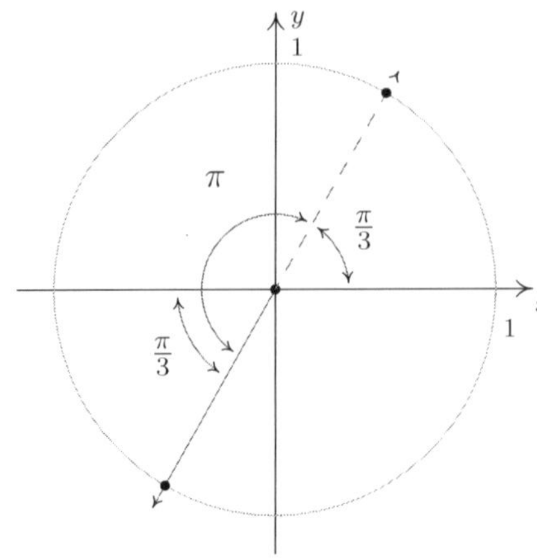

3. From the table of common values, we see that $\frac{\pi}{4}$ has a cotangent of 1, which means the solutions to $\cot(\theta) = -1$ have a reference angle of $\frac{\pi}{4}$. To find the quadrants in which our solutions lie, we note that $\cot(\theta) = \frac{x}{y}$ for a point (x, y) on the Unit Circle where $y \neq 0$. If $\cot(\theta)$ is negative, then x and y must have different signs (i.e., one positive and one negative.) Hence, our solutions lie in Quadrants II and IV. Our Quadrant II solution is $\theta = \frac{3\pi}{4} + 2\pi k$, and for Quadrant IV, we get $\theta = \frac{7\pi}{4} + 2\pi k$ for integers k. Can these lists be combined? Indeed they can - one such way to capture all the solutions is: $\theta = \frac{3\pi}{4} + \pi k$ for integers k.

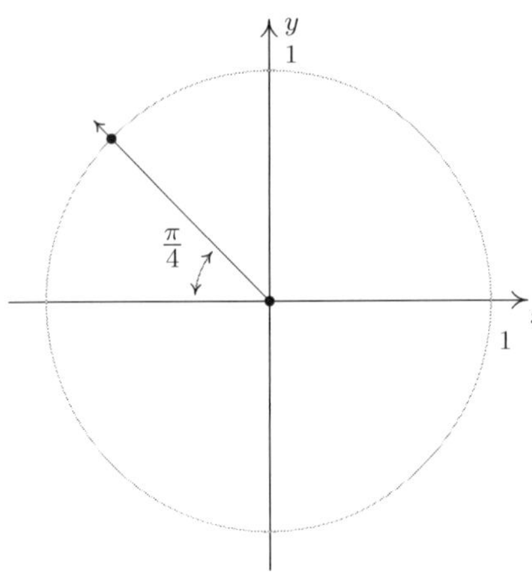

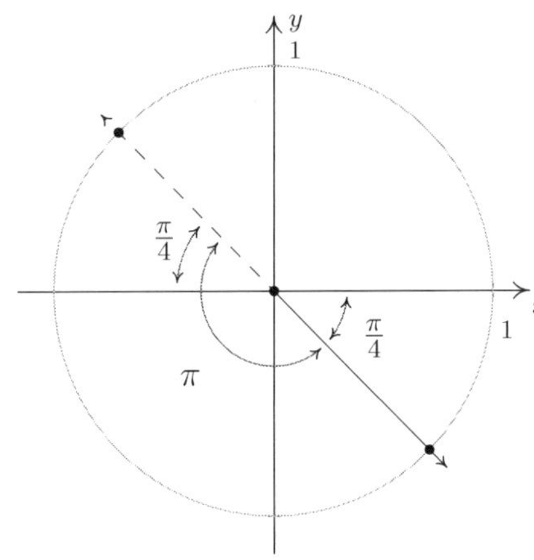

□

We have already seen the importance of identities in trigonometry. Our next task is to use use the Reciprocal and Quotient Identities found in Theorem 10.6 coupled with the Pythagorean Identity found in Theorem 10.1 to derive new Pythagorean-like identities for the remaining four circular functions. Assuming $\cos(\theta) \neq 0$, we may start with $\cos^2(\theta) + \sin^2(\theta) = 1$ and divide both sides by $\cos^2(\theta)$ to obtain $1 + \frac{\sin^2(\theta)}{\cos^2(\theta)} = \frac{1}{\cos^2(\theta)}$. Using properties of exponents along with the Reciprocal and Quotient Identities, this reduces to $1 + \tan^2(\theta) = \sec^2(\theta)$. If $\sin(\theta) \neq 0$, we can divide both sides of the identity $\cos^2(\theta) + \sin^2(\theta) = 1$ by $\sin^2(\theta)$, apply Theorem 10.6 once again, and obtain $\cot^2(\theta) + 1 = \csc^2(\theta)$. These three Pythagorean Identities are worth memorizing and they, along with some of their other common forms, are summarized in the following theorem.

Theorem 10.8. The Pythagorean Identities:

1. $\cos^2(\theta)+\sin^2(\theta)=1$.

   Common Alternate Forms:

   - $1-\sin^2(\theta)=\cos^2(\theta)$
   - $1-\cos^2(\theta)=\sin^2(\theta)$

2. $1+\tan^2(\theta)=\sec^2(\theta)$, provided $\cos(\theta)\neq 0$.

   Common Alternate Forms:

   - $\sec^2(\theta)-\tan^2(\theta)=1$
   - $\sec^2(\theta)-1=\tan^2(\theta)$

3. $1+\cot^2(\theta)=\csc^2(\theta)$, provided $\sin(\theta)\neq 0$.

   Common Alternate Forms:

   - $\csc^2(\theta)-\cot^2(\theta)=1$
   - $\csc^2(\theta)-1=\cot^2(\theta)$

Trigonometric identities play an important role in not just Trigonometry, but in Calculus as well. We'll use them in this book to find the values of the circular functions of an angle and solve equations and inequalities. In Calculus, they are needed to simplify otherwise complicated expressions. In the next example, we make good use of the Theorems 10.6 and 10.8.

Example 10.3.3. Verify the following identities. Assume that all quantities are defined.

1. $\dfrac{1}{\csc(\theta)}=\sin(\theta)$
2. $\tan(\theta)=\sin(\theta)\sec(\theta)$
3. $(\sec(\theta)-\tan(\theta))(\sec(\theta)+\tan(\theta))=1$
4. $\dfrac{\sec(\theta)}{1-\tan(\theta)}=\dfrac{1}{\cos(\theta)-\sin(\theta)}$
5. $6\sec(\theta)\tan(\theta)=\dfrac{3}{1-\sin(\theta)}-\dfrac{3}{1+\sin(\theta)}$
6. $\dfrac{\sin(\theta)}{1-\cos(\theta)}=\dfrac{1+\cos(\theta)}{\sin(\theta)}$

Solution. In verifying identities, we typically start with the more complicated side of the equation and use known identities to *transform* it into the other side of the equation.

1. To verify $\frac{1}{\csc(\theta)}=\sin(\theta)$, we start with the left side. Using $\csc(\theta)=\frac{1}{\sin(\theta)}$, we get:

   $$\frac{1}{\csc(\theta)}=\frac{1}{\frac{1}{\sin(\theta)}}=\sin(\theta),$$

   which is what we were trying to prove.

2. Starting with the right hand side of $\tan(\theta) = \sin(\theta)\sec(\theta)$, we use $\sec(\theta) = \frac{1}{\cos(\theta)}$ and find:

$$\sin(\theta)\sec(\theta) = \sin(\theta)\frac{1}{\cos(\theta)} = \frac{\sin(\theta)}{\cos(\theta)} = \tan(\theta),$$

where the last equality is courtesy of Theorem 10.6.

3. Expanding the left hand side of the equation gives: $(\sec(\theta) - \tan(\theta))(\sec(\theta) + \tan(\theta)) = \sec^2(\theta) - \tan^2(\theta)$. According to Theorem 10.8, $\sec^2(\theta) - \tan^2(\theta) = 1$. Putting it all together,

$$(\sec(\theta) - \tan(\theta))(\sec(\theta) + \tan(\theta)) = \sec^2(\theta) - \tan^2(\theta) = 1.$$

4. While both sides of our last identity contain fractions, the left side affords us more opportunities to use our identities.[5] Substituting $\sec(\theta) = \frac{1}{\cos(\theta)}$ and $\tan(\theta) = \frac{\sin(\theta)}{\cos(\theta)}$, we get:

$$\begin{aligned}
\frac{\sec(\theta)}{1 - \tan(\theta)} &= \frac{\frac{1}{\cos(\theta)}}{1 - \frac{\sin(\theta)}{\cos(\theta)}} = \frac{\frac{1}{\cos(\theta)}}{1 - \frac{\sin(\theta)}{\cos(\theta)}} \cdot \frac{\cos(\theta)}{\cos(\theta)} \\
&= \frac{\left(\frac{1}{\cos(\theta)}\right)(\cos(\theta))}{\left(1 - \frac{\sin(\theta)}{\cos(\theta)}\right)(\cos(\theta))} = \frac{1}{(1)(\cos(\theta)) - \left(\frac{\sin(\theta)}{\cos(\theta)}\right)(\cos(\theta))} \\
&= \frac{1}{\cos(\theta) - \sin(\theta)},
\end{aligned}$$

which is exactly what we had set out to show.

5. The right hand side of the equation seems to hold more promise. We get common denominators and add:

$$\begin{aligned}
\frac{3}{1 - \sin(\theta)} - \frac{3}{1 + \sin(\theta)} &= \frac{3(1 + \sin(\theta))}{(1 - \sin(\theta))(1 + \sin(\theta))} - \frac{3(1 - \sin(\theta))}{(1 + \sin(\theta))(1 - \sin(\theta))} \\
&= \frac{3 + 3\sin(\theta)}{1 - \sin^2(\theta)} - \frac{3 - 3\sin(\theta)}{1 - \sin^2(\theta)} \\
&= \frac{(3 + 3\sin(\theta)) - (3 - 3\sin(\theta))}{1 - \sin^2(\theta)} \\
&= \frac{6\sin(\theta)}{1 - \sin^2(\theta)}
\end{aligned}$$

[5] Or, to put to another way, earn more partial credit if this were an exam question!

At this point, it is worth pausing to remind ourselves of our goal. We wish to transform this expression into $6\sec(\theta)\tan(\theta)$. Using a reciprocal and quotient identity, we find $6\sec(\theta)\tan(\theta) = 6\left(\frac{1}{\cos(\theta)}\right)\left(\frac{\sin(\theta)}{\cos(\theta)}\right)$. In other words, we need to get cosines in our denominator. Theorem 10.8 tells us $1-\sin^2(\theta) = \cos^2(\theta)$ so we get:

$$\begin{aligned}\frac{3}{1-\sin(\theta)} - \frac{3}{1+\sin(\theta)} &= \frac{6\sin(\theta)}{1-\sin^2(\theta)} = \frac{6\sin(\theta)}{\cos^2(\theta)} \\ &= 6\left(\frac{1}{\cos(\theta)}\right)\left(\frac{\sin(\theta)}{\cos(\theta)}\right) = 6\sec(\theta)\tan(\theta)\end{aligned}$$

6. It is debatable which side of the identity is more complicated. One thing which stands out is that the denominator on the left hand side is $1-\cos(\theta)$, while the numerator of the right hand side is $1+\cos(\theta)$. This suggests the strategy of starting with the left hand side and multiplying the numerator and denominator by the quantity $1+\cos(\theta)$:

$$\begin{aligned}\frac{\sin(\theta)}{1-\cos(\theta)} &= \frac{\sin(\theta)}{(1-\cos(\theta))}\cdot\frac{(1+\cos(\theta))}{(1+\cos(\theta))} = \frac{\sin(\theta)(1+\cos(\theta))}{(1-\cos(\theta))(1+\cos(\theta))} \\ &= \frac{\sin(\theta)(1+\cos(\theta))}{1-\cos^2(\theta)} = \frac{\sin(\theta)(1+\cos(\theta))}{\sin^2(\theta)} \\ &= \frac{\cancel{\sin(\theta)}(1+\cos(\theta))}{\cancel{\sin(\theta)}\sin(\theta)} = \frac{1+\cos(\theta)}{\sin(\theta)}\end{aligned}$$

□

In Example 10.3.3 number 6 above, we see that multiplying $1-\cos(\theta)$ by $1+\cos(\theta)$ produces a difference of squares that can be simplified to one term using Theorem 10.8. This is exactly the same kind of phenomenon that occurs when we multiply expressions such as $1-\sqrt{2}$ by $1+\sqrt{2}$ or $3-4i$ by $3+4i$. (Can you recall instances from Algebra where we did such things?) For this reason, the quantities $(1-\cos(\theta))$ and $(1+\cos(\theta))$ are called 'Pythagorean Conjugates.' Below is a list of other common Pythagorean Conjugates.

Pythagorean Conjugates

- $1-\cos(\theta)$ and $1+\cos(\theta)$: $(1-\cos(\theta))(1+\cos(\theta)) = 1-\cos^2(\theta) = \sin^2(\theta)$
- $1-\sin(\theta)$ and $1+\sin(\theta)$: $(1-\sin(\theta))(1+\sin(\theta)) = 1-\sin^2(\theta) = \cos^2(\theta)$
- $\sec(\theta)-1$ and $\sec(\theta)+1$: $(\sec(\theta)-1)(\sec(\theta)+1) = \sec^2(\theta)-1 = \tan^2(\theta)$
- $\sec(\theta)-\tan(\theta)$ and $\sec(\theta)+\tan(\theta)$: $(\sec(\theta)-\tan(\theta))(\sec(\theta)+\tan(\theta)) = \sec^2(\theta)-\tan^2(\theta) = 1$
- $\csc(\theta)-1$ and $\csc(\theta)+1$: $(\csc(\theta)-1)(\csc(\theta)+1) = \csc^2(\theta)-1 = \cot^2(\theta)$
- $\csc(\theta)-\cot(\theta)$ and $\csc(\theta)+\cot(\theta)$: $(\csc(\theta)-\cot(\theta))(\csc(\theta)+\cot(\theta)) = \csc^2(\theta)-\cot^2(\theta) = 1$

Verifying trigonometric identities requires a healthy mix of tenacity and inspiration. You will need to spend many hours struggling with them just to become proficient in the basics. Like many things in life, there is no short-cut here – there is no complete algorithm for verifying identities. Nevertheless, a summary of some strategies which may be helpful (depending on the situation) is provided below and ample practice is provided for you in the Exercises.

Strategies for Verifying Identities

- Try working on the more complicated side of the identity.
- Use the Reciprocal and Quotient Identities in Theorem 10.6 to write functions on one side of the identity in terms of the functions on the other side of the identity. Simplify the resulting complex fractions.
- Add rational expressions with unlike denominators by obtaining common denominators.
- Use the Pythagorean Identities in Theorem 10.8 to 'exchange' sines and cosines, secants and tangents, cosecants and cotangents, and simplify sums or differences of squares to one term.
- Multiply numerator **and** denominator by Pythagorean Conjugates in order to take advantage of the Pythagorean Identities in Theorem 10.8.
- If you find yourself stuck working with one side of the identity, try starting with the other side of the identity and see if you can find a way to bridge the two parts of your work.

10.3.1 Beyond the Unit Circle

In Section 10.2, we generalized the cosine and sine functions from coordinates on the Unit Circle to coordinates on circles of radius r. Using Theorem 10.3 in conjunction with Theorem 10.8, we generalize the remaining circular functions in kind.

Theorem 10.9. Suppose $Q(x, y)$ is the point on the terminal side of an angle θ (plotted in standard position) which lies on the circle of radius r, $x^2 + y^2 = r^2$. Then:

- $\sec(\theta) = \dfrac{r}{x} = \dfrac{\sqrt{x^2+y^2}}{x}$, provided $x \neq 0$.
- $\csc(\theta) = \dfrac{r}{y} = \dfrac{\sqrt{x^2+y^2}}{y}$, provided $y \neq 0$.
- $\tan(\theta) = \dfrac{y}{x}$, provided $x \neq 0$.
- $\cot(\theta) = \dfrac{x}{y}$, provided $y \neq 0$.

Example 10.3.4.

1. Suppose the terminal side of θ, when plotted in standard position, contains the point $Q(3, -4)$. Find the values of the six circular functions of θ.

2. Suppose θ is a Quadrant IV angle with $\cot(\theta) = -4$. Find the values of the five remaining circular functions of θ.

Solution.

1. Since $x = 3$ and $y = -4$, $r = \sqrt{x^2 + y^2} = \sqrt{(3)^2 + (-4)^2} = \sqrt{25} = 5$. Theorem 10.9 tells us $\cos(\theta) = \frac{3}{5}$, $\sin(\theta) = -\frac{4}{5}$, $\sec(\theta) = \frac{5}{3}$, $\csc(\theta) = -\frac{5}{4}$, $\tan(\theta) = -\frac{4}{3}$ and $\cot(\theta) = -\frac{3}{4}$.

2. In order to use Theorem 10.9, we need to find a point $Q(x, y)$ which lies on the terminal side of θ, when θ is plotted in standard position. We have that $\cot(\theta) = -4 = \frac{x}{y}$, and since θ is a Quadrant IV angle, we also know $x > 0$ and $y < 0$. Viewing $-4 = \frac{4}{-1}$, we may choose[6] $x = 4$ and $y = -1$ so that $r = \sqrt{x^2 + y^2} = \sqrt{(4)^2 + (-1)^2} = \sqrt{17}$. Applying Theorem 10.9 once more, we find $\cos(\theta) = \frac{4}{\sqrt{17}} = \frac{4\sqrt{17}}{17}$, $\sin(\theta) = -\frac{1}{\sqrt{17}} = -\frac{\sqrt{17}}{17}$, $\sec(\theta) = \frac{\sqrt{17}}{4}$, $\csc(\theta) = -\sqrt{17}$ and $\tan(\theta) = -\frac{1}{4}$. □

We may also specialize Theorem 10.9 to the case of acute angles θ which reside in a right triangle, as visualized below.

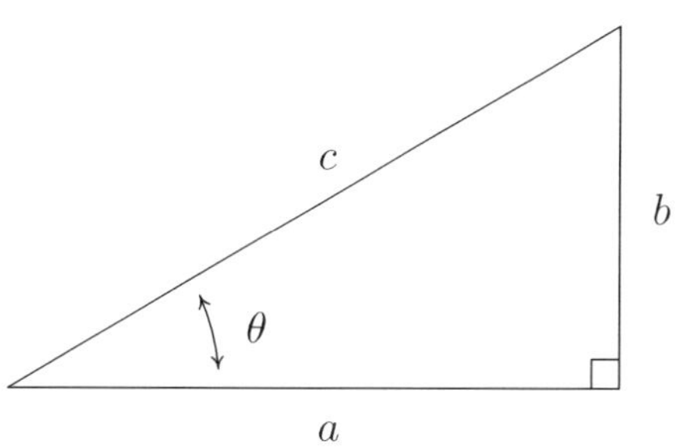

Theorem 10.10. Suppose θ is an acute angle residing in a right triangle. If the length of the side adjacent to θ is a, the length of the side opposite θ is b, and the length of the hypotenuse is c, then

$$\tan(\theta) = \frac{b}{a} \qquad \sec(\theta) = \frac{c}{a} \qquad \csc(\theta) = \frac{c}{b} \qquad \cot(\theta) = \frac{a}{b}$$

The following example uses Theorem 10.10 as well as the concept of an 'angle of inclination.' The angle of inclination (or angle of elevation) of an object refers to the angle whose initial side is some kind of base-line (say, the ground), and whose terminal side is the line-of-sight to an object above the base-line. This is represented schematically below.

[6] We may choose *any* values x and y so long as $x > 0$, $y < 0$ and $\frac{x}{y} = -4$. For example, we could choose $x = 8$ and $y = -2$. The fact that all such points lie on the terminal side of θ is a consequence of the fact that the terminal side of θ is the portion of the line with slope $-\frac{1}{4}$ which extends from the origin into Quadrant IV.

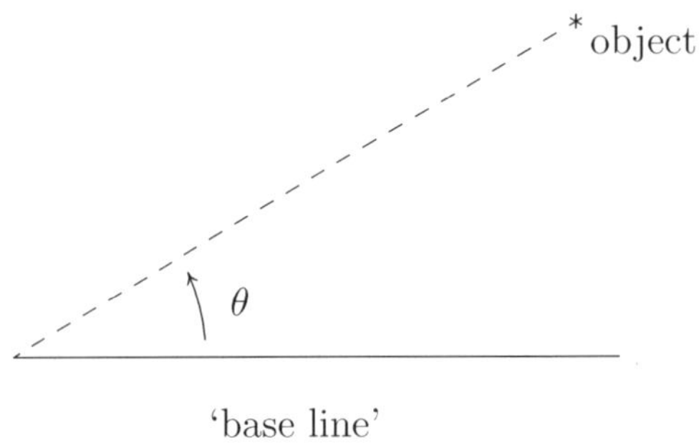

The angle of inclination from the base line to the object is θ

Example 10.3.5.

1. The angle of inclination from a point on the ground 30 feet away to the top of Lakeland's Armington Clocktower[7] is 60°. Find the height of the Clocktower to the nearest foot.

2. In order to determine the height of a California Redwood tree, two sightings from the ground, one 200 feet directly behind the other, are made. If the angles of inclination were 45° and 30°, respectively, how tall is the tree to the nearest foot?

Solution.

1. We can represent the problem situation using a right triangle as shown below. If we let h denote the height of the tower, then Theorem 10.10 gives $\tan\left(60^\circ\right) = \frac{h}{30}$. From this we get $h = 30\tan\left(60^\circ\right) = 30\sqrt{3} \approx 51.96$. Hence, the Clocktower is approximately 52 feet tall.

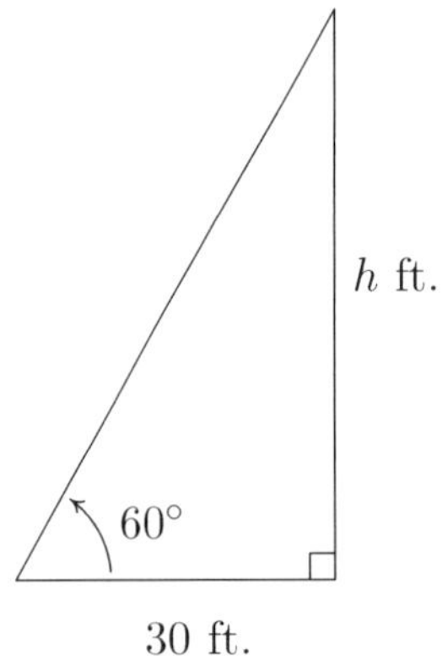

Finding the height of the Clocktower

2. Sketching the problem situation below, we find ourselves with two unknowns: the height h of the tree and the distance x from the base of the tree to the first observation point.

[7] Named in honor of Raymond Q. Armington, Lakeland's Clocktower has been a part of campus since 1972.

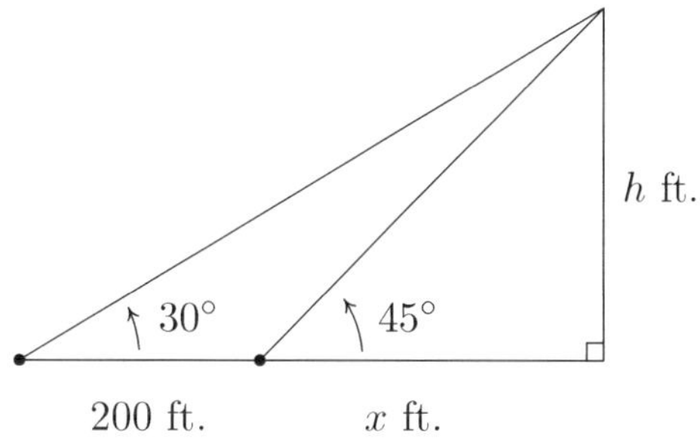

Finding the height of a California Redwood

Using Theorem 10.10, we get a pair of equations: $\tan(45^\circ) = \frac{h}{x}$ and $\tan(30^\circ) = \frac{h}{x+200}$. Since $\tan(45^\circ) = 1$, the first equation gives $\frac{h}{x} = 1$, or $x = h$. Substituting this into the second equation gives $\frac{h}{h+200} = \tan(30^\circ) = \frac{\sqrt{3}}{3}$. Clearing fractions, we get $3h = (h+200)\sqrt{3}$. The result is a linear equation for h, so we proceed to expand the right hand side and gather all the terms involving h to one side.

$$\begin{aligned}
3h &= (h+200)\sqrt{3} \\
3h &= h\sqrt{3} + 200\sqrt{3} \\
3h - h\sqrt{3} &= 200\sqrt{3} \\
(3-\sqrt{3})h &= 200\sqrt{3} \\
h &= \frac{200\sqrt{3}}{3-\sqrt{3}} \approx 273.20
\end{aligned}$$

Hence, the tree is approximately 273 feet tall. □

As we did in Section 10.2.1, we may consider all six circular functions as functions of real numbers. At this stage, there are three equivalent ways to define the functions $\sec(t)$, $\csc(t)$, $\tan(t)$ and $\cot(t)$ for real numbers t. First, we could go through the formality of the wrapping function on page 704 and define these functions as the appropriate ratios of x and y coordinates of points on the Unit Circle; second, we could define them by associating the real number t with the angle $\theta = t$ radians so that the value of the trigonometric function of t coincides with that of θ; lastly, we could simply define them using the Reciprocal and Quotient Identities as combinations of the functions $f(t) = \cos(t)$ and $g(t) = \sin(t)$. Presently, we adopt the last approach. We now set about determining the domains and ranges of the remaining four circular functions. Consider the function $F(t) = \sec(t)$ defined as $F(t) = \sec(t) = \frac{1}{\cos(t)}$. We know F is undefined whenever $\cos(t) = 0$. From Example 10.2.5 number 3, we know $\cos(t) = 0$ whenever $t = \frac{\pi}{2} + \pi k$ for integers k. Hence, our domain for $F(t) = \sec(t)$, in set builder notation is $\{t : t \neq \frac{\pi}{2} + \pi k, \text{for integers } k\}$. To get a better understanding what set of real numbers we're dealing with, it pays to write out and graph this set. Running through a few values of k, we find the domain to be $\{t : t \neq \pm\frac{\pi}{2}, \pm\frac{3\pi}{2}, \pm\frac{5\pi}{2}, \ldots\}$. Graphing this set on the number line we get

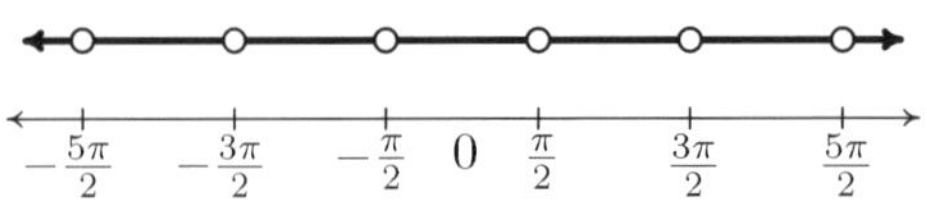

Using interval notation to describe this set, we get

$$\ldots \cup \left(-\frac{5\pi}{2}, -\frac{3\pi}{2}\right) \cup \left(-\frac{3\pi}{2}, -\frac{\pi}{2}\right) \cup \left(-\frac{\pi}{2}, \frac{\pi}{2}\right) \cup \left(\frac{\pi}{2}, \frac{3\pi}{2}\right) \cup \left(\frac{3\pi}{2}, \frac{5\pi}{2}\right) \cup \ldots$$

This is cumbersome, to say the least! In order to write this in a more compact way, we note that from the set-builder description of the domain, the kth point excluded from the domain, which we'll call x_k, can be found by the formula $x_k = \frac{\pi}{2} + \pi k$. (We are using sequence notation from Chapter 9.) Getting a common denominator and factoring out the π in the numerator, we get $x_k = \frac{(2k+1)\pi}{2}$. The domain consists of the intervals determined by successive points x_k: $(x_k, x_{k+1}) = \left(\frac{(2k+1)\pi}{2}, \frac{(2k+3)\pi}{2}\right)$. In order to capture all of the intervals in the domain, k must run through all of the integers, that is, $k = 0, \pm 1, \pm 2, \ldots$. The way we denote taking the union of infinitely many intervals like this is to use what we call in this text **extended interval notation**. The domain of $F(t) = \sec(t)$ can now be written as

$$\bigcup_{k=-\infty}^{\infty} \left(\frac{(2k+1)\pi}{2}, \frac{(2k+3)\pi}{2}\right)$$

The reader should compare this notation with summation notation introduced in Section 9.2, in particular the notation used to describe geometric series in Theorem 9.2. In the same way the index k in the series

$$\sum_{k=1}^{\infty} ar^{k-1}$$

can never equal the upper limit ∞, but rather, ranges through all of the natural numbers, the index k in the union

$$\bigcup_{k=-\infty}^{\infty} \left(\frac{(2k+1)\pi}{2}, \frac{(2k+3)\pi}{2}\right)$$

can never actually be ∞ or $-\infty$, but rather, this conveys the idea that k ranges through all of the integers. Now that we have painstakingly determined the domain of $F(t) = \sec(t)$, it is time to discuss the range. Once again, we appeal to the definition $F(t) = \sec(t) = \frac{1}{\cos(t)}$. The range of $f(t) = \cos(t)$ is $[-1, 1]$, and since $F(t) = \sec(t)$ is undefined when $\cos(t) = 0$, we split our discussion into two cases: when $0 < \cos(t) \leq 1$ and when $-1 \leq \cos(t) < 0$. If $0 < \cos(t) \leq 1$, then we can divide the inequality $\cos(t) \leq 1$ by $\cos(t)$ to obtain $\sec(t) = \frac{1}{\cos(t)} \geq 1$. Moreover, using the notation introduced in Section 4.2, we have that as $\cos(t) \to 0^+$, $\sec(t) = \frac{1}{\cos(t)} \approx \frac{1}{\text{very small } (+)} \approx \text{very big } (+)$. In other words, as $\cos(t) \to 0^+, \sec(t) \to \infty$. If, on the other hand, if $-1 \leq \cos(t) < 0$, then dividing by $\cos(t)$ causes a reversal of the inequality so that $\sec(t) = \frac{1}{\sec(t)} \leq -1$. In this case, as $\cos(t) \to 0^-$, $\sec(t) = \frac{1}{\cos(t)} \approx \frac{1}{\text{very small } (-)} \approx \text{very big } (-)$, so that as $\cos(t) \to 0^-$, we get $\sec(t) \to -\infty$. Since

$f(t) = \cos(t)$ admits all of the values in $[-1, 1]$, the function $F(t) = \sec(t)$ admits all of the values in $(-\infty, -1] \cup [1, \infty)$. Using set-builder notation, the range of $F(t) = \sec(t)$ can be written as $\{u : u \leq -1 \text{ or } u \geq 1\}$, or, more succinctly,[8] as $\{u : |u| \geq 1\}$.[9] Similar arguments can be used to determine the domains and ranges of the remaining three circular functions: $\csc(t)$, $\tan(t)$ and $\cot(t)$. The reader is encouraged to do so. (See the Exercises.) For now, we gather these facts into the theorem below.

Theorem 10.11. Domains and Ranges of the Circular Functions

- The function $f(t) = \cos(t)$
  - has domain $(-\infty, \infty)$
  - has range $[-1, 1]$
- The function $g(t) = \sin(t)$
  - has domain $(-\infty, \infty)$
  - has range $[-1, 1]$
- The function $F(t) = \sec(t) = \dfrac{1}{\cos(t)}$
  - has domain $\{t : t \neq \frac{\pi}{2} + \pi k, \text{for integers } k\} = \displaystyle\bigcup_{k=-\infty}^{\infty} \left(\frac{(2k+1)\pi}{2}, \frac{(2k+3)\pi}{2} \right)$
  - has range $\{u : |u| \geq 1\} = (-\infty, -1] \cup [1, \infty)$
- The function $G(t) = \csc(t) = \dfrac{1}{\sin(t)}$
  - has domain $\{t : t \neq \pi k, \text{for integers } k\} = \displaystyle\bigcup_{k=-\infty}^{\infty} (k\pi, (k+1)\pi)$
  - has range $\{u : |u| \geq 1\} = (-\infty, -1] \cup [1, \infty)$
- The function $J(t) = \tan(t) = \dfrac{\sin(t)}{\cos(t)}$
  - has domain $\{t : t \neq \frac{\pi}{2} + \pi k, \text{for integers } k\} = \displaystyle\bigcup_{k=-\infty}^{\infty} \left(\frac{(2k+1)\pi}{2}, \frac{(2k+3)\pi}{2} \right)$
  - has range $(-\infty, \infty)$
- The function $K(t) = \cot(t) = \dfrac{\cos(t)}{\sin(t)}$
  - has domain $\{t : t \neq \pi k, \text{for integers } k\} = \displaystyle\bigcup_{k=-\infty}^{\infty} (k\pi, (k+1)\pi)$
  - has range $(-\infty, \infty)$

---

[8] Using Theorem 2.4 from Section 2.4.

[9] Notice we have used the variable 'u' as the 'dummy variable' to describe the range elements. While there is no mathematical reason to do this (we are describing a set of real numbers, and, as such, could use t again) we choose u to help solidify the idea that these real numbers are the outputs from the inputs, which we have been calling t.

We close this section with a few notes about solving equations which involve the circular functions. First, the discussion on page 735 in Section 10.2.1 concerning solving equations applies to all six circular functions, not just $f(t) = \cos(t)$ and $g(t) = \sin(t)$. In particular, to solve the equation $\cot(t) = -1$ for real numbers t, we can use the same thought process we used in Example 10.3.2, number 3 to solve $\cot(\theta) = -1$ for angles θ in radian measure – we just need to remember to write our answers using the variable t as opposed to θ. Next, it is critical that you know the domains and ranges of the six circular functions so that you know which equations have no solutions. For example, $\sec(t) = \frac{1}{2}$ has no solution because $\frac{1}{2}$ is not in the range of secant. Finally, you will need to review the notions of reference angles and coterminal angles so that you can see why $\csc(t) = -42$ has an infinite set of solutions in Quadrant III and another infinite set of solutions in Quadrant IV.

10.3.2 EXERCISES

In Exercises 1 - 20, find the exact value or state that it is undefined.

1. $\tan\left(\frac{\pi}{4}\right)$
2. $\sec\left(\frac{\pi}{6}\right)$
3. $\csc\left(\frac{5\pi}{6}\right)$
4. $\cot\left(\frac{4\pi}{3}\right)$
5. $\tan\left(-\frac{11\pi}{6}\right)$
6. $\sec\left(-\frac{3\pi}{2}\right)$
7. $\csc\left(-\frac{\pi}{3}\right)$
8. $\cot\left(\frac{13\pi}{2}\right)$
9. $\tan(117\pi)$
10. $\sec\left(-\frac{5\pi}{3}\right)$
11. $\csc(3\pi)$
12. $\cot(-5\pi)$
13. $\tan\left(\frac{31\pi}{2}\right)$
14. $\sec\left(\frac{\pi}{4}\right)$
15. $\csc\left(-\frac{7\pi}{4}\right)$
16. $\cot\left(\frac{7\pi}{6}\right)$
17. $\tan\left(\frac{2\pi}{3}\right)$
18. $\sec(-7\pi)$
19. $\csc\left(\frac{\pi}{2}\right)$
20. $\cot\left(\frac{3\pi}{4}\right)$

In Exercises 21 - 34, use the given the information to find the exact values of the remaining circular functions of θ.

21. $\sin(\theta) = \frac{3}{5}$ with θ in Quadrant II
22. $\tan(\theta) = \frac{12}{5}$ with θ in Quadrant III
23. $\csc(\theta) = \frac{25}{24}$ with θ in Quadrant I
24. $\sec(\theta) = 7$ with θ in Quadrant IV
25. $\csc(\theta) = -\frac{10\sqrt{91}}{91}$ with θ in Quadrant III
26. $\cot(\theta) = -23$ with θ in Quadrant II
27. $\tan(\theta) = -2$ with θ in Quadrant IV.
28. $\sec(\theta) = -4$ with θ in Quadrant II.
29. $\cot(\theta) = \sqrt{5}$ with θ in Quadrant III.
30. $\cos(\theta) = \frac{1}{3}$ with θ in Quadrant I.
31. $\cot(\theta) = 2$ with $0 < \theta < \frac{\pi}{2}$.
32. $\csc(\theta) = 5$ with $\frac{\pi}{2} < \theta < \pi$.
33. $\tan(\theta) = \sqrt{10}$ with $\pi < \theta < \frac{3\pi}{2}$.
34. $\sec(\theta) = 2\sqrt{5}$ with $\frac{3\pi}{2} < \theta < 2\pi$.

In Exercises 35 - 42, use your calculator to approximate the given value to three decimal places. Make sure your calculator is in the proper angle measurement mode!

35. $\csc(78.95^\circ)$
36. $\tan(-2.01)$
37. $\cot(392.994)$
38. $\sec(207^\circ)$
39. $\csc(5.902)$
40. $\tan(39.672^\circ)$
41. $\cot(3^\circ)$
42. $\sec(0.45)$

In Exercises 43 - 57, find all of the angles which satisfy the equation.

43. $\tan(\theta) = \sqrt{3}$
44. $\sec(\theta) = 2$
45. $\csc(\theta) = -1$
46. $\cot(\theta) = \dfrac{\sqrt{3}}{3}$
47. $\tan(\theta) = 0$
48. $\sec(\theta) = 1$
49. $\csc(\theta) = 2$
50. $\cot(\theta) = 0$
51. $\tan(\theta) = -1$
52. $\sec(\theta) = 0$
53. $\csc(\theta) = -\dfrac{1}{2}$
54. $\sec(\theta) = -1$
55. $\tan(\theta) = -\sqrt{3}$
56. $\csc(\theta) = -2$
57. $\cot(\theta) = -1$

In Exercises 58 - 65, solve the equation for t. Give exact values.

58. $\cot(t) = 1$
59. $\tan(t) = \dfrac{\sqrt{3}}{3}$
60. $\sec(t) = -\dfrac{2\sqrt{3}}{3}$
61. $\csc(t) = 0$
62. $\cot(t) = -\sqrt{3}$
63. $\tan(t) = -\dfrac{\sqrt{3}}{3}$
64. $\sec(t) = \dfrac{2\sqrt{3}}{3}$
65. $\csc(t) = \dfrac{2\sqrt{3}}{3}$

In Exercises 66 - 69, use Theorem 10.10 to find the requested quantities.

66. Find θ, a, and c.

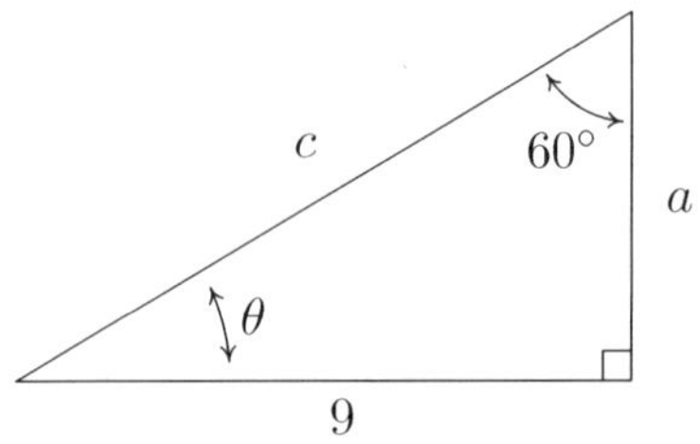

67. Find α, b, and c.

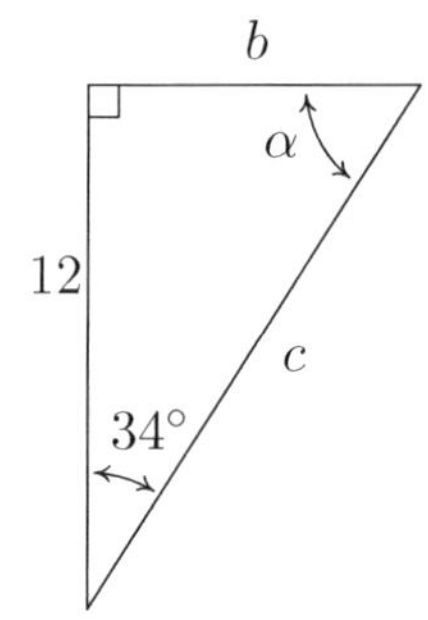

68. Find θ, a, and c.

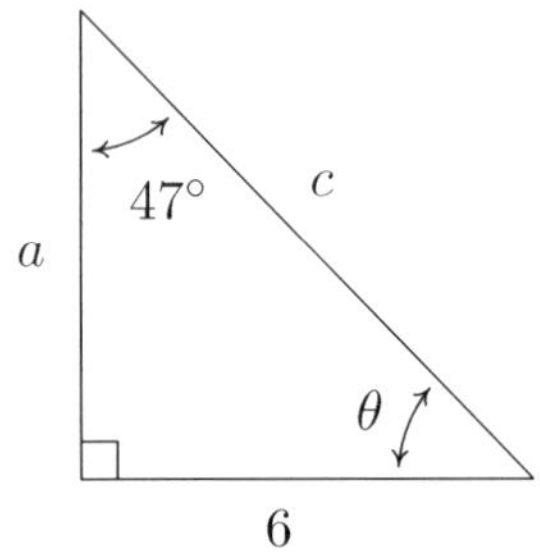

69. Find β, b, and c.

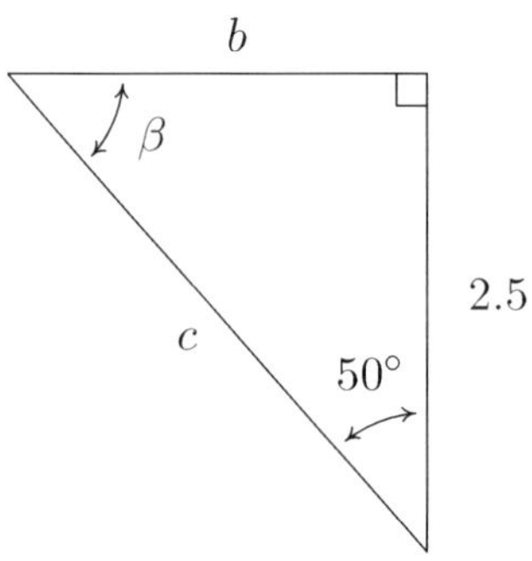

In Exercises 70 - 75, use Theorem 10.10 to answer the question. Assume that θ is an angle in a right triangle.

70. If $\theta = 30^\circ$ and the side opposite θ has length 4, how long is the side adjacent to θ?

71. If $\theta = 15^\circ$ and the hypotenuse has length 10, how long is the side opposite θ?

72. If $\theta = 87^\circ$ and the side adjacent to θ has length 2, how long is the side opposite θ?

73. If $\theta = 38.2^\circ$ and the side opposite θ has lengh 14, how long is the hypoteneuse?

74. If $\theta = 2.05^\circ$ and the hypotenuse has length 3.98, how long is the side adjacent to θ?

75. If $\theta = 42^\circ$ and the side adjacent to θ has length 31, how long is the side opposite θ?

76. A tree standing vertically on level ground casts a 120 foot long shadow. The angle of elevation from the end of the shadow to the top of the tree is 21.4°. Find the height of the tree to the nearest foot. With the help of your classmates, research the term *umbra versa* and see what it has to do with the shadow in this problem.

77. The broadcast tower for radio station WSAZ (Home of "Algebra in the Morning with Carl and Jeff") has two enormous flashing red lights on it: one at the very top and one a few feet below the top. From a point 5000 feet away from the base of the tower on level ground the angle of elevation to the top light is 7.970° and to the second light is 7.125°. Find the distance between the lights to the nearest foot.

78. On page 753 we defined the angle of inclination (also known as the angle of elevation) and in this exercise we introduce a related angle - the angle of depression (also known as the angle of declination). The angle of depression of an object refers to the angle whose initial side is a horizontal line above the object and whose terminal side is the line-of-sight to the object below the horizontal. This is represented schematically below.

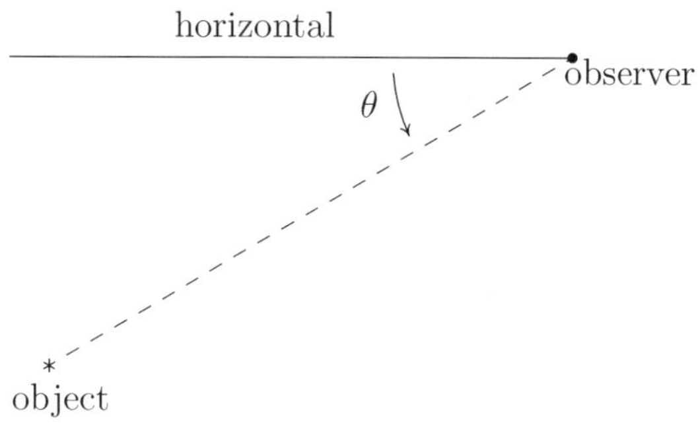

The angle of depression from the horizontal to the object is θ

(a) Show that if the horizontal is above and parallel to level ground then the angle of depression (from observer to object) and the angle of inclination (from object to observer) will be congruent because they are alternate interior angles.

(b) From a firetower 200 feet above level ground in the Sasquatch National Forest, a ranger spots a fire off in the distance. The angle of depression to the fire is 2.5°. How far away from the base of the tower is the fire?

(c) The ranger in part 78b sees a Sasquatch running directly from the fire towards the firetower. The ranger takes two sightings. At the first sighting, the angle of depression from the tower to the Sasquatch is 6°. The second sighting, taken just 10 seconds later, gives the the angle of depression as 6.5°. How far did the Saquatch travel in those 10 seconds? Round your answer to the nearest foot. How fast is it running in miles per hour? Round your answer to the nearest mile per hour. If the Sasquatch keeps up this pace, how long will it take for the Sasquatch to reach the firetower from his location at the second sighting? Round your answer to the nearest minute.

79. When I stand 30 feet away from a tree at home, the angle of elevation to the top of the tree is 50° and the angle of depression to the base of the tree is 10°. What is the height of the tree? Round your answer to the nearest foot.

80. From the observation deck of the lighthouse at Sasquatch Point 50 feet above the surface of Lake Ippizuti, a lifeguard spots a boat out on the lake sailing directly toward the lighthouse. The first sighting had an angle of depression of 8.2° and the second sighting had an angle of depression of 25.9°. How far had the boat traveled between the sightings?

81. A guy wire 1000 feet long is attached to the top of a tower. When pulled taut it makes a 43° angle with the ground. How tall is the tower? How far away from the base of the tower does the wire hit the ground?

In Exercises 82 - 128, verify the identity. Assume that all quantities are defined.

82. $\cos(\theta)\sec(\theta) = 1$

83. $\tan(\theta)\cos(\theta) = \sin(\theta)$

84. $\sin(\theta)\csc(\theta) = 1$

85. $\tan(\theta)\cot(\theta) = 1$

86. $\csc(\theta)\cos(\theta) = \cot(\theta)$

87. $\dfrac{\sin(\theta)}{\cos^2(\theta)} = \sec(\theta)\tan(\theta)$

88. $\dfrac{\cos(\theta)}{\sin^2(\theta)} = \csc(\theta)\cot(\theta)$

89. $\dfrac{1+\sin(\theta)}{\cos(\theta)} = \sec(\theta) + \tan(\theta)$

90. $\dfrac{1-\cos(\theta)}{\sin(\theta)} = \csc(\theta) - \cot(\theta)$

91. $\dfrac{\cos(\theta)}{1-\sin^2(\theta)} = \sec(\theta)$

92. $\dfrac{\sin(\theta)}{1-\cos^2(\theta)} = \csc(\theta)$

93. $\dfrac{\sec(\theta)}{1+\tan^2(\theta)} = \cos(\theta)$

94. $\dfrac{\csc(\theta)}{1+\cot^2(\theta)} = \sin(\theta)$

95. $\dfrac{\tan(\theta)}{\sec^2(\theta)-1} = \cot(\theta)$

96. $\dfrac{\cot(\theta)}{\csc^2(\theta)-1} = \tan(\theta)$

97. $4\cos^2(\theta) + 4\sin^2(\theta) = 4$

98. $9 - \cos^2(\theta) - \sin^2(\theta) = 8$

99. $\tan^3(\theta) = \tan(\theta)\sec^2(\theta) - \tan(\theta)$

100. $\sin^5(\theta) = \left(1-\cos^2(\theta)\right)^2 \sin(\theta)$

101. $\sec^{10}(\theta) = \left(1+\tan^2(\theta)\right)^4 \sec^2(\theta)$

102. $\cos^2(\theta)\tan^3(\theta) = \tan(\theta) - \sin(\theta)\cos(\theta)$

103. $\sec^4(\theta) - \sec^2(\theta) = \tan^2(\theta) + \tan^4(\theta)$

104. $\dfrac{\cos(\theta)+1}{\cos(\theta)-1} = \dfrac{1+\sec(\theta)}{1-\sec(\theta)}$

105. $\dfrac{\sin(\theta)+1}{\sin(\theta)-1} = \dfrac{1+\csc(\theta)}{1-\csc(\theta)}$

106. $\dfrac{1-\cot(\theta)}{1+\cot(\theta)} = \dfrac{\tan(\theta)-1}{\tan(\theta)+1}$

107. $\dfrac{1-\tan(\theta)}{1+\tan(\theta)} = \dfrac{\cos(\theta)-\sin(\theta)}{\cos(\theta)+\sin(\theta)}$

108. $\tan(\theta) + \cot(\theta) = \sec(\theta)\csc(\theta)$

109. $\csc(\theta) - \sin(\theta) = \cot(\theta)\cos(\theta)$

110. $\cos(\theta) - \sec(\theta) = -\tan(\theta)\sin(\theta)$

111. $\cos(\theta)(\tan(\theta) + \cot(\theta)) = \csc(\theta)$

112. $\sin(\theta)(\tan(\theta) + \cot(\theta)) = \sec(\theta)$

113. $\dfrac{1}{1-\cos(\theta)} + \dfrac{1}{1+\cos(\theta)} = 2\csc^2(\theta)$

114. $\dfrac{1}{\sec(\theta)+1} + \dfrac{1}{\sec(\theta)-1} = 2\csc(\theta)\cot(\theta)$

115. $\dfrac{1}{\csc(\theta)+1} + \dfrac{1}{\csc(\theta)-1} = 2\sec(\theta)\tan(\theta)$

116. $\dfrac{1}{\csc(\theta)-\cot(\theta)} - \dfrac{1}{\csc(\theta)+\cot(\theta)} = 2\cot(\theta)$

117. $\dfrac{\cos(\theta)}{1-\tan(\theta)} + \dfrac{\sin(\theta)}{1-\cot(\theta)} = \sin(\theta) + \cos(\theta)$

118. $\dfrac{1}{\sec(\theta)+\tan(\theta)} = \sec(\theta) - \tan(\theta)$

119. $\dfrac{1}{\sec(\theta)-\tan(\theta)} = \sec(\theta) + \tan(\theta)$

120. $\dfrac{1}{\csc(\theta)-\cot(\theta)} = \csc(\theta) + \cot(\theta)$

121. $\dfrac{1}{\csc(\theta)+\cot(\theta)} = \csc(\theta) - \cot(\theta)$

122. $\dfrac{1}{1-\sin(\theta)} = \sec^2(\theta) + \sec(\theta)\tan(\theta)$

123. $\dfrac{1}{1+\sin(\theta)} = \sec^2(\theta) - \sec(\theta)\tan(\theta)$

124. $\dfrac{1}{1-\cos(\theta)} = \csc^2(\theta) + \csc(\theta)\cot(\theta)$

125. $\dfrac{1}{1+\cos(\theta)} = \csc^2(\theta) - \csc(\theta)\cot(\theta)$

126. $\dfrac{\cos(\theta)}{1+\sin(\theta)} = \dfrac{1-\sin(\theta)}{\cos(\theta)}$

127. $\csc(\theta) - \cot(\theta) = \dfrac{\sin(\theta)}{1+\cos(\theta)}$

128. $\dfrac{1-\sin(\theta)}{1+\sin(\theta)} = (\sec(\theta) - \tan(\theta))^2$

In Exercises 129 - 132, verify the identity. You may need to consult Sections 2.2 and 6.2 for a review of the properties of absolute value and logarithms before proceeding.

129. $\ln|\sec(\theta)| = -\ln|\cos(\theta)|$

130. $-\ln|\csc(\theta)| = \ln|\sin(\theta)|$

131. $-\ln|\sec(\theta) - \tan(\theta)| = \ln|\sec(\theta) + \tan(\theta)|$

132. $-\ln|\csc(\theta) + \cot(\theta)| = \ln|\csc(\theta) - \cot(\theta)|$

133. Verify the domains and ranges of the tangent, cosecant and cotangent functions as presented in Theorem 10.11.

134. As we did in Exercise 74 in Section 10.2, let α and β be the two acute angles of a right triangle. (Thus α and β are complementary angles.) Show that $\sec(\alpha) = \csc(\beta)$ and $\tan(\alpha) = \cot(\beta)$. The fact that co-functions of complementary angles are equal in this case is not an accident and a more general result will be given in Section 10.4.

135. We wish to establish the inequality $\cos(\theta) < \dfrac{\sin(\theta)}{\theta} < 1$ for $0 < \theta < \dfrac{\pi}{2}$. Use the diagram from the beginning of the section, partially reproduced below, to answer the following.

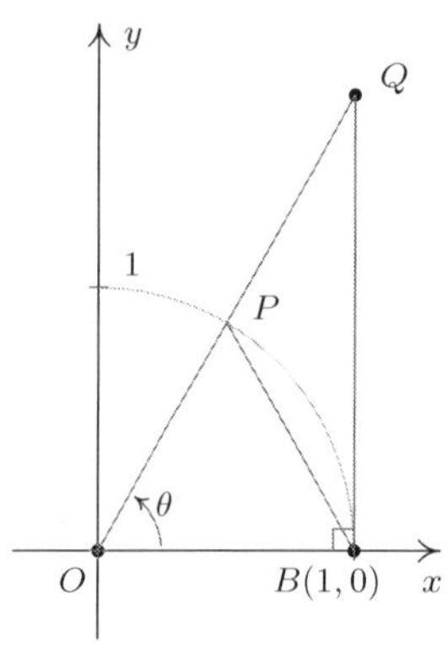

(a) Show that triangle OPB has area $\dfrac{1}{2}\sin(\theta)$.

(b) Show that the circular sector OPB with central angle θ has area $\dfrac{1}{2}\theta$.

(c) Show that triangle OQB has area $\dfrac{1}{2}\tan(\theta)$.

(d) Comparing areas, show that $\sin(\theta) < \theta < \tan(\theta)$ for $0 < \theta < \dfrac{\pi}{2}$.

(e) Use the inequality $\sin(\theta) < \theta$ to show that $\dfrac{\sin(\theta)}{\theta} < 1$ for $0 < \theta < \dfrac{\pi}{2}$.

(f) Use the inequality $\theta < \tan(\theta)$ to show that $\cos(\theta) < \dfrac{\sin(\theta)}{\theta}$ for $0 < \theta < \dfrac{\pi}{2}$. Combine this with the previous part to complete the proof.

136. Show that $\cos(\theta) < \dfrac{\sin(\theta)}{\theta} < 1$ also holds for $-\dfrac{\pi}{2} < \theta < 0$.

137. Explain why the fact that $\tan(\theta) = 3 = \frac{3}{1}$ does not mean $\sin(\theta) = 3$ and $\cos(\theta) = 1$? (See the solution to number 6 in Example 10.3.1.)

10.3.3 Answers

1. $\tan\left(\frac{\pi}{4}\right) = 1$
2. $\sec\left(\frac{\pi}{6}\right) = \frac{2\sqrt{3}}{3}$
3. $\csc\left(\frac{5\pi}{6}\right) = 2$
4. $\cot\left(\frac{4\pi}{3}\right) = \frac{\sqrt{3}}{3}$
5. $\tan\left(-\frac{11\pi}{6}\right) = \frac{\sqrt{3}}{3}$
6. $\sec\left(-\frac{3\pi}{2}\right)$ is undefined
7. $\csc\left(-\frac{\pi}{3}\right) = -\frac{2\sqrt{3}}{3}$
8. $\cot\left(\frac{13\pi}{2}\right) = 0$
9. $\tan(117\pi) = 0$
10. $\sec\left(-\frac{5\pi}{3}\right) = 2$
11. $\csc(3\pi)$ is undefined
12. $\cot(-5\pi)$ is undefined
13. $\tan\left(\frac{31\pi}{2}\right)$ is undefined
14. $\sec\left(\frac{\pi}{4}\right) = \sqrt{2}$
15. $\csc\left(-\frac{7\pi}{4}\right) = \sqrt{2}$
16. $\cot\left(\frac{7\pi}{6}\right) = \sqrt{3}$
17. $\tan\left(\frac{2\pi}{3}\right) = -\sqrt{3}$
18. $\sec(-7\pi) = -1$
19. $\csc\left(\frac{\pi}{2}\right) = 1$
20. $\cot\left(\frac{3\pi}{4}\right) = -1$
21. $\sin(\theta) = \frac{3}{5}, \cos(\theta) = -\frac{4}{5}, \tan(\theta) = -\frac{3}{4}, \csc(\theta) = \frac{5}{3}, \sec(\theta) = -\frac{5}{4}, \cot(\theta) = -\frac{4}{3}$
22. $\sin(\theta) = -\frac{12}{13}, \cos(\theta) = -\frac{5}{13}, \tan(\theta) = \frac{12}{5}, \csc(\theta) = -\frac{13}{12}, \sec(\theta) = -\frac{13}{5}, \cot(\theta) = \frac{5}{12}$
23. $\sin(\theta) = \frac{24}{25}, \cos(\theta) = \frac{7}{25}, \tan(\theta) = \frac{24}{7}, \csc(\theta) = \frac{25}{24}, \sec(\theta) = \frac{25}{7}, \cot(\theta) = \frac{7}{24}$
24. $\sin(\theta) = \frac{-4\sqrt{3}}{7}, \cos(\theta) = \frac{1}{7}, \tan(\theta) = -4\sqrt{3}, \csc(\theta) = -\frac{7\sqrt{3}}{12}, \sec(\theta) = 7, \cot(\theta) = -\frac{\sqrt{3}}{12}$
25. $\sin(\theta) = -\frac{\sqrt{91}}{10}, \cos(\theta) = -\frac{3}{10}, \tan(\theta) = \frac{\sqrt{91}}{3}, \csc(\theta) = -\frac{10\sqrt{91}}{91}, \sec(\theta) = -\frac{10}{3}, \cot(\theta) = \frac{3\sqrt{91}}{91}$
26. $\sin(\theta) = \frac{\sqrt{530}}{530}, \cos(\theta) = -\frac{23\sqrt{530}}{530}, \tan(\theta) = -\frac{1}{23}, \csc(\theta) = \sqrt{530}, \sec(\theta) = -\frac{\sqrt{530}}{23}, \cot(\theta) = -23$
27. $\sin(\theta) = -\frac{2\sqrt{5}}{5}, \cos(\theta) = \frac{\sqrt{5}}{5}, \tan(\theta) = -2, \csc(\theta) = -\frac{\sqrt{5}}{2}, \sec(\theta) = \sqrt{5}, \cot(\theta) = -\frac{1}{2}$
28. $\sin(\theta) = \frac{\sqrt{15}}{4}, \cos(\theta) = -\frac{1}{4}, \tan(\theta) = -\sqrt{15}, \csc(\theta) = \frac{4\sqrt{15}}{15}, \sec(\theta) = -4, \cot(\theta) = -\frac{\sqrt{15}}{15}$
29. $\sin(\theta) = -\frac{\sqrt{6}}{6}, \cos(\theta) = -\frac{\sqrt{30}}{6}, \tan(\theta) = \frac{\sqrt{5}}{5}, \csc(\theta) = -\sqrt{6}, \sec(\theta) = -\frac{\sqrt{30}}{5}, \cot(\theta) = \sqrt{5}$
30. $\sin(\theta) = \frac{2\sqrt{2}}{3}, \cos(\theta) = \frac{1}{3}, \tan(\theta) = 2\sqrt{2}, \csc(\theta) = \frac{3\sqrt{2}}{4}, \sec(\theta) = 3, \cot(\theta) = \frac{\sqrt{2}}{4}$
31. $\sin(\theta) = \frac{\sqrt{5}}{5}, \cos(\theta) = \frac{2\sqrt{5}}{5}, \tan(\theta) = \frac{1}{2}, \csc(\theta) = \sqrt{5}, \sec(\theta) = \frac{\sqrt{5}}{2}, \cot(\theta) = 2$
32. $\sin(\theta) = \frac{1}{5}, \cos(\theta) = -\frac{2\sqrt{6}}{5}, \tan(\theta) = -\frac{\sqrt{6}}{12}, \csc(\theta) = 5, \sec(\theta) = -\frac{5\sqrt{6}}{12}, \cot(\theta) = -2\sqrt{6}$
33. $\sin(\theta) = -\frac{\sqrt{110}}{11}, \cos(\theta) = -\frac{\sqrt{11}}{11}, \tan(\theta) = \sqrt{10}, \csc(\theta) = -\frac{\sqrt{110}}{10}, \sec(\theta) = -\sqrt{11}, \cot(\theta) = \frac{\sqrt{10}}{10}$
34. $\sin(\theta) = -\frac{\sqrt{95}}{10}, \cos(\theta) = \frac{\sqrt{5}}{10}, \tan(\theta) = -\sqrt{19}, \csc(\theta) = -\frac{2\sqrt{95}}{19}, \sec(\theta) = 2\sqrt{5}, \cot(\theta) = -\frac{\sqrt{19}}{19}$

35. $\csc(78.95^\circ) \approx 1.019$

36. $\tan(-2.01) \approx 2.129$

37. $\cot(392.994) \approx 3.292$

38. $\sec(207^\circ) \approx -1.122$

39. $\csc(5.902) \approx -2.688$

40. $\tan(39.672^\circ) \approx 0.829$

41. $\cot(3^\circ) \approx 19.081$

42. $\sec(0.45) \approx 1.111$

43. $\tan(\theta) = \sqrt{3}$ when $\theta = \frac{\pi}{3} + \pi k$ for any integer k

44. $\sec(\theta) = 2$ when $\theta = \frac{\pi}{3} + 2\pi k$ or $\theta = \frac{5\pi}{3} + 2\pi k$ for any integer k

45. $\csc(\theta) = -1$ when $\theta = \frac{3\pi}{2} + 2\pi k$ for any integer k.

46. $\cot(\theta) = \frac{\sqrt{3}}{3}$ when $\theta = \frac{\pi}{3} + \pi k$ for any integer k

47. $\tan(\theta) = 0$ when $\theta = \pi k$ for any integer k

48. $\sec(\theta) = 1$ when $\theta = 2\pi k$ for any integer k

49. $\csc(\theta) = 2$ when $\theta = \frac{\pi}{6} + 2\pi k$ or $\theta = \frac{5\pi}{6} + 2\pi k$ for any integer k.

50. $\cot(\theta) = 0$ when $\theta = \frac{\pi}{2} + \pi k$ for any integer k

51. $\tan(\theta) = -1$ when $\theta = \frac{3\pi}{4} + \pi k$ for any integer k

52. $\sec(\theta) = 0$ never happens

53. $\csc(\theta) = -\frac{1}{2}$ never happens

54. $\sec(\theta) = -1$ when $\theta = \pi + 2\pi k = (2k+1)\pi$ for any integer k

55. $\tan(\theta) = -\sqrt{3}$ when $\theta = \frac{2\pi}{3} + \pi k$ for any integer k

56. $\csc(\theta) = -2$ when $\theta = \frac{7\pi}{6} + 2\pi k$ or $\theta = \frac{11\pi}{6} + 2\pi k$ for any integer k

57. $\cot(\theta) = -1$ when $\theta = \frac{3\pi}{4} + \pi k$ for any integer k

58. $\cot(t) = 1$ when $t = \frac{\pi}{4} + \pi k$ for any integer k

59. $\tan(t) = \frac{\sqrt{3}}{3}$ when $t = \frac{\pi}{6} + \pi k$ for any integer k

60. $\sec(t) = -\frac{2\sqrt{3}}{3}$ when $t = \frac{5\pi}{6} + 2\pi k$ or $t = \frac{7\pi}{6} + 2\pi k$ for any integer k

61. $\csc(t) = 0$ never happens

62. $\cot(t) = -\sqrt{3}$ when $t = \frac{5\pi}{6} + \pi k$ for any integer k

63. $\tan(t) = -\frac{\sqrt{3}}{3}$ when $t = \frac{5\pi}{6} + \pi k$ for any integer k

64. $\sec(t) = \frac{2\sqrt{3}}{3}$ when $t = \frac{\pi}{6} + 2\pi k$ or $t = \frac{11\pi}{6} + 2\pi k$ for any integer k

65. $\csc(t) = \frac{2\sqrt{3}}{3}$ when $t = \frac{\pi}{3} + 2\pi k$ or $t = \frac{2\pi}{3} + 2\pi k$ for any integer k

66. $\theta = 30^\circ$, $a = 3\sqrt{3}$, $c = \sqrt{108} = 6\sqrt{3}$

67. $\alpha = 56^\circ$, $b = 12\tan(34^\circ) = 8.094$, $c = 12\sec(34^\circ) = \frac{12}{\cos(34^\circ)} \approx 14.475$

68. $\theta = 43^\circ$, $a = 6\cot(47^\circ) = \frac{6}{\tan(47^\circ)} \approx 5.595$, $c = 6\csc(47^\circ) = \frac{6}{\sin(47^\circ)} \approx 8.204$

69. $\beta = 40^\circ$, $b = 2.5\tan(50^\circ) \approx 2.979$, $c = 2.5\sec(50^\circ) = \frac{2.5}{\cos(50^\circ)} \approx 3.889$

70. The side adjacent to θ has length $4\sqrt{3} \approx 6.928$

71. The side opposite θ has length $10\sin(15^\circ) \approx 2.588$

72. The side opposite θ is $2\tan(87^\circ) \approx 38.162$

73. The hypoteneuse has length $14\csc(38.2^\circ) = \frac{14}{\sin(38.2^\circ)} \approx 22.639$

74. The side adjacent to θ has length $3.98\cos(2.05^\circ) \approx 3.977$

75. The side opposite θ has length $31\tan(42^\circ) \approx 27.912$

76. The tree is about 47 feet tall.

77. The lights are about 75 feet apart.

78. (b) The fire is about 4581 feet from the base of the tower.

    (c) The Sasquatch ran $200\cot(6^\circ) - 200\cot(6.5^\circ) \approx 147$ feet in those 10 seconds. This translates to ≈ 10 miles per hour. At the scene of the second sighting, the Sasquatch was ≈ 1755 feet from the tower, which means, if it keeps up this pace, it will reach the tower in about 2 minutes.

79. The tree is about 41 feet tall.

80. The boat has traveled about 244 feet.

81. The tower is about 682 feet tall. The guy wire hits the ground about 731 feet away from the base of the tower.

10.4 TRIGONOMETRIC IDENTITIES

In Section 10.3, we saw the utility of the Pythagorean Identities in Theorem 10.8 along with the Quotient and Reciprocal Identities in Theorem 10.6. Not only did these identities help us compute the values of the circular functions for angles, they were also useful in simplifying expressions involving the circular functions. In this section, we introduce several collections of identities which have uses in this course and beyond. Our first set of identities is the 'Even / Odd' identities.[1]

Theorem 10.12. Even / Odd Identities: For all applicable angles θ,

- $\cos(-\theta) = \cos(\theta)$
- $\sin(-\theta) = -\sin(\theta)$
- $\tan(-\theta) = -\tan(\theta)$
- $\sec(-\theta) = \sec(\theta)$
- $\csc(-\theta) = -\csc(\theta)$
- $\cot(-\theta) = -\cot(\theta)$

In light of the Quotient and Reciprocal Identities, Theorem 10.6, it suffices to show $\cos(-\theta) = \cos(\theta)$ and $\sin(-\theta) = -\sin(\theta)$. The remaining four circular functions can be expressed in terms of $\cos(\theta)$ and $\sin(\theta)$ so the proofs of their Even / Odd Identities are left as exercises. Consider an angle θ plotted in standard position. Let θ_0 be the angle coterminal with θ with $0 \leq \theta_0 < 2\pi$. (We can construct the angle θ_0 by rotating counter-clockwise from the positive x-axis to the terminal side of θ as pictured below.) Since θ and θ_0 are coterminal, $\cos(\theta) = \cos(\theta_0)$ and $\sin(\theta) = \sin(\theta_0)$.

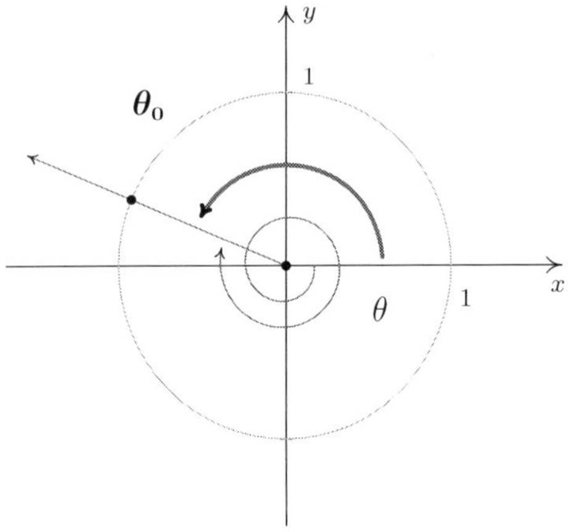

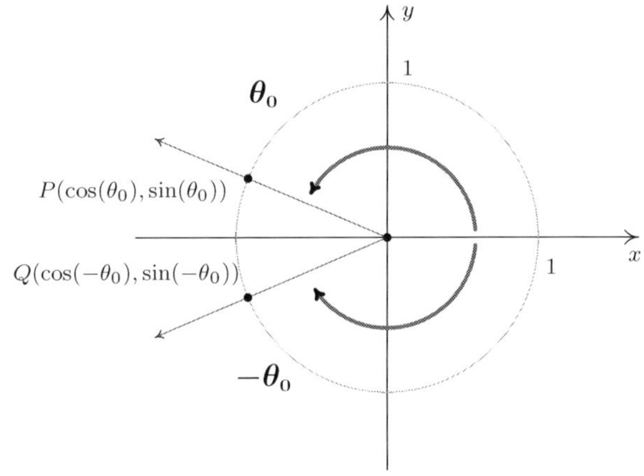

We now consider the angles $-\theta$ and $-\theta_0$. Since θ is coterminal with θ_0, there is some integer k so that $\theta = \theta_0 + 2\pi \cdot k$. Therefore, $-\theta = -\theta_0 - 2\pi \cdot k = -\theta_0 + 2\pi \cdot (-k)$. Since k is an integer, so is $(-k)$, which means $-\theta$ is coterminal with $-\theta_0$. Hence, $\cos(-\theta) = \cos(-\theta_0)$ and $\sin(-\theta) = \sin(-\theta_0)$. Let P and Q denote the points on the terminal sides of θ_0 and $-\theta_0$, respectively, which lie on the Unit Circle. By definition, the coordinates of P are $(\cos(\theta_0), \sin(\theta_0))$ and the coordinates of Q are $(\cos(-\theta_0), \sin(-\theta_0))$. Since θ_0 and $-\theta_0$ sweep out congruent central sectors of the Unit Circle, it

[1]As mentioned at the end of Section 10.2, properties of the circular functions when thought of as functions of angles in radian measure hold equally well if we view these functions as functions of real numbers. Not surprisingly, the Even / Odd properties of the circular functions are so named because they identify cosine and secant as even functions, while the remaining four circular functions are odd. (See Section 1.6.)

follows that the points P and Q are symmetric about the x-axis. Thus, $\cos(-\theta_0) = \cos(\theta_0)$ and $\sin(-\theta_0) = -\sin(\theta_0)$. Since the cosines and sines of θ_0 and $-\theta_0$ are the same as those for θ and $-\theta$, respectively, we get $\cos(-\theta) = \cos(\theta)$ and $\sin(-\theta) = -\sin(\theta)$, as required. The Even / Odd Identities are readily demonstrated using any of the 'common angles' noted in Section 10.2. Their true utility, however, lies not in computation, but in simplifying expressions involving the circular functions. In fact, our next batch of identities makes heavy use of the Even / Odd Identities.

Theorem 10.13. Sum and Difference Identities for Cosine: For all angles α and β,

- $\cos(\alpha + \beta) = \cos(\alpha)\cos(\beta) - \sin(\alpha)\sin(\beta)$
- $\cos(\alpha - \beta) = \cos(\alpha)\cos(\beta) + \sin(\alpha)\sin(\beta)$

We first prove the result for differences. As in the proof of the Even / Odd Identities, we can reduce the proof for general angles α and β to angles α_0 and β_0, coterminal with α and β, respectively, each of which measure between 0 and 2π radians. Since α and α_0 are coterminal, as are β and β_0, it follows that $\alpha - \beta$ is coterminal with $\alpha_0 - \beta_0$. Consider the case below where $\alpha_0 \geq \beta_0$.

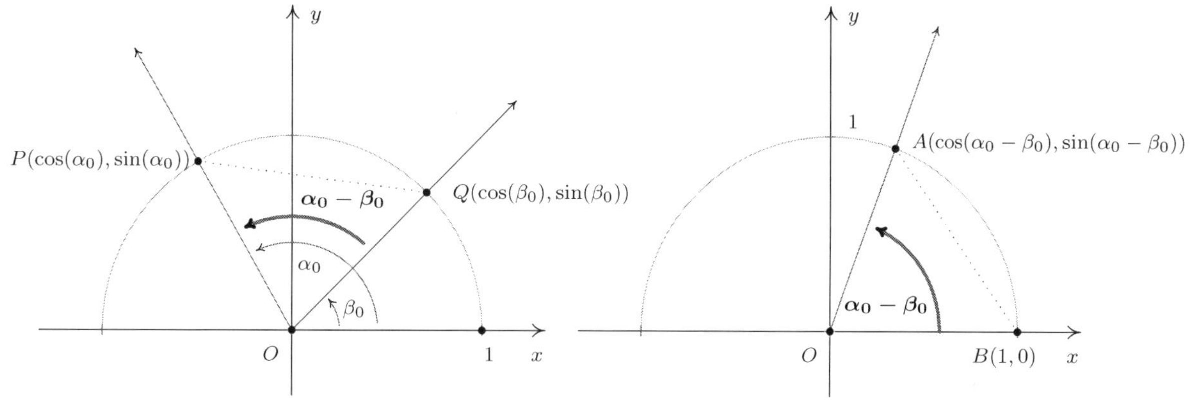

Since the angles POQ and AOB are congruent, the distance between P and Q is equal to the distance between A and B.[2] The distance formula, Equation 1.1, yields

$$\sqrt{(\cos(\alpha_0) - \cos(\beta_0))^2 + (\sin(\alpha_0) - \sin(\beta_0))^2} = \sqrt{(\cos(\alpha_0 - \beta_0) - 1)^2 + (\sin(\alpha_0 - \beta_0) - 0)^2}$$

Squaring both sides, we expand the left hand side of this equation as

$$\begin{aligned} (\cos(\alpha_0) - \cos(\beta_0))^2 + (\sin(\alpha_0) - \sin(\beta_0))^2 &= \cos^2(\alpha_0) - 2\cos(\alpha_0)\cos(\beta_0) + \cos^2(\beta_0) \\ &\quad + \sin^2(\alpha_0) - 2\sin(\alpha_0)\sin(\beta_0) + \sin^2(\beta_0) \\ &= \cos^2(\alpha_0) + \sin^2(\alpha_0) + \cos^2(\beta_0) + \sin^2(\beta_0) \\ &\quad - 2\cos(\alpha_0)\cos(\beta_0) - 2\sin(\alpha_0)\sin(\beta_0) \end{aligned}$$

[2] In the picture we've drawn, the triangles POQ and AOB are congruent, which is even better. However, $\alpha_0 - \beta_0$ could be 0 or it could be π, neither of which makes a triangle. It could also be larger than π, which makes a triangle, just not the one we've drawn. You should think about those three cases.

From the Pythagorean Identities, $\cos^2(\alpha_0)+\sin^2(\alpha_0)=1$ and $\cos^2(\beta_0)+\sin^2(\beta_0)=1$, so

$$(\cos(\alpha_0)-\cos(\beta_0))^2+(\sin(\alpha_0)-\sin(\beta_0))^2 \quad = \quad 2-2\cos(\alpha_0)\cos(\beta_0)-2\sin(\alpha_0)\sin(\beta_0)$$

Turning our attention to the right hand side of our equation, we find

$$\begin{aligned}(\cos(\alpha_0-\beta_0)-1)^2+(\sin(\alpha_0-\beta_0)-0)^2 &= \cos^2(\alpha_0-\beta_0)-2\cos(\alpha_0-\beta_0)+1+\sin^2(\alpha_0-\beta_0)\\ &= 1+\cos^2(\alpha_0-\beta_0)+\sin^2(\alpha_0-\beta_0)-2\cos(\alpha_0-\beta_0)\end{aligned}$$

Once again, we simplify $\cos^2(\alpha_0-\beta_0)+\sin^2(\alpha_0-\beta_0)=1$, so that

$$(\cos(\alpha_0-\beta_0)-1)^2+(\sin(\alpha_0-\beta_0)-0)^2 \quad = \quad 2-2\cos(\alpha_0-\beta_0)$$

Putting it all together, we get $2-2\cos(\alpha_0)\cos(\beta_0)-2\sin(\alpha_0)\sin(\beta_0)=2-2\cos(\alpha_0-\beta_0)$, which simplifies to: $\cos(\alpha_0-\beta_0)=\cos(\alpha_0)\cos(\beta_0)+\sin(\alpha_0)\sin(\beta_0)$. Since α and α_0, β and β_0 and $\alpha-\beta$ and $\alpha_0-\beta_0$ are all coterminal pairs of angles, we have $\cos(\alpha-\beta)=\cos(\alpha)\cos(\beta)+\sin(\alpha)\sin(\beta)$. For the case where $\alpha_0\leq\beta_0$, we can apply the above argument to the angle $\beta_0-\alpha_0$ to obtain the identity $\cos(\beta_0-\alpha_0)=\cos(\beta_0)\cos(\alpha_0)+\sin(\beta_0)\sin(\alpha_0)$. Applying the Even Identity of cosine, we get $\cos(\beta_0-\alpha_0)=\cos(-(\alpha_0-\beta_0))=\cos(\alpha_0-\beta_0)$, and we get the identity in this case, too.

To get the sum identity for cosine, we use the difference formula along with the Even/Odd Identities

$$\cos(\alpha+\beta)=\cos(\alpha-(-\beta))=\cos(\alpha)\cos(-\beta)+\sin(\alpha)\sin(-\beta)=\cos(\alpha)\cos(\beta)-\sin(\alpha)\sin(\beta)$$

We put these newfound identities to good use in the following example.

Example 10.4.1.

1. Find the exact value of $\cos(15^\circ)$.
2. Verify the identity: $\cos\left(\frac{\pi}{2}-\theta\right)=\sin(\theta)$.

Solution.

1. In order to use Theorem 10.13 to find $\cos(15^\circ)$, we need to write 15° as a sum or difference of angles whose cosines and sines we know. One way to do so is to write $15^\circ=45^\circ-30^\circ$.

$$\begin{aligned}\cos(15^\circ) &= \cos(45^\circ-30^\circ)\\ &= \cos(45^\circ)\cos(30^\circ)+\sin(45^\circ)\sin(30^\circ)\\ &= \left(\frac{\sqrt{2}}{2}\right)\left(\frac{\sqrt{3}}{2}\right)+\left(\frac{\sqrt{2}}{2}\right)\left(\frac{1}{2}\right)\\ &= \frac{\sqrt{6}+\sqrt{2}}{4}\end{aligned}$$

2. In a straightforward application of Theorem 10.13, we find

$$\begin{aligned}
\cos\left(\frac{\pi}{2}-\theta\right) &= \cos\left(\frac{\pi}{2}\right)\cos(\theta)+\sin\left(\frac{\pi}{2}\right)\sin(\theta) \\
&= (0)(\cos(\theta))+(1)(\sin(\theta)) \\
&= \sin(\theta)
\end{aligned}$$

□

The identity verified in Example 10.4.1, namely, $\cos\left(\frac{\pi}{2}-\theta\right)=\sin(\theta)$, is the first of the celebrated 'cofunction' identities. These identities were first hinted at in Exercise 74 in Section 10.2. From $\sin(\theta)=\cos\left(\frac{\pi}{2}-\theta\right)$, we get:

$$\sin\left(\frac{\pi}{2}-\theta\right)=\cos\left(\frac{\pi}{2}-\left[\frac{\pi}{2}-\theta\right]\right)=\cos(\theta),$$

which says, in words, that the 'co'sine of an angle is the sine of its 'co'mplement. Now that these identities have been established for cosine and sine, the remaining circular functions follow suit. The remaining proofs are left as exercises.

Theorem 10.14. Cofunction Identities: For all applicable angles θ,

- $\cos\left(\frac{\pi}{2}-\theta\right)=\sin(\theta)$
- $\sin\left(\frac{\pi}{2}-\theta\right)=\cos(\theta)$
- $\sec\left(\frac{\pi}{2}-\theta\right)=\csc(\theta)$
- $\csc\left(\frac{\pi}{2}-\theta\right)=\sec(\theta)$
- $\tan\left(\frac{\pi}{2}-\theta\right)=\cot(\theta)$
- $\cot\left(\frac{\pi}{2}-\theta\right)=\tan(\theta)$

With the Cofunction Identities in place, we are now in the position to derive the sum and difference formulas for sine. To derive the sum formula for sine, we convert to cosines using a cofunction identity, then expand using the difference formula for cosine

$$\begin{aligned}
\sin(\alpha+\beta) &= \cos\left(\frac{\pi}{2}-(\alpha+\beta)\right) \\
&= \cos\left(\left[\frac{\pi}{2}-\alpha\right]-\beta\right) \\
&= \cos\left(\frac{\pi}{2}-\alpha\right)\cos(\beta)+\sin\left(\frac{\pi}{2}-\alpha\right)\sin(\beta) \\
&= \sin(\alpha)\cos(\beta)+\cos(\alpha)\sin(\beta)
\end{aligned}$$

We can derive the difference formula for sine by rewriting $\sin(\alpha-\beta)$ as $\sin(\alpha+(-\beta))$ and using the sum formula and the Even / Odd Identities. Again, we leave the details to the reader.

Theorem 10.15. Sum and Difference Identities for Sine: For all angles α and β,

- $\sin(\alpha+\beta)=\sin(\alpha)\cos(\beta)+\cos(\alpha)\sin(\beta)$
- $\sin(\alpha-\beta)=\sin(\alpha)\cos(\beta)-\cos(\alpha)\sin(\beta)$

Example 10.4.2.

1. Find the exact value of $\sin\left(\frac{19\pi}{12}\right)$

2. If α is a Quadrant II angle with $\sin(\alpha) = \frac{5}{13}$, and β is a Quadrant III angle with $\tan(\beta) = 2$, find $\sin(\alpha - \beta)$.

3. Derive a formula for $\tan(\alpha + \beta)$ in terms of $\tan(\alpha)$ and $\tan(\beta)$.

Solution.

1. As in Example 10.4.1, we need to write the angle $\frac{19\pi}{12}$ as a sum or difference of common angles. The denominator of 12 suggests a combination of angles with denominators 3 and 4. One such combination is $\frac{19\pi}{12} = \frac{4\pi}{3} + \frac{\pi}{4}$. Applying Theorem 10.15, we get

$$\begin{aligned}\sin\left(\frac{19\pi}{12}\right) &= \sin\left(\frac{4\pi}{3} + \frac{\pi}{4}\right)\\ &= \sin\left(\frac{4\pi}{3}\right)\cos\left(\frac{\pi}{4}\right) + \cos\left(\frac{4\pi}{3}\right)\sin\left(\frac{\pi}{4}\right)\\ &= \left(-\frac{\sqrt{3}}{2}\right)\left(\frac{\sqrt{2}}{2}\right) + \left(-\frac{1}{2}\right)\left(\frac{\sqrt{2}}{2}\right)\\ &= \frac{-\sqrt{6} - \sqrt{2}}{4}\end{aligned}$$

2. In order to find $\sin(\alpha - \beta)$ using Theorem 10.15, we need to find $\cos(\alpha)$ and both $\cos(\beta)$ and $\sin(\beta)$. To find $\cos(\alpha)$, we use the Pythagorean Identity $\cos^2(\alpha) + \sin^2(\alpha) = 1$. Since $\sin(\alpha) = \frac{5}{13}$, we have $\cos^2(\alpha) + \left(\frac{5}{13}\right)^2 = 1$, or $\cos(\alpha) = \pm\frac{12}{13}$. Since α is a Quadrant II angle, $\cos(\alpha) = -\frac{12}{13}$. We now set about finding $\cos(\beta)$ and $\sin(\beta)$. We have several ways to proceed, but the Pythagorean Identity $1 + \tan^2(\beta) = \sec^2(\beta)$ is a quick way to get $\sec(\beta)$, and hence, $\cos(\beta)$. With $\tan(\beta) = 2$, we get $1 + 2^2 = \sec^2(\beta)$ so that $\sec(\beta) = \pm\sqrt{5}$. Since β is a Quadrant III angle, we choose $\sec(\beta) = -\sqrt{5}$ so $\cos(\beta) = \frac{1}{\sec(\beta)} = \frac{1}{-\sqrt{5}} = -\frac{\sqrt{5}}{5}$. We now need to determine $\sin(\beta)$. We could use The Pythagorean Identity $\cos^2(\beta) + \sin^2(\beta) = 1$, but we opt instead to use a quotient identity. From $\tan(\beta) = \frac{\sin(\beta)}{\cos(\beta)}$, we have $\sin(\beta) = \tan(\beta)\cos(\beta)$ so we get $\sin(\beta) = (2)\left(-\frac{\sqrt{5}}{5}\right) = -\frac{2\sqrt{5}}{5}$. We now have all the pieces needed to find $\sin(\alpha - \beta)$:

$$\begin{aligned}\sin(\alpha - \beta) &= \sin(\alpha)\cos(\beta) - \cos(\alpha)\sin(\beta)\\ &= \left(\frac{5}{13}\right)\left(-\frac{\sqrt{5}}{5}\right) - \left(-\frac{12}{13}\right)\left(-\frac{2\sqrt{5}}{5}\right)\\ &= -\frac{29\sqrt{5}}{65}\end{aligned}$$

3. We can start expanding $\tan(\alpha+\beta)$ using a quotient identity and our sum formulas

$$\begin{aligned}\tan(\alpha+\beta) &= \frac{\sin(\alpha+\beta)}{\cos(\alpha+\beta)} \\ &= \frac{\sin(\alpha)\cos(\beta)+\cos(\alpha)\sin(\beta)}{\cos(\alpha)\cos(\beta)-\sin(\alpha)\sin(\beta)}\end{aligned}$$

Since $\tan(\alpha) = \frac{\sin(\alpha)}{\cos(\alpha)}$ and $\tan(\beta) = \frac{\sin(\beta)}{\cos(\beta)}$, it looks as though if we divide both numerator and denominator by $\cos(\alpha)\cos(\beta)$ we will have what we want

$$\begin{aligned}\tan(\alpha+\beta) &= \frac{\sin(\alpha)\cos(\beta)+\cos(\alpha)\sin(\beta)}{\cos(\alpha)\cos(\beta)-\sin(\alpha)\sin(\beta)} \cdot \frac{\dfrac{1}{\cos(\alpha)\cos(\beta)}}{\dfrac{1}{\cos(\alpha)\cos(\beta)}} \\ &= \frac{\dfrac{\sin(\alpha)\cos(\beta)}{\cos(\alpha)\cos(\beta)}+\dfrac{\cos(\alpha)\sin(\beta)}{\cos(\alpha)\cos(\beta)}}{\dfrac{\cos(\alpha)\cos(\beta)}{\cos(\alpha)\cos(\beta)}-\dfrac{\sin(\alpha)\sin(\beta)}{\cos(\alpha)\cos(\beta)}} \\ &= \frac{\dfrac{\sin(\alpha)\cancel{\cos(\beta)}}{\cos(\alpha)\cancel{\cos(\beta)}}+\dfrac{\cancel{\cos(\alpha)}\sin(\beta)}{\cancel{\cos(\alpha)}\cos(\beta)}}{\dfrac{\cancel{\cos(\alpha)}\cancel{\cos(\beta)}}{\cancel{\cos(\alpha)}\cancel{\cos(\beta)}}-\dfrac{\sin(\alpha)\sin(\beta)}{\cos(\alpha)\cos(\beta)}} \\ &= \frac{\tan(\alpha)+\tan(\beta)}{1-\tan(\alpha)\tan(\beta)}\end{aligned}$$

Naturally, this formula is limited to those cases where all of the tangents are defined. □

The formula developed in Exercise 10.4.2 for $\tan(\alpha+\beta)$ can be used to find a formula for $\tan(\alpha-\beta)$ by rewriting the difference as a sum, $\tan(\alpha+(-\beta))$, and the reader is encouraged to fill in the details. Below we summarize all of the sum and difference formulas for cosine, sine and tangent.

Theorem 10.16. Sum and Difference Identities: For all applicable angles α and β,

- $\cos(\alpha\pm\beta) = \cos(\alpha)\cos(\beta)\mp\sin(\alpha)\sin(\beta)$
- $\sin(\alpha\pm\beta) = \sin(\alpha)\cos(\beta)\pm\cos(\alpha)\sin(\beta)$
- $\tan(\alpha\pm\beta) = \dfrac{\tan(\alpha)\pm\tan(\beta)}{1\mp\tan(\alpha)\tan(\beta)}$

In the statement of Theorem 10.16, we have combined the cases for the sum '+' and difference '−' of angles into one formula. The convention here is that if you want the formula for the sum '+' of two angles, you use the top sign in the formula; for the difference, '−', use the bottom sign. For example,

$$\tan(\alpha-\beta)=\frac{\tan(\alpha)-\tan(\beta)}{1+\tan(\alpha)\tan(\beta)}$$

If we specialize the sum formulas in Theorem 10.16 to the case when $\alpha=\beta$, we obtain the following 'Double Angle' Identities.

Theorem 10.17. Double Angle Identities: For all applicable angles θ,

- $\cos(2\theta)=\begin{cases}\cos^2(\theta)-\sin^2(\theta)\\ 2\cos^2(\theta)-1\\ 1-2\sin^2(\theta)\end{cases}$
- $\sin(2\theta)=2\sin(\theta)\cos(\theta)$
- $\tan(2\theta)=\dfrac{2\tan(\theta)}{1-\tan^2(\theta)}$

The three different forms for $\cos(2\theta)$ can be explained by our ability to 'exchange' squares of cosine and sine via the Pythagorean Identity $\cos^2(\theta)+\sin^2(\theta)=1$ and we leave the details to the reader. It is interesting to note that to determine the value of $\cos(2\theta)$, only *one* piece of information is required: either $\cos(\theta)$ or $\sin(\theta)$. To determine $\sin(2\theta)$, however, it appears that we must know both $\sin(\theta)$ and $\cos(\theta)$. In the next example, we show how we can find $\sin(2\theta)$ knowing just one piece of information, namely $\tan(\theta)$.

Example 10.4.3.

1. Suppose $P(-3,4)$ lies on the terminal side of θ when θ is plotted in standard position. Find $\cos(2\theta)$ and $\sin(2\theta)$ and determine the quadrant in which the terminal side of the angle 2θ lies when it is plotted in standard position.
2. If $\sin(\theta)=x$ for $-\frac{\pi}{2}\leq\theta\leq\frac{\pi}{2}$, find an expression for $\sin(2\theta)$ in terms of x.
3. Verify the identity: $\sin(2\theta)=\dfrac{2\tan(\theta)}{1+\tan^2(\theta)}$.
4. Express $\cos(3\theta)$ as a polynomial in terms of $\cos(\theta)$.

Solution.

1. Using Theorem 10.3 from Section 10.2 with $x=-3$ and $y=4$, we find $r=\sqrt{x^2+y^2}=5$. Hence, $\cos(\theta)=-\frac{3}{5}$ and $\sin(\theta)=\frac{4}{5}$. Applying Theorem 10.17, we get $\cos(2\theta)=\cos^2(\theta)-\sin^2(\theta)=\left(-\frac{3}{5}\right)^2-\left(\frac{4}{5}\right)^2=-\frac{7}{25}$, and $\sin(2\theta)=2\sin(\theta)\cos(\theta)=2\left(\frac{4}{5}\right)\left(-\frac{3}{5}\right)=-\frac{24}{25}$. Since both cosine and sine of 2θ are negative, the terminal side of 2θ, when plotted in standard position, lies in Quadrant III.

2. If your first reaction to '$\sin(\theta) = x$' is 'No it's not, $\cos(\theta) = x$!' then you have indeed learned something, and we take comfort in that. However, context is everything. Here, 'x' is just a variable - it does not necessarily represent the x-coordinate of the point on The Unit Circle which lies on the terminal side of θ, assuming θ is drawn in standard position. Here, x represents the quantity $\sin(\theta)$, and what we wish to know is how to express $\sin(2\theta)$ in terms of x. We will see more of this kind of thing in Section 10.6, and, as usual, this is something we need for Calculus. Since $\sin(2\theta) = 2\sin(\theta)\cos(\theta)$, we need to write $\cos(\theta)$ in terms of x to finish the problem. We substitute $x = \sin(\theta)$ into the Pythagorean Identity, $\cos^2(\theta) + \sin^2(\theta) = 1$, to get $\cos^2(\theta) + x^2 = 1$, or $\cos(\theta) = \pm\sqrt{1-x^2}$. Since $-\frac{\pi}{2} \leq \theta \leq \frac{\pi}{2}$, $\cos(\theta) \geq 0$, and thus $\cos(\theta) = \sqrt{1-x^2}$. Our final answer is $\sin(2\theta) = 2\sin(\theta)\cos(\theta) = 2x\sqrt{1-x^2}$.

3. We start with the right hand side of the identity and note that $1 + \tan^2(\theta) = \sec^2(\theta)$. From this point, we use the Reciprocal and Quotient Identities to rewrite $\tan(\theta)$ and $\sec(\theta)$ in terms of $\cos(\theta)$ and $\sin(\theta)$:

$$\begin{aligned} \frac{2\tan(\theta)}{1+\tan^2(\theta)} &= \frac{2\tan(\theta)}{\sec^2(\theta)} = \frac{2\left(\frac{\sin(\theta)}{\cos(\theta)}\right)}{\frac{1}{\cos^2(\theta)}} = 2\left(\frac{\sin(\theta)}{\cos(\theta)}\right)\cos^2(\theta) \\ &= 2\left(\frac{\sin(\theta)}{\cancel{\cos(\theta)}}\right)\cancel{\cos(\theta)}\cos(\theta) = 2\sin(\theta)\cos(\theta) = \sin(2\theta) \end{aligned}$$

4. In Theorem 10.17, one of the formulas for $\cos(2\theta)$, namely $\cos(2\theta) = 2\cos^2(\theta) - 1$, expresses $\cos(2\theta)$ as a polynomial in terms of $\cos(\theta)$. We are now asked to find such an identity for $\cos(3\theta)$. Using the sum formula for cosine, we begin with

$$\begin{aligned} \cos(3\theta) &= \cos(2\theta + \theta) \\ &= \cos(2\theta)\cos(\theta) - \sin(2\theta)\sin(\theta) \end{aligned}$$

Our ultimate goal is to express the right hand side in terms of $\cos(\theta)$ only. We substitute $\cos(2\theta) = 2\cos^2(\theta) - 1$ and $\sin(2\theta) = 2\sin(\theta)\cos(\theta)$ which yields

$$\begin{aligned} \cos(3\theta) &= \cos(2\theta)\cos(\theta) - \sin(2\theta)\sin(\theta) \\ &= \left(2\cos^2(\theta) - 1\right)\cos(\theta) - \left(2\sin(\theta)\cos(\theta)\right)\sin(\theta) \\ &= 2\cos^3(\theta) - \cos(\theta) - 2\sin^2(\theta)\cos(\theta) \end{aligned}$$

Finally, we exchange $\sin^2(\theta)$ for $1 - \cos^2(\theta)$ courtesy of the Pythagorean Identity, and get

$$\begin{aligned} \cos(3\theta) &= 2\cos^3(\theta) - \cos(\theta) - 2\sin^2(\theta)\cos(\theta) \\ &= 2\cos^3(\theta) - \cos(\theta) - 2\left(1 - \cos^2(\theta)\right)\cos(\theta) \\ &= 2\cos^3(\theta) - \cos(\theta) - 2\cos(\theta) + 2\cos^3(\theta) \\ &= 4\cos^3(\theta) - 3\cos(\theta) \end{aligned}$$

and we are done. □

In the last problem in Example 10.4.3, we saw how we could rewrite $\cos(3\theta)$ as sums of powers of $\cos(\theta)$. In Calculus, we have occasion to do the reverse; that is, reduce the power of cosine and sine. Solving the identity $\cos(2\theta) = 2\cos^2(\theta) - 1$ for $\cos^2(\theta)$ and the identity $\cos(2\theta) = 1 - 2\sin^2(\theta)$ for $\sin^2(\theta)$ results in the aptly-named 'Power Reduction' formulas below.

Theorem 10.18. Power Reduction Formulas: For all angles θ,

- $\cos^2(\theta) = \dfrac{1+\cos(2\theta)}{2}$
- $\sin^2(\theta) = \dfrac{1-\cos(2\theta)}{2}$

Example 10.4.4. Rewrite $\sin^2(\theta)\cos^2(\theta)$ as a sum and difference of cosines to the first power.

Solution. We begin with a straightforward application of Theorem 10.18

$$\begin{aligned} \sin^2(\theta)\cos^2(\theta) &= \left(\frac{1-\cos(2\theta)}{2}\right)\left(\frac{1+\cos(2\theta)}{2}\right) \\ &= \frac{1}{4}\left(1-\cos^2(2\theta)\right) \\ &= \frac{1}{4} - \frac{1}{4}\cos^2(2\theta) \end{aligned}$$

Next, we apply the power reduction formula to $\cos^2(2\theta)$ to finish the reduction

$$\begin{aligned} \sin^2(\theta)\cos^2(\theta) &= \frac{1}{4} - \frac{1}{4}\cos^2(2\theta) \\ &= \frac{1}{4} - \frac{1}{4}\left(\frac{1+\cos(2(2\theta))}{2}\right) \\ &= \frac{1}{4} - \frac{1}{8} - \frac{1}{8}\cos(4\theta) \\ &= \frac{1}{8} - \frac{1}{8}\cos(4\theta) \end{aligned}$$

□

Another application of the Power Reduction Formulas is the Half Angle Formulas. To start, we apply the Power Reduction Formula to $\cos^2\left(\frac{\theta}{2}\right)$

$$\cos^2\left(\frac{\theta}{2}\right) = \frac{1+\cos\left(2\left(\frac{\theta}{2}\right)\right)}{2} = \frac{1+\cos(\theta)}{2}.$$

We can obtain a formula for $\cos\left(\frac{\theta}{2}\right)$ by extracting square roots. In a similar fashion, we may obtain a half angle formula for sine, and by using a quotient formula, obtain a half angle formula for tangent. We summarize these formulas below.

Theorem 10.19. Half Angle Formulas: For all applicable angles θ,

- $\cos\left(\dfrac{\theta}{2}\right) = \pm\sqrt{\dfrac{1+\cos(\theta)}{2}}$
- $\sin\left(\dfrac{\theta}{2}\right) = \pm\sqrt{\dfrac{1-\cos(\theta)}{2}}$
- $\tan\left(\dfrac{\theta}{2}\right) = \pm\sqrt{\dfrac{1-\cos(\theta)}{1+\cos(\theta)}}$

where the choice of $\pm$ depends on the quadrant in which the terminal side of $\dfrac{\theta}{2}$ lies.

Example 10.4.5.

1. Use a half angle formula to find the exact value of $\cos(15^\circ)$.
2. Suppose $-\pi \leq \theta \leq 0$ with $\cos(\theta) = -\frac{3}{5}$. Find $\sin\left(\frac{\theta}{2}\right)$.
3. Use the identity given in number 3 of Example 10.4.3 to derive the identity
$$\tan\left(\frac{\theta}{2}\right) = \frac{\sin(\theta)}{1+\cos(\theta)}$$

Solution.

1. To use the half angle formula, we note that $15^\circ = \frac{30^\circ}{2}$ and since 15° is a Quadrant I angle, its cosine is positive. Thus we have
$$\begin{aligned} \cos(15^\circ) &= +\sqrt{\frac{1+\cos(30^\circ)}{2}} = \sqrt{\frac{1+\frac{\sqrt{3}}{2}}{2}} \\ &= \sqrt{\frac{1+\frac{\sqrt{3}}{2}}{2}\cdot\frac{2}{2}} = \sqrt{\frac{2+\sqrt{3}}{4}} = \frac{\sqrt{2+\sqrt{3}}}{2} \end{aligned}$$
Back in Example 10.4.1, we found $\cos(15^\circ)$ by using the difference formula for cosine. In that case, we determined $\cos(15^\circ) = \frac{\sqrt{6}+\sqrt{2}}{4}$. The reader is encouraged to prove that these two expressions are equal.
2. If $-\pi \leq \theta \leq 0$, then $-\frac{\pi}{2} \leq \frac{\theta}{2} \leq 0$, which means $\sin\left(\frac{\theta}{2}\right) < 0$. Theorem 10.19 gives
$$\begin{aligned} \sin\left(\frac{\theta}{2}\right) &= -\sqrt{\frac{1-\cos(\theta)}{2}} = -\sqrt{\frac{1-\left(-\frac{3}{5}\right)}{2}} \\ &= -\sqrt{\frac{1+\frac{3}{5}}{2}\cdot\frac{5}{5}} = -\sqrt{\frac{8}{10}} = -\frac{2\sqrt{5}}{5} \end{aligned}$$

3. Instead of our usual approach to verifying identities, namely starting with one side of the equation and trying to transform it into the other, we will start with the identity we proved in number 3 of Example 10.4.3 and manipulate it into the identity we are asked to prove. The identity we are asked to start with is $\sin(2\theta) = \frac{2\tan(\theta)}{1+\tan^2(\theta)}$. If we are to use this to derive an identity for $\tan\left(\frac{\theta}{2}\right)$, it seems reasonable to proceed by replacing each occurrence of θ with $\frac{\theta}{2}$

$$\begin{aligned} \sin\left(2\left(\tfrac{\theta}{2}\right)\right) &= \frac{2\tan\left(\frac{\theta}{2}\right)}{1+\tan^2\left(\frac{\theta}{2}\right)} \\ \sin(\theta) &= \frac{2\tan\left(\frac{\theta}{2}\right)}{1+\tan^2\left(\frac{\theta}{2}\right)} \end{aligned}$$

We now have the $\sin(\theta)$ we need, but we somehow need to get a factor of $1+\cos(\theta)$ involved. To get cosines involved, recall that $1+\tan^2\left(\frac{\theta}{2}\right) = \sec^2\left(\frac{\theta}{2}\right)$. We continue to manipulate our given identity by converting secants to cosines and using a power reduction formula

$$\begin{aligned} \sin(\theta) &= \frac{2\tan\left(\frac{\theta}{2}\right)}{1+\tan^2\left(\frac{\theta}{2}\right)} \\ \sin(\theta) &= \frac{2\tan\left(\frac{\theta}{2}\right)}{\sec^2\left(\frac{\theta}{2}\right)} \\ \sin(\theta) &= 2\tan\left(\tfrac{\theta}{2}\right)\cos^2\left(\tfrac{\theta}{2}\right) \\ \sin(\theta) &= 2\tan\left(\tfrac{\theta}{2}\right)\left(\frac{1+\cos\left(2\left(\frac{\theta}{2}\right)\right)}{2}\right) \\ \sin(\theta) &= \tan\left(\tfrac{\theta}{2}\right)(1+\cos(\theta)) \\ \tan\left(\frac{\theta}{2}\right) &= \frac{\sin(\theta)}{1+\cos(\theta)} \end{aligned}$$

□

Our next batch of identities, the Product to Sum Formulas,[3] are easily verified by expanding each of the right hand sides in accordance with Theorem 10.16 and as you should expect by now we leave the details as exercises. They are of particular use in Calculus, and we list them here for reference.

Theorem 10.20. Product to Sum Formulas: For all angles α and β,

- $\cos(\alpha)\cos(\beta) = \frac{1}{2}\left[\cos(\alpha-\beta)+\cos(\alpha+\beta)\right]$
- $\sin(\alpha)\sin(\beta) = \frac{1}{2}\left[\cos(\alpha-\beta)-\cos(\alpha+\beta)\right]$
- $\sin(\alpha)\cos(\beta) = \frac{1}{2}\left[\sin(\alpha-\beta)+\sin(\alpha+\beta)\right]$

[3]These are also known as the Prosthaphaeresis Formulas and have a rich history. The authors recommend that you conduct some research on them as your schedule allows.

Related to the Product to Sum Formulas are the Sum to Product Formulas, which we will have need of in Section 10.7. These are easily verified using the Product to Sum Formulas, and as such, their proofs are left as exercises.

Theorem 10.21. Sum to Product Formulas: For all angles α and β,

- $\cos(\alpha) + \cos(\beta) = 2\cos\left(\dfrac{\alpha+\beta}{2}\right)\cos\left(\dfrac{\alpha-\beta}{2}\right)$
- $\cos(\alpha) - \cos(\beta) = -2\sin\left(\dfrac{\alpha+\beta}{2}\right)\sin\left(\dfrac{\alpha-\beta}{2}\right)$
- $\sin(\alpha) \pm \sin(\beta) = 2\sin\left(\dfrac{\alpha\pm\beta}{2}\right)\cos\left(\dfrac{\alpha\mp\beta}{2}\right)$

Example 10.4.6.

1. Write $\cos(2\theta)\cos(6\theta)$ as a sum.
2. Write $\sin(\theta) - \sin(3\theta)$ as a product.

Solution.

1. Identifying $\alpha = 2\theta$ and $\beta = 6\theta$, we find

$$\begin{aligned}\cos(2\theta)\cos(6\theta) &= \tfrac{1}{2}\left[\cos(2\theta-6\theta)+\cos(2\theta+6\theta)\right]\\ &= \tfrac{1}{2}\cos(-4\theta)+\tfrac{1}{2}\cos(8\theta)\\ &= \tfrac{1}{2}\cos(4\theta)+\tfrac{1}{2}\cos(8\theta),\end{aligned}$$

where the last equality is courtesy of the even identity for cosine, $\cos(-4\theta) = \cos(4\theta)$.

2. Identifying $\alpha = \theta$ and $\beta = 3\theta$ yields

$$\begin{aligned}\sin(\theta)-\sin(3\theta) &= 2\sin\left(\frac{\theta-3\theta}{2}\right)\cos\left(\frac{\theta+3\theta}{2}\right)\\ &= 2\sin\left(-\theta\right)\cos\left(2\theta\right)\\ &= -2\sin\left(\theta\right)\cos\left(2\theta\right),\end{aligned}$$

where the last equality is courtesy of the odd identity for sine, $\sin(-\theta) = -\sin(\theta)$. □

The reader is reminded that all of the identities presented in this section which regard the circular functions as functions of angles (in radian measure) apply equally well to the circular (trigonometric) functions regarded as functions of real numbers. In Exercises 38 - 43 in Section 10.5, we see how some of these identities manifest themselves geometrically as we study the graphs of the these functions. In the upcoming Exercises, however, you need to do all of your work analytically without graphs.

10.4.1 Exercises

In Exercises 1 - 6, use the Even / Odd Identities to verify the identity. Assume all quantities are defined.

1. $\sin(3\pi - 2\theta) = -\sin(2\theta - 3\pi)$
2. $\cos\left(-\frac{\pi}{4} - 5t\right) = \cos\left(5t + \frac{\pi}{4}\right)$
3. $\tan(-t^2 + 1) = -\tan(t^2 - 1)$
4. $\csc(-\theta - 5) = -\csc(\theta + 5)$
5. $\sec(-6t) = \sec(6t)$
6. $\cot(9 - 7\theta) = -\cot(7\theta - 9)$

In Exercises 7 - 21, use the Sum and Difference Identities to find the exact value. You may have need of the Quotient, Reciprocal or Even / Odd Identities as well.

7. $\cos(75^\circ)$
8. $\sec(165^\circ)$
9. $\sin(105^\circ)$
10. $\csc(195^\circ)$
11. $\cot(255^\circ)$
12. $\tan(375^\circ)$
13. $\cos\left(\frac{13\pi}{12}\right)$
14. $\sin\left(\frac{11\pi}{12}\right)$
15. $\tan\left(\frac{13\pi}{12}\right)$
16. $\cos\left(\frac{7\pi}{12}\right)$
17. $\tan\left(\frac{17\pi}{12}\right)$
18. $\sin\left(\frac{\pi}{12}\right)$
19. $\cot\left(\frac{11\pi}{12}\right)$
20. $\csc\left(\frac{5\pi}{12}\right)$
21. $\sec\left(-\frac{\pi}{12}\right)$
22. If α is a Quadrant IV angle with $\cos(\alpha) = \frac{\sqrt{5}}{5}$, and $\sin(\beta) = \frac{\sqrt{10}}{10}$, where $\frac{\pi}{2} < \beta < \pi$, find
    (a) $\cos(\alpha + \beta)$
    (b) $\sin(\alpha + \beta)$
    (c) $\tan(\alpha + \beta)$
    (d) $\cos(\alpha - \beta)$
    (e) $\sin(\alpha - \beta)$
    (f) $\tan(\alpha - \beta)$
23. If $\csc(\alpha) = 3$, where $0 < \alpha < \frac{\pi}{2}$, and β is a Quadrant II angle with $\tan(\beta) = -7$, find
    (a) $\cos(\alpha + \beta)$
    (b) $\sin(\alpha + \beta)$
    (c) $\tan(\alpha + \beta)$
    (d) $\cos(\alpha - \beta)$
    (e) $\sin(\alpha - \beta)$
    (f) $\tan(\alpha - \beta)$
24. If $\sin(\alpha) = \frac{3}{5}$, where $0 < \alpha < \frac{\pi}{2}$, and $\cos(\beta) = \frac{12}{13}$ where $\frac{3\pi}{2} < \beta < 2\pi$, find
    (a) $\sin(\alpha + \beta)$
    (b) $\cos(\alpha - \beta)$
    (c) $\tan(\alpha - \beta)$

25. If $\sec(\alpha) = -\frac{5}{3}$, where $\frac{\pi}{2} < \alpha < \pi$, and $\tan(\beta) = \frac{24}{7}$, where $\pi < \beta < \frac{3\pi}{2}$, find

(a) $\csc(\alpha - \beta)$ (b) $\sec(\alpha + \beta)$ (c) $\cot(\alpha + \beta)$

In Exercises 26 - 38, verify the identity.

26. $\cos(\theta - \pi) = -\cos(\theta)$

27. $\sin(\pi - \theta) = \sin(\theta)$

28. $\tan\left(\theta + \frac{\pi}{2}\right) = -\cot(\theta)$

29. $\sin(\alpha + \beta) + \sin(\alpha - \beta) = 2\sin(\alpha)\cos(\beta)$

30. $\sin(\alpha + \beta) - \sin(\alpha - \beta) = 2\cos(\alpha)\sin(\beta)$

31. $\cos(\alpha + \beta) + \cos(\alpha - \beta) = 2\cos(\alpha)\cos(\beta)$

32. $\cos(\alpha + \beta) - \cos(\alpha - \beta) = -2\sin(\alpha)\sin(\beta)$

33. $\dfrac{\sin(\alpha + \beta)}{\sin(\alpha - \beta)} = \dfrac{1 + \cot(\alpha)\tan(\beta)}{1 - \cot(\alpha)\tan(\beta)}$

34. $\dfrac{\cos(\alpha + \beta)}{\cos(\alpha - \beta)} = \dfrac{1 - \tan(\alpha)\tan(\beta)}{1 + \tan(\alpha)\tan(\beta)}$

35. $\dfrac{\tan(\alpha + \beta)}{\tan(\alpha - \beta)} = \dfrac{\sin(\alpha)\cos(\alpha) + \sin(\beta)\cos(\beta)}{\sin(\alpha)\cos(\alpha) - \sin(\beta)\cos(\beta)}$

36. $\dfrac{\sin(t + h) - \sin(t)}{h} = \cos(t)\left(\dfrac{\sin(h)}{h}\right) + \sin(t)\left(\dfrac{\cos(h) - 1}{h}\right)$

37. $\dfrac{\cos(t + h) - \cos(t)}{h} = \cos(t)\left(\dfrac{\cos(h) - 1}{h}\right) - \sin(t)\left(\dfrac{\sin(h)}{h}\right)$

38. $\dfrac{\tan(t + h) - \tan(t)}{h} = \left(\dfrac{\tan(h)}{h}\right)\left(\dfrac{\sec^2(t)}{1 - \tan(t)\tan(h)}\right)$

In Exercises 39 - 48, use the Half Angle Formulas to find the exact value. You may have need of the Quotient, Reciprocal or Even / Odd Identities as well.

39. $\cos(75^\circ)$ (compare with Exercise 7)

40. $\sin(105^\circ)$ (compare with Exercise 9)

41. $\cos(67.5^\circ)$

42. $\sin(157.5^\circ)$

43. $\tan(112.5^\circ)$

44. $\cos\left(\frac{7\pi}{12}\right)$ (compare with Exercise 16)

45. $\sin\left(\frac{\pi}{12}\right)$ (compare with Exercise 18)

46. $\cos\left(\frac{\pi}{8}\right)$

47. $\sin\left(\frac{5\pi}{8}\right)$

48. $\tan\left(\frac{7\pi}{8}\right)$

In Exercises 49 - 58, use the given information about θ to find the exact values of

- $\sin(2\theta)$
- $\cos(2\theta)$
- $\tan(2\theta)$
- $\sin\left(\frac{\theta}{2}\right)$
- $\cos\left(\frac{\theta}{2}\right)$
- $\tan\left(\frac{\theta}{2}\right)$

49. $\sin(\theta) = -\frac{7}{25}$ where $\frac{3\pi}{2} < \theta < 2\pi$
50. $\cos(\theta) = \frac{28}{53}$ where $0 < \theta < \frac{\pi}{2}$
51. $\tan(\theta) = \frac{12}{5}$ where $\pi < \theta < \frac{3\pi}{2}$
52. $\csc(\theta) = 4$ where $\frac{\pi}{2} < \theta < \pi$
53. $\cos(\theta) = \frac{3}{5}$ where $0 < \theta < \frac{\pi}{2}$
54. $\sin(\theta) = -\frac{4}{5}$ where $\pi < \theta < \frac{3\pi}{2}$
55. $\cos(\theta) = \frac{12}{13}$ where $\frac{3\pi}{2} < \theta < 2\pi$
56. $\sin(\theta) = \frac{5}{13}$ where $\frac{\pi}{2} < \theta < \pi$
57. $\sec(\theta) = \sqrt{5}$ where $\frac{3\pi}{2} < \theta < 2\pi$
58. $\tan(\theta) = -2$ where $\frac{\pi}{2} < \theta < \pi$

In Exercises 59 - 73, verify the identity. Assume all quantities are defined.

59. $(\cos(\theta) + \sin(\theta))^2 = 1 + \sin(2\theta)$
60. $(\cos(\theta) - \sin(\theta))^2 = 1 - \sin(2\theta)$
61. $\tan(2\theta) = \dfrac{1}{1 - \tan(\theta)} - \dfrac{1}{1 + \tan(\theta)}$
62. $\csc(2\theta) = \dfrac{\cot(\theta) + \tan(\theta)}{2}$
63. $8\sin^4(\theta) = \cos(4\theta) - 4\cos(2\theta) + 3$
64. $8\cos^4(\theta) = \cos(4\theta) + 4\cos(2\theta) + 3$
65. $\sin(3\theta) = 3\sin(\theta) - 4\sin^3(\theta)$
66. $\sin(4\theta) = 4\sin(\theta)\cos^3(\theta) - 4\sin^3(\theta)\cos(\theta)$
67. $32\sin^2(\theta)\cos^4(\theta) = 2 + \cos(2\theta) - 2\cos(4\theta) - \cos(6\theta)$
68. $32\sin^4(\theta)\cos^2(\theta) = 2 - \cos(2\theta) - 2\cos(4\theta) + \cos(6\theta)$
69. $\cos(4\theta) = 8\cos^4(\theta) - 8\cos^2(\theta) + 1$
70. $\cos(8\theta) = 128\cos^8(\theta) - 256\cos^6(\theta) + 160\cos^4(\theta) - 32\cos^2(\theta) + 1$ (HINT: Use the result to 69.)
71. $\sec(2\theta) = \dfrac{\cos(\theta)}{\cos(\theta) + \sin(\theta)} + \dfrac{\sin(\theta)}{\cos(\theta) - \sin(\theta)}$
72. $\dfrac{1}{\cos(\theta) - \sin(\theta)} + \dfrac{1}{\cos(\theta) + \sin(\theta)} = \dfrac{2\cos(\theta)}{\cos(2\theta)}$
73. $\dfrac{1}{\cos(\theta) - \sin(\theta)} - \dfrac{1}{\cos(\theta) + \sin(\theta)} = \dfrac{2\sin(\theta)}{\cos(2\theta)}$

In Exercises 74 - 79, write the given product as a sum. You may need to use an Even/Odd Identity.

74. $\cos(3\theta)\cos(5\theta)$

75. $\sin(2\theta)\sin(7\theta)$

76. $\sin(9\theta)\cos(\theta)$

77. $\cos(2\theta)\cos(6\theta)$

78. $\sin(3\theta)\sin(2\theta)$

79. $\cos(\theta)\sin(3\theta)$

In Exercises 80 - 85, write the given sum as a product. You may need to use an Even/Odd or Cofunction Identity.

80. $\cos(3\theta) + \cos(5\theta)$

81. $\sin(2\theta) - \sin(7\theta)$

82. $\cos(5\theta) - \cos(6\theta)$

83. $\sin(9\theta) - \sin(-\theta)$

84. $\sin(\theta) + \cos(\theta)$

85. $\cos(\theta) - \sin(\theta)$

86. Suppose θ is a Quadrant I angle with $\sin(\theta) = x$. Verify the following formulas

    (a) $\cos(\theta) = \sqrt{1 - x^2}$

    (b) $\sin(2\theta) = 2x\sqrt{1 - x^2}$

    (c) $\cos(2\theta) = 1 - 2x^2$

87. Discuss with your classmates how each of the formulas, if any, in Exercise 86 change if we change assume θ is a Quadrant II, III, or IV angle.

88. Suppose θ is a Quadrant I angle with $\tan(\theta) = x$. Verify the following formulas

    (a) $\cos(\theta) = \dfrac{1}{\sqrt{x^2 + 1}}$

    (b) $\sin(\theta) = \dfrac{x}{\sqrt{x^2 + 1}}$

    (c) $\sin(2\theta) = \dfrac{2x}{x^2 + 1}$

    (d) $\cos(2\theta) = \dfrac{1 - x^2}{x^2 + 1}$

89. Discuss with your classmates how each of the formulas, if any, in Exercise 88 change if we change assume θ is a Quadrant II, III, or IV angle.

90. If $\sin(\theta) = \dfrac{x}{2}$ for $-\dfrac{\pi}{2} < \theta < \dfrac{\pi}{2}$, find an expression for $\cos(2\theta)$ in terms of x.

91. If $\tan(\theta) = \dfrac{x}{7}$ for $-\dfrac{\pi}{2} < \theta < \dfrac{\pi}{2}$, find an expression for $\sin(2\theta)$ in terms of x.

92. If $\sec(\theta) = \dfrac{x}{4}$ for $0 < \theta < \dfrac{\pi}{2}$, find an expression for $\ln|\sec(\theta) + \tan(\theta)|$ in terms of x.

93. Show that $\cos^2(\theta) - \sin^2(\theta) = 2\cos^2(\theta) - 1 = 1 - 2\sin^2(\theta)$ for all θ.

94. Let θ be a Quadrant III angle with $\cos(\theta) = -\dfrac{1}{5}$. Show that this is not enough information to determine the sign of $\sin\left(\dfrac{\theta}{2}\right)$ by first assuming $3\pi < \theta < \dfrac{7\pi}{2}$ and then assuming $\pi < \theta < \dfrac{3\pi}{2}$ and computing $\sin\left(\dfrac{\theta}{2}\right)$ in both cases.

95. Without using your calculator, show that $\dfrac{\sqrt{2+\sqrt{3}}}{2} = \dfrac{\sqrt{6}+\sqrt{2}}{4}$

96. In part 4 of Example 10.4.3, we wrote $\cos(3\theta)$ as a polynomial in terms of $\cos(\theta)$. In Exercise 69, we had you verify an identity which expresses $\cos(4\theta)$ as a polynomial in terms of $\cos(\theta)$. Can you find a polynomial in terms of $\cos(\theta)$ for $\cos(5\theta)$? $\cos(6\theta)$? Can you find a pattern so that $\cos(n\theta)$ could be written as a polynomial in cosine for any natural number n?

97. In Exercise 65, we has you verify an identity which expresses $\sin(3\theta)$ as a polynomial in terms of $\sin(\theta)$. Can you do the same for $\sin(5\theta)$? What about for $\sin(4\theta)$? If not, what goes wrong?

98. Verify the Even / Odd Identities for tangent, secant, cosecant and cotangent.

99. Verify the Cofunction Identities for tangent, secant, cosecant and cotangent.

100. Verify the Difference Identities for sine and tangent.

101. Verify the Product to Sum Identities.

102. Verify the Sum to Product Identities.

10.4.2 Answers

7. $\cos(75^\circ) = \dfrac{\sqrt{6}-\sqrt{2}}{4}$

8. $\sec(165^\circ) = -\dfrac{4}{\sqrt{2}+\sqrt{6}} = \sqrt{2}-\sqrt{6}$

9. $\sin(105^\circ) = \dfrac{\sqrt{6}+\sqrt{2}}{4}$

10. $\csc(195^\circ) = \dfrac{4}{\sqrt{2}-\sqrt{6}} = -(\sqrt{2}+\sqrt{6})$

11. $\cot(255^\circ) = \dfrac{\sqrt{3}-1}{\sqrt{3}+1} = 2-\sqrt{3}$

12. $\tan(375^\circ) = \dfrac{3-\sqrt{3}}{3+\sqrt{3}} = 2-\sqrt{3}$

13. $\cos\left(\dfrac{13\pi}{12}\right) = -\dfrac{\sqrt{6}+\sqrt{2}}{4}$

14. $\sin\left(\dfrac{11\pi}{12}\right) = \dfrac{\sqrt{6}-\sqrt{2}}{4}$

15. $\tan\left(\dfrac{13\pi}{12}\right) = \dfrac{3-\sqrt{3}}{3+\sqrt{3}} = 2-\sqrt{3}$

16. $\cos\left(\dfrac{7\pi}{12}\right) = \dfrac{\sqrt{2}-\sqrt{6}}{4}$

17. $\tan\left(\dfrac{17\pi}{12}\right) = 2+\sqrt{3}$

18. $\sin\left(\dfrac{\pi}{12}\right) = \dfrac{\sqrt{6}-\sqrt{2}}{4}$

19. $\cot\left(\dfrac{11\pi}{12}\right) = -(2+\sqrt{3})$

20. $\csc\left(\dfrac{5\pi}{12}\right) = \sqrt{6}-\sqrt{2}$

21. $\sec\left(-\dfrac{\pi}{12}\right) = \sqrt{6}-\sqrt{2}$

22. (a) $\cos(\alpha+\beta) = -\dfrac{\sqrt{2}}{10}$
    (b) $\sin(\alpha+\beta) = \dfrac{7\sqrt{2}}{10}$
    (c) $\tan(\alpha+\beta) = -7$
    (d) $\cos(\alpha-\beta) = -\dfrac{\sqrt{2}}{2}$
    (e) $\sin(\alpha-\beta) = \dfrac{\sqrt{2}}{2}$
    (f) $\tan(\alpha-\beta) = -1$

23. (a) $\cos(\alpha+\beta) = -\dfrac{4+7\sqrt{2}}{30}$
    (b) $\sin(\alpha+\beta) = \dfrac{28-\sqrt{2}}{30}$
    (c) $\tan(\alpha+\beta) = \dfrac{-28+\sqrt{2}}{4+7\sqrt{2}} = \dfrac{63-100\sqrt{2}}{41}$
    (d) $\cos(\alpha-\beta) = \dfrac{-4+7\sqrt{2}}{30}$
    (e) $\sin(\alpha-\beta) = -\dfrac{28+\sqrt{2}}{30}$
    (f) $\tan(\alpha-\beta) = \dfrac{28+\sqrt{2}}{4-7\sqrt{2}} = -\dfrac{63+100\sqrt{2}}{41}$

24. (a) $\sin(\alpha+\beta) = \dfrac{16}{65}$
    (b) $\cos(\alpha-\beta) = \dfrac{33}{65}$
    (c) $\tan(\alpha-\beta) = \dfrac{56}{33}$

25. (a) $\csc(\alpha - \beta) = -\dfrac{5}{4}$ (b) $\sec(\alpha + \beta) = \dfrac{125}{117}$ (c) $\cot(\alpha + \beta) = \dfrac{117}{44}$

39. $\cos(75^\circ) = \dfrac{\sqrt{2 - \sqrt{3}}}{2}$

40. $\sin(105^\circ) = \dfrac{\sqrt{2 + \sqrt{3}}}{2}$

41. $\cos(67.5^\circ) = \dfrac{\sqrt{2 - \sqrt{2}}}{2}$

42. $\sin(157.5^\circ) = \dfrac{\sqrt{2 - \sqrt{2}}}{2}$

43. $\tan(112.5^\circ) = -\sqrt{\dfrac{2 + \sqrt{2}}{2 - \sqrt{2}}} = -1 - \sqrt{2}$

44. $\cos\left(\dfrac{7\pi}{12}\right) = -\dfrac{\sqrt{2 - \sqrt{3}}}{2}$

45. $\sin\left(\dfrac{\pi}{12}\right) = \dfrac{\sqrt{2 - \sqrt{3}}}{2}$

46. $\cos\left(\dfrac{\pi}{8}\right) = \dfrac{\sqrt{2 + \sqrt{2}}}{2}$

47. $\sin\left(\dfrac{5\pi}{8}\right) = \dfrac{\sqrt{2 + \sqrt{2}}}{2}$

48. $\tan\left(\dfrac{7\pi}{8}\right) = -\sqrt{\dfrac{2 - \sqrt{2}}{2 + \sqrt{2}}} = 1 - \sqrt{2}$

49.
- $\sin(2\theta) = -\dfrac{336}{625}$
- $\cos(2\theta) = \dfrac{527}{625}$
- $\tan(2\theta) = -\dfrac{336}{527}$
- $\sin\left(\frac{\theta}{2}\right) = \dfrac{\sqrt{2}}{10}$
- $\cos\left(\frac{\theta}{2}\right) = -\dfrac{7\sqrt{2}}{10}$
- $\tan\left(\frac{\theta}{2}\right) = -\dfrac{1}{7}$

50.
- $\sin(2\theta) = \dfrac{2520}{2809}$
- $\cos(2\theta) = -\dfrac{1241}{2809}$
- $\tan(2\theta) = -\dfrac{2520}{1241}$
- $\sin\left(\frac{\theta}{2}\right) = \dfrac{5\sqrt{106}}{106}$
- $\cos\left(\frac{\theta}{2}\right) = \dfrac{9\sqrt{106}}{106}$
- $\tan\left(\frac{\theta}{2}\right) = \dfrac{5}{9}$

51.
- $\sin(2\theta) = \dfrac{120}{169}$
- $\cos(2\theta) = -\dfrac{119}{169}$
- $\tan(2\theta) = -\dfrac{120}{119}$
- $\sin\left(\frac{\theta}{2}\right) = \dfrac{3\sqrt{13}}{13}$
- $\cos\left(\frac{\theta}{2}\right) = -\dfrac{2\sqrt{13}}{13}$
- $\tan\left(\frac{\theta}{2}\right) = -\dfrac{3}{2}$

52.
- $\sin(2\theta) = -\dfrac{\sqrt{15}}{8}$
- $\cos(2\theta) = \dfrac{7}{8}$
- $\tan(2\theta) = -\dfrac{\sqrt{15}}{7}$
- $\sin\left(\frac{\theta}{2}\right) = \dfrac{\sqrt{8 + 2\sqrt{15}}}{4}$
- $\cos\left(\frac{\theta}{2}\right) = \dfrac{\sqrt{8 - 2\sqrt{15}}}{4}$
- $\tan\left(\frac{\theta}{2}\right) = \sqrt{\dfrac{8 + 2\sqrt{15}}{8 - 2\sqrt{15}}}$
  $\tan\left(\frac{\theta}{2}\right) = 4 + \sqrt{15}$

53.
- $\sin(2\theta) = \dfrac{24}{25}$
- $\cos(2\theta) = -\dfrac{7}{25}$
- $\tan(2\theta) = -\dfrac{24}{7}$
- $\sin\left(\frac{\theta}{2}\right) = \dfrac{\sqrt{5}}{5}$
- $\cos\left(\frac{\theta}{2}\right) = \dfrac{2\sqrt{5}}{5}$
- $\tan\left(\frac{\theta}{2}\right) = \dfrac{1}{2}$

54.
- $\sin(2\theta) = \frac{24}{25}$
- $\cos(2\theta) = -\frac{7}{25}$
- $\tan(2\theta) = -\frac{24}{7}$
- $\sin\left(\frac{\theta}{2}\right) = \frac{2\sqrt{5}}{5}$
- $\cos\left(\frac{\theta}{2}\right) = -\frac{\sqrt{5}}{5}$
- $\tan\left(\frac{\theta}{2}\right) = -2$

55.
- $\sin(2\theta) = -\frac{120}{169}$
- $\cos(2\theta) = \frac{119}{169}$
- $\tan(2\theta) = -\frac{120}{119}$
- $\sin\left(\frac{\theta}{2}\right) = \frac{\sqrt{26}}{26}$
- $\cos\left(\frac{\theta}{2}\right) = -\frac{5\sqrt{26}}{26}$
- $\tan\left(\frac{\theta}{2}\right) = -\frac{1}{5}$

56.
- $\sin(2\theta) = -\frac{120}{169}$
- $\cos(2\theta) = \frac{119}{169}$
- $\tan(2\theta) = -\frac{120}{119}$
- $\sin\left(\frac{\theta}{2}\right) = \frac{5\sqrt{26}}{26}$
- $\cos\left(\frac{\theta}{2}\right) = \frac{\sqrt{26}}{26}$
- $\tan\left(\frac{\theta}{2}\right) = 5$

57.
- $\sin(2\theta) = -\frac{4}{5}$
- $\cos(2\theta) = -\frac{3}{5}$
- $\tan(2\theta) = \frac{4}{3}$
- $\sin\left(\frac{\theta}{2}\right) = \frac{\sqrt{50 - 10\sqrt{5}}}{10}$
- $\cos\left(\frac{\theta}{2}\right) = -\frac{\sqrt{50 + 10\sqrt{5}}}{10}$
- $\tan\left(\frac{\theta}{2}\right) = -\sqrt{\frac{5 - \sqrt{5}}{5 + \sqrt{5}}}$

  $\tan\left(\frac{\theta}{2}\right) = \frac{5 - 5\sqrt{5}}{10}$

58.
- $\sin(2\theta) = -\frac{4}{5}$
- $\cos(2\theta) = -\frac{3}{5}$
- $\tan(2\theta) = \frac{4}{3}$
- $\sin\left(\frac{\theta}{2}\right) = \frac{\sqrt{50 + 10\sqrt{5}}}{10}$
- $\cos\left(\frac{\theta}{2}\right) = \frac{\sqrt{50 - 10\sqrt{5}}}{10}$
- $\tan\left(\frac{\theta}{2}\right) = \sqrt{\frac{5 + \sqrt{5}}{5 - \sqrt{5}}}$

  $\tan\left(\frac{\theta}{2}\right) = \frac{5 + 5\sqrt{5}}{10}$

74. $\frac{\cos(2\theta) + \cos(8\theta)}{2}$

75. $\frac{\cos(5\theta) - \cos(9\theta)}{2}$

76. $\frac{\sin(8\theta) + \sin(10\theta)}{2}$

77. $\frac{\cos(4\theta) + \cos(8\theta)}{2}$

78. $\frac{\cos(\theta) - \cos(5\theta)}{2}$

79. $\frac{\sin(2\theta) + \sin(4\theta)}{2}$

80. $2\cos(4\theta)\cos(\theta)$

81. $-2\cos\left(\frac{9}{2}\theta\right)\sin\left(\frac{5}{2}\theta\right)$

82. $2\sin\left(\frac{11}{2}\theta\right)\sin\left(\frac{1}{2}\theta\right)$

83. $2\cos(4\theta)\sin(5\theta)$

84. $\sqrt{2}\cos\left(\theta - \frac{\pi}{4}\right)$

85. $-\sqrt{2}\sin\left(\theta - \frac{\pi}{4}\right)$

90. $1 - \frac{x^2}{2}$

91. $\frac{14x}{x^2 + 49}$

92. $\ln|x + \sqrt{x^2 + 16}| - \ln(4)$

10.5 GRAPHS OF THE TRIGONOMETRIC FUNCTIONS

In this section, we return to our discussion of the circular (trigonometric) functions as functions of real numbers and pick up where we left off in Sections 10.2.1 and 10.3.1. As usual, we begin our study with the functions $f(t) = \cos(t)$ and $g(t) = \sin(t)$.

10.5.1 GRAPHS OF THE COSINE AND SINE FUNCTIONS

From Theorem 10.5 in Section 10.2.1, we know that the domain of $f(t) = \cos(t)$ and of $g(t) = \sin(t)$ is all real numbers, $(-\infty, \infty)$, and the range of both functions is $[-1, 1]$. The Even / Odd Identities in Theorem 10.12 tell us $\cos(-t) = \cos(t)$ for all real numbers t and $\sin(-t) = -\sin(t)$ for all real numbers t. This means $f(t) = \cos(t)$ is an even function, while $g(t) = \sin(t)$ is an odd function.[1] Another important property of these functions is that for coterminal angles α and β, $\cos(\alpha) = \cos(\beta)$ and $\sin(\alpha) = \sin(\beta)$. Said differently, $\cos(t+2\pi k) = \cos(t)$ and $\sin(t+2\pi k) = \sin(t)$ for all real numbers t and any integer k. This last property is given a special name.

Definition 10.3. Periodic Functions: A function f is said to be **periodic** if there is a real number c so that $f(t+c) = f(t)$ for all real numbers t in the domain of f. The smallest positive number p for which $f(t+p) = f(t)$ for all real numbers t in the domain of f, if it exists, is called the **period** of f.

We have already seen a family of periodic functions in Section 2.1: the constant functions. However, despite being periodic a constant function has no period. (We'll leave that odd gem as an exercise for you.) Returning to the circular functions, we see that by Definition 10.3, $f(t) = \cos(t)$ is periodic, since $\cos(t + 2\pi k) = \cos(t)$ for any integer k. To determine the period of f, we need to find the smallest real number p so that $f(t+p) = f(t)$ for all real numbers t or, said differently, the smallest positive real number p such that $\cos(t+p) = \cos(t)$ for all real numbers t. We know that $\cos(t+2\pi) = \cos(t)$ for all real numbers t but the question remains if any smaller real number will do the trick. Suppose $p > 0$ and $\cos(t+p) = \cos(t)$ for all real numbers t. Then, in particular, $\cos(0+p) = \cos(0)$ so that $\cos(p) = 1$. From this we know p is a multiple of 2π and, since the smallest positive multiple of 2π is 2π itself, we have the result. Similarly, we can show $g(t) = \sin(t)$ is also periodic with 2π as its period.[2] Having period 2π essentially means that we can completely understand everything about the functions $f(t) = \cos(t)$ and $g(t) = \sin(t)$ by studying one interval of length 2π, say $[0, 2\pi]$.[3]

One last property of the functions $f(t) = \cos(t)$ and $g(t) = \sin(t)$ is worth pointing out: both of these functions are continuous and smooth. Recall from Section 3.1 that geometrically this means the graphs of the cosine and sine functions have no jumps, gaps, holes in the graph, asymptotes,

[1] See section 1.6 for a review of these concepts.

[2] Alternatively, we can use the Cofunction Identities in Theorem 10.14 to show that $g(t) = \sin(t)$ is periodic with period 2π since $g(t) = \sin(t) = \cos\left(\frac{\pi}{2} - t\right) = f\left(\frac{\pi}{2} - t\right)$.

[3] Technically, we should study the interval $[0, 2\pi)$,[4] since whatever happens at $t = 2\pi$ is the same as what happens at $t = 0$. As we will see shortly, $t = 2\pi$ gives us an extra 'check' when we go to graph these functions.

[4] In some advanced texts, the interval of choice is $[-\pi, \pi)$.

corners or cusps. As we shall see, the graphs of both $f(t) = \cos(t)$ and $g(t) = \sin(t)$ meander nicely and don't cause any trouble. We summarize these facts in the following theorem.

Theorem 10.22. Properties of the Cosine and Sine Functions

- The function $f(x) = \cos(x)$
  - has domain $(-\infty, \infty)$
  - has range $[-1, 1]$
  - is continuous and smooth
  - is even
  - has period 2π
- The function $g(x) = \sin(x)$
  - has domain $(-\infty, \infty)$
  - has range $[-1, 1]$
  - is continuous and smooth
  - is odd
  - has period 2π

In the chart above, we followed the convention established in Section 1.6 and used x as the independent variable and y as the dependent variable.[5] This allows us to turn our attention to graphing the cosine and sine functions in the Cartesian Plane. To graph $y = \cos(x)$, we make a table as we did in Section 1.6 using some of the 'common values' of x in the interval $[0, 2\pi]$. This generates a *portion* of the cosine graph, which we call the '**fundamental cycle**' of $y = \cos(x)$.

x	$\cos(x)$	$(x, \cos(x))$
0	1	$(0, 1)$
$\frac{\pi}{4}$	$\frac{\sqrt{2}}{2}$	$\left(\frac{\pi}{4}, \frac{\sqrt{2}}{2}\right)$
$\frac{\pi}{2}$	0	$\left(\frac{\pi}{2}, 0\right)$
$\frac{3\pi}{4}$	$-\frac{\sqrt{2}}{2}$	$\left(\frac{3\pi}{4}, -\frac{\sqrt{2}}{2}\right)$
π	-1	$(\pi, -1)$
$\frac{5\pi}{4}$	$-\frac{\sqrt{2}}{2}$	$\left(\frac{5\pi}{4}, -\frac{\sqrt{2}}{2}\right)$
$\frac{3\pi}{2}$	0	$\left(\frac{3\pi}{2}, 0\right)$
$\frac{7\pi}{4}$	$\frac{\sqrt{2}}{2}$	$\left(\frac{7\pi}{4}, \frac{\sqrt{2}}{2}\right)$
2π	1	$(2\pi, 1)$

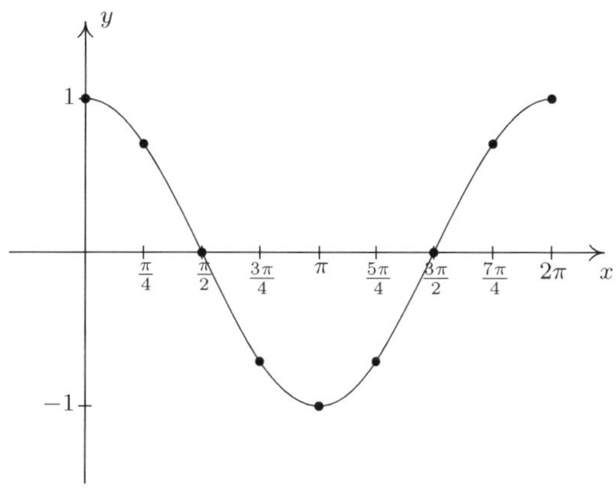

The 'fundamental cycle' of $y = \cos(x)$.

A few things about the graph above are worth mentioning. First, this graph represents only part of the graph of $y = \cos(x)$. To get the entire graph, we imagine 'copying and pasting' this graph end to end infinitely in both directions (left and right) on the x-axis. Secondly, the vertical scale here has been greatly exaggerated for clarity and aesthetics. Below is an accurate-to-scale graph of $y = \cos(x)$ showing several cycles with the 'fundamental cycle' plotted thicker than the others. The

[5]The use of x and y in this context is not to be confused with the x- and y-coordinates of points on the Unit Circle which define cosine and sine. Using the term 'trigonometric function' as opposed to 'circular function' can help with that, but one could then ask, "Hey, where's the triangle?"

graph of $y = \cos(x)$ is usually described as 'wavelike' – indeed, many of the applications involving the cosine and sine functions feature modeling wavelike phenomena.

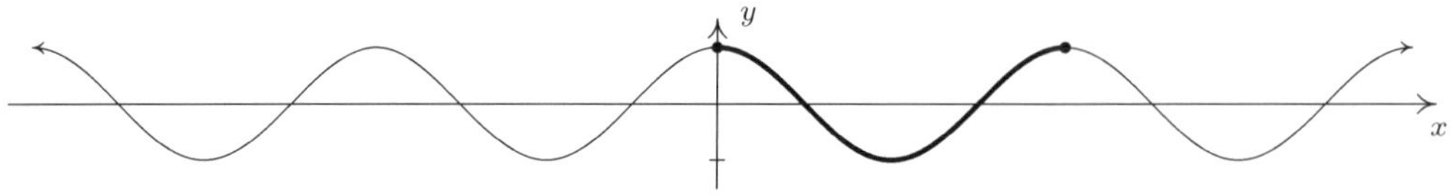

An accurately scaled graph of $y = \cos(x)$.

We can plot the fundamental cycle of the graph of $y = \sin(x)$ similarly, with similar results.

x	$\sin(x)$	$(x, \sin(x))$
0	0	$(0, 0)$
$\frac{\pi}{4}$	$\frac{\sqrt{2}}{2}$	$\left(\frac{\pi}{4}, \frac{\sqrt{2}}{2}\right)$
$\frac{\pi}{2}$	1	$\left(\frac{\pi}{2}, 1\right)$
$\frac{3\pi}{4}$	$\frac{\sqrt{2}}{2}$	$\left(\frac{3\pi}{4}, \frac{\sqrt{2}}{2}\right)$
π	0	$(\pi, 0)$
$\frac{5\pi}{4}$	$-\frac{\sqrt{2}}{2}$	$\left(\frac{5\pi}{4}, -\frac{\sqrt{2}}{2}\right)$
$\frac{3\pi}{2}$	-1	$\left(\frac{3\pi}{2}, -1\right)$
$\frac{7\pi}{4}$	$-\frac{\sqrt{2}}{2}$	$\left(\frac{7\pi}{4}, -\frac{\sqrt{2}}{2}\right)$
2π	0	$(2\pi, 0)$

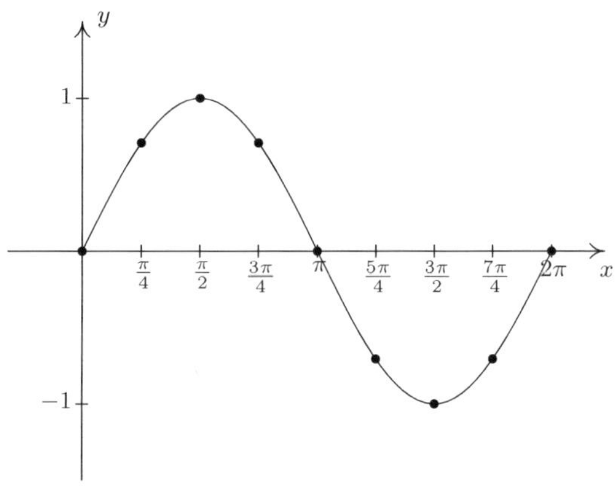

The 'fundamental cycle' of $y = \sin(x)$.

As with the graph of $y = \cos(x)$, we provide an accurately scaled graph of $y = \sin(x)$ below with the fundamental cycle highlighted.

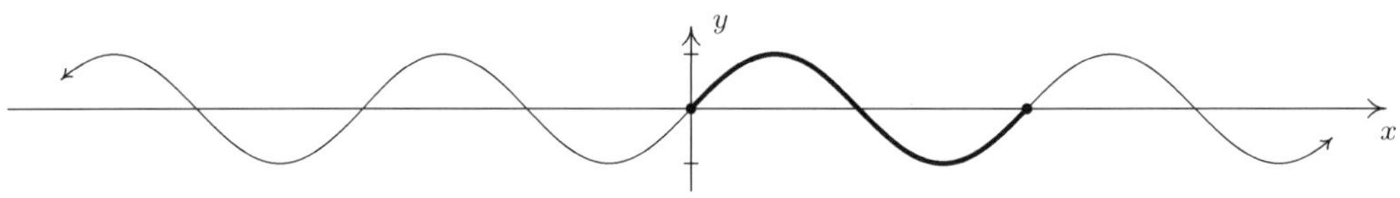

An accurately scaled graph of $y = \sin(x)$.

It is no accident that the graphs of $y = \cos(x)$ and $y = \sin(x)$ are so similar. Using a cofunction identity along with the even property of cosine, we have

$$\sin(x) = \cos\left(\frac{\pi}{2} - x\right) = \cos\left(-\left(x - \frac{\pi}{2}\right)\right) = \cos\left(x - \frac{\pi}{2}\right)$$

Recalling Section 1.7, we see from this formula that the graph of $y = \sin(x)$ is the result of shifting the graph of $y = \cos(x)$ to the right $\frac{\pi}{2}$ units. A visual inspection confirms this.

Now that we know the basic shapes of the graphs of $y = \cos(x)$ and $y = \sin(x)$, we can use Theorem 1.7 in Section 1.7 to graph more complicated curves. To do so, we need to keep track of the movement of some key points on the original graphs. We choose to track the values $x = 0$, $\frac{\pi}{2}$, π, $\frac{3\pi}{2}$ and 2π. These 'quarter marks' correspond to quadrantal angles, and as such, mark the location of the zeros and the local extrema of these functions over exactly one period. Before we begin our next example, we need to review the concept of the 'argument' of a function as first introduced in Section 1.4. For the function $f(x) = 1 - 5\cos(2x - \pi)$, the argument of f is x. We shall have occasion, however, to refer to the argument of the *cosine*, which in this case is $2x - \pi$. Loosely stated, the argument of a trigonometric function is the expression 'inside' the function.

Example 10.5.1. Graph one cycle of the following functions. State the period of each.

1. $f(x) = 3\cos\left(\frac{\pi x - \pi}{2}\right) + 1$
2. $g(x) = \frac{1}{2}\sin(\pi - 2x) + \frac{3}{2}$

Solution.

1. We set the argument of the cosine, $\frac{\pi x - \pi}{2}$, equal to each of the values: 0, $\frac{\pi}{2}$, π, $\frac{3\pi}{2}$, 2π and solve for x. We summarize the results below.

a	$\frac{\pi x-\pi}{2} = a$	x
0	$\frac{\pi x-\pi}{2} = 0$	1
$\frac{\pi}{2}$	$\frac{\pi x-\pi}{2} = \frac{\pi}{2}$	2
π	$\frac{\pi x-\pi}{2} = \pi$	3
$\frac{3\pi}{2}$	$\frac{\pi x-\pi}{2} = \frac{3\pi}{2}$	4
2π	$\frac{\pi x-\pi}{2} = 2\pi$	5

Next, we substitute each of these x values into $f(x) = 3\cos\left(\frac{\pi x-\pi}{2}\right) + 1$ to determine the corresponding y-values and connect the dots in a pleasing wavelike fashion.

x	$f(x)$	$(x, f(x))$
1	4	$(1, 4)$
2	1	$(2, 1)$
3	-2	$(3, -2)$
4	1	$(4, 1)$
5	4	$(5, 4)$

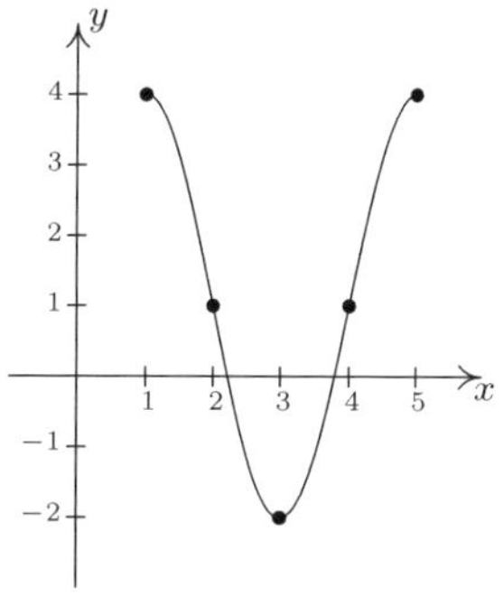

One cycle of $y = f(x)$.

One cycle is graphed on $[1, 5]$ so the period is the length of that interval which is 4.

2. Proceeding as above, we set the argument of the sine, $\pi - 2x$, equal to each of our quarter marks and solve for x.

a	$\pi - 2x = a$	x
0	$\pi - 2x = 0$	$\frac{\pi}{2}$
$\frac{\pi}{2}$	$\pi - 2x = \frac{\pi}{2}$	$\frac{\pi}{4}$
π	$\pi - 2x = \pi$	0
$\frac{3\pi}{2}$	$\pi - 2x = \frac{3\pi}{2}$	$-\frac{\pi}{4}$
2π	$\pi - 2x = 2\pi$	$-\frac{\pi}{2}$

We now find the corresponding y-values on the graph by substituting each of these x-values into $g(x) = \frac{1}{2}\sin(\pi - 2x) + \frac{3}{2}$. Once again, we connect the dots in a wavelike fashion.

x	$g(x)$	$(x, g(x))$
$\frac{\pi}{2}$	$\frac{3}{2}$	$\left(\frac{\pi}{2}, \frac{3}{2}\right)$
$\frac{\pi}{4}$	2	$\left(\frac{\pi}{4}, 2\right)$
0	$\frac{3}{2}$	$\left(0, \frac{3}{2}\right)$
$-\frac{\pi}{4}$	1	$\left(-\frac{\pi}{4}, 1\right)$
$-\frac{\pi}{2}$	$\frac{3}{2}$	$\left(-\frac{\pi}{2}, \frac{3}{2}\right)$

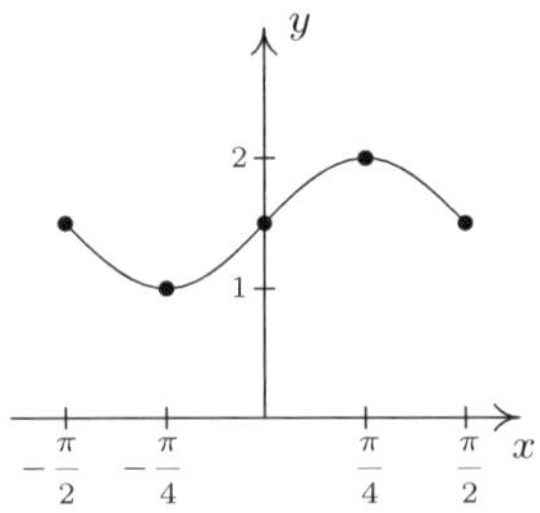

One cycle of $y = g(x)$.

One cycle was graphed on the interval $\left[-\frac{\pi}{2}, \frac{\pi}{2}\right]$ so the period is $\frac{\pi}{2} - \left(-\frac{\pi}{2}\right) = \pi$. □

The functions in Example 10.5.1 are examples of **sinusoids**. Roughly speaking, a sinusoid is the result of taking the basic graph of $f(x) = \cos(x)$ or $g(x) = \sin(x)$ and performing any of the transformations[6] mentioned in Section 1.7. Sinusoids can be characterized by four properties: period, amplitude, phase shift and vertical shift. We have already discussed period, that is, how long it takes for the sinusoid to complete one cycle. The standard period of both $f(x) = \cos(x)$ and $g(x) = \sin(x)$ is 2π, but horizontal scalings will change the period of the resulting sinusoid. The **amplitude** of the sinusoid is a measure of how 'tall' the wave is, as indicated in the figure below. The amplitude of the standard cosine and sine functions is 1, but vertical scalings can alter this.

[6] We have already seen how the Even/Odd and Cofunction Identities can be used to rewrite $g(x) = \sin(x)$ as a transformed version of $f(x) = \cos(x)$, so of course, the reverse is true: $f(x) = \cos(x)$ can be written as a transformed version of $g(x) = \sin(x)$. The authors have seen some instances where sinusoids are always converted to cosine functions while in other disciplines, the sinusoids are always written in terms of sine functions. We will discuss the applications of sinusoids in greater detail in Chapter 11. Until then, we will keep our options open.

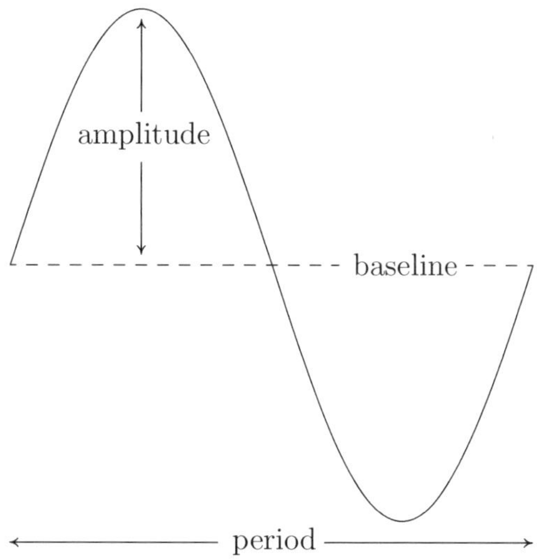

The **phase shift** of the sinusoid is the horizontal shift experienced by the fundamental cycle. We have seen that a phase (horizontal) shift of $\frac{\pi}{2}$ to the right takes $f(x) = \cos(x)$ to $g(x) = \sin(x)$ since $\cos\left(x - \frac{\pi}{2}\right) = \sin(x)$. As the reader can verify, a phase shift of $\frac{\pi}{2}$ to the left takes $g(x) = \sin(x)$ to $f(x) = \cos(x)$. The vertical shift of a sinusoid is exactly the same as the vertical shifts in Section 1.7. In most contexts, the vertical shift of a sinusoid is assumed to be 0, but we state the more general case below. The following theorem, which is reminiscent of Theorem 1.7 in Section 1.7, shows how to find these four fundamental quantities from the formula of the given sinusoid.

Theorem 10.23. For $\omega > 0$, the functions

$$C(x) = A\cos(\omega x + \phi) + B \quad \text{and} \quad S(x) = A\sin(\omega x + \phi) + B$$

- have period $\dfrac{2\pi}{\omega}$
- have phase shift $-\dfrac{\phi}{\omega}$
- have amplitude $|A|$
- have vertical shift B

We note that in some scientific and engineering circles, the quantity ϕ mentioned in Theorem 10.23 is called the **phase** of the sinusoid. Since our interest in this book is primarily with *graphing* sinusoids, we focus our attention on the horizontal shift $-\frac{\phi}{\omega}$ induced by ϕ.

The proof of Theorem 10.23 is a direct application of Theorem 1.7 in Section 1.7 and is left to the reader. The parameter ω, which is stipulated to be positive, is called the (**angular**) **frequency** of the sinusoid and is the number of cycles the sinusoid completes over a 2π interval. We can always ensure $\omega > 0$ using the Even/Odd Identities.[7] We now test out Theorem 10.23 using the functions f and g featured in Example 10.5.1. First, we write $f(x)$ in the form prescribed in Theorem 10.23,

$$f(x) = 3\cos\left(\frac{\pi x - \pi}{2}\right) + 1 = 3\cos\left(\frac{\pi}{2}x + \left(-\frac{\pi}{2}\right)\right) + 1,$$

[7]Try using the formulas in Theorem 10.23 applied to $C(x) = \cos(-x + \pi)$ to see why we need $\omega > 0$.

so that $A = 3$, $\omega = \frac{\pi}{2}$, $\phi = -\frac{\pi}{2}$ and $B = 1$. According to Theorem 10.23, the period of f is $\frac{2\pi}{\omega} = \frac{2\pi}{\pi/2} = 4$, the amplitude is $|A| = |3| = 3$, the phase shift is $-\frac{\phi}{\omega} = -\frac{-\pi/2}{\pi/2} = 1$ (indicating a shift to the *right* 1 unit) and the vertical shift is $B = 1$ (indicating a shift *up* 1 unit.) All of these match with our graph of $y = f(x)$. Moreover, if we start with the basic shape of the cosine graph, shift it 1 unit to the right, 1 unit up, stretch the amplitude to 3 and shrink the period to 4, we will have reconstructed one period of the graph of $y = f(x)$. In other words, instead of tracking the five 'quarter marks' through the transformations to plot $y = f(x)$, we can use five other pieces of information: the phase shift, vertical shift, amplitude, period and basic shape of the cosine curve. Turning our attention now to the function g in Example 10.5.1, we first need to use the odd property of the sine function to write it in the form required by Theorem 10.23

$$g(x) = \frac{1}{2}\sin(\pi - 2x) + \frac{3}{2} = \frac{1}{2}\sin(-(2x - \pi)) + \frac{3}{2} = -\frac{1}{2}\sin(2x - \pi) + \frac{3}{2} = -\frac{1}{2}\sin(2x + (-\pi)) + \frac{3}{2}$$

We find $A = -\frac{1}{2}$, $\omega = 2$, $\phi = -\pi$ and $B = \frac{3}{2}$. The period is then $\frac{2\pi}{2} = \pi$, the amplitude is $\left|-\frac{1}{2}\right| = \frac{1}{2}$, the phase shift is $-\frac{-\pi}{2} = \frac{\pi}{2}$ (indicating a shift *right* $\frac{\pi}{2}$ units) and the vertical shift is *up* $\frac{3}{2}$. Note that, in this case, all of the data match our graph of $y = g(x)$ with the exception of the phase shift. Instead of the graph *starting* at $x = \frac{\pi}{2}$, it ends there. Remember, however, that the graph presented in Example 10.5.1 is only one portion of the graph of $y = g(x)$. Indeed, another complete cycle begins at $x = \frac{\pi}{2}$, and this is the cycle Theorem 10.23 is detecting. The reason for the discrepancy is that, in order to apply Theorem 10.23, we had to rewrite the formula for $g(x)$ using the odd property of the sine function. Note that whether we graph $y = g(x)$ using the 'quarter marks' approach or using the Theorem 10.23, we get one complete cycle of the graph, which means we have completely determined the sinusoid.

Example 10.5.2. Below is the graph of one complete cycle of a sinusoid $y = f(x)$.

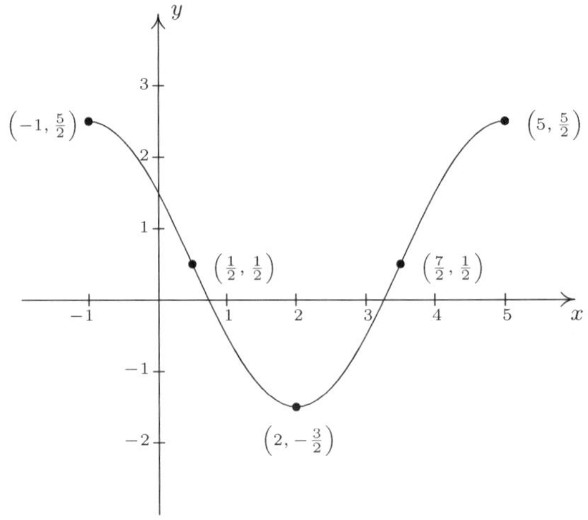

One cycle of $y = f(x)$.

1. Find a cosine function whose graph matches the graph of $y = f(x)$.

2. Find a sine function whose graph matches the graph of $y = f(x)$.

Solution.

1. We fit the data to a function of the form $C(x) = A\cos(\omega x + \phi) + B$. Since one cycle is graphed over the interval $[-1, 5]$, its period is $5 - (-1) = 6$. According to Theorem 10.23, $6 = \frac{2\pi}{\omega}$, so that $\omega = \frac{\pi}{3}$. Next, we see that the phase shift is -1, so we have $-\frac{\phi}{\omega} = -1$, or $\phi = \omega = \frac{\pi}{3}$. To find the amplitude, note that the range of the sinusoid is $\left[-\frac{3}{2}, \frac{5}{2}\right]$. As a result, the amplitude $A = \frac{1}{2}\left[\frac{5}{2} - \left(-\frac{3}{2}\right)\right] = \frac{1}{2}(4) = 2$. Finally, to determine the vertical shift, we average the endpoints of the range to find $B = \frac{1}{2}\left[\frac{5}{2} + \left(-\frac{3}{2}\right)\right] = \frac{1}{2}(1) = \frac{1}{2}$. Our final answer is $C(x) = 2\cos\left(\frac{\pi}{3}x + \frac{\pi}{3}\right) + \frac{1}{2}$.

2. Most of the work to fit the data to a function of the form $S(x) = A\sin(\omega x + \phi) + B$ is done. The period, amplitude and vertical shift are the same as before with $\omega = \frac{\pi}{3}$, $A = 2$ and $B = \frac{1}{2}$. The trickier part is finding the phase shift. To that end, we imagine extending the graph of the given sinusoid as in the figure below so that we can identify a cycle beginning at $\left(\frac{7}{2}, \frac{1}{2}\right)$. Taking the phase shift to be $\frac{7}{2}$, we get $-\frac{\phi}{\omega} = \frac{7}{2}$, or $\phi = -\frac{7}{2}\omega = -\frac{7}{2}\left(\frac{\pi}{3}\right) = -\frac{7\pi}{6}$. Hence, our answer is $S(x) = 2\sin\left(\frac{\pi}{3}x - \frac{7\pi}{6}\right) + \frac{1}{2}$.

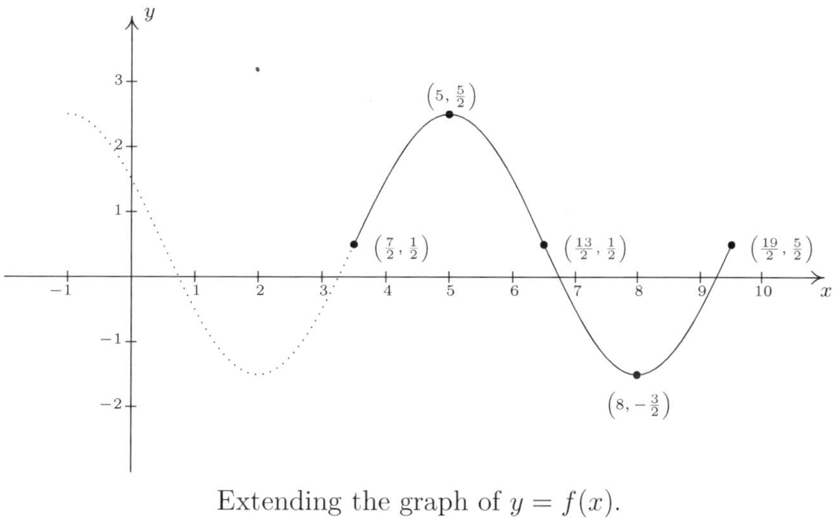

Extending the graph of $y = f(x)$.

□

Note that each of the answers given in Example 10.5.2 is one choice out of many possible answers. For example, when fitting a sine function to the data, we could have chosen to start at $\left(\frac{1}{2}, \frac{1}{2}\right)$ taking $A = -2$. In this case, the phase shift is $\frac{1}{2}$ so $\phi = -\frac{\pi}{6}$ for an answer of $S(x) = -2\sin\left(\frac{\pi}{3}x - \frac{\pi}{6}\right) + \frac{1}{2}$. Alternatively, we could have extended the graph of $y = f(x)$ to the left and considered a sine function starting at $\left(-\frac{5}{2}, \frac{1}{2}\right)$, and so on. Each of these formulas determine the same sinusoid curve and their formulas are all equivalent using identities. Speaking of identities, if we use the sum identity for cosine, we can expand the formula to yield

$$C(x) = A\cos(\omega x + \phi) + B = A\cos(\omega x)\cos(\phi) - A\sin(\omega x)\sin(\phi) + B.$$

Similarly, using the sum identity for sine, we get

$$S(x) = A\sin(\omega x + \phi) + B = A\sin(\omega x)\cos(\phi) + A\cos(\omega x)\sin(\phi) + B.$$

Making these observations allows us to recognize (and graph) functions as sinusoids which, at first glance, don't appear to fit the forms of either $C(x)$ or $S(x)$.

Example 10.5.3. Consider the function $f(x) = \cos(2x) - \sqrt{3}\sin(2x)$. Find a formula for $f(x)$:

1. in the form $C(x) = A\cos(\omega x + \phi) + B$ for $\omega > 0$

2. in the form $S(x) = A\sin(\omega x + \phi) + B$ for $\omega > 0$

Check your answers analytically using identities and graphically using a calculator.

Solution.

1. The key to this problem is to use the expanded forms of the sinusoid formulas and match up corresponding coefficients. Equating $f(x) = \cos(2x) - \sqrt{3}\sin(2x)$ with the expanded form of $C(x) = A\cos(\omega x + \phi) + B$, we get

$$\cos(2x) - \sqrt{3}\sin(2x) = A\cos(\omega x)\cos(\phi) - A\sin(\omega x)\sin(\phi) + B$$

It should be clear that we can take $\omega = 2$ and $B = 0$ to get

$$\cos(2x) - \sqrt{3}\sin(2x) = A\cos(2x)\cos(\phi) - A\sin(2x)\sin(\phi)$$

To determine A and ϕ, a bit more work is involved. We get started by equating the coefficients of the trigonometric functions on either side of the equation. On the left hand side, the coefficient of $\cos(2x)$ is 1, while on the right hand side, it is $A\cos(\phi)$. Since this equation is to hold for all real numbers, we must have[8] that $A\cos(\phi) = 1$. Similarly, we find by equating the coefficients of $\sin(2x)$ that $A\sin(\phi) = \sqrt{3}$. What we have here is a system of nonlinear equations! We can temporarily eliminate the dependence on ϕ by using the Pythagorean Identity. We know $\cos^2(\phi) + \sin^2(\phi) = 1$, so multiplying this by A^2 gives $A^2\cos^2(\phi) + A^2\sin^2(\phi) = A^2$. Since $A\cos(\phi) = 1$ and $A\sin(\phi) = \sqrt{3}$, we get $A^2 = 1^2 + (\sqrt{3})^2 = 4$ or $A = \pm 2$. Choosing $A = 2$, we have $2\cos(\phi) = 1$ and $2\sin(\phi) = \sqrt{3}$ or, after some rearrangement, $\cos(\phi) = \frac{1}{2}$ and $\sin(\phi) = \frac{\sqrt{3}}{2}$. One such angle ϕ which satisfies this criteria is $\phi = \frac{\pi}{3}$. Hence, one way to write $f(x)$ as a sinusoid is $f(x) = 2\cos\left(2x + \frac{\pi}{3}\right)$. We can easily check our answer using the sum formula for cosine

$$\begin{aligned} f(x) &= 2\cos\left(2x + \frac{\pi}{3}\right) \\ &= 2\left[\cos(2x)\cos\left(\frac{\pi}{3}\right) - \sin(2x)\sin\left(\frac{\pi}{3}\right)\right] \\ &= 2\left[\cos(2x)\left(\frac{1}{2}\right) - \sin(2x)\left(\frac{\sqrt{3}}{2}\right)\right] \\ &= \cos(2x) - \sqrt{3}\sin(2x) \end{aligned}$$

[8] This should remind you of equation coefficients of like powers of x in Section 8.6.

2. Proceeding as before, we equate $f(x) = \cos(2x) - \sqrt{3}\sin(2x)$ with the expanded form of $S(x) = A\sin(\omega x + \phi) + B$ to get

$$\cos(2x) - \sqrt{3}\sin(2x) = A\sin(\omega x)\cos(\phi) + A\cos(\omega x)\sin(\phi) + B$$

Once again, we may take $\omega = 2$ and $B = 0$ so that

$$\cos(2x) - \sqrt{3}\sin(2x) = A\sin(2x)\cos(\phi) + A\cos(2x)\sin(\phi)$$

We equate[9] the coefficients of $\cos(2x)$ on either side and get $A\sin(\phi) = 1$ and $A\cos(\phi) = -\sqrt{3}$. Using $A^2\cos^2(\phi) + A^2\sin^2(\phi) = A^2$ as before, we get $A = \pm 2$, and again we choose $A = 2$. This means $2\sin(\phi) = 1$, or $\sin(\phi) = \frac{1}{2}$, and $2\cos(\phi) = -\sqrt{3}$, which means $\cos(\phi) = -\frac{\sqrt{3}}{2}$. One such angle which meets these criteria is $\phi = \frac{5\pi}{6}$. Hence, we have $f(x) = 2\sin\left(2x + \frac{5\pi}{6}\right)$. Checking our work analytically, we have

$$\begin{aligned} f(x) &= 2\sin\left(2x + \tfrac{5\pi}{6}\right) \\ &= 2\left[\sin(2x)\cos\left(\tfrac{5\pi}{6}\right) + \cos(2x)\sin\left(\tfrac{5\pi}{6}\right)\right] \\ &= 2\left[\sin(2x)\left(-\tfrac{\sqrt{3}}{2}\right) + \cos(2x)\left(\tfrac{1}{2}\right)\right] \\ &= \cos(2x) - \sqrt{3}\sin(2x) \end{aligned}$$

Graphing the three formulas for $f(x)$ result in the identical curve, verifying our analytic work.

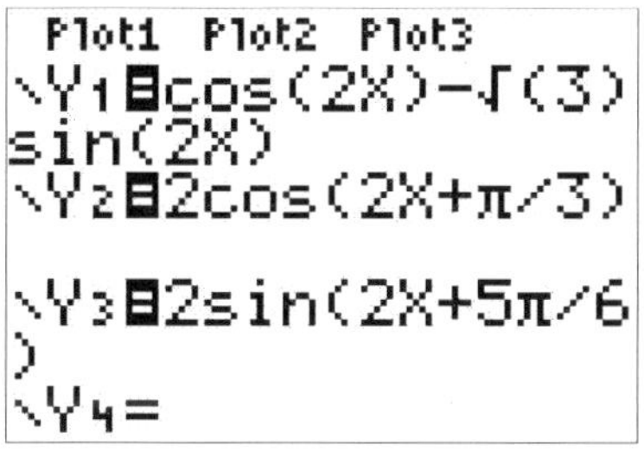

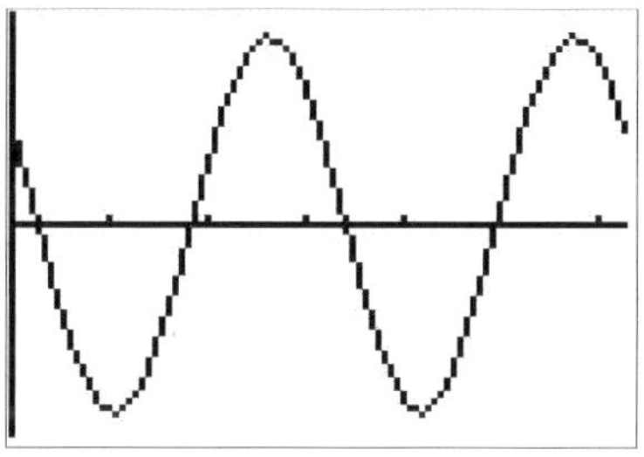

□

It is important to note that in order for the technique presented in Example 10.5.3 to fit a function into one of the forms in Theorem 10.23, the arguments of the cosine and sine function much match. That is, while $f(x) = \cos(2x) - \sqrt{3}\sin(2x)$ is a sinusoid, $g(x) = \cos(2x) - \sqrt{3}\sin(3x)$ is not.[10] It is also worth mentioning that, had we chosen $A = -2$ instead of $A = 2$ as we worked through Example 10.5.3, our final answers would have *looked* different. The reader is encouraged to rework Example 10.5.3 using $A = -2$ to see what these differences are, and then for a challenging exercise, use identities to show that the formulas are all equivalent. The general equations to fit a function of the form $f(x) = a\cos(\omega x) + b\sin(\omega x) + B$ into one of the forms in Theorem 10.23 are explored in Exercise 35.

[9] Be careful here!

[10] This graph does, however, exhibit sinusoid-like characteristics! Check it out!

10.5.2 Graphs of the Secant and Cosecant Functions

We now turn our attention to graphing $y = \sec(x)$. Since $\sec(x) = \frac{1}{\cos(x)}$, we can use our table of values for the graph of $y = \cos(x)$ and take reciprocals. We know from Section 10.3.1 that the domain of $F(x) = \sec(x)$ excludes all odd multiples of $\frac{\pi}{2}$, and sure enough, we run into trouble at $x = \frac{\pi}{2}$ and $x = \frac{3\pi}{2}$ since $\cos(x) = 0$ at these values. Using the notation introduced in Section 4.2, we have that as $x \to \frac{\pi}{2}^{-}$, $\cos(x) \to 0^{+}$, so $\sec(x) \to \infty$. (See Section 10.3.1 for a more detailed analysis.) Similarly, we find that as $x \to \frac{\pi}{2}^{+}$, $\sec(x) \to -\infty$; as $x \to \frac{3\pi}{2}^{-}$, $\sec(x) \to -\infty$; and as $x \to \frac{3\pi}{2}^{+}$, $\sec(x) \to \infty$. This means we have a pair of vertical asymptotes to the graph of $y = \sec(x)$, $x = \frac{\pi}{2}$ and $x = \frac{3\pi}{2}$. Since $\cos(x)$ is periodic with period 2π, it follows that $\sec(x)$ is also.[11] Below we graph a fundamental cycle of $y = \sec(x)$ along with a more complete graph obtained by the usual 'copying and pasting.'[12]

x	$\cos(x)$	$\sec(x)$	$(x, \sec(x))$
0	1	1	$(0, 1)$
$\frac{\pi}{4}$	$\frac{\sqrt{2}}{2}$	$\sqrt{2}$	$\left(\frac{\pi}{4}, \sqrt{2}\right)$
$\frac{\pi}{2}$	0	undefined	
$\frac{3\pi}{4}$	$-\frac{\sqrt{2}}{2}$	$-\sqrt{2}$	$\left(\frac{3\pi}{4}, -\sqrt{2}\right)$
π	-1	-1	$(\pi, -1)$
$\frac{5\pi}{4}$	$-\frac{\sqrt{2}}{2}$	$-\sqrt{2}$	$\left(\frac{5\pi}{4}, -\sqrt{2}\right)$
$\frac{3\pi}{2}$	0	undefined	
$\frac{7\pi}{4}$	$\frac{\sqrt{2}}{2}$	$\sqrt{2}$	$\left(\frac{7\pi}{4}, \sqrt{2}\right)$
2π	1	1	$(2\pi, 1)$

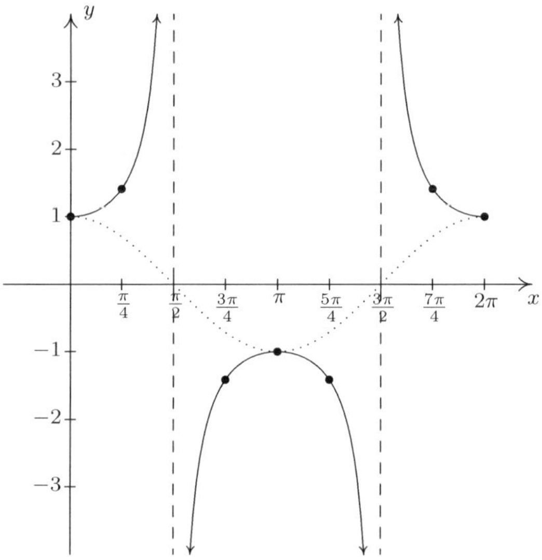

The 'fundamental cycle' of $y = \sec(x)$.

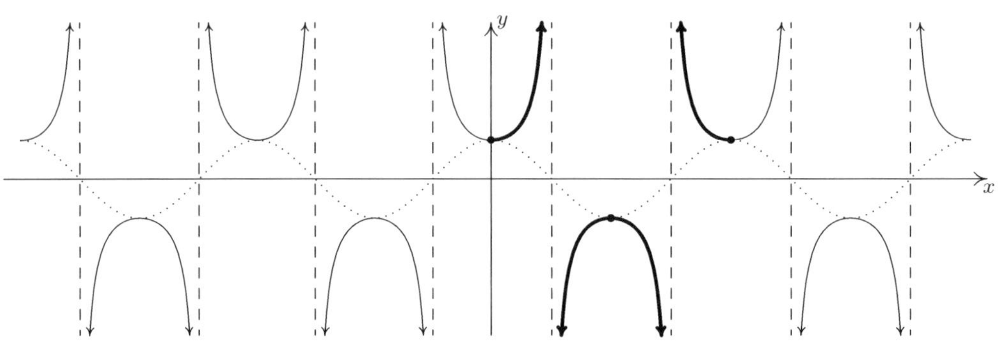

The graph of $y = \sec(x)$.

[11] Provided $\sec(\alpha)$ and $\sec(\beta)$ are defined, $\sec(\alpha) = \sec(\beta)$ if and only if $\cos(\alpha) = \cos(\beta)$. Hence, $\sec(x)$ inherits its period from $\cos(x)$.

[12] In Section 10.3.1, we argued the range of $F(x) = \sec(x)$ is $(-\infty, -1] \cup [1, \infty)$. We can now see this graphically.

As one would expect, to graph $y = \csc(x)$ we begin with $y = \sin(x)$ and take reciprocals of the corresponding y-values. Here, we encounter issues at $x = 0$, $x = \pi$ and $x = 2\pi$. Proceeding with the usual analysis, we graph the fundamental cycle of $y = \csc(x)$ below along with the dotted graph of $y = \sin(x)$ for reference. Since $y = \sin(x)$ and $y = \cos(x)$ are merely phase shifts of each other, so too are $y = \csc(x)$ and $y = \sec(x)$.

x	$\sin(x)$	$\csc(x)$	$(x, \csc(x))$
0	0	undefined	
$\frac{\pi}{4}$	$\frac{\sqrt{2}}{2}$	$\sqrt{2}$	$\left(\frac{\pi}{4}, \sqrt{2}\right)$
$\frac{\pi}{2}$	1	1	$\left(\frac{\pi}{2}, 1\right)$
$\frac{3\pi}{4}$	$\frac{\sqrt{2}}{2}$	$\sqrt{2}$	$\left(\frac{3\pi}{4}, \sqrt{2}\right)$
π	0	undefined	
$\frac{5\pi}{4}$	$-\frac{\sqrt{2}}{2}$	$-\sqrt{2}$	$\left(\frac{5\pi}{4}, -\sqrt{2}\right)$
$\frac{3\pi}{2}$	-1	-1	$\left(\frac{3\pi}{2}, -1\right)$
$\frac{7\pi}{4}$	$-\frac{\sqrt{2}}{2}$	$-\sqrt{2}$	$\left(\frac{7\pi}{4}, -\sqrt{2}\right)$
2π	0	undefined	

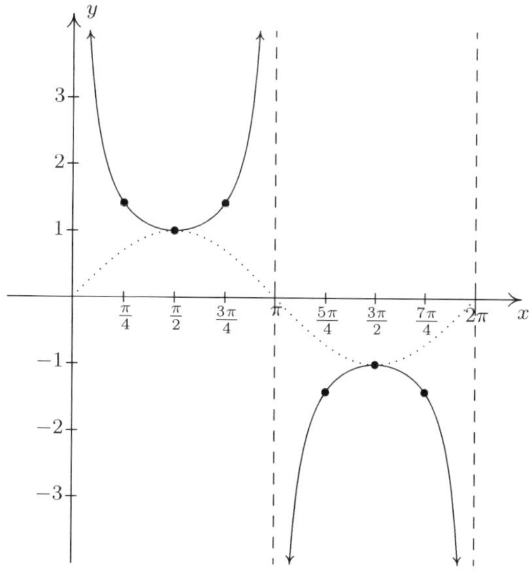

The 'fundamental cycle' of $y = \csc(x)$.

Once again, our domain and range work in Section 10.3.1 is verified geometrically in the graph of $y = G(x) = \csc(x)$.

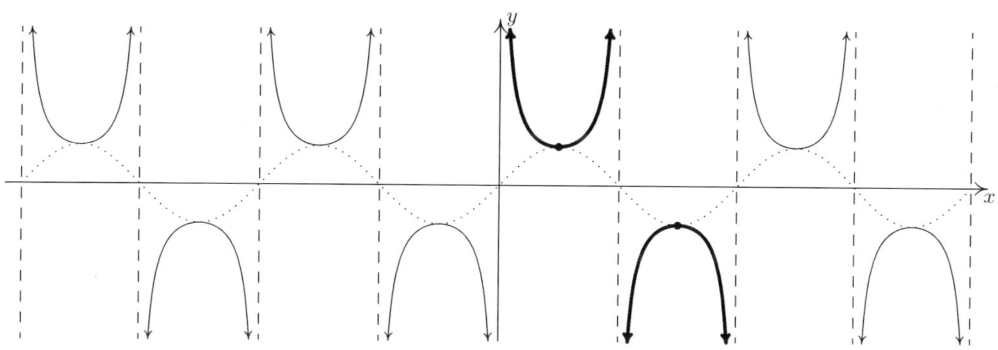

The graph of $y = \csc(x)$.

Note that, on the intervals between the vertical asymptotes, both $F(x) = \sec(x)$ and $G(x) = \csc(x)$ are continuous and smooth. In other words, they are continuous and smooth *on their domains*.[13] The following theorem summarizes the properties of the secant and cosecant functions. Note that

[13] Just like the rational functions in Chapter 4 are continuous and smooth on their domains because polynomials are continuous and smooth everywhere, the secant and cosecant functions are continuous and smooth on their domains since the cosine and sine functions are continuous and smooth everywhere.

all of these properties are direct results of them being reciprocals of the cosine and sine functions, respectively.

Theorem 10.24. Properties of the Secant and Cosecant Functions

- The function $F(x) = \sec(x)$
  - has domain $\left\{x : x \neq \frac{\pi}{2} + \pi k, \; k \text{ is an integer}\right\} = \displaystyle\bigcup_{k=-\infty}^{\infty} \left(\frac{(2k+1)\pi}{2}, \frac{(2k+3)\pi}{2}\right)$
  - has range $\{y : |y| \geq 1\} = (-\infty, -1] \cup [1, \infty)$
  - is continuous and smooth on its domain
  - is even
  - has period 2π
- The function $G(x) = \csc(x)$
  - has domain $\{x : x \neq \pi k, \; k \text{ is an integer}\} = \displaystyle\bigcup_{k=-\infty}^{\infty} (k\pi, (k+1)\pi)$
  - has range $\{y : |y| \geq 1\} = (-\infty, -1] \cup [1, \infty)$
  - is continuous and smooth on its domain
  - is odd
  - has period 2π

In the next example, we discuss graphing more general secant and cosecant curves.

Example 10.5.4. Graph one cycle of the following functions. State the period of each.

1. $f(x) = 1 - 2\sec(2x)$
2. $g(x) = \dfrac{\csc(\pi - \pi x) - 5}{3}$

Solution.

1. To graph $y = 1 - 2\sec(2x)$, we follow the same procedure as in Example 10.5.1. First, we set the argument of secant, $2x$, equal to the 'quarter marks' 0, $\frac{\pi}{2}$, π, $\frac{3\pi}{2}$ and 2π and solve for x.

a	$2x = a$	x
0	$2x = 0$	0
$\frac{\pi}{2}$	$2x = \frac{\pi}{2}$	$\frac{\pi}{4}$
π	$2x = \pi$	$\frac{\pi}{2}$
$\frac{3\pi}{2}$	$2x = \frac{3\pi}{2}$	$\frac{3\pi}{4}$
2π	$2x = 2\pi$	π

Next, we substitute these x values into $f(x)$. If $f(x)$ exists, we have a point on the graph; otherwise, we have found a vertical asymptote. In addition to these points and asymptotes, we have graphed the associated cosine curve – in this case $y = 1 - 2\cos(2x)$ – dotted in the picture below. Since one cycle is graphed over the interval $[0, \pi]$, the period is $\pi - 0 = \pi$.

x	$f(x)$	$(x, f(x))$
0	-1	$(0, -1)$
$\frac{\pi}{4}$	undefined	
$\frac{\pi}{2}$	3	$\left(\frac{\pi}{2}, 3\right)$
$\frac{3\pi}{4}$	undefined	
π	-1	$(\pi, -1)$

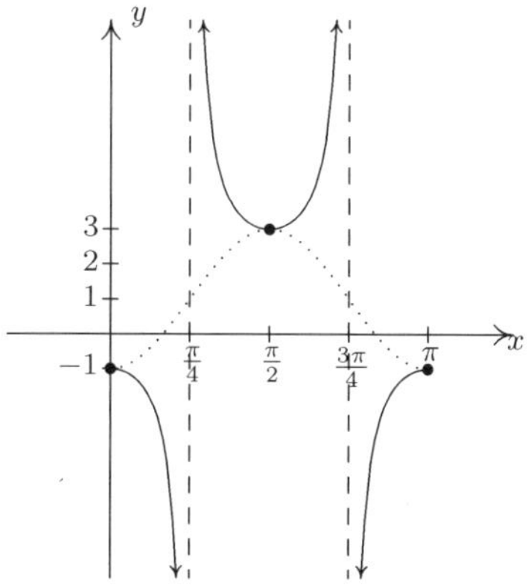

One cycle of $y = 1 - 2\sec(2x)$.

2. Proceeding as before, we set the argument of cosecant in $g(x) = \frac{\csc(\pi - \pi x) - 5}{3}$ equal to the quarter marks and solve for x.

a	$\pi - \pi x = a$	x
0	$\pi - \pi x = 0$	1
$\frac{\pi}{2}$	$\pi - \pi x = \frac{\pi}{2}$	$\frac{1}{2}$
π	$\pi - \pi x = \pi$	0
$\frac{3\pi}{2}$	$\pi - \pi x = \frac{3\pi}{2}$	$-\frac{1}{2}$
2π	$\pi - \pi x = 2\pi$	-1

Substituting these x-values into $g(x)$, we generate the graph below and find the period to be $1 - (-1) = 2$. The associated sine curve, $y = \frac{\sin(\pi - \pi x) - 5}{3}$, is dotted in as a reference.

x	$g(x)$	$(x, g(x))$
1	undefined	
$\frac{1}{2}$	$-\frac{4}{3}$	$\left(\frac{1}{2}, -\frac{4}{3}\right)$
0	undefined	
$-\frac{1}{2}$	-2	$\left(-\frac{1}{2}, -2\right)$
-1	undefined	

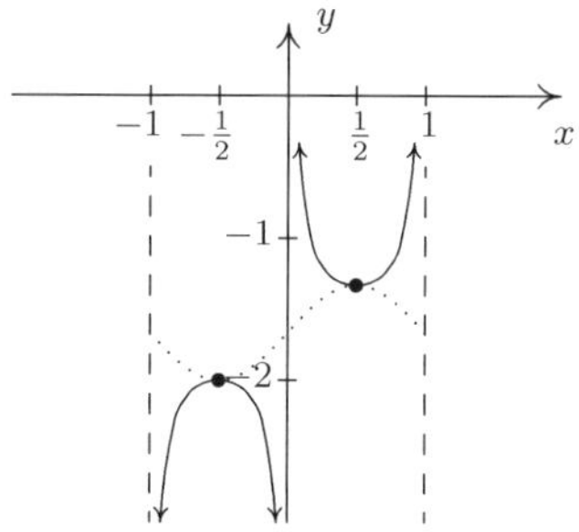

One cycle of $y = \frac{\csc(\pi - \pi x) - 5}{3}$. □

Before moving on, we note that it is possible to speak of the period, phase shift and vertical shift of secant and cosecant graphs and use even/odd identities to put them in a form similar to the sinusoid forms mentioned in Theorem 10.23. Since these quantities match those of the corresponding cosine and sine curves, we do not spell this out explicitly. Finally, since the ranges of secant and cosecant are unbounded, there is no amplitude associated with these curves.

10.5.3 Graphs of the Tangent and Cotangent Functions

Finally, we turn our attention to the graphs of the tangent and cotangent functions. When constructing a table of values for the tangent function, we see that $J(x) = \tan(x)$ is undefined at $x = \frac{\pi}{2}$ and $x = \frac{3\pi}{2}$, in accordance with our findings in Section 10.3.1. As $x \to \frac{\pi}{2}^-$, $\sin(x) \to 1^-$ and $\cos(x) \to 0^+$, so that $\tan(x) = \frac{\sin(x)}{\cos(x)} \to \infty$ producing a vertical asymptote at $x = \frac{\pi}{2}$. Using a similar analysis, we get that as $x \to \frac{\pi}{2}^+$, $\tan(x) \to -\infty$; as $x \to \frac{3\pi}{2}^-$, $\tan(x) \to \infty$; and as $x \to \frac{3\pi}{2}^+$, $\tan(x) \to -\infty$. Plotting this information and performing the usual 'copy and paste' produces:

x	$\tan(x)$	$(x, \tan(x))$
0	0	$(0,0)$
$\frac{\pi}{4}$	1	$\left(\frac{\pi}{4}, 1\right)$
$\frac{\pi}{2}$	undefined	
$\frac{3\pi}{4}$	-1	$\left(\frac{3\pi}{4}, -1\right)$
π	0	$(\pi, 0)$
$\frac{5\pi}{4}$	1	$\left(\frac{5\pi}{4}, 1\right)$
$\frac{3\pi}{2}$	undefined	
$\frac{7\pi}{4}$	-1	$\left(\frac{7\pi}{4}, -1\right)$
2π	0	$(2\pi, 0)$

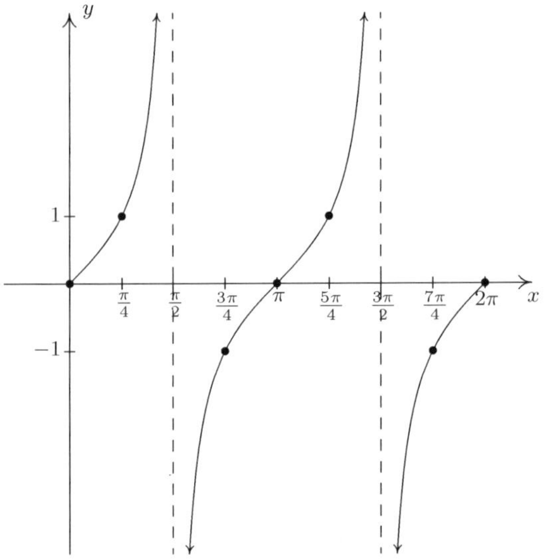

The graph of $y = \tan(x)$ over $[0, 2\pi]$.

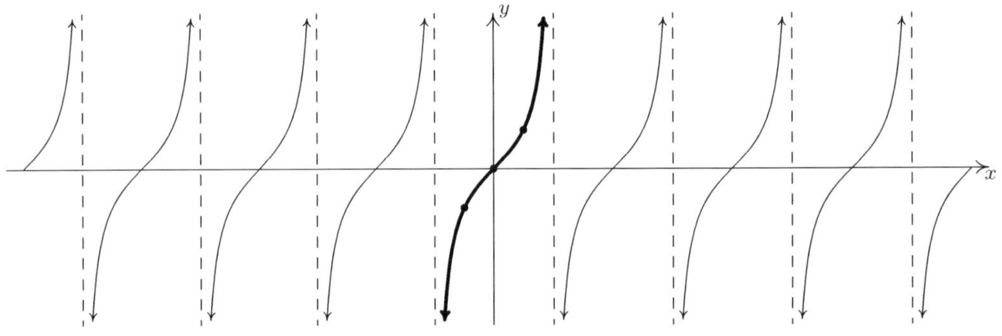

The graph of $y = \tan(x)$.

From the graph, it appears as if the tangent function is periodic with period π. To prove that this is the case, we appeal to the sum formula for tangents. We have:

$$\tan(x+\pi) = \frac{\tan(x)+\tan(\pi)}{1-\tan(x)\tan(\pi)} = \frac{\tan(x)+0}{1-(\tan(x))(0)} = \tan(x),$$

which tells us the period of $\tan(x)$ is at most π. To show that it is exactly π, suppose p is a positive real number so that $\tan(x+p) = \tan(x)$ for all real numbers x. For $x = 0$, we have $\tan(p) = \tan(0+p) = \tan(0) = 0$, which means p is a multiple of π. The smallest positive multiple of π is π itself, so we have established the result. We take as our fundamental cycle for $y = \tan(x)$ the interval $\left(-\frac{\pi}{2}, \frac{\pi}{2}\right)$, and use as our 'quarter marks' $x = -\frac{\pi}{2}$, $-\frac{\pi}{4}$, 0, $\frac{\pi}{4}$ and $\frac{\pi}{2}$. From the graph, we see confirmation of our domain and range work in Section 10.3.1.

It should be no surprise that $K(x) = \cot(x)$ behaves similarly to $J(x) = \tan(x)$. Plotting $\cot(x)$ over the interval $[0, 2\pi]$ results in the graph below.

x	$\cot(x)$	$(x, \cot(x))$
0	undefined	
$\frac{\pi}{4}$	1	$\left(\frac{\pi}{4}, 1\right)$
$\frac{\pi}{2}$	0	$\left(\frac{\pi}{2}, 0\right)$
$\frac{3\pi}{4}$	-1	$\left(\frac{3\pi}{4}, -1\right)$
π	undefined	
$\frac{5\pi}{4}$	1	$\left(\frac{5\pi}{4}, 1\right)$
$\frac{3\pi}{2}$	0	$\left(\frac{3\pi}{2}, 0\right)$
$\frac{7\pi}{4}$	-1	$\left(\frac{7\pi}{4}, -1\right)$
2π	undefined	

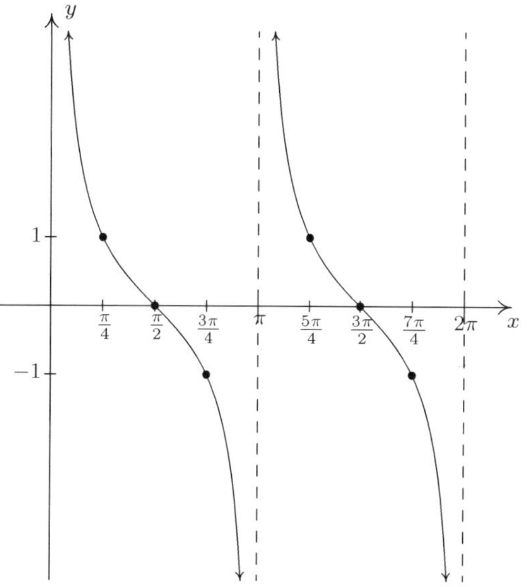

The graph of $y = \cot(x)$ over $[0, 2\pi]$.

From these data, it clearly appears as if the period of $\cot(x)$ is π, and we leave it to the reader to prove this.[14] We take as one fundamental cycle the interval $(0, \pi)$ with quarter marks: $x = 0$, $\frac{\pi}{4}$, $\frac{\pi}{2}$, $\frac{3\pi}{4}$ and π. A more complete graph of $y = \cot(x)$ is below, along with the fundamental cycle highlighted as usual. Once again, we see the domain and range of $K(x) = \cot(x)$ as read from the graph matches with what we found analytically in Section 10.3.1.

[14] Certainly, mimicking the proof that the period of $\tan(x)$ is an option; for another approach, consider transforming $\tan(x)$ to $\cot(x)$ using identities.

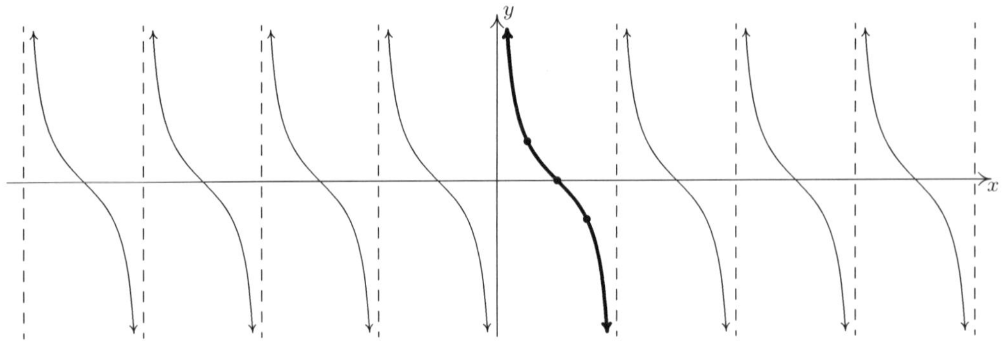

The graph of $y = \cot(x)$.

The properties of the tangent and cotangent functions are summarized below. As with Theorem 10.24, each of the results below can be traced back to properties of the cosine and sine functions and the definition of the tangent and cotangent functions as quotients thereof.

Theorem 10.25. Properties of the Tangent and Cotangent Functions

- The function $J(x) = \tan(x)$
  - has domain $\left\{x : x \neq \frac{\pi}{2} + \pi k,\ k \text{ is an integer}\right\} = \displaystyle\bigcup_{k=-\infty}^{\infty} \left(\frac{(2k+1)\pi}{2}, \frac{(2k+3)\pi}{2}\right)$
  - has range $(-\infty, \infty)$
  - is continuous and smooth on its domain
  - is odd
  - has period π
- The function $K(x) = \cot(x)$
  - has domain $\{x : x \neq \pi k,\ k \text{ is an integer}\} = \displaystyle\bigcup_{k=-\infty}^{\infty} (k\pi, (k+1)\pi)$
  - has range $(-\infty, \infty)$
  - is continuous and smooth on its domain
  - is odd
  - has period π

Example 10.5.5. Graph one cycle of the following functions. Find the period.

1. $f(x) = 1 - \tan\left(\frac{x}{2}\right)$.
2. $g(x) = 2\cot\left(\frac{\pi}{2}x + \pi\right) + 1$.

Solution.

1. We proceed as we have in all of the previous graphing examples by setting the argument of tangent in $f(x) = 1 - \tan\left(\frac{x}{2}\right)$, namely $\frac{x}{2}$, equal to each of the 'quarter marks' $-\frac{\pi}{2}$, $-\frac{\pi}{4}$, 0, $\frac{\pi}{4}$ and $\frac{\pi}{2}$, and solving for x.

a	$\frac{x}{2} = a$	x
$-\frac{\pi}{2}$	$\frac{x}{2} = -\frac{\pi}{2}$	$-\pi$
$-\frac{\pi}{4}$	$\frac{x}{2} = -\frac{\pi}{4}$	$-\frac{\pi}{2}$
0	$\frac{x}{2} = 0$	0
$\frac{\pi}{4}$	$\frac{x}{2} = \frac{\pi}{4}$	$\frac{\pi}{2}$
$\frac{\pi}{2}$	$\frac{x}{2} = \frac{\pi}{2}$	π

Substituting these x-values into $f(x)$, we find points on the graph and the vertical asymptotes.

x	$f(x)$	$(x, f(x))$
$-\pi$	undefined	
$-\frac{\pi}{2}$	2	$\left(-\frac{\pi}{2}, 2\right)$
0	1	$(0, 1)$
$\frac{\pi}{2}$	0	$\left(\frac{\pi}{2}, 0\right)$
π	undefined	

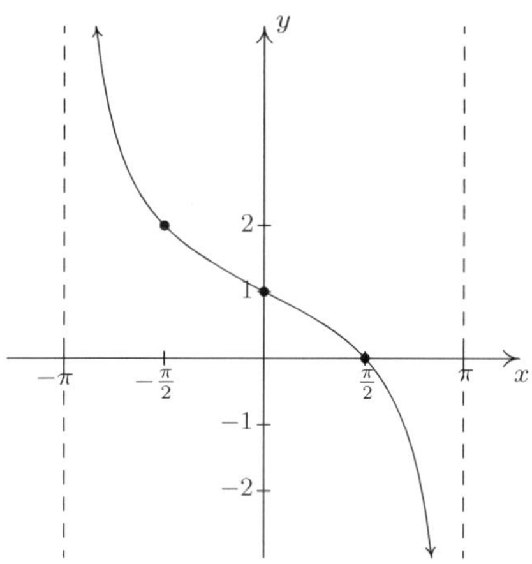

One cycle of $y = 1 - \tan\left(\frac{x}{2}\right)$.

We see that the period is $\pi - (-\pi) = 2\pi$.

2. The 'quarter marks' for the fundamental cycle of the cotangent curve are 0, $\frac{\pi}{4}$, $\frac{\pi}{2}$, $\frac{3\pi}{4}$ and π. To graph $g(x) = 2\cot\left(\frac{\pi}{2}x + \pi\right) + 1$, we begin by setting $\frac{\pi}{2}x + \pi$ equal to each quarter mark and solving for x.

a	$\frac{\pi}{2}x + \pi = a$	x
0	$\frac{\pi}{2}x + \pi = 0$	-2
$\frac{\pi}{4}$	$\frac{\pi}{2}x + \pi = \frac{\pi}{4}$	$-\frac{3}{2}$
$\frac{\pi}{2}$	$\frac{\pi}{2}x + \pi = \frac{\pi}{2}$	-1
$\frac{3\pi}{4}$	$\frac{\pi}{2}x + \pi = \frac{3\pi}{4}$	$-\frac{1}{2}$
π	$\frac{\pi}{2}x + \pi = \pi$	0

We now use these x-values to generate our graph.

x	$g(x)$	$(x, g(x))$
-2	undefined	
$-\frac{3}{2}$	3	$\left(-\frac{3}{2}, 3\right)$
-1	1	$(-1, 1)$
$-\frac{1}{2}$	-1	$\left(-\frac{1}{2}, -1\right)$
0	undefined	

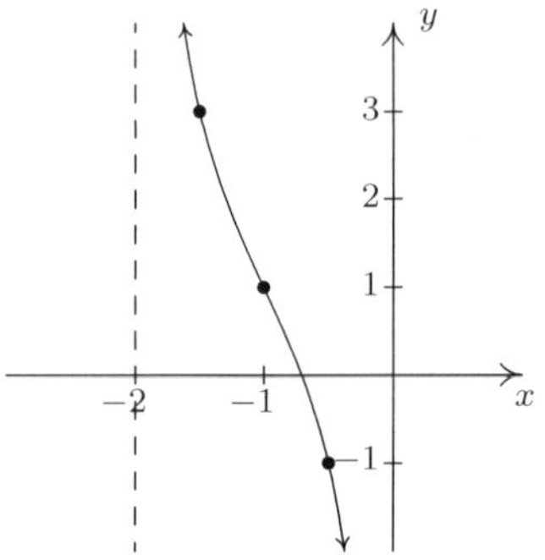

One cycle of $y = 2\cot\left(\frac{\pi}{2}x + \pi\right) + 1$.

We find the period to be $0 - (-2) = 2$. □

As with the secant and cosecant functions, it is possible to extend the notion of period, phase shift and vertical shift to the tangent and cotangent functions as we did for the cosine and sine functions in Theorem 10.23. Since the number of classical applications involving sinusoids far outnumber those involving tangent and cotangent functions, we omit this. The ambitious reader is invited to formulate such a theorem, however.

10.5.4 Exercises

In Exercises 1 - 12, graph one cycle of the given function. State the period, amplitude, phase shift and vertical shift of the function.

1. $y = 3\sin(x)$
2. $y = \sin(3x)$
3. $y = -2\cos(x)$
4. $y = \cos\left(x - \frac{\pi}{2}\right)$
5. $y = -\sin\left(x + \frac{\pi}{3}\right)$
6. $y = \sin(2x - \pi)$
7. $y = -\frac{1}{3}\cos\left(\frac{1}{2}x + \frac{\pi}{3}\right)$
8. $y = \cos(3x - 2\pi) + 4$
9. $y = \sin\left(-x - \frac{\pi}{4}\right) - 2$
10. $y = \frac{2}{3}\cos\left(\frac{\pi}{2} - 4x\right) + 1$
11. $y = -\frac{3}{2}\cos\left(2x + \frac{\pi}{3}\right) - \frac{1}{2}$
12. $y = 4\sin(-2\pi x + \pi)$

In Exercises 13 - 24, graph one cycle of the given function. State the period of the function.

13. $y = \tan\left(x - \frac{\pi}{3}\right)$
14. $y = 2\tan\left(\frac{1}{4}x\right) - 3$
15. $y = \frac{1}{3}\tan(-2x - \pi) + 1$
16. $y = \sec\left(x - \frac{\pi}{2}\right)$
17. $y = -\csc\left(x + \frac{\pi}{3}\right)$
18. $y = -\frac{1}{3}\sec\left(\frac{1}{2}x + \frac{\pi}{3}\right)$
19. $y = \csc(2x - \pi)$
20. $y = \sec(3x - 2\pi) + 4$
21. $y = \csc\left(-x - \frac{\pi}{4}\right) - 2$
22. $y = \cot\left(x + \frac{\pi}{6}\right)$
23. $y = -11\cot\left(\frac{1}{5}x\right)$
24. $y = \frac{1}{3}\cot\left(2x + \frac{3\pi}{2}\right) + 1$

In Exercises 25 - 34, use Example 10.5.3 as a guide to show that the function is a sinusoid by rewriting it in the forms $C(x) = A\cos(\omega x + \phi) + B$ and $S(x) = A\sin(\omega x + \phi) + B$ for $\omega > 0$ and $0 \leq \phi < 2\pi$.

25. $f(x) = \sqrt{2}\sin(x) + \sqrt{2}\cos(x) + 1$
26. $f(x) = 3\sqrt{3}\sin(3x) - 3\cos(3x)$
27. $f(x) = -\sin(x) + \cos(x) - 2$
28. $f(x) = -\frac{1}{2}\sin(2x) - \frac{\sqrt{3}}{2}\cos(2x)$
29. $f(x) = 2\sqrt{3}\cos(x) - 2\sin(x)$
30. $f(x) = \frac{3}{2}\cos(2x) - \frac{3\sqrt{3}}{2}\sin(2x) + 6$
31. $f(x) = -\frac{1}{2}\cos(5x) - \frac{\sqrt{3}}{2}\sin(5x)$
32. $f(x) = -6\sqrt{3}\cos(3x) - 6\sin(3x) - 3$

33. $f(x) = \frac{5\sqrt{2}}{2}\sin(x) - \frac{5\sqrt{2}}{2}\cos(x)$

34. $f(x) = 3\sin\left(\frac{x}{6}\right) - 3\sqrt{3}\cos\left(\frac{x}{6}\right)$

35. In Exercises 25 - 34, you should have noticed a relationship between the phases ϕ for the $S(x)$ and $C(x)$. Show that if $f(x) = A\sin(\omega x + \alpha) + B$, then $f(x) = A\cos(\omega x + \beta) + B$ where $\beta = \alpha - \frac{\pi}{2}$.

36. Let ϕ be an angle measured in radians and let $P(a, b)$ be a point on the terminal side of ϕ when it is drawn in standard position. Use Theorem 10.3 and the sum identity for sine in Theorem 10.15 to show that $f(x) = a\sin(\omega x) + b\cos(\omega x) + B$ (with $\omega > 0$) can be rewritten as $f(x) = \sqrt{a^2 + b^2}\sin(\omega x + \phi) + B$.

37. With the help of your classmates, express the domains of the functions in Examples 10.5.4 and 10.5.5 using extended interval notation. (We will revisit this in Section 10.7.)

In Exercises 38 - 43, verify the identity by graphing the right and left hand sides on a calculator.

38. $\sin^2(x) + \cos^2(x) = 1$

39. $\sec^2(x) - \tan^2(x) = 1$

40. $\cos(x) = \sin\left(\frac{\pi}{2} - x\right)$

41. $\tan(x + \pi) = \tan(x)$

42. $\sin(2x) = 2\sin(x)\cos(x)$

43. $\tan\left(\frac{x}{2}\right) = \frac{\sin(x)}{1 + \cos(x)}$

In Exercises 44 - 50, graph the function with the help of your calculator and discuss the given questions with your classmates.

44. $f(x) = \cos(3x) + \sin(x)$. Is this function periodic? If so, what is the period?

45. $f(x) = \frac{\sin(x)}{x}$. What appears to be the horizontal asymptote of the graph?

46. $f(x) = x\sin(x)$. Graph $y = \pm x$ on the same set of axes and describe the behavior of f.

47. $f(x) = \sin\left(\frac{1}{x}\right)$. What's happening as $x \to 0$?

48. $f(x) = x - \tan(x)$. Graph $y = x$ on the same set of axes and describe the behavior of f.

49. $f(x) = e^{-0.1x}(\cos(2x) + \sin(2x))$. Graph $y = \pm e^{-0.1x}$ on the same set of axes and describe the behavior of f.

50. $f(x) = e^{-0.1x}(\cos(2x) + 2\sin(x))$. Graph $y = \pm e^{-0.1x}$ on the same set of axes and describe the behavior of f.

51. Show that a constant function f is periodic by showing that $f(x + 117) = f(x)$ for all real numbers x. Then show that f has no period by showing that you cannot find a *smallest* number p such that $f(x + p) = f(x)$ for all real numbers x. Said another way, show that $f(x + p) = f(x)$ for all real numbers x for ALL values of $p > 0$, so no smallest value exists to satisfy the definition of 'period'.

10.5.5 Answers

1. $y = 3\sin(x)$
   Period: 2π
   Amplitude: 3
   Phase Shift: 0
   Vertical Shift: 0

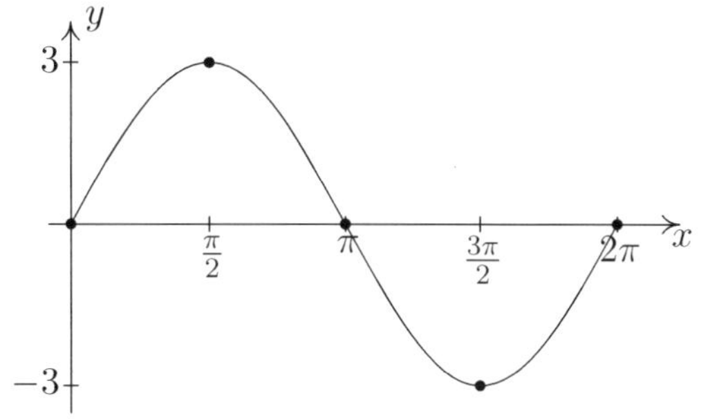

2. $y = \sin(3x)$
   Period: $\dfrac{2\pi}{3}$
   Amplitude: 1
   Phase Shift: 0
   Vertical Shift: 0

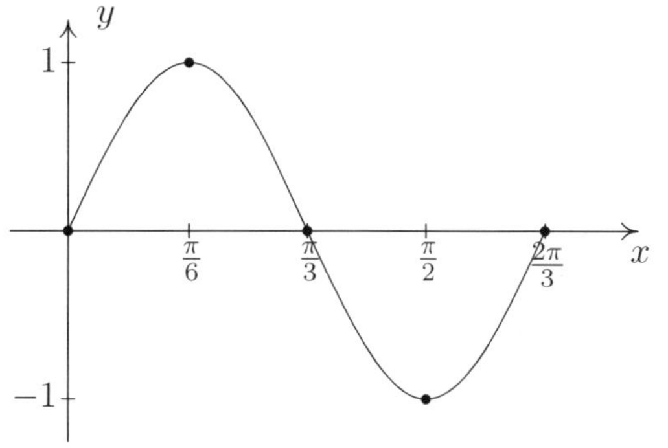

3. $y = -2\cos(x)$
   Period: 2π
   Amplitude: 2
   Phase Shift: 0
   Vertical Shift: 0

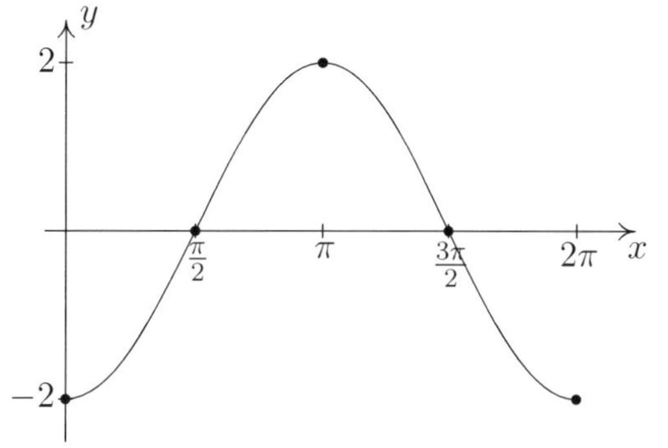

4. $y = \cos\left(x - \dfrac{\pi}{2}\right)$
   Period: 2π
   Amplitude: 1
   Phase Shift: $\dfrac{\pi}{2}$
   Vertical Shift: 0

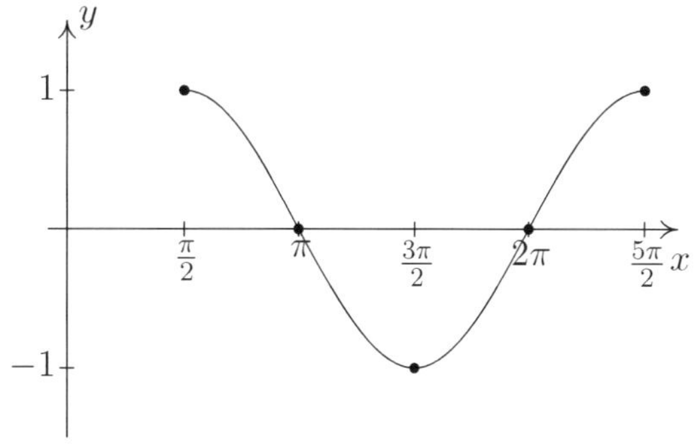

5. $y = -\sin\left(x + \frac{\pi}{3}\right)$
Period: 2π
Amplitude: 1
Phase Shift: $-\frac{\pi}{3}$
Vertical Shift: 0

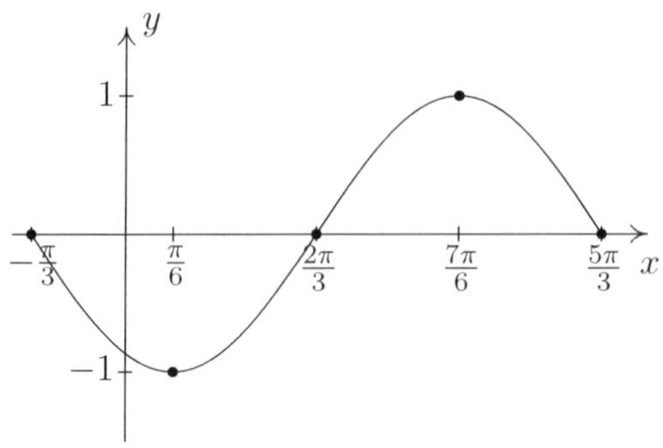

6. $y = \sin(2x - \pi)$
Period: π
Amplitude: 1
Phase Shift: $\frac{\pi}{2}$
Vertical Shift: 0

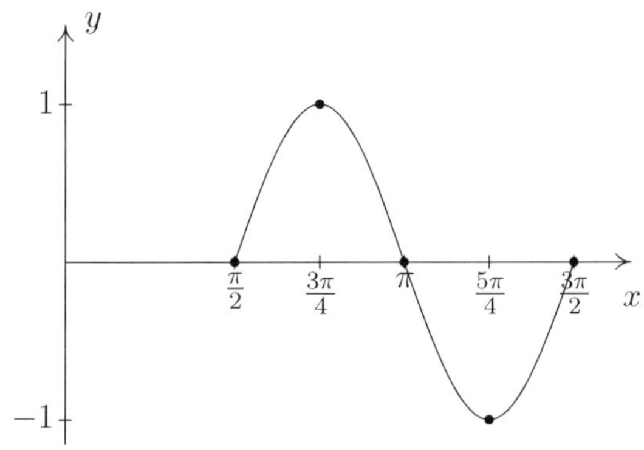

7. $y = -\frac{1}{3}\cos\left(\frac{1}{2}x + \frac{\pi}{3}\right)$
Period: 4π
Amplitude: $\frac{1}{3}$
Phase Shift: $-\frac{2\pi}{3}$
Vertical Shift: 0

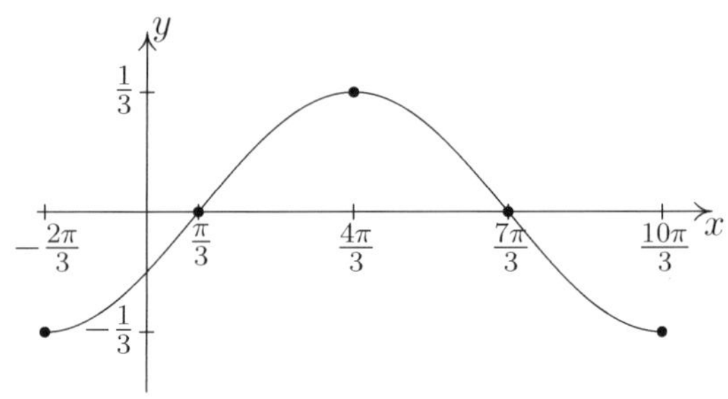

8. $y = \cos(3x - 2\pi) + 4$
Period: $\frac{2\pi}{3}$
Amplitude: 1
Phase Shift: $\frac{2\pi}{3}$
Vertical Shift: 4

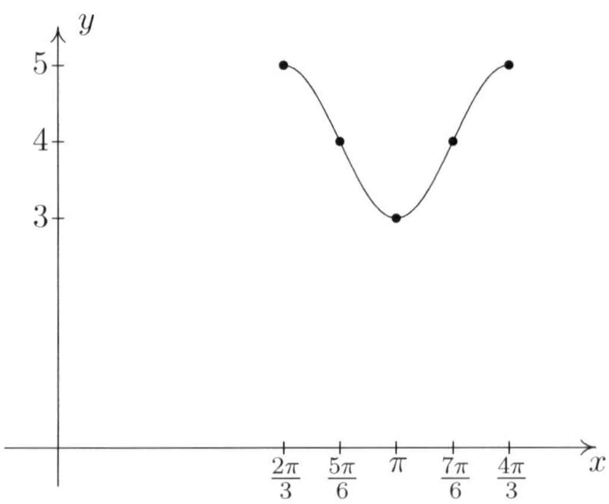

9. $y = \sin\left(-x - \frac{\pi}{4}\right) - 2$
Period: 2π
Amplitude: 1
Phase Shift: $-\frac{\pi}{4}$ (You need to use $y = -\sin\left(x + \frac{\pi}{4}\right) - 2$ to find this.)[15]
Vertical Shift: -2

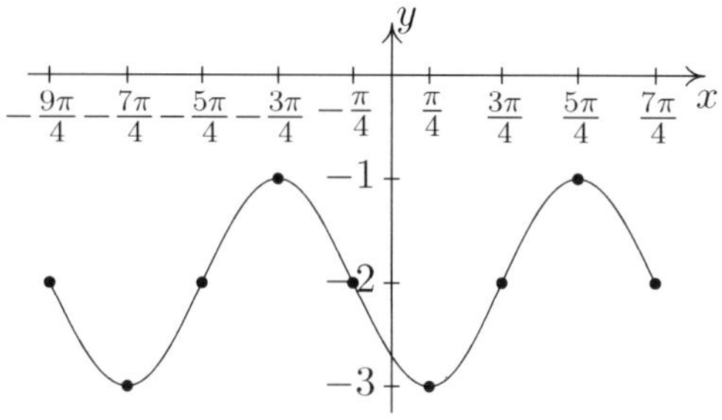

10. $y = \frac{2}{3}\cos\left(\frac{\pi}{2} - 4x\right) + 1$
Period: $\frac{\pi}{2}$
Amplitude: $\frac{2}{3}$
Phase Shift: $\frac{\pi}{8}$ (You need to use $y = \frac{2}{3}\cos\left(4x - \frac{\pi}{2}\right) + 1$ to find this.)[16]
Vertical Shift: 1

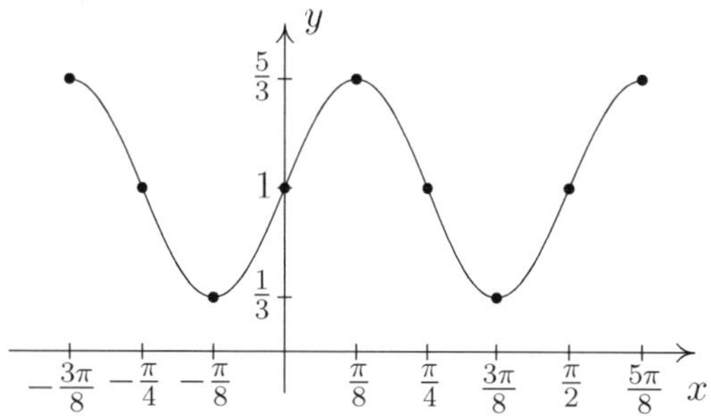

11. $y = -\frac{3}{2}\cos\left(2x + \frac{\pi}{3}\right) - \frac{1}{2}$
Period: π
Amplitude: $\frac{3}{2}$
Phase Shift: $-\frac{\pi}{6}$
Vertical Shift: $-\frac{1}{2}$

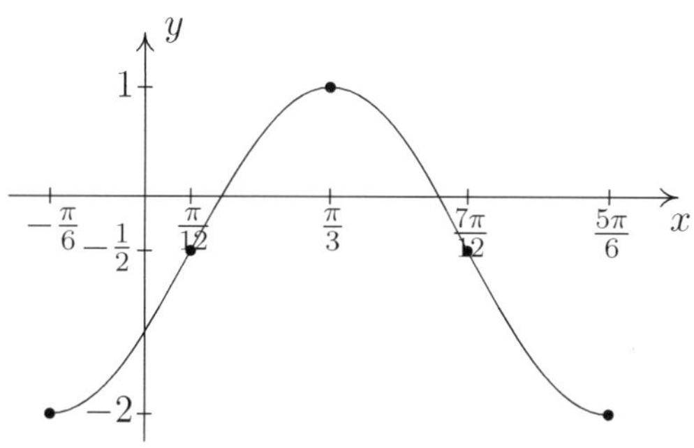

12. $y = 4\sin(-2\pi x + \pi)$
Period: 1
Amplitude: 4
Phase Shift: $\frac{1}{2}$ (You need to use $y = -4\sin(2\pi x - \pi)$ to find this.)[17]
Vertical Shift: 0

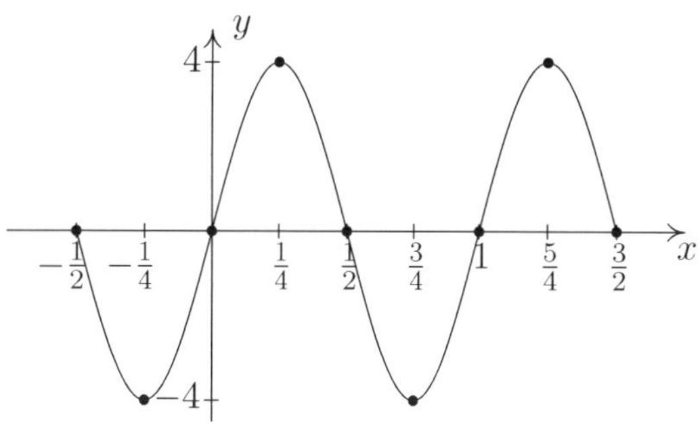

[15]Two cycles of the graph are shown to illustrate the discrepancy discussed on page 796.
[16]Again, we graph two cycles to illustrate the discrepancy discussed on page 796.
[17]This will be the last time we graph two cycles to illustrate the discrepancy discussed on page 796.

13. $y = \tan\left(x - \frac{\pi}{3}\right)$
Period: π

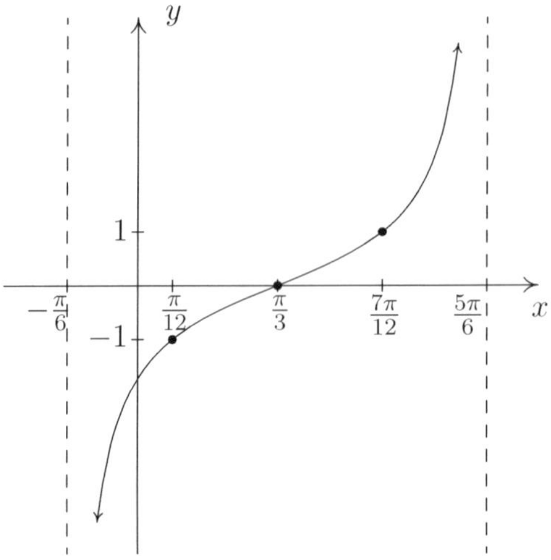

14. $y = 2\tan\left(\frac{1}{4}x\right) - 3$
Period: 4π

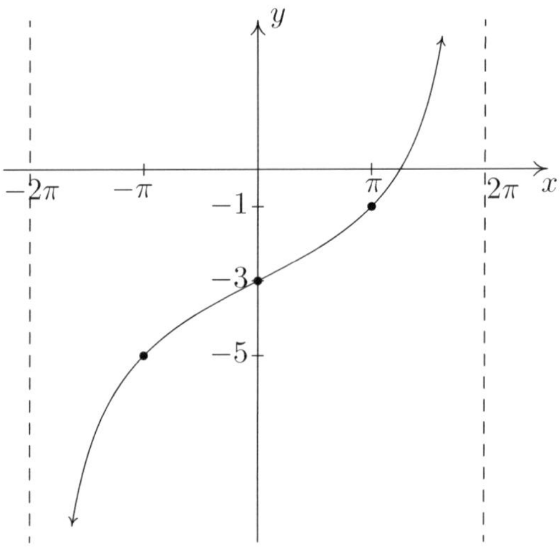

15. $y = \frac{1}{3}\tan(-2x - \pi) + 1$
is equivalent to
$y = -\frac{1}{3}\tan(2x + \pi) + 1$
via the Even / Odd identity for tangent.
Period: $\frac{\pi}{2}$

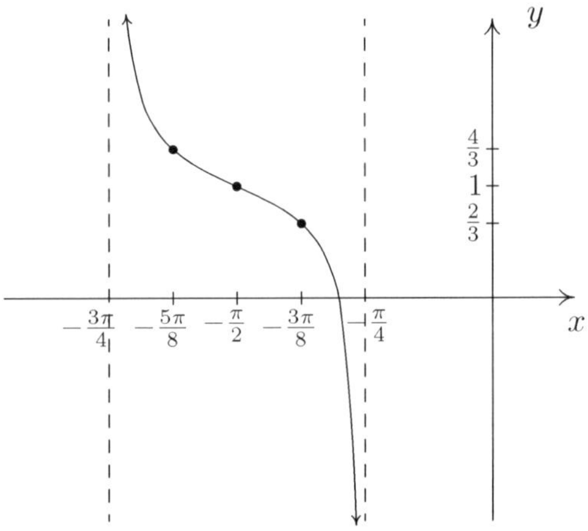

16. $y = \sec\left(x - \frac{\pi}{2}\right)$
Start with $y = \cos\left(x - \frac{\pi}{2}\right)$
Period: 2π

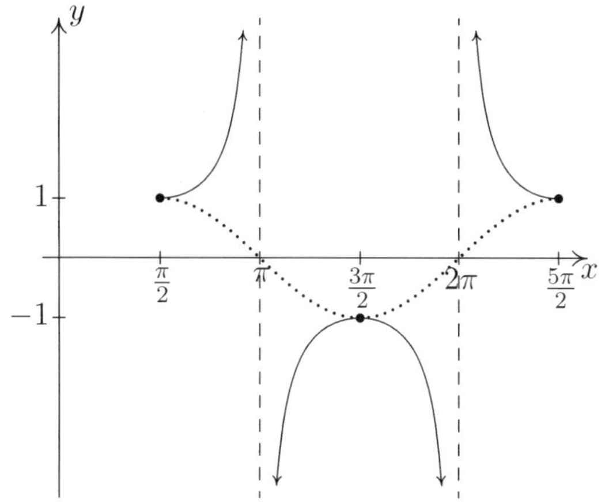

17. $y = -\csc\left(x + \dfrac{\pi}{3}\right)$
Start with $y = -\sin\left(x + \dfrac{\pi}{3}\right)$
Period: 2π

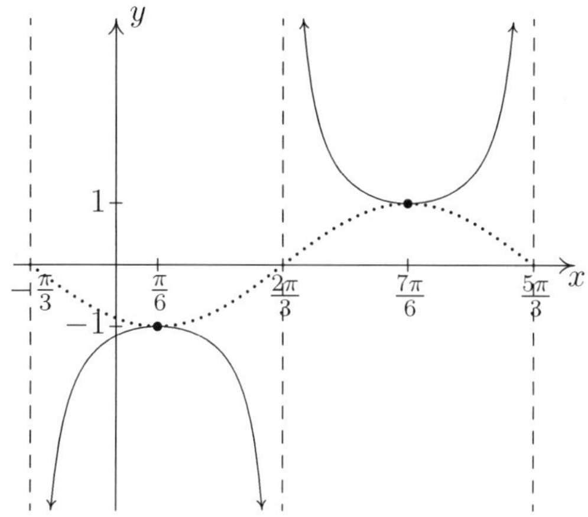

18. $y = -\dfrac{1}{3}\sec\left(\dfrac{1}{2}x + \dfrac{\pi}{3}\right)$
Start with $y = -\dfrac{1}{3}\cos\left(\dfrac{1}{2}x + \dfrac{\pi}{3}\right)$
Period: 4π

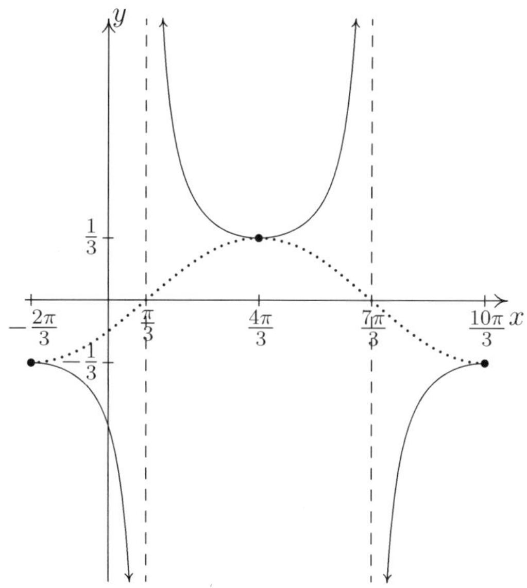

19. $y = \csc(2x - \pi)$
Start with $y = \sin(2x - \pi)$
Period: π

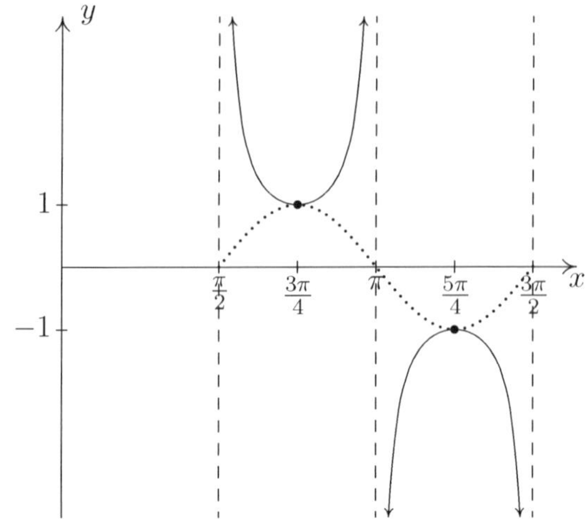

20. $y = \sec(3x - 2\pi) + 4$
Start with $y = \cos(3x - 2\pi) + 4$
Period: $\dfrac{2\pi}{3}$

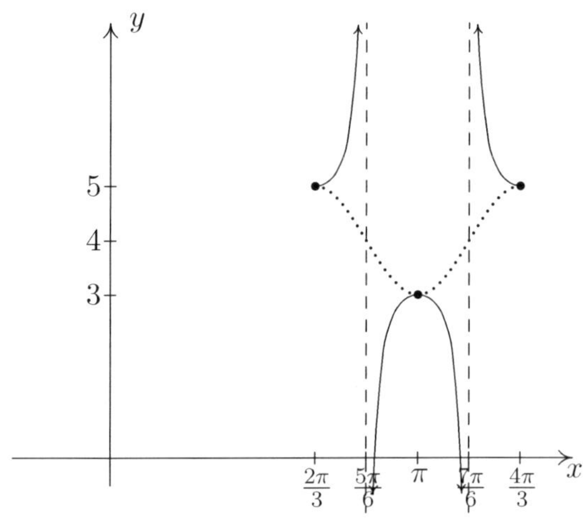

21. $y = \csc\left(-x - \dfrac{\pi}{4}\right) - 2$
Start with $y = \sin\left(-x - \dfrac{\pi}{4}\right) - 2$
Period: 2π

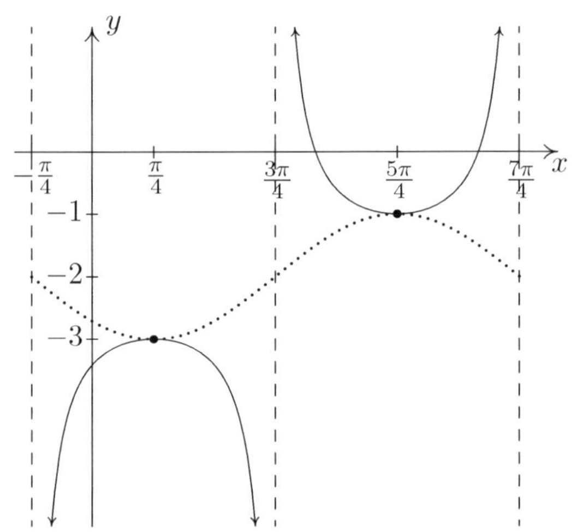

22. $y = \cot\left(x + \frac{\pi}{6}\right)$
Period: π

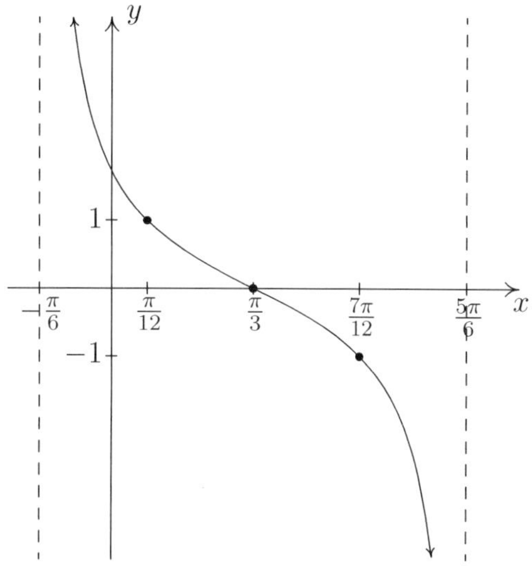

23. $y = -11\cot\left(\frac{1}{5}x\right)$
Period: 5π

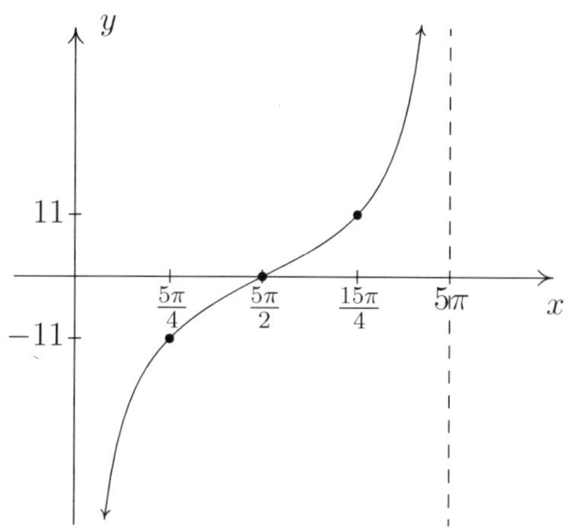

24. $y = \frac{1}{3}\cot\left(2x + \frac{3\pi}{2}\right) + 1$
Period: $\frac{\pi}{2}$

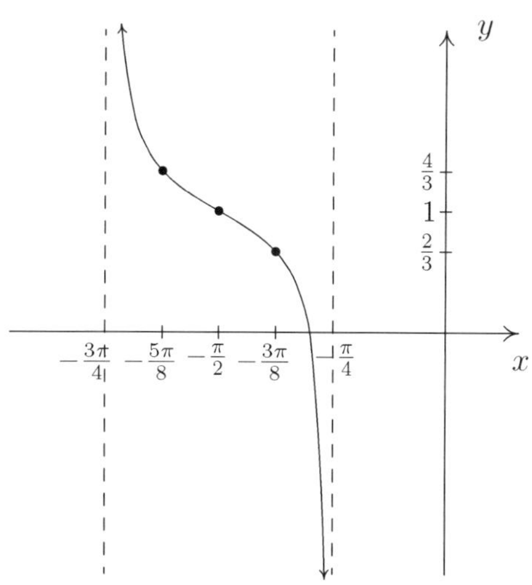

25. $f(x) = \sqrt{2}\sin(x) + \sqrt{2}\cos(x) + 1 = 2\sin\left(x + \frac{\pi}{4}\right) + 1 = 2\cos\left(x + \frac{7\pi}{4}\right) + 1$

26. $f(x) = 3\sqrt{3}\sin(3x) - 3\cos(3x) = 6\sin\left(3x + \frac{11\pi}{6}\right) = 6\cos\left(3x + \frac{4\pi}{3}\right)$

27. $f(x) = -\sin(x) + \cos(x) - 2 = \sqrt{2}\sin\left(x + \frac{3\pi}{4}\right) - 2 = \sqrt{2}\cos\left(x + \frac{\pi}{4}\right) - 2$

28. $f(x) = -\frac{1}{2}\sin(2x) - \frac{\sqrt{3}}{2}\cos(2x) = \sin\left(2x + \frac{4\pi}{3}\right) = \cos\left(2x + \frac{5\pi}{6}\right)$

29. $f(x) = 2\sqrt{3}\cos(x) - 2\sin(x) = 4\sin\left(x + \frac{2\pi}{3}\right) = 4\cos\left(x + \frac{\pi}{6}\right)$

30. $f(x) = \frac{3}{2}\cos(2x) - \frac{3\sqrt{3}}{2}\sin(2x) + 6 = 3\sin\left(2x + \frac{5\pi}{6}\right) + 6 = 3\cos\left(2x + \frac{\pi}{3}\right) + 6$

31. $f(x) = -\frac{1}{2}\cos(5x) - \frac{\sqrt{3}}{2}\sin(5x) = \sin\left(5x + \frac{7\pi}{6}\right) = \cos\left(5x + \frac{2\pi}{3}\right)$

32. $f(x) = -6\sqrt{3}\cos(3x) - 6\sin(3x) - 3 = 12\sin\left(3x + \frac{4\pi}{3}\right) - 3 = 12\cos\left(3x + \frac{5\pi}{6}\right) - 3$

33. $f(x) = \frac{5\sqrt{2}}{2}\sin(x) - \frac{5\sqrt{2}}{2}\cos(x) = 5\sin\left(x + \frac{7\pi}{4}\right) = 5\cos\left(x + \frac{5\pi}{4}\right)$

34. $f(x) = 3\sin\left(\frac{x}{6}\right) - 3\sqrt{3}\cos\left(\frac{x}{6}\right) = 6\sin\left(\frac{x}{6} + \frac{5\pi}{3}\right) = 6\cos\left(\frac{x}{6} + \frac{7\pi}{6}\right)$

10.6 The Inverse Trigonometric Functions

As the title indicates, in this section we concern ourselves with finding inverses of the (circular) trigonometric functions. Our immediate problem is that, owing to their periodic nature, none of the six circular functions is one-to-one. To remedy this, we restrict the domains of the circular functions in the same way we restricted the domain of the quadratic function in Example 5.2.3 in Section 5.2 to obtain a one-to-one function. We first consider $f(x) = \cos(x)$. Choosing the interval $[0, \pi]$ allows us to keep the range as $[-1, 1]$ as well as the properties of being smooth and continuous.

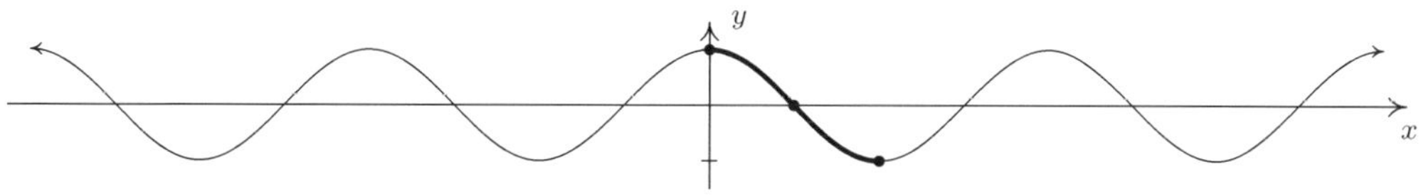

Restricting the domain of $f(x) = \cos(x)$ to $[0, \pi]$.

Recall from Section 5.2 that the inverse of a function f is typically denoted f^{-1}. For this reason, some textbooks use the notation $f^{-1}(x) = \cos^{-1}(x)$ for the inverse of $f(x) = \cos(x)$. The obvious pitfall here is our convention of writing $(\cos(x))^2$ as $\cos^2(x)$, $(\cos(x))^3$ as $\cos^3(x)$ and so on. It is far too easy to confuse $\cos^{-1}(x)$ with $\frac{1}{\cos(x)} = \sec(x)$ so we will not use this notation in our text.[1] Instead, we use the notation $f^{-1}(x) = \arccos(x)$, read 'arc-cosine of x'. To understand the 'arc' in 'arccosine', recall that an inverse function, by definition, reverses the process of the original function. The function $f(t) = \cos(t)$ takes a real number input t, associates it with the angle $\theta = t$ radians, and returns the value $\cos(\theta)$. Digging deeper,[2] we have that $\cos(\theta) = \cos(t)$ is the x-coordinate of the terminal point on the Unit Circle of an oriented arc of length $|t|$ whose initial point is $(1, 0)$. Hence, we may view the inputs to $f(t) = \cos(t)$ as oriented arcs and the outputs as x-coordinates on the Unit Circle. The function f^{-1}, then, would take x-coordinates on the Unit Circle and return oriented arcs, hence the 'arc' in arccosine. Below are the graphs of $f(x) = \cos(x)$ and $f^{-1}(x) = \arccos(x)$, where we obtain the latter from the former by reflecting it across the line $y = x$, in accordance with Theorem 5.3.

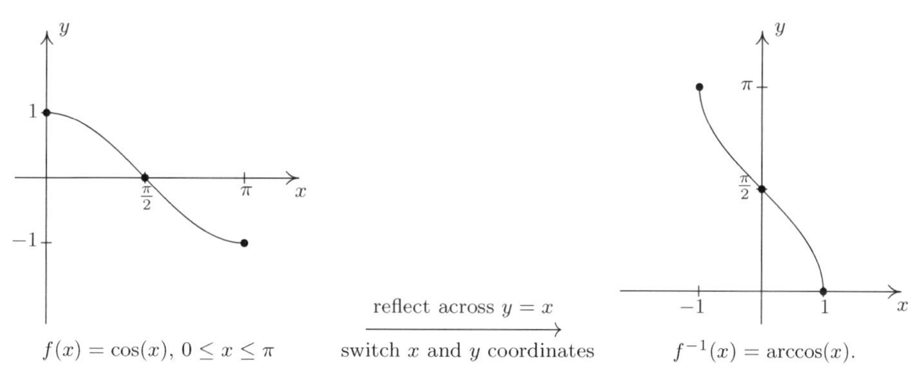

[1] But be aware that many books do! As always, be sure to check the context!

[2] See page 704 if you need a review of how we associate real numbers with angles in radian measure.

We restrict $g(x) = \sin(x)$ in a similar manner, although the interval of choice is $\left[-\frac{\pi}{2}, \frac{\pi}{2}\right]$.

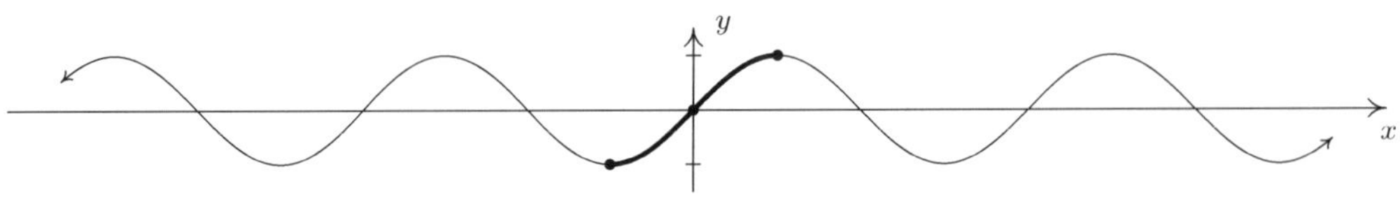

Restricting the domain of $f(x) = \sin(x)$ to $\left[-\frac{\pi}{2}, \frac{\pi}{2}\right]$.

It should be no surprise that we call $g^{-1}(x) = \arcsin(x)$, which is read 'arc-sine of x'.

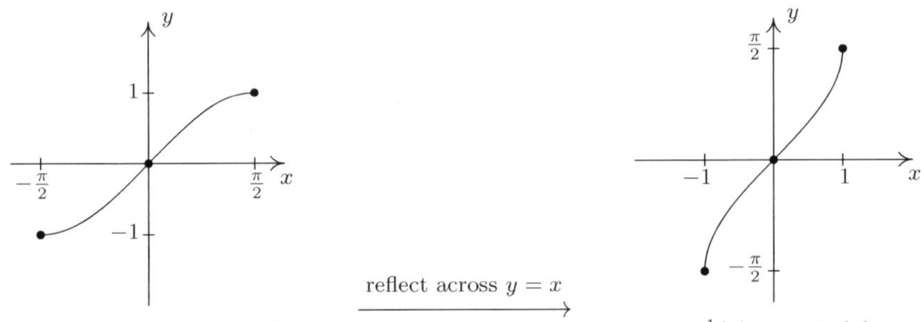

$g(x) = \sin(x)$, $-\frac{\pi}{2} \leq x \leq \frac{\pi}{2}$. switch x and y coordinates $g^{-1}(x) = \arcsin(x)$.

We list some important facts about the arccosine and arcsine functions in the following theorem.

Theorem 10.26. Properties of the Arccosine and Arcsine Functions

- Properties of $F(x) = \arccos(x)$
  - Domain: $[-1, 1]$
  - Range: $[0, \pi]$
  - $\arccos(x) = t$ if and only if $0 \leq t \leq \pi$ and $\cos(t) = x$
  - $\cos(\arccos(x)) = x$ provided $-1 \leq x \leq 1$
  - $\arccos(\cos(x)) = x$ provided $0 \leq x \leq \pi$
- Properties of $G(x) = \arcsin(x)$
  - Domain: $[-1, 1]$
  - Range: $\left[-\frac{\pi}{2}, \frac{\pi}{2}\right]$
  - $\arcsin(x) = t$ if and only if $-\frac{\pi}{2} \leq t \leq \frac{\pi}{2}$ and $\sin(t) = x$
  - $\sin(\arcsin(x)) = x$ provided $-1 \leq x \leq 1$
  - $\arcsin(\sin(x)) = x$ provided $-\frac{\pi}{2} \leq x \leq \frac{\pi}{2}$
  - additionally, arcsine is odd

Everything in Theorem 10.26 is a direct consequence of the facts that $f(x) = \cos(x)$ for $0 \leq x \leq \pi$ and $F(x) = \arccos(x)$ are inverses of each other as are $g(x) = \sin(x)$ for $-\frac{\pi}{2} \leq x \leq \frac{\pi}{2}$ and $G(x) = \arcsin(x)$. It's about time for an example.

Example 10.6.1.

1. Find the exact values of the following.

   (a) $\arccos\left(\frac{1}{2}\right)$

   (b) $\arcsin\left(\frac{\sqrt{2}}{2}\right)$

   (c) $\arccos\left(-\frac{\sqrt{2}}{2}\right)$

   (d) $\arcsin\left(-\frac{1}{2}\right)$

   (e) $\arccos\left(\cos\left(\frac{\pi}{6}\right)\right)$

   (f) $\arccos\left(\cos\left(\frac{11\pi}{6}\right)\right)$

   (g) $\cos\left(\arccos\left(-\frac{3}{5}\right)\right)$

   (h) $\sin\left(\arccos\left(-\frac{3}{5}\right)\right)$

2. Rewrite the following as algebraic expressions of x and state the domain on which the equivalence is valid.

   (a) $\tan(\arccos(x))$

   (b) $\cos(2\arcsin(x))$

Solution.

1. (a) To find $\arccos\left(\frac{1}{2}\right)$, we need to find the real number t (or, equivalently, an angle measuring t radians) which lies between 0 and π with $\cos(t) = \frac{1}{2}$. We know $t = \frac{\pi}{3}$ meets these criteria, so $\arccos\left(\frac{1}{2}\right) = \frac{\pi}{3}$.

   (b) The value of $\arcsin\left(\frac{\sqrt{2}}{2}\right)$ is a real number t between $-\frac{\pi}{2}$ and $\frac{\pi}{2}$ with $\sin(t) = \frac{\sqrt{2}}{2}$. The number we seek is $t = \frac{\pi}{4}$. Hence, $\arcsin\left(\frac{\sqrt{2}}{2}\right) = \frac{\pi}{4}$.

   (c) The number $t = \arccos\left(-\frac{\sqrt{2}}{2}\right)$ lies in the interval $[0, \pi]$ with $\cos(t) = -\frac{\sqrt{2}}{2}$. Our answer is $\arccos\left(-\frac{\sqrt{2}}{2}\right) = \frac{3\pi}{4}$.

   (d) To find $\arcsin\left(-\frac{1}{2}\right)$, we seek the number t in the interval $\left[-\frac{\pi}{2}, \frac{\pi}{2}\right]$ with $\sin(t) = -\frac{1}{2}$. The answer is $t = -\frac{\pi}{6}$ so that $\arcsin\left(-\frac{1}{2}\right) = -\frac{\pi}{6}$.

   (e) Since $0 \leq \frac{\pi}{6} \leq \pi$, we could simply invoke Theorem 10.26 to get $\arccos\left(\cos\left(\frac{\pi}{6}\right)\right) = \frac{\pi}{6}$. However, in order to make sure we understand *why* this is the case, we choose to work the example through using the definition of arccosine. Working from the inside out, $\arccos\left(\cos\left(\frac{\pi}{6}\right)\right) = \arccos\left(\frac{\sqrt{3}}{2}\right)$. Now, $\arccos\left(\frac{\sqrt{3}}{2}\right)$ is the real number t with $0 \leq t \leq \pi$ and $\cos(t) = \frac{\sqrt{3}}{2}$. We find $t = \frac{\pi}{6}$, so that $\arccos\left(\cos\left(\frac{\pi}{6}\right)\right) = \frac{\pi}{6}$.

(f) Since $\frac{11\pi}{6}$ does not fall between 0 and π, Theorem 10.26 does not apply. We are forced to work through from the inside out starting with $\arccos\left(\cos\left(\frac{11\pi}{6}\right)\right) = \arccos\left(\frac{\sqrt{3}}{2}\right)$. From the previous problem, we know $\arccos\left(\frac{\sqrt{3}}{2}\right) = \frac{\pi}{6}$. Hence, $\arccos\left(\cos\left(\frac{11\pi}{6}\right)\right) = \frac{\pi}{6}$.

(g) One way to simplify $\cos\left(\arccos\left(-\frac{3}{5}\right)\right)$ is to use Theorem 10.26 directly. Since $-\frac{3}{5}$ is between -1 and 1, we have that $\cos\left(\arccos\left(-\frac{3}{5}\right)\right) = -\frac{3}{5}$ and we are done. However, as before, to really understand *why* this cancellation occurs, we let $t = \arccos\left(-\frac{3}{5}\right)$. Then, by definition, $\cos(t) = -\frac{3}{5}$. Hence, $\cos\left(\arccos\left(-\frac{3}{5}\right)\right) = \cos(t) = -\frac{3}{5}$, and we are finished in (nearly) the same amount of time.

(h) As in the previous example, we let $t = \arccos\left(-\frac{3}{5}\right)$ so that $\cos(t) = -\frac{3}{5}$ for some t where $0 \leq t \leq \pi$. Since $\cos(t) < 0$, we can narrow this down a bit and conclude that $\frac{\pi}{2} < t < \pi$, so that t corresponds to an angle in Quadrant II. In terms of t, then, we need to find $\sin\left(\arccos\left(-\frac{3}{5}\right)\right) = \sin(t)$. Using the Pythagorean Identity $\cos^2(t) + \sin^2(t) = 1$, we get $\left(-\frac{3}{5}\right)^2 + \sin^2(t) = 1$ or $\sin(t) = \pm\frac{4}{5}$. Since t corresponds to a Quadrants II angle, we choose $\sin(t) = \frac{4}{5}$. Hence, $\sin\left(\arccos\left(-\frac{3}{5}\right)\right) = \frac{4}{5}$.

2. (a) We begin this problem in the same manner we began the previous two problems. To help us see the forest for the trees, we let $t = \arccos(x)$, so our goal is to find a way to express $\tan(\arccos(x)) = \tan(t)$ in terms of x. Since $t = \arccos(x)$, we know $\cos(t) = x$ where $0 \leq t \leq \pi$, but since we are after an expression for $\tan(t)$, we know we need to throw out $t = \frac{\pi}{2}$ from consideration. Hence, either $0 \leq t < \frac{\pi}{2}$ or $\frac{\pi}{2} < t \leq \pi$ so that, geometrically, t corresponds to an angle in Quadrant I or Quadrant II. One approach[3] to finding $\tan(t)$ is to use the quotient identity $\tan(t) = \frac{\sin(t)}{\cos(t)}$. Substituting $\cos(t) = x$ into the Pythagorean Identity $\cos^2(t) + \sin^2(t) = 1$ gives $x^2 + \sin^2(t) = 1$, from which we get $\sin(t) = \pm\sqrt{1 - x^2}$. Since t corresponds to angles in Quadrants I and II, $\sin(t) \geq 0$, so we choose $\sin(t) = \sqrt{1 - x^2}$. Thus,

$$\tan(t) = \frac{\sin(t)}{\cos(t)} = \frac{\sqrt{1 - x^2}}{x}$$

To determine the values of x for which this equivalence is valid, we consider our substitution $t = \arccos(x)$. Since the domain of $\arccos(x)$ is $[-1, 1]$, we know we must restrict $-1 \leq x \leq 1$. Additionally, since we had to discard $t = \frac{\pi}{2}$, we need to discard $x = \cos\left(\frac{\pi}{2}\right) = 0$. Hence, $\tan(\arccos(x)) = \frac{\sqrt{1-x^2}}{x}$ is valid for x in $[-1, 0) \cup (0, 1]$.

(b) We proceed as in the previous problem by writing $t = \arcsin(x)$ so that t lies in the interval $\left[-\frac{\pi}{2}, \frac{\pi}{2}\right]$ with $\sin(t) = x$. We aim to express $\cos(2\arcsin(x)) = \cos(2t)$ in terms of x. Since $\cos(2t)$ is defined everywhere, we get no additional restrictions on t as we did in the previous problem. We have three choices for rewriting $\cos(2t)$: $\cos^2(t) - \sin^2(t)$, $2\cos^2(t) - 1$ and $1 - 2\sin^2(t)$. Since we know $x = \sin(t)$, it is easiest to use the last form:

$$\cos(2\arcsin(x)) = \cos(2t) = 1 - 2\sin^2(t) = 1 - 2x^2$$

[3]Alternatively, we could use the identity: $1 + \tan^2(t) = \sec^2(t)$. Since $x = \cos(t)$, $\sec(t) = \frac{1}{\cos(t)} = \frac{1}{x}$. The reader is invited to work through this approach to see what, if any, difficulties arise.

To find the restrictions on x, we once again appeal to our substitution $t = \arcsin(x)$. Since $\arcsin(x)$ is defined only for $-1 \leq x \leq 1$, the equivalence $\cos(2\arcsin(x)) = 1-2x^2$ is valid only on $[-1, 1]$. □

A few remarks about Example 10.6.1 are in order. Most of the common errors encountered in dealing with the inverse circular functions come from the need to restrict the domains of the original functions so that they are one-to-one. One instance of this phenomenon is the fact that $\arccos\left(\cos\left(\frac{11\pi}{6}\right)\right) = \frac{\pi}{6}$ as opposed to $\frac{11\pi}{6}$. This is the exact same phenomenon discussed in Section 5.2 when we saw $\sqrt{(-2)^2} = 2$ as opposed to -2. Additionally, even though the expression we arrived at in part 2b above, namely $1 - 2x^2$, is defined for all real numbers, the equivalence $\cos(2\arcsin(x)) = 1 - 2x^2$ is valid for only $-1 \leq x \leq 1$. This is akin to the fact that while the expression x is defined for all real numbers, the equivalence $\left(\sqrt{x}\right)^2 = x$ is valid only for $x \geq 0$. For this reason, it pays to be careful when we determine the intervals where such equivalences are valid.

The next pair of functions we wish to discuss are the inverses of tangent and cotangent, which are named arctangent and arccotangent, respectively. First, we restrict $f(x) = \tan(x)$ to its fundamental cycle on $\left(-\frac{\pi}{2}, \frac{\pi}{2}\right)$ to obtain $f^{-1}(x) = \arctan(x)$. Among other things, note that the *vertical* asymptotes $x = -\frac{\pi}{2}$ and $x = \frac{\pi}{2}$ of the graph of $f(x) = \tan(x)$ become the *horizontal* asymptotes $y = -\frac{\pi}{2}$ and $y = \frac{\pi}{2}$ of the graph of $f^{-1}(x) = \arctan(x)$.

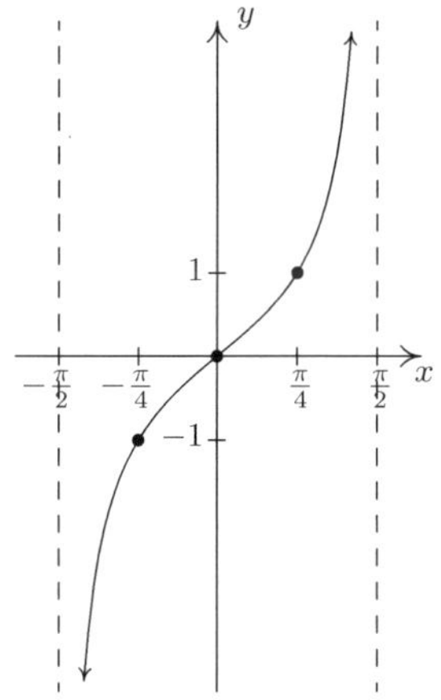

$f(x) = \tan(x)$, $-\frac{\pi}{2} < x < \frac{\pi}{2}$.

reflect across $y = x$ → switch x and y coordinates

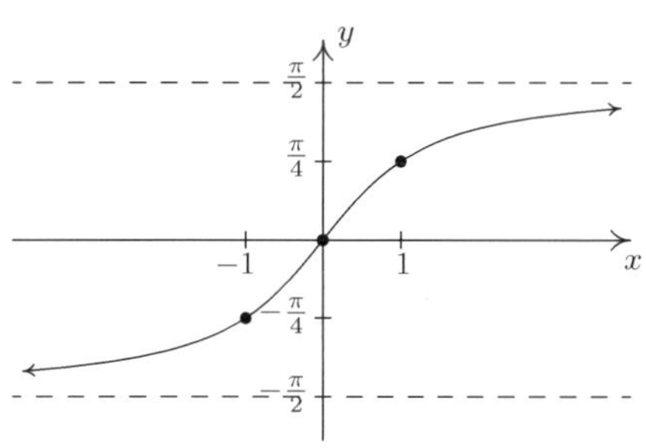

$f^{-1}(x) = \arctan(x)$.

Next, we restrict $g(x) = \cot(x)$ to its fundamental cycle on $(0, \pi)$ to obtain $g^{-1}(x) = \operatorname{arccot}(x)$. Once again, the vertical asymptotes $x = 0$ and $x = \pi$ of the graph of $g(x) = \cot(x)$ become the horizontal asymptotes $y = 0$ and $y = \pi$ of the graph of $g^{-1}(x) = \operatorname{arccot}(x)$. We show these graphs on the next page and list some of the basic properties of the arctangent and arccotangent functions.

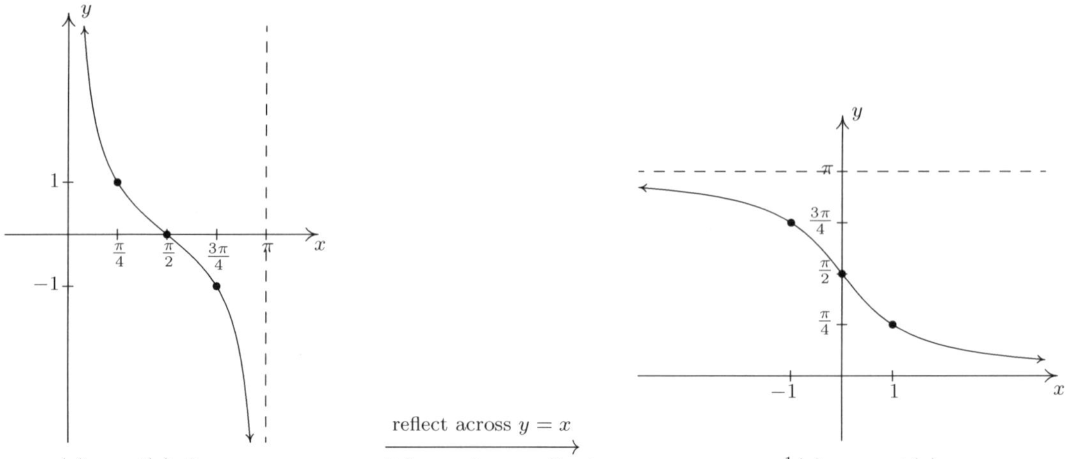

$g(x) = \cot(x)$, $0 < x < \pi$. $\qquad$ $g^{-1}(x) = \operatorname{arccot}(x)$.

Theorem 10.27. Properties of the Arctangent and Arccotangent Functions

- Properties of $F(x) = \arctan(x)$
  - Domain: $(-\infty, \infty)$
  - Range: $\left(-\frac{\pi}{2}, \frac{\pi}{2}\right)$
  - as $x \to -\infty$, $\arctan(x) \to -\frac{\pi}{2}^{+}$; as $x \to \infty$, $\arctan(x) \to \frac{\pi}{2}^{-}$
  - $\arctan(x) = t$ if and only if $-\frac{\pi}{2} < t < \frac{\pi}{2}$ and $\tan(t) = x$
  - $\arctan(x) = \operatorname{arccot}\left(\frac{1}{x}\right)$ for $x > 0$
  - $\tan\left(\arctan(x)\right) = x$ for all real numbers x
  - $\arctan(\tan(x)) = x$ provided $-\frac{\pi}{2} < x < \frac{\pi}{2}$
  - additionally, arctangent is odd
- Properties of $G(x) = \operatorname{arccot}(x)$
  - Domain: $(-\infty, \infty)$
  - Range: $(0, \pi)$
  - as $x \to -\infty$, $\operatorname{arccot}(x) \to \pi^{-}$; as $x \to \infty$, $\operatorname{arccot}(x) \to 0^{+}$
  - $\operatorname{arccot}(x) = t$ if and only if $0 < t < \pi$ and $\cot(t) = x$
  - $\operatorname{arccot}(x) = \arctan\left(\frac{1}{x}\right)$ for $x > 0$
  - $\cot\left(\operatorname{arccot}(x)\right) = x$ for all real numbers x
  - $\operatorname{arccot}(\cot(x)) = x$ provided $0 < x < \pi$

Example 10.6.2.

1. Find the exact values of the following.

   (a) $\arctan(\sqrt{3})$ (b) $\operatorname{arccot}(-\sqrt{3})$

   (c) $\cot(\operatorname{arccot}(-5))$ (d) $\sin\left(\arctan\left(-\frac{3}{4}\right)\right)$

2. Rewrite the following as algebraic expressions of x and state the domain on which the equivalence is valid.

   (a) $\tan(2\arctan(x))$ (b) $\cos(\operatorname{arccot}(2x))$

Solution.

1. (a) We know $\arctan(\sqrt{3})$ is the real number t between $-\frac{\pi}{2}$ and $\frac{\pi}{2}$ with $\tan(t) = \sqrt{3}$. We find $t = \frac{\pi}{3}$, so $\arctan(\sqrt{3}) = \frac{\pi}{3}$.

   (b) The real number $t = \operatorname{arccot}(-\sqrt{3})$ lies in the interval $(0, \pi)$ with $\cot(t) = -\sqrt{3}$. We get $\operatorname{arccot}(-\sqrt{3}) = \frac{5\pi}{6}$.

   (c) We can apply Theorem 10.27 directly and obtain $\cot(\operatorname{arccot}(-5)) = -5$. However, working it through provides us with yet another opportunity to understand why this is the case. Letting $t = \operatorname{arccot}(-5)$, we have that t belongs to the interval $(0, \pi)$ and $\cot(t) = -5$. Hence, $\cot(\operatorname{arccot}(-5)) = \cot(t) = -5$.

   (d) We start simplifying $\sin\left(\arctan\left(-\frac{3}{4}\right)\right)$ by letting $t = \arctan\left(-\frac{3}{4}\right)$. Then $\tan(t) = -\frac{3}{4}$ for some $-\frac{\pi}{2} < t < \frac{\pi}{2}$. Since $\tan(t) < 0$, we know, in fact, $-\frac{\pi}{2} < t < 0$. One way to proceed is to use The Pythagorean Identity, $1+\cot^2(t) = \csc^2(t)$, since this relates the reciprocals of $\tan(t)$ and $\sin(t)$ and is valid for all t under consideration.[4] From $\tan(t) = -\frac{3}{4}$, we get $\cot(t) = -\frac{4}{3}$. Substituting, we get $1 + \left(-\frac{4}{3}\right)^2 = \csc^2(t)$ so that $\csc(t) = \pm\frac{5}{3}$. Since $-\frac{\pi}{2} < t < 0$, we choose $\csc(t) = -\frac{5}{3}$, so $\sin(t) = -\frac{3}{5}$. Hence, $\sin\left(\arctan\left(-\frac{3}{4}\right)\right) = -\frac{3}{5}$.

2. (a) If we let $t = \arctan(x)$, then $-\frac{\pi}{2} < t < \frac{\pi}{2}$ and $\tan(t) = x$. We look for a way to express $\tan(2\arctan(x)) = \tan(2t)$ in terms of x. Before we get started using identities, we note that $\tan(2t)$ is undefined when $2t = \frac{\pi}{2} + \pi k$ for integers k. Dividing both sides of this equation by 2 tells us we need to exclude values of t where $t = \frac{\pi}{4} + \frac{\pi}{2}k$, where k is an integer. The only members of this family which lie in $\left(-\frac{\pi}{2}, \frac{\pi}{2}\right)$ are $t = \pm\frac{\pi}{4}$, which means the values of t under consideration are $\left(-\frac{\pi}{2}, -\frac{\pi}{4}\right) \cup \left(-\frac{\pi}{4}, \frac{\pi}{4}\right) \cup \left(\frac{\pi}{4}, \frac{\pi}{2}\right)$. Returning to $\arctan(2t)$, we note the double angle identity $\tan(2t) = \frac{2\tan(t)}{1-\tan^2(t)}$, is valid for all the values of t under consideration, hence we get

$$\tan(2\arctan(x)) = \tan(2t) = \frac{2\tan(t)}{1-\tan^2(t)} = \frac{2x}{1-x^2}$$

[4]It's always a good idea to make sure the identities used in these situations are valid for all values t under consideration. Check our work back in Example 10.6.1. Were the identities we used there valid for all t under consideration? A pedantic point, to be sure, but what else do you expect from this book?

To find where this equivalence is valid we check back with our substitution $t = \arctan(x)$. Since the domain of $\arctan(x)$ is all real numbers, the only exclusions come from the values of t we discarded earlier, $t = \pm\frac{\pi}{4}$. Since $x = \tan(t)$, this means we exclude $x = \tan\left(\pm\frac{\pi}{4}\right) = \pm 1$. Hence, the equivalence $\tan(2\arctan(x)) = \frac{2x}{1-x^2}$ holds for all x in $(-\infty, -1) \cup (-1, 1) \cup (1, \infty)$.

(b) To get started, we let $t = \operatorname{arccot}(2x)$ so that $\cot(t) = 2x$ where $0 < t < \pi$. In terms of t, $\cos(\operatorname{arccot}(2x)) = \cos(t)$, and our goal is to express the latter in terms of x. Since $\cos(t)$ is always defined, there are no additional restrictions on t, so we can begin using identities to relate $\cot(t)$ to $\cos(t)$. The identity $\cot(t) = \frac{\cos(t)}{\sin(t)}$ is valid for t in $(0, \pi)$, so our strategy is to obtain $\sin(t)$ in terms of x, then write $\cos(t) = \cot(t)\sin(t)$. The identity $1 + \cot^2(t) = \csc^2(t)$ holds for all t in $(0, \pi)$ and relates $\cot(t)$ and $\csc(t) = \frac{1}{\sin(t)}$. Substituting $\cot(t) = 2x$, we get $1 + (2x)^2 = \csc^2(t)$, or $\csc(t) = \pm\sqrt{4x^2+1}$. Since t is between 0 and π, $\csc(t) > 0$, so $\csc(t) = \sqrt{4x^2+1}$ which gives $\sin(t) = \frac{1}{\sqrt{4x^2+1}}$. Hence,

$$\cos(\operatorname{arccot}(2x)) = \cos(t) = \cot(t)\sin(t) = \frac{2x}{\sqrt{4x^2+1}}$$

Since $\operatorname{arccot}(2x)$ is defined for all real numbers x and we encountered no additional restrictions on t, we have $\cos\left(\operatorname{arccot}(2x)\right) = \frac{2x}{\sqrt{4x^2+1}}$ for all real numbers x. □

The last two functions to invert are secant and cosecant. A portion of each of their graphs, which were first discussed in Subsection 10.5.2, are given below with the fundamental cycles highlighted.

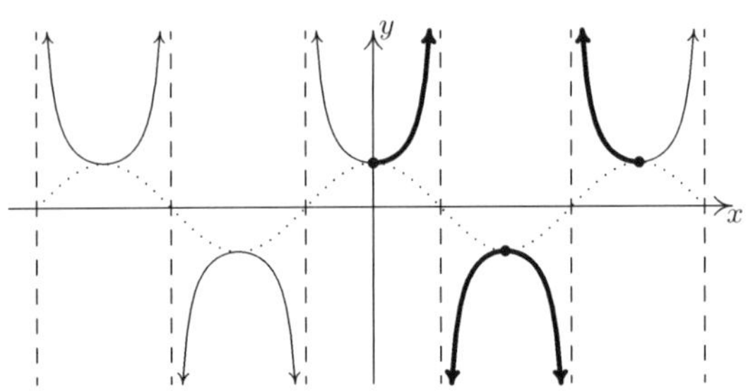

The graph of $y = \sec(x)$.

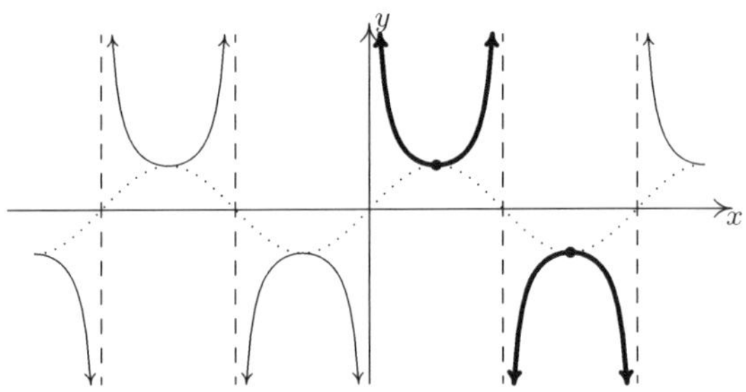

The graph of $y = \csc(x)$.

It is clear from the graph of secant that we cannot find one single continuous piece of its graph which covers its entire range of $(-\infty, -1] \cup [1, \infty)$ and restricts the domain of the function so that it is one-to-one. The same is true for cosecant. Thus in order to define the arcsecant and arccosecant functions, we must settle for a piecewise approach wherein we choose one piece to cover the top of the range, namely $[1, \infty)$, and another piece to cover the bottom, namely $(-\infty, -1]$. There are two generally accepted ways make these choices which restrict the domains of these functions so that they are one-to-one. One approach simplifies the Trigonometry associated with the inverse functions, but complicates the Calculus; the other makes the Calculus easier, but the Trigonometry less so. We present both points of view.

10.6.1 Inverses of Secant and Cosecant: Trigonometry Friendly Approach

In this subsection, we restrict the secant and cosecant functions to coincide with the restrictions on cosine and sine, respectively. For $f(x) = \sec(x)$, we restrict the domain to $\left[0, \frac{\pi}{2}\right) \cup \left(\frac{\pi}{2}, \pi\right]$

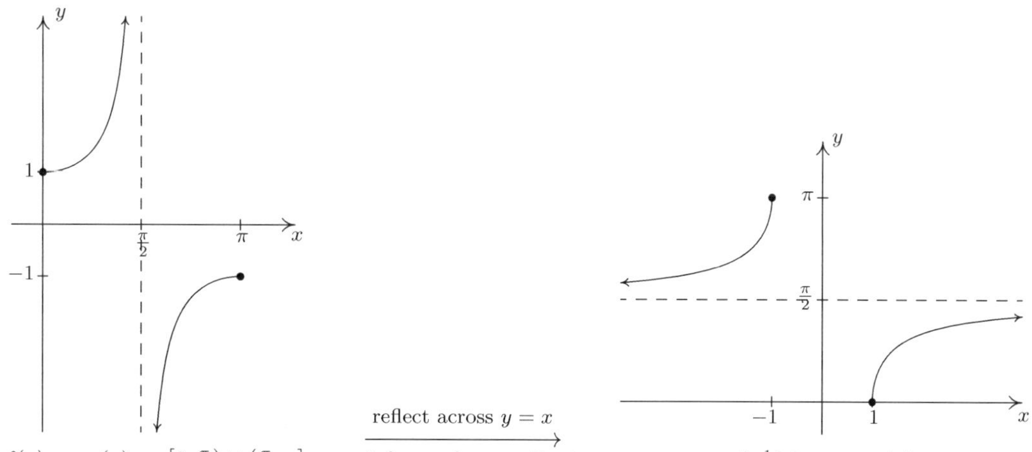

$f(x) = \sec(x)$ on $\left[0, \frac{\pi}{2}\right) \cup \left(\frac{\pi}{2}, \pi\right]$ switch x and y coordinates $f^{-1}(x) = \operatorname{arcsec}(x)$

and we restrict $g(x) = \csc(x)$ to $\left[-\frac{\pi}{2}, 0\right) \cup \left(0, \frac{\pi}{2}\right]$.

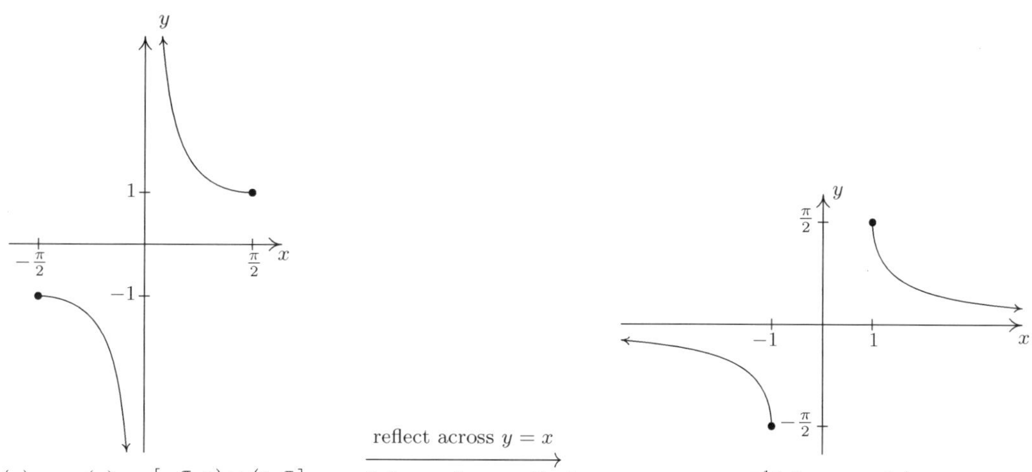

$g(x) = \csc(x)$ on $\left[-\frac{\pi}{2}, 0\right) \cup \left(0, \frac{\pi}{2}\right]$ switch x and y coordinates $g^{-1}(x) = \operatorname{arccsc}(x)$

Note that for both arcsecant and arccosecant, the domain is $(-\infty, -1] \cup [1, \infty)$. Taking a page from Section 2.2, we can rewrite this as $\{x : |x| \geq 1\}$. This is often done in Calculus textbooks, so we include it here for completeness. Using these definitions, we get the following properties of the arcsecant and arccosecant functions.

Theorem 10.28. Properties of the Arcsecant and Arccosecant Functions[a]

- Properties of $F(x) = \operatorname{arcsec}(x)$
  - Domain: $\{x : |x| \geq 1\} = (-\infty, -1] \cup [1, \infty)$
  - Range: $\left[0, \frac{\pi}{2}\right) \cup \left(\frac{\pi}{2}, \pi\right]$
  - as $x \to -\infty$, $\operatorname{arcsec}(x) \to \frac{\pi}{2}^+$; as $x \to \infty$, $\operatorname{arcsec}(x) \to \frac{\pi}{2}^-$
  - $\operatorname{arcsec}(x) = t$ if and only if $0 \leq t < \frac{\pi}{2}$ or $\frac{\pi}{2} < t \leq \pi$ and $\sec(t) = x$
  - $\operatorname{arcsec}(x) = \arccos\left(\frac{1}{x}\right)$ provided $|x| \geq 1$
  - $\sec(\operatorname{arcsec}(x)) = x$ provided $|x| \geq 1$
  - $\operatorname{arcsec}(\sec(x)) = x$ provided $0 \leq x < \frac{\pi}{2}$ or $\frac{\pi}{2} < x \leq \pi$
- Properties of $G(x) = \operatorname{arccsc}(x)$
  - Domain: $\{x : |x| \geq 1\} = (-\infty, -1] \cup [1, \infty)$
  - Range: $\left[-\frac{\pi}{2}, 0\right) \cup \left(0, \frac{\pi}{2}\right]$
  - as $x \to -\infty$, $\operatorname{arccsc}(x) \to 0^-$; as $x \to \infty$, $\operatorname{arccsc}(x) \to 0^+$
  - $\operatorname{arccsc}(x) = t$ if and only if $-\frac{\pi}{2} \leq t < 0$ or $0 < t \leq \frac{\pi}{2}$ and $\csc(t) = x$
  - $\operatorname{arccsc}(x) = \arcsin\left(\frac{1}{x}\right)$ provided $|x| \geq 1$
  - $\csc(\operatorname{arccsc}(x)) = x$ provided $|x| \geq 1$
  - $\operatorname{arccsc}(\csc(x)) = x$ provided $-\frac{\pi}{2} \leq x < 0$ or $0 < x \leq \frac{\pi}{2}$
  - additionally, arccosecant is odd

[a]... assuming the "Trigonometry Friendly" ranges are used.

Example 10.6.3.

1. Find the exact values of the following.

   (a) $\operatorname{arcsec}(2)$ (b) $\operatorname{arccsc}(-2)$ (c) $\operatorname{arcsec}\left(\sec\left(\frac{5\pi}{4}\right)\right)$ (d) $\cot\left(\operatorname{arccsc}(-3)\right)$

2. Rewrite the following as algebraic expressions of x and state the domain on which the equivalence is valid.

   (a) $\tan(\operatorname{arcsec}(x))$ (b) $\cos(\operatorname{arccsc}(4x))$

Solution.

1. (a) Using Theorem 10.28, we have $\text{arcsec}(2) = \arccos\left(\frac{1}{2}\right) = \frac{\pi}{3}$.
   (b) Once again, Theorem 10.28 comes to our aid giving $\text{arccsc}(-2) = \arcsin\left(-\frac{1}{2}\right) = -\frac{\pi}{6}$.
   (c) Since $\frac{5\pi}{4}$ doesn't fall between 0 and $\frac{\pi}{2}$ or $\frac{\pi}{2}$ and π, we cannot use the inverse property stated in Theorem 10.28. We can, nevertheless, begin by working 'inside out' which yields $\text{arcsec}\left(\sec\left(\frac{5\pi}{4}\right)\right) = \text{arcsec}(-\sqrt{2}) = \arccos\left(-\frac{\sqrt{2}}{2}\right) = \frac{3\pi}{4}$.
   (d) One way to begin to simplify $\cot\left(\text{arccsc}\left(-3\right)\right)$ is to let $t = \text{arccsc}(-3)$. Then, $\csc(t) = -3$ and, since this is negative, we have that t lies in the interval $\left[-\frac{\pi}{2}, 0\right)$. We are after $\cot\left(\text{arccsc}\left(-3\right)\right) = \cot(t)$, so we use the Pythagorean Identity $1 + \cot^2(t) = \csc^2(t)$. Substituting, we have $1 + \cot^2(t) = (-3)^2$, or $\cot(t) = \pm\sqrt{8} = \pm 2\sqrt{2}$. Since $-\frac{\pi}{2} \leq t < 0$, $\cot(t) < 0$, so we get $\cot\left(\text{arccsc}\left(-3\right)\right) = -2\sqrt{2}$.

2. (a) We begin simplifying $\tan(\text{arcsec}(x))$ by letting $t = \text{arcsec}(x)$. Then, $\sec(t) = x$ for t in $\left[0, \frac{\pi}{2}\right) \cup \left(\frac{\pi}{2}, \pi\right]$, and we seek a formula for $\tan(t)$. Since $\tan(t)$ is defined for all t values under consideration, we have no additional restrictions on t. To relate $\sec(t)$ to $\tan(t)$, we use the identity $1 + \tan^2(t) = \sec^2(t)$. This is valid for all values of t under consideration, and when we substitute $\sec(t) = x$, we get $1 + \tan^2(t) = x^2$. Hence, $\tan(t) = \pm\sqrt{x^2 - 1}$. If t belongs to $\left[0, \frac{\pi}{2}\right)$ then $\tan(t) \geq 0$; if, on the the other hand, t belongs to $\left(\frac{\pi}{2}, \pi\right]$ then $\tan(t) \leq 0$. As a result, we get a piecewise defined function for $\tan(t)$

$$\tan(t) = \begin{cases} \sqrt{x^2 - 1}, & \text{if } 0 \leq t < \frac{\pi}{2} \\ -\sqrt{x^2 - 1}, & \text{if } \frac{\pi}{2} < t \leq \pi \end{cases}$$

   Now we need to determine what these conditions on t mean for x. Since $x = \sec(t)$, when $0 \leq t < \frac{\pi}{2}$, $x \geq 1$, and when $\frac{\pi}{2} < t \leq \pi$, $x \leq -1$. Since we encountered no further restrictions on t, the equivalence below holds for all x in $(-\infty, -1] \cup [1, \infty)$.

$$\tan(\text{arcsec}(x)) = \begin{cases} \sqrt{x^2 - 1}, & \text{if } x \geq 1 \\ -\sqrt{x^2 - 1}, & \text{if } x \leq -1 \end{cases}$$

   (b) To simplify $\cos(\text{arccsc}(4x))$, we start by letting $t = \text{arccsc}(4x)$. Then $\csc(t) = 4x$ for t in $\left[-\frac{\pi}{2}, 0\right) \cup \left(0, \frac{\pi}{2}\right]$, and we now set about finding an expression for $\cos(\text{arccsc}(4x)) = \cos(t)$. Since $\cos(t)$ is defined for all t, we do not encounter any additional restrictions on t. From $\csc(t) = 4x$, we get $\sin(t) = \frac{1}{4x}$, so to find $\cos(t)$, we can make use if the identity $\cos^2(t) + \sin^2(t) = 1$. Substituting $\sin(t) = \frac{1}{4x}$ gives $\cos^2(t) + \left(\frac{1}{4x}\right)^2 = 1$. Solving, we get

$$\cos(t) = \pm\sqrt{\frac{16x^2 - 1}{16x^2}} = \pm\frac{\sqrt{16x^2 - 1}}{4|x|}$$

   Since t belongs to $\left[-\frac{\pi}{2}, 0\right) \cup \left(0, \frac{\pi}{2}\right]$, we know $\cos(t) \geq 0$, so we choose $\cos(t) = \frac{\sqrt{16 - x^2}}{4|x|}$. (The absolute values here are necessary, since x could be negative.) To find the values for which this equivalence is valid, we look back at our original substution, $t = \operatorname{arccsc}(4x)$. Since the domain of $\operatorname{arccsc}(x)$ requires its argument x to satisfy $|x| \geq 1$, the domain of $\operatorname{arccsc}(4x)$ requires $|4x| \geq 1$. Using Theorem 2.4, we rewrite this inequality and solve to get $x \leq -\frac{1}{4}$ or $x \geq \frac{1}{4}$. Since we had no additional restrictions on t, the equivalence $\cos(\operatorname{arccsc}(4x)) = \frac{\sqrt{16x^2-1}}{4|x|}$ holds for all x in $\left(-\infty, -\frac{1}{4}\right] \cup \left[\frac{1}{4}, \infty\right)$. $\square$

10.6.2 Inverses of Secant and Cosecant: Calculus Friendly Approach

In this subsection, we restrict $f(x) = \sec(x)$ to $\left[0, \frac{\pi}{2}\right) \cup \left[\pi, \frac{3\pi}{2}\right)$

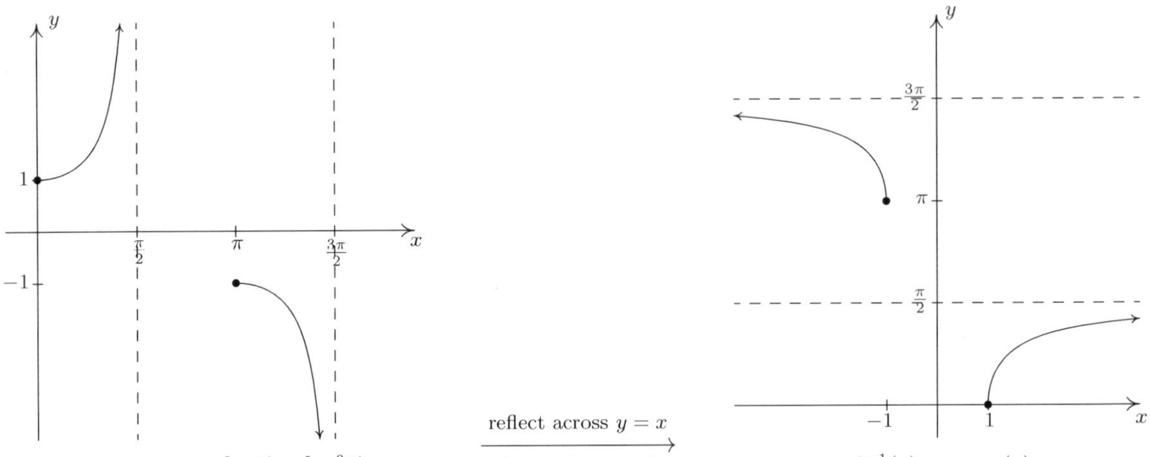

$f(x) = \sec(x)$ on $\left[0, \frac{\pi}{2}\right) \cup \left[\pi, \frac{3\pi}{2}\right)$ $\qquad$ $f^{-1}(x) = \operatorname{arcsec}(x)$

and we restrict $g(x) = \csc(x)$ to $\left(0, \frac{\pi}{2}\right] \cup \left(\pi, \frac{3\pi}{2}\right]$.

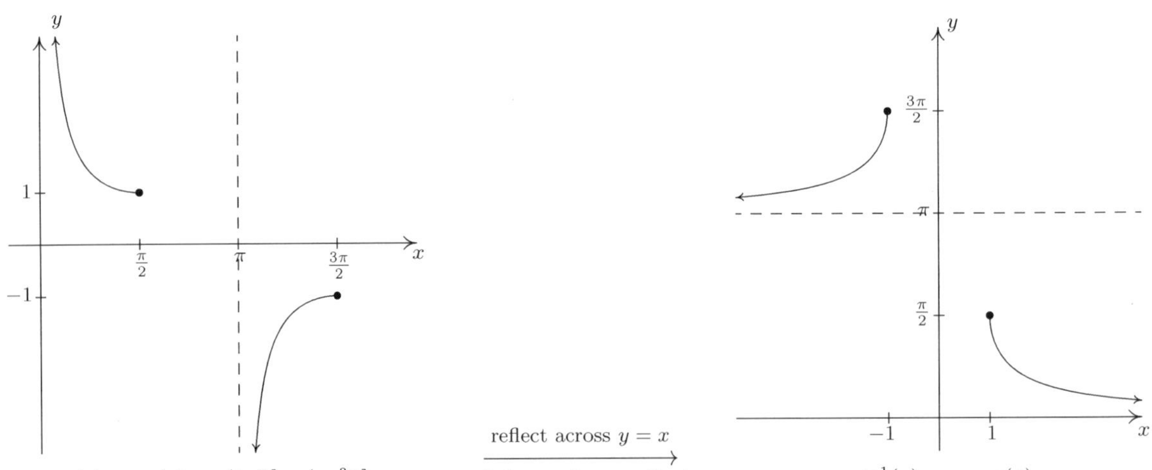

$g(x) = \csc(x)$ on $\left(0, \frac{\pi}{2}\right] \cup \left(\pi, \frac{3\pi}{2}\right]$ $\qquad$ $g^{-1}(x) = \operatorname{arccsc}(x)$

Using these definitions, we get the following result.

Theorem 10.29. Properties of the Arcsecant and Arccosecant Functions[a]

- Properties of $F(x) = \operatorname{arcsec}(x)$
  - Domain: $\{x : |x| \geq 1\} = (-\infty, -1] \cup [1, \infty)$
  - Range: $\left[0, \frac{\pi}{2}\right) \cup \left[\pi, \frac{3\pi}{2}\right)$
  - as $x \to -\infty$, $\operatorname{arcsec}(x) \to \frac{3\pi}{2}^-$; as $x \to \infty$, $\operatorname{arcsec}(x) \to \frac{\pi}{2}^-$
  - $\operatorname{arcsec}(x) = t$ if and only if $0 \leq t < \frac{\pi}{2}$ or $\pi \leq t < \frac{3\pi}{2}$ and $\sec(t) = x$
  - $\operatorname{arcsec}(x) = \arccos\left(\frac{1}{x}\right)$ for $x \geq 1$ only[b]
  - $\sec\left(\operatorname{arcsec}(x)\right) = x$ provided $|x| \geq 1$
  - $\operatorname{arcsec}(\sec(x)) = x$ provided $0 \leq x < \frac{\pi}{2}$ or $\pi \leq x < \frac{3\pi}{2}$
- Properties of $G(x) = \operatorname{arccsc}(x)$
  - Domain: $\{x : |x| \geq 1\} = (-\infty, -1] \cup [1, \infty)$
  - Range: $\left(0, \frac{\pi}{2}\right] \cup \left(\pi, \frac{3\pi}{2}\right]$
  - as $x \to -\infty$, $\operatorname{arccsc}(x) \to \pi^+$; as $x \to \infty$, $\operatorname{arccsc}(x) \to 0^+$
  - $\operatorname{arccsc}(x) = t$ if and only if $0 < t \leq \frac{\pi}{2}$ or $\pi < t \leq \frac{3\pi}{2}$ and $\csc(t) = x$
  - $\operatorname{arccsc}(x) = \arcsin\left(\frac{1}{x}\right)$ for $x \geq 1$ only[c]
  - $\csc\left(\operatorname{arccsc}(x)\right) = x$ provided $|x| \geq 1$
  - $\operatorname{arccsc}(\csc(x)) = x$ provided $0 < x \leq \frac{\pi}{2}$ or $\pi < x \leq \frac{3\pi}{2}$

[a] ...assuming the "Calculus Friendly" ranges are used.
[b] Compare this with the similar result in Theorem 10.28.
[c] Compare this with the similar result in Theorem 10.28.

Our next example is a duplicate of Example 10.6.3. The interested reader is invited to compare and contrast the solution to each.

Example 10.6.4.

1. Find the exact values of the following.

   (a) $\operatorname{arcsec}(2)$ (b) $\operatorname{arccsc}(-2)$ (c) $\operatorname{arcsec}\left(\sec\left(\frac{5\pi}{4}\right)\right)$ (d) $\cot\left(\operatorname{arccsc}(-3)\right)$

2. Rewrite the following as algebraic expressions of x and state the domain on which the equivalence is valid.

   (a) $\tan(\operatorname{arcsec}(x))$ (b) $\cos(\operatorname{arccsc}(4x))$

Solution.

1. (a) Since $2 \geq 1$, we may invoke Theorem 10.29 to get $\operatorname{arcsec}(2) = \arccos\left(\frac{1}{2}\right) = \frac{\pi}{3}$.

   (b) Unfortunately, -2 is not greater to or equal to 1, so we cannot apply Theorem 10.29 to $\operatorname{arccsc}(-2)$ and convert this into an arcsine problem. Instead, we appeal to the definition. The real number $t = \operatorname{arccsc}(-2)$ lies in $\left(0, \frac{\pi}{2}\right] \cup \left(\pi, \frac{3\pi}{2}\right]$ and satisfies $\csc(t) = -2$. The t we're after is $t = \frac{7\pi}{6}$, so $\operatorname{arccsc}(-2) = \frac{7\pi}{6}$.

   (c) Since $\frac{5\pi}{4}$ lies between π and $\frac{3\pi}{2}$, we may apply Theorem 10.29 directly to simplify $\operatorname{arcsec}\left(\sec\left(\frac{5\pi}{4}\right)\right) = \frac{5\pi}{4}$. We encourage the reader to work this through using the definition as we have done in the previous examples to see how it goes.

   (d) To simplify $\cot\left(\operatorname{arccsc}(-3)\right)$ we let $t = \operatorname{arccsc}(-3)$ so that $\cot\left(\operatorname{arccsc}(-3)\right) = \cot(t)$. We know $\csc(t) = -3$, and since this is negative, t lies in $\left(\pi, \frac{3\pi}{2}\right]$. Using the identity $1 + \cot^2(t) = \csc^2(t)$, we find $1 + \cot^2(t) = (-3)^2$ so that $\cot(t) = \pm\sqrt{8} = \pm 2\sqrt{2}$. Since t is in the interval $\left(\pi, \frac{3\pi}{2}\right]$, we know $\cot(t) > 0$. Our answer is $\cot\left(\operatorname{arccsc}(-3)\right) = 2\sqrt{2}$.

2. (a) We begin simplifying $\tan(\operatorname{arcsec}(x))$ by letting $t = \operatorname{arcsec}(x)$. Then, $\sec(t) = x$ for t in $\left[0, \frac{\pi}{2}\right) \cup \left[\pi, \frac{3\pi}{2}\right)$, and we seek a formula for $\tan(t)$. Since $\tan(t)$ is defined for all t values under consideration, we have no additional restrictions on t. To relate $\sec(t)$ to $\tan(t)$, we use the identity $1 + \tan^2(t) = \sec^2(t)$. This is valid for all values of t under consideration, and when we substitute $\sec(t) = x$, we get $1 + \tan^2(t) = x^2$. Hence, $\tan(t) = \pm\sqrt{x^2 - 1}$. Since t lies in $\left[0, \frac{\pi}{2}\right) \cup \left[\pi, \frac{3\pi}{2}\right)$, $\tan(t) \geq 0$, so we choose $\tan(t) = \sqrt{x^2 - 1}$. Since we found no additional restrictions on t, the equivalence $\tan(\operatorname{arcsec}(x)) = \sqrt{x^2 - 1}$ holds for all x in the domain of $t = \operatorname{arcsec}(x)$, namely $(-\infty, -1] \cup [1, \infty)$.

   (b) To simplify $\cos(\operatorname{arccsc}(4x))$, we start by letting $t = \operatorname{arccsc}(4x)$. Then $\csc(t) = 4x$ for t in $\left(0, \frac{\pi}{2}\right] \cup \left(\pi, \frac{3\pi}{2}\right]$, and we now set about finding an expression for $\cos(\operatorname{arccsc}(4x)) = \cos(t)$. Since $\cos(t)$ is defined for all t, we do not encounter any additional restrictions on t. From $\csc(t) = 4x$, we get $\sin(t) = \frac{1}{4x}$, so to find $\cos(t)$, we can make use if the identity $\cos^2(t) + \sin^2(t) = 1$. Substituting $\sin(t) = \frac{1}{4x}$ gives $\cos^2(t) + \left(\frac{1}{4x}\right)^2 = 1$. Solving, we get

   $$\cos(t) = \pm\sqrt{\frac{16x^2 - 1}{16x^2}} = \pm\frac{\sqrt{16x^2 - 1}}{4|x|}$$

   If t lies in $\left(0, \frac{\pi}{2}\right]$, then $\cos(t) \geq 0$, and we choose $\cos(t) = \frac{\sqrt{16x^2-1}}{4|x|}$. Otherwise, t belongs to $\left(\pi, \frac{3\pi}{2}\right]$ in which case $\cos(t) \leq 0$, so, we choose $\cos(t) = -\frac{\sqrt{16x^2-1}}{4|x|}$ This leads us to a (momentarily) piecewise defined function for $\cos(t)$

   $$\cos(t) = \begin{cases} \dfrac{\sqrt{16x^2 - 1}}{4|x|}, & \text{if } 0 \leq t \leq \frac{\pi}{2} \\ -\dfrac{\sqrt{16x^2 - 1}}{4|x|}, & \text{if } \pi < t \leq \frac{3\pi}{2} \end{cases}$$

We now see what these restrictions mean in terms of x. Since $4x = \csc(t)$, we get that for $0 \leq t \leq \frac{\pi}{2}$, $4x \geq 1$, or $x \geq \frac{1}{4}$. In this case, we can simplify $|x| = x$ so

$$\cos(t) = \frac{\sqrt{16x^2-1}}{4|x|} = \frac{\sqrt{16x^2-1}}{4x}$$

Similarly, for $\pi < t \leq \frac{3\pi}{2}$, we get $4x \leq -1$, or $x \leq -\frac{1}{4}$. In this case, $|x| = -x$, so we also get

$$\cos(t) = -\frac{\sqrt{16x^2-1}}{4|x|} = -\frac{\sqrt{16x^2-1}}{4(-x)} = \frac{\sqrt{16x^2-1}}{4x}$$

Hence, in all cases, $\cos(\operatorname{arccsc}(4x)) = \frac{\sqrt{16x^2-1}}{4x}$, and this equivalence is valid for all x in the domain of $t = \operatorname{arccsc}(4x)$, namely $\left(-\infty, -\frac{1}{4}\right] \cup \left[\frac{1}{4}, \infty\right)$ □

10.6.3 Calculators and the Inverse Circular Functions.

In the sections to come, we will have need to approximate the values of the inverse circular functions. On most calculators, only the arcsine, arccosine and arctangent functions are available and they are usually labeled as $\sin^{-1}, \cos^{-1}$ and $\tan^{-1}$, respectively. If we are asked to approximate these values, it is a simple matter to punch up the appropriate decimal on the calculator. If we are asked for an arccotangent, arcsecant or arccosecant, however, we often need to employ some ingenuity, as our next example illustrates.

Example 10.6.5.

1. Use a calculator to approximate the following values to four decimal places.

   (a) $\operatorname{arccot}(2)$ (b) $\operatorname{arcsec}(5)$ (c) $\operatorname{arccot}(-2)$ (d) $\operatorname{arccsc}\left(-\frac{3}{2}\right)$

2. Find the domain and range of the following functions. Check your answers using a calculator.

   (a) $f(x) = \frac{\pi}{2} - \arccos\left(\frac{x}{5}\right)$ (b) $f(x) = 3\arctan(4x)$. (c) $f(x) = \operatorname{arccot}\left(\frac{x}{2}\right) + \pi$

Solution.

1. (a) Since $2 > 0$, we can use the property listed in Theorem 10.27 to rewrite $\operatorname{arccot}(2)$ as $\operatorname{arccot}(2) = \arctan\left(\frac{1}{2}\right)$. In 'radian' mode, we find $\operatorname{arccot}(2) = \arctan\left(\frac{1}{2}\right) \approx 0.4636$.

   (b) Since $5 \geq 1$, we can use the property from either Theorem 10.28 or Theorem 10.29 to write $\operatorname{arcsec}(5) = \arccos\left(\frac{1}{5}\right) \approx 1.3694$.

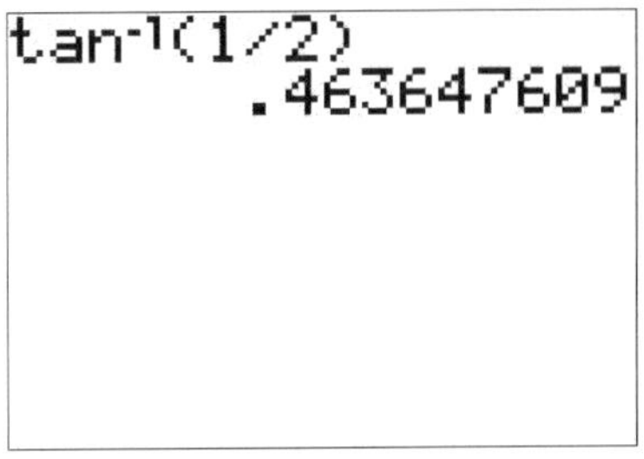

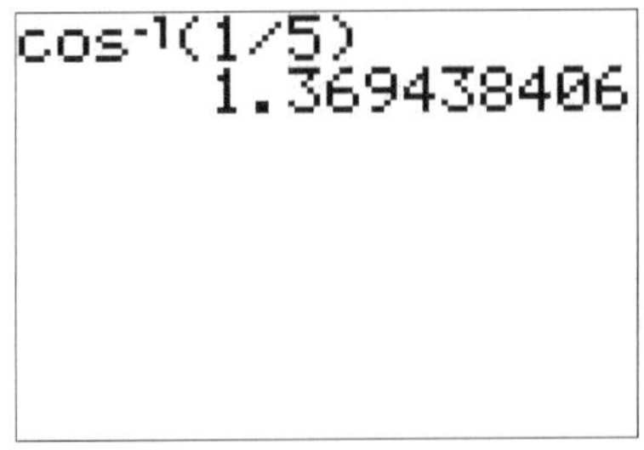

(c) Since the argument -2 is negative, we cannot directly apply Theorem 10.27 to help us find $\text{arccot}(-2)$. Let $t = \text{arccot}(-2)$. Then t is a real number such that $0 < t < \pi$ and $\cot(t) = -2$. Moreover, since $\cot(t) < 0$, we know $\frac{\pi}{2} < t < \pi$. Geometrically, this means t corresponds to a Quadrant II angle $\theta = t$ radians. This allows us to proceed using a 'reference angle' approach. Consider α, the reference angle for θ, as pictured below. By definition, α is an acute angle so $0 < \alpha < \frac{\pi}{2}$, and the Reference Angle Theorem, Theorem 10.2, tells us that $\cot(\alpha) = 2$. This means $\alpha = \text{arccot}(2)$ radians. Since the argument of arccotangent is now a *positive* 2, we can use Theorem 10.27 to get $\alpha = \text{arccot}(2) = \arctan\left(\frac{1}{2}\right)$ radians. Since $\theta = \pi - \alpha = \pi - \arctan\left(\frac{1}{2}\right) \approx 2.6779$ radians, we get $\text{arccot}(-2) \approx 2.6779$.

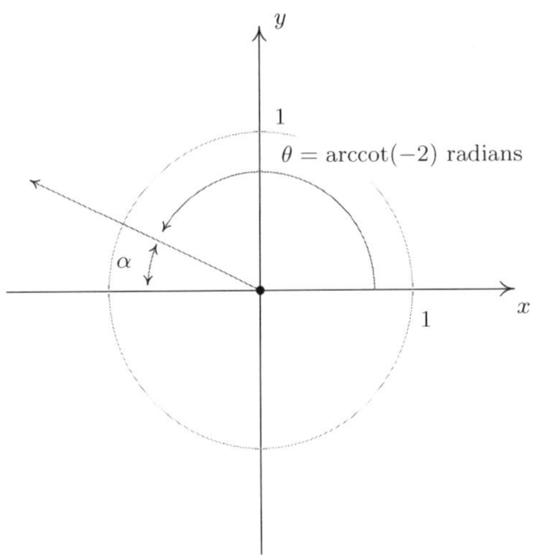

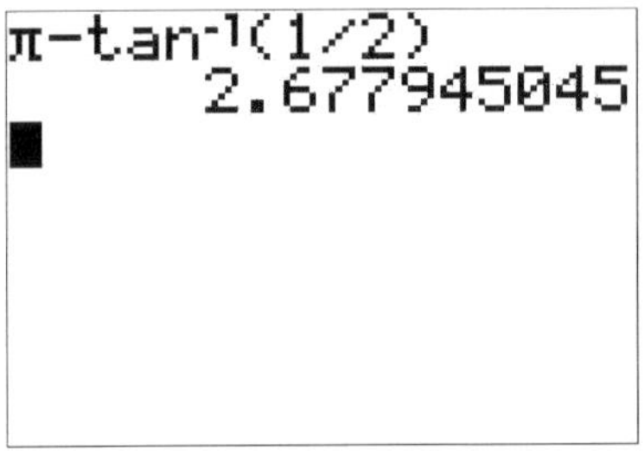

Another way to attack the problem is to use $\arctan\left(-\frac{1}{2}\right)$. By definition, the real number $t = \arctan\left(-\frac{1}{2}\right)$ satisfies $\tan(t) = -\frac{1}{2}$ with $-\frac{\pi}{2} < t < \frac{\pi}{2}$. Since $\tan(t) < 0$, we know more specifically that $-\frac{\pi}{2} < t < 0$, so t corresponds to an angle β in Quadrant IV. To find the value of $\text{arccot}(-2)$, we once again visualize the angle $\theta = \text{arccot}(-2)$ radians and note that it is a Quadrant II angle with $\tan(\theta) = -\frac{1}{2}$. This means it is exactly π units away from β, and we get $\theta = \pi + \beta = \pi + \arctan\left(-\frac{1}{2}\right) \approx 2.6779$ radians. Hence, as before, $\text{arccot}(-2) \approx 2.6779$.

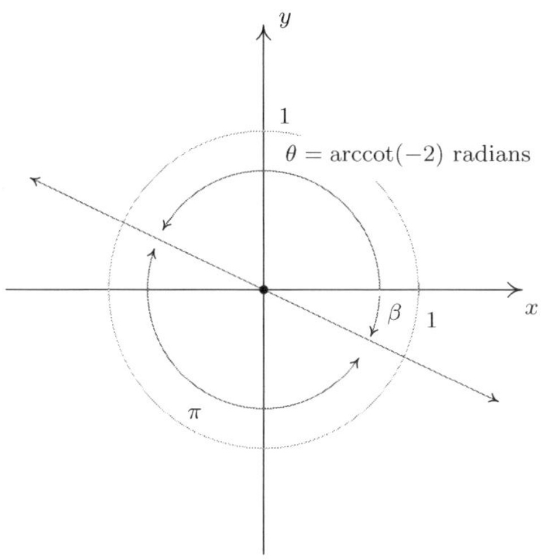

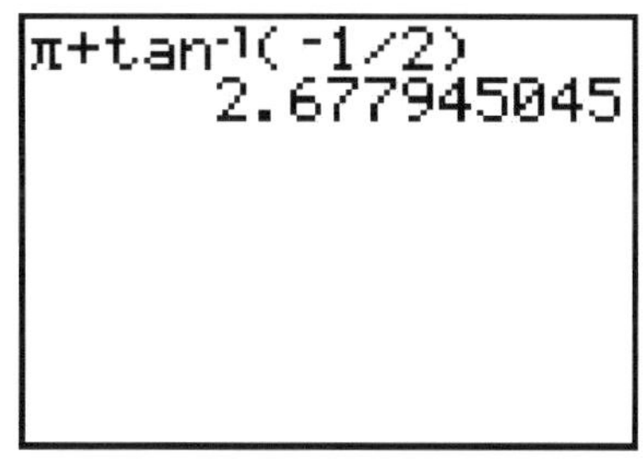

(d) If the range of arccosecant is taken to be $\left[-\frac{\pi}{2}, 0\right) \cup \left(0, \frac{\pi}{2}\right]$, we can use Theorem 10.28 to get $\text{arccsc}\left(-\frac{3}{2}\right) = \arcsin\left(-\frac{2}{3}\right) \approx -0.7297$. If, on the other hand, the range of arccosecant is taken to be $\left(0, \frac{\pi}{2}\right] \cup \left(\pi, \frac{3\pi}{2}\right]$, then we proceed as in the previous problem by letting $t = \text{arccsc}\left(-\frac{3}{2}\right)$. Then t is a real number with $\csc(t) = -\frac{3}{2}$. Since $\csc(t) < 0$, we have that $\pi < \theta \leq \frac{3\pi}{2}$, so t corresponds to a Quadrant III angle, θ. As above, we let α be the reference angle for θ. Then $0 < \alpha < \frac{\pi}{2}$ and $\csc(\alpha) = \frac{3}{2}$, which means $\alpha = \text{arccsc}\left(\frac{3}{2}\right)$ radians. Since the argument of arccosecant is now positive, we may use Theorem 10.29 to get $\alpha = \text{arccsc}\left(\frac{3}{2}\right) = \arcsin\left(\frac{2}{3}\right)$ radians. Since $\theta = \pi + \alpha = \pi + \arcsin\left(\frac{2}{3}\right) \approx 3.8713$ radians, $\text{arccsc}\left(-\frac{3}{2}\right) \approx 3.8713$.

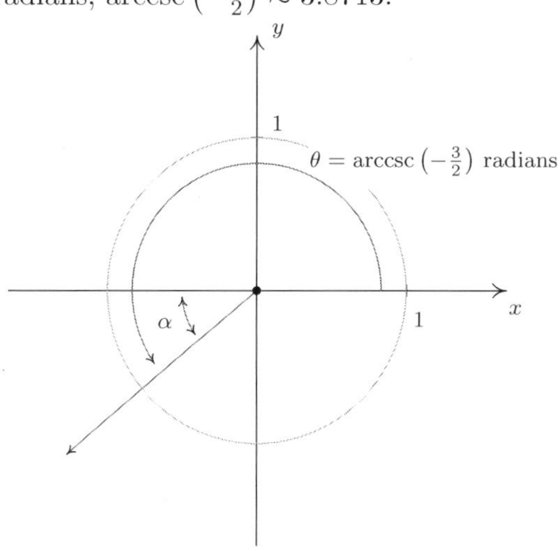

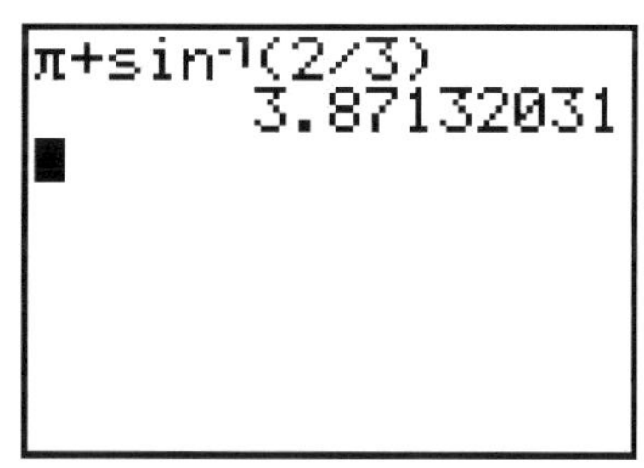

2. (a) Since the domain of $F(x) = \arccos(x)$ is $-1 \leq x \leq 1$, we can find the domain of $f(x) = \frac{\pi}{2} - \arccos\left(\frac{x}{5}\right)$ by setting the argument of the arccosine, in this case $\frac{x}{5}$, between -1 and 1. Solving $-1 \leq \frac{x}{5} \leq 1$ gives $-5 \leq x \leq 5$, so the domain is $[-5, 5]$. To determine the range of f, we take a cue from Section 1.7. Three 'key' points on the graph of $F(x) = \arccos(x)$ are $(-1, \pi)$, $\left(0, \frac{\pi}{2}\right)$ and $(1, 0)$. Following the procedure outlined in Theorem 1.7, we track these points to $\left(-5, -\frac{\pi}{2}\right)$, $(0, 0)$ and $\left(5, \frac{\pi}{2}\right)$. Plotting these values tells us that the range[5] of f is $\left[-\frac{\pi}{2}, \frac{\pi}{2}\right]$. Our graph confirms our results.

   (b) To find the domain and range of $f(x) = 3\arctan(4x)$, we note that since the domain of $F(x) = \arctan(x)$ is all real numbers, the only restrictions, if any, on the domain of $f(x) = 3\arctan(4x)$ come from the argument of the arctangent, in this case, $4x$. Since $4x$ is defined for all real numbers, we have established that the domain of f is all real numbers. To determine the range of f, we can, once again, appeal to Theorem 1.7. Choosing our 'key' point to be $(0, 0)$ and tracking the horizontal asymptotes $y = -\frac{\pi}{2}$ and $y = \frac{\pi}{2}$, we find that the graph of $y = f(x) = 3\arctan(4x)$ differs from the graph of $y = F(x) = \arctan(x)$ by a horizontal compression by a factor of 4 and a vertical stretch by a factor of 3. It is the latter which affects the range, producing a range of $\left(-\frac{3\pi}{2}, \frac{3\pi}{2}\right)$. We confirm our findings on the calculator below.

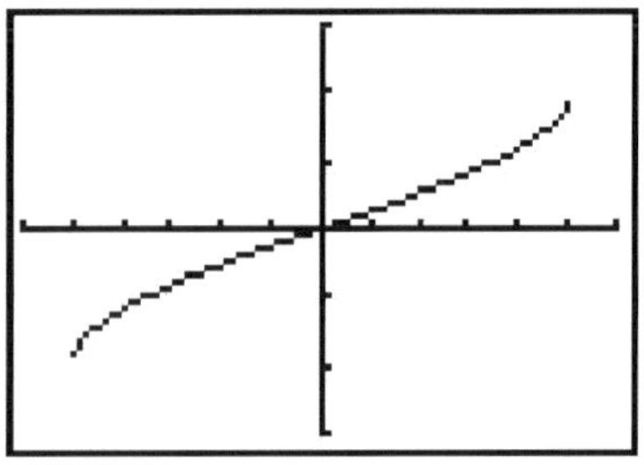

$y = f(x) = \dfrac{\pi}{2} - \arccos\left(\dfrac{x}{5}\right)$

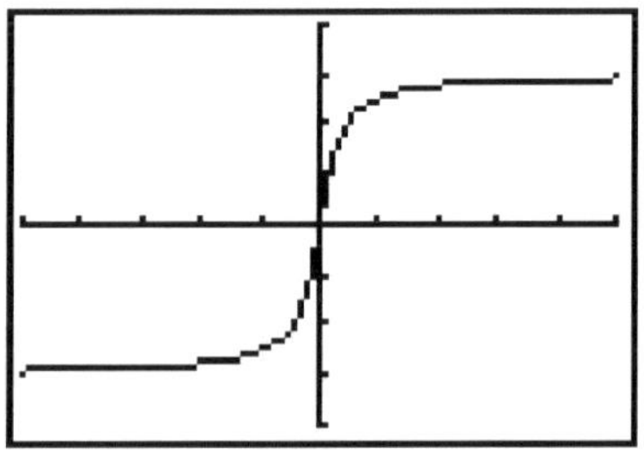

$y = f(x) = 3\arctan(4x)$

   (c) To find the domain of $g(x) = \operatorname{arccot}\left(\frac{x}{2}\right) + \pi$, we proceed as above. Since the domain of $G(x) = \operatorname{arccot}(x)$ is $(-\infty, \infty)$, and $\frac{x}{2}$ is defined for all x, we get that the domain of g is $(-\infty, \infty)$ as well. As for the range, we note that the range of $G(x) = \operatorname{arccot}(x)$, like that of $F(x) = \arctan(x)$, is limited by a pair of horizontal asymptotes, in this case $y = 0$ and $y = \pi$. Following Theorem 1.7, we graph $y = g(x) = \operatorname{arccot}\left(\frac{x}{2}\right) + \pi$ starting with $y = G(x) = \operatorname{arccot}(x)$ and first performing a horizontal expansion by a factor of 2 and following that with a vertical shift upwards by π. This latter transformation is the one which affects the range, making it now $(\pi, 2\pi)$. To check this graphically, we encounter a bit of a problem, since on many calculators, there is no shortcut button corresponding to the arccotangent function. Taking a cue from number 1c, we attempt to rewrite $g(x) = \operatorname{arccot}\left(\frac{x}{2}\right) + \pi$ in terms of the arctangent function. Using Theorem 10.27, we have that $\operatorname{arccot}\left(\frac{x}{2}\right) = \arctan\left(\frac{2}{x}\right)$ when $\frac{x}{2} > 0$, or, in this case, when $x > 0$. Hence, for $x > 0$, we have $g(x) = \arctan\left(\frac{2}{x}\right) + \pi$. When $\frac{x}{2} < 0$, we can use the same argument in number 1c that gave us $\operatorname{arccot}(-2) = \pi + \arctan\left(-\frac{1}{2}\right)$ to give us $\operatorname{arccot}\left(\frac{x}{2}\right) = \pi + \arctan\left(\frac{2}{x}\right)$.

[5] It also confirms our domain!

Hence, for $x < 0$, $g(x) = \pi + \arctan\left(\frac{2}{x}\right) + \pi = \arctan\left(\frac{2}{x}\right) + 2\pi$. What about $x = 0$? We know $g(0) = \operatorname{arccot}(0) + \pi = \pi$, and neither of the formulas for g involving arctangent will produce this result.[6] Hence, in order to graph $y = g(x)$ on our calculators, we need to write it as a piecewise defined function:

$$g(x) = \operatorname{arccot}\left(\frac{x}{2}\right) + \pi = \begin{cases} \arctan\left(\frac{2}{x}\right) + 2\pi, & \text{if } x < 0 \\ \pi, & \text{if } x = 0 \\ \arctan\left(\frac{2}{x}\right) + \pi, & \text{if } x > 0 \end{cases}$$

We show the input and the result below.

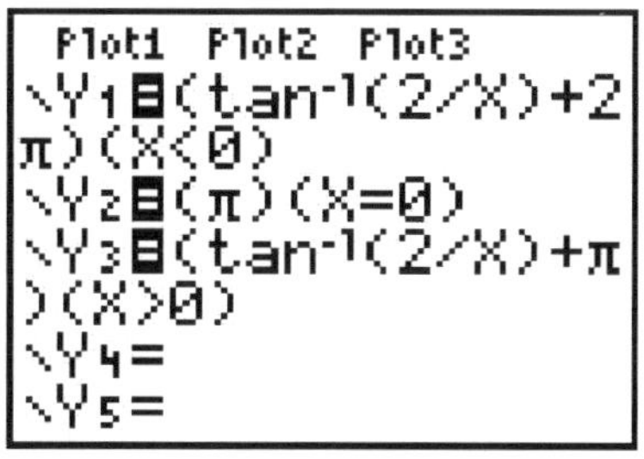

$y = g(x)$ in terms of arctangent

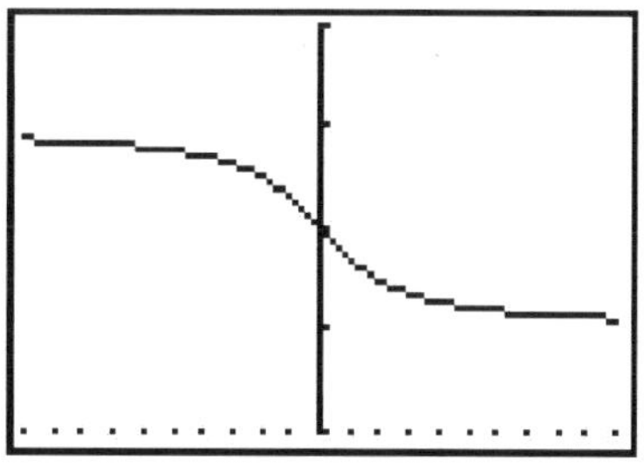
$y = g(x) = \operatorname{arccot}\left(\frac{x}{2}\right) + \pi$

□

The inverse trigonometric functions are typically found in applications whenever the measure of an angle is required. One such scenario is presented in the following example.

Example 10.6.6. [7] The roof on the house below has a '6/12 pitch'. This means that when viewed from the side, the roof line has a rise of 6 feet over a run of 12 feet. Find the angle of inclination from the bottom of the roof to the top of the roof. Express your answer in decimal degrees, rounded to the nearest hundredth of a degree.

Front View

Side View

Solution. If we divide the side view of the house down the middle, we find that the roof line forms the hypotenuse of a right triangle with legs of length 6 feet and 12 feet. Using Theorem 10.10, we

[6] Without Calculus, of course ...

[7] The authors would like to thank Dan Stitz for this problem and associated graphics.

find the angle of inclination, labeled θ below, satisfies $\tan(\theta) = \frac{6}{12} = \frac{1}{2}$. Since θ is an acute angle, we can use the arctangent function and we find $\theta = \arctan\left(\frac{1}{2}\right)$ radians $\approx 26.56°$.

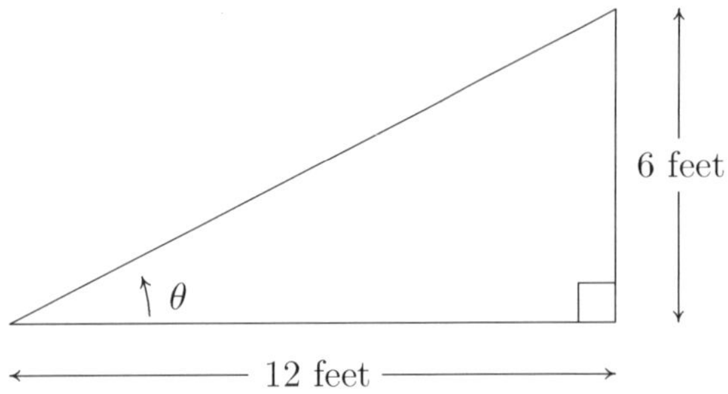

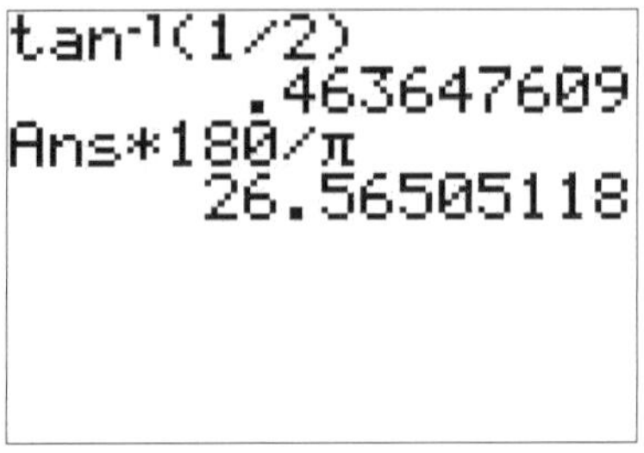

□

10.6.4 Solving Equations Using the Inverse Trigonometric Functions.

In Sections 10.2 and 10.3, we learned how to solve equations like $\sin(\theta) = \frac{1}{2}$ for angles θ and $\tan(t) = -1$ for real numbers t. In each case, we ultimately appealed to the Unit Circle and relied on the fact that the answers corresponded to a set of 'common angles' listed on page 724. If, on the other hand, we had been asked to find all angles with $\sin(\theta) = \frac{1}{3}$ or solve $\tan(t) = -2$ for real numbers t, we would have been hard-pressed to do so. With the introduction of the inverse trigonometric functions, however, we are now in a position to solve these equations. A good parallel to keep in mind is how the square root function can be used to solve certain quadratic equations. The equation $x^2 = 4$ is a lot like $\sin(\theta) = \frac{1}{2}$ in that it has friendly, 'common value' answers $x = \pm 2$. The equation $x^2 = 7$, on the other hand, is a lot like $\sin(\theta) = \frac{1}{3}$. We know[8] there are answers, but we can't express them using 'friendly' numbers.[9] To solve $x^2 = 7$, we make use of the square root function and write $x = \pm\sqrt{7}$. We can certainly *approximate* these answers using a calculator, but as far as exact answers go, we leave them as $x = \pm\sqrt{7}$. In the same way, we will use the arcsine function to solve $\sin(\theta) = \frac{1}{3}$, as seen in the following example.

Example 10.6.7. Solve the following equations.

1. Find all angles θ for which $\sin(\theta) = \frac{1}{3}$.
2. Find all real numbers t for which $\tan(t) = -2$
3. Solve $\sec(x) = -\frac{5}{3}$ for x.

Solution.

1. If $\sin(\theta) = \frac{1}{3}$, then the terminal side of θ, when plotted in standard position, intersects the Unit Circle at $y = \frac{1}{3}$. Geometrically, we see that this happens at two places: in Quadrant I and Quadrant II. If we let α denote the acute solution to the equation, then all the solutions

[8] How do we know this again?

[9] This is all, of course, a matter of opinion. For the record, the authors find $\pm\sqrt{7}$ just as 'nice' as ± 2.

to this equation in Quadrant I are coterminal with α, and α serves as the reference angle for all of the solutions to this equation in Quadrant II.

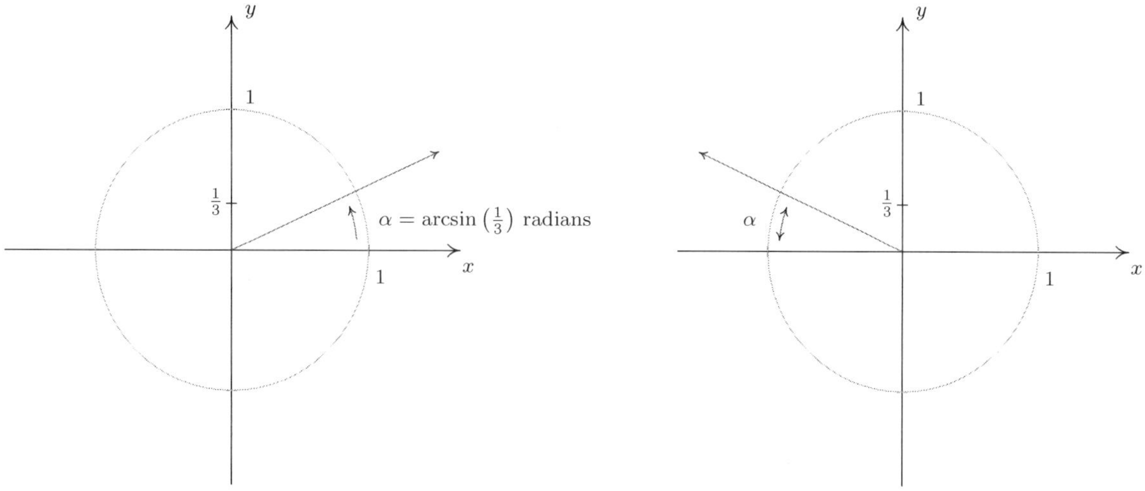

Since $\frac{1}{3}$ isn't the sine of any of the 'common angles' discussed earlier, we use the arcsine functions to express our answers. The real number $t = \arcsin\left(\frac{1}{3}\right)$ is defined so it satisfies $0 < t < \frac{\pi}{2}$ with $\sin(t) = \frac{1}{3}$. Hence, $\alpha = \arcsin\left(\frac{1}{3}\right)$ radians. Since the solutions in Quadrant I are all coterminal with α, we get part of our solution to be $\theta = \alpha + 2\pi k = \arcsin\left(\frac{1}{3}\right) + 2\pi k$ for integers k. Turning our attention to Quadrant II, we get one solution to be $\pi - \alpha$. Hence, the Quadrant II solutions are $\theta = \pi - \alpha + 2\pi k = \pi - \arcsin\left(\frac{1}{3}\right) + 2\pi k$, for integers k.

2. We may visualize the solutions to $\tan(t) = -2$ as angles θ with $\tan(\theta) = -2$. Since tangent is negative only in Quadrants II and IV, we focus our efforts there.

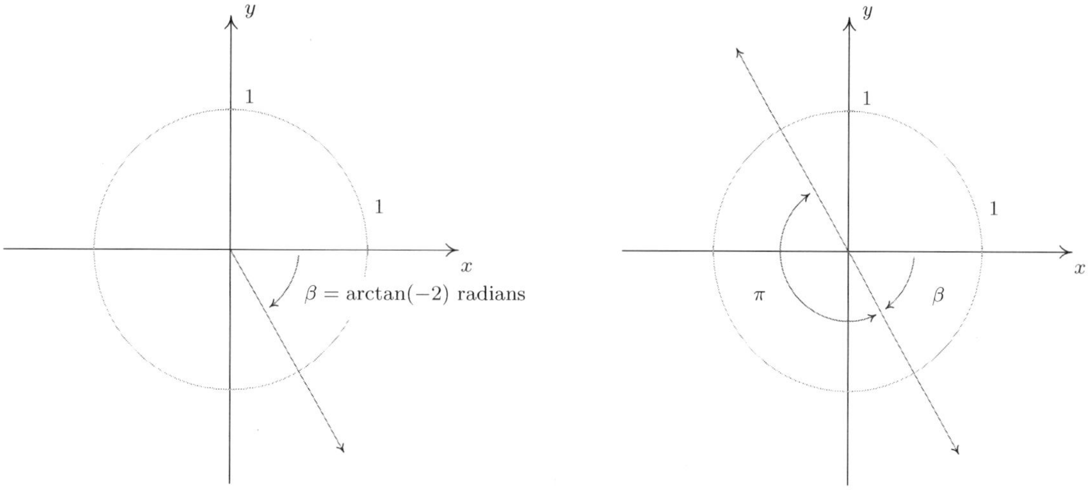

Since -2 isn't the tangent of any of the 'common angles', we need to use the arctangent function to express our answers. The real number $t = \arctan(-2)$ satisfies $\tan(t) = -2$ and $-\frac{\pi}{2} < t < 0$. If we let $\beta = \arctan(-2)$ radians, we see that all of the Quadrant IV solutions to $\tan(\theta) = -2$ are coterminal with β. Moreover, the solutions from Quadrant II differ by exactly π units from the solutions in Quadrant IV, so all the solutions to $\tan(\theta) = -2$ are of the form $\theta = \beta + \pi k = \arctan(-2) + \pi k$ for some integer k. Switching back to the variable t, we record our final answer to $\tan(t) = -2$ as $t = \arctan(-2) + \pi k$ for integers k.

3. The last equation we are asked to solve, $\sec(x) = -\frac{5}{3}$, poses two immediate problems. First, we are not told whether or not x represents an angle or a real number. We assume the latter, but note that we will use angles and the Unit Circle to solve the equation regardless. Second, as we have mentioned, there is no universally accepted range of the arcsecant function. For that reason, we adopt the advice given in Section 10.3 and convert this to the cosine problem $\cos(x) = -\frac{3}{5}$. Adopting an angle approach, we consider the equation $\cos(\theta) = -\frac{3}{5}$ and note that our solutions lie in Quadrants II and III. Since $-\frac{3}{5}$ isn't the cosine of any of the 'common angles', we'll need to express our solutions in terms of the arccosine function. The real number $t = \arccos\left(-\frac{3}{5}\right)$ is defined so that $\frac{\pi}{2} < t < \pi$ with $\cos(t) = -\frac{3}{5}$. If we let $\beta = \arccos\left(-\frac{3}{5}\right)$ radians, we see that β is a Quadrant II angle. To obtain a Quadrant III angle solution, we may simply use $-\beta = -\arccos\left(-\frac{3}{5}\right)$. Since all angle solutions are coterminal with β or $-\beta$, we get our solutions to $\cos(\theta) = -\frac{3}{5}$ to be $\theta = \beta + 2\pi k = \arccos\left(-\frac{3}{5}\right) + 2\pi k$ or $\theta = -\beta + 2\pi k = -\arccos\left(-\frac{3}{5}\right) + 2\pi k$ for integers k. Switching back to the variable x, we record our final answer to $\sec(x) = -\frac{5}{3}$ as $x = \arccos\left(-\frac{3}{5}\right) + 2\pi k$ or $x = -\arccos\left(-\frac{3}{5}\right) + 2\pi k$ for integers k.

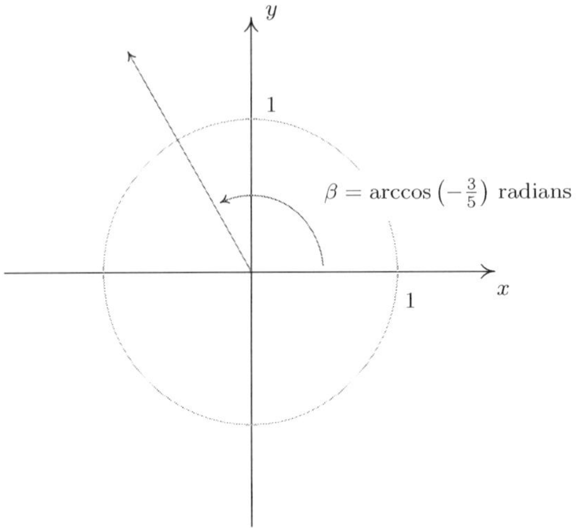

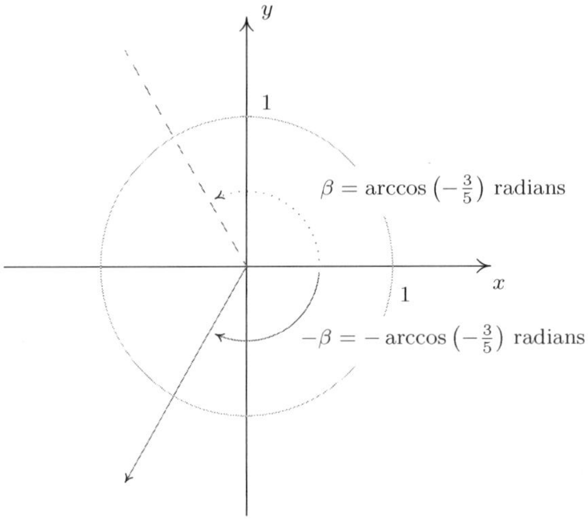

□

The reader is encouraged to check the answers found in Example 10.6.7 - both analytically and with the calculator (see Section 10.6.3). With practice, the inverse trigonometric functions will become as familiar to you as the square root function. Speaking of practice ...

10.6.5 Exercises

In Exercises 1 - 40, find the exact value.

1. $\arcsin(-1)$
2. $\arcsin\left(-\frac{\sqrt{3}}{2}\right)$
3. $\arcsin\left(-\frac{\sqrt{2}}{2}\right)$
4. $\arcsin\left(-\frac{1}{2}\right)$
5. $\arcsin(0)$
6. $\arcsin\left(\frac{1}{2}\right)$
7. $\arcsin\left(\frac{\sqrt{2}}{2}\right)$
8. $\arcsin\left(\frac{\sqrt{3}}{2}\right)$
9. $\arcsin(1)$
10. $\arccos(-1)$
11. $\arccos\left(-\frac{\sqrt{3}}{2}\right)$
12. $\arccos\left(-\frac{\sqrt{2}}{2}\right)$
13. $\arccos\left(-\frac{1}{2}\right)$
14. $\arccos(0)$
15. $\arccos\left(\frac{1}{2}\right)$
16. $\arccos\left(\frac{\sqrt{2}}{2}\right)$
17. $\arccos\left(\frac{\sqrt{3}}{2}\right)$
18. $\arccos(1)$
19. $\arctan\left(-\sqrt{3}\right)$
20. $\arctan(-1)$
21. $\arctan\left(-\frac{\sqrt{3}}{3}\right)$
22. $\arctan(0)$
23. $\arctan\left(\frac{\sqrt{3}}{3}\right)$
24. $\arctan(1)$
25. $\arctan\left(\sqrt{3}\right)$
26. $\operatorname{arccot}\left(-\sqrt{3}\right)$
27. $\operatorname{arccot}(-1)$
28. $\operatorname{arccot}\left(-\frac{\sqrt{3}}{3}\right)$
29. $\operatorname{arccot}(0)$
30. $\operatorname{arccot}\left(\frac{\sqrt{3}}{3}\right)$
31. $\operatorname{arccot}(1)$
32. $\operatorname{arccot}\left(\sqrt{3}\right)$
33. $\operatorname{arcsec}(2)$
34. $\operatorname{arccsc}(2)$
35. $\operatorname{arcsec}\left(\sqrt{2}\right)$
36. $\operatorname{arccsc}\left(\sqrt{2}\right)$
37. $\operatorname{arcsec}\left(\frac{2\sqrt{3}}{3}\right)$
38. $\operatorname{arccsc}\left(\frac{2\sqrt{3}}{3}\right)$
39. $\operatorname{arcsec}(1)$
40. $\operatorname{arccsc}(1)$

In Exercises 41 - 48, assume that the range of arcsecant is $\left[0, \frac{\pi}{2}\right) \cup \left[\pi, \frac{3\pi}{2}\right)$ and that the range of arccosecant is $\left(0, \frac{\pi}{2}\right] \cup \left(\pi, \frac{3\pi}{2}\right]$ when finding the exact value.

41. $\operatorname{arcsec}(-2)$
42. $\operatorname{arcsec}\left(-\sqrt{2}\right)$
43. $\operatorname{arcsec}\left(-\frac{2\sqrt{3}}{3}\right)$
44. $\operatorname{arcsec}(-1)$
45. $\operatorname{arccsc}(-2)$
46. $\operatorname{arccsc}\left(-\sqrt{2}\right)$
47. $\operatorname{arccsc}\left(-\frac{2\sqrt{3}}{3}\right)$
48. $\operatorname{arccsc}(-1)$

In Exercises 49 - 56, assume that the range of arcsecant is $\left[0, \frac{\pi}{2}\right) \cup \left(\frac{\pi}{2}, \pi\right]$ and that the range of arccosecant is $\left[-\frac{\pi}{2}, 0\right) \cup \left(0, \frac{\pi}{2}\right]$ when finding the exact value.

49. $\text{arcsec}(-2)$
50. $\text{arcsec}(-\sqrt{2})$
51. $\text{arcsec}\left(-\frac{2\sqrt{3}}{3}\right)$
52. $\text{arcsec}(-1)$
53. $\text{arccsc}(-2)$
54. $\text{arccsc}(-\sqrt{2})$
55. $\text{arccsc}\left(-\frac{2\sqrt{3}}{3}\right)$
56. $\text{arccsc}(-1)$

In Exercises 57 - 86, find the exact value or state that it is undefined.

57. $\sin\left(\arcsin\left(\frac{1}{2}\right)\right)$
58. $\sin\left(\arcsin\left(-\frac{\sqrt{2}}{2}\right)\right)$
59. $\sin\left(\arcsin\left(\frac{3}{5}\right)\right)$
60. $\sin(\arcsin(-0.42))$
61. $\sin\left(\arcsin\left(\frac{5}{4}\right)\right)$
62. $\cos\left(\arccos\left(\frac{\sqrt{2}}{2}\right)\right)$
63. $\cos\left(\arccos\left(-\frac{1}{2}\right)\right)$
64. $\cos\left(\arccos\left(\frac{5}{13}\right)\right)$
65. $\cos(\arccos(-0.998))$
66. $\cos(\arccos(\pi))$
67. $\tan(\arctan(-1))$
68. $\tan(\arctan(\sqrt{3}))$
69. $\tan\left(\arctan\left(\frac{5}{12}\right)\right)$
70. $\tan(\arctan(0.965))$
71. $\tan(\arctan(3\pi))$
72. $\cot(\text{arccot}(1))$
73. $\cot(\text{arccot}(-\sqrt{3}))$
74. $\cot\left(\text{arccot}\left(-\frac{7}{24}\right)\right)$
75. $\cot(\text{arccot}(-0.001))$
76. $\cot\left(\text{arccot}\left(\frac{17\pi}{4}\right)\right)$
77. $\sec(\text{arcsec}(2))$
78. $\sec(\text{arcsec}(-1))$
79. $\sec\left(\text{arcsec}\left(\frac{1}{2}\right)\right)$
80. $\sec(\text{arcsec}(0.75))$
81. $\sec(\text{arcsec}(117\pi))$
82. $\csc(\text{arccsc}(\sqrt{2}))$
83. $\csc\left(\text{arccsc}\left(-\frac{2\sqrt{3}}{3}\right)\right)$
84. $\csc\left(\text{arccsc}\left(\frac{\sqrt{2}}{2}\right)\right)$
85. $\csc(\text{arccsc}(1.0001))$
86. $\csc\left(\text{arccsc}\left(\frac{\pi}{4}\right)\right)$

In Exercises 87 - 106, find the exact value or state that it is undefined.

87. $\arcsin\left(\sin\left(\frac{\pi}{6}\right)\right)$
88. $\arcsin\left(\sin\left(-\frac{\pi}{3}\right)\right)$
89. $\arcsin\left(\sin\left(\frac{3\pi}{4}\right)\right)$

90. $\arcsin\left(\sin\left(\frac{11\pi}{6}\right)\right)$
91. $\arcsin\left(\sin\left(\frac{4\pi}{3}\right)\right)$
92. $\arccos\left(\cos\left(\frac{\pi}{4}\right)\right)$
93. $\arccos\left(\cos\left(\frac{2\pi}{3}\right)\right)$
94. $\arccos\left(\cos\left(\frac{3\pi}{2}\right)\right)$
95. $\arccos\left(\cos\left(-\frac{\pi}{6}\right)\right)$
96. $\arccos\left(\cos\left(\frac{5\pi}{4}\right)\right)$
97. $\arctan\left(\tan\left(\frac{\pi}{3}\right)\right)$
98. $\arctan\left(\tan\left(-\frac{\pi}{4}\right)\right)$
99. $\arctan(\tan(\pi))$
100. $\arctan\left(\tan\left(\frac{\pi}{2}\right)\right)$
101. $\arctan\left(\tan\left(\frac{2\pi}{3}\right)\right)$
102. $\text{arccot}\left(\cot\left(\frac{\pi}{3}\right)\right)$
103. $\text{arccot}\left(\cot\left(-\frac{\pi}{4}\right)\right)$
104. $\text{arccot}(\cot(\pi))$
105. $\text{arccot}\left(\cot\left(\frac{\pi}{2}\right)\right)$
106. $\text{arccot}\left(\cot\left(\frac{2\pi}{3}\right)\right)$

In Exercises 107 - 118, assume that the range of arcsecant is $\left[0, \frac{\pi}{2}\right) \cup \left[\pi, \frac{3\pi}{2}\right)$ and that the range of arccosecant is $\left(0, \frac{\pi}{2}\right] \cup \left(\pi, \frac{3\pi}{2}\right]$ when finding the exact value.

107. $\text{arcsec}\left(\sec\left(\frac{\pi}{4}\right)\right)$
108. $\text{arcsec}\left(\sec\left(\frac{4\pi}{3}\right)\right)$
109. $\text{arcsec}\left(\sec\left(\frac{5\pi}{6}\right)\right)$
110. $\text{arcsec}\left(\sec\left(-\frac{\pi}{2}\right)\right)$
111. $\text{arcsec}\left(\sec\left(\frac{5\pi}{3}\right)\right)$
112. $\text{arccsc}\left(\csc\left(\frac{\pi}{6}\right)\right)$
113. $\text{arccsc}\left(\csc\left(\frac{5\pi}{4}\right)\right)$
114. $\text{arccsc}\left(\csc\left(\frac{2\pi}{3}\right)\right)$
115. $\text{arccsc}\left(\csc\left(-\frac{\pi}{2}\right)\right)$
116. $\text{arccsc}\left(\csc\left(\frac{11\pi}{6}\right)\right)$
117. $\text{arcsec}\left(\sec\left(\frac{11\pi}{12}\right)\right)$
118. $\text{arccsc}\left(\csc\left(\frac{9\pi}{8}\right)\right)$

In Exercises 119 - 130, assume that the range of arcsecant is $\left[0, \frac{\pi}{2}\right) \cup \left(\frac{\pi}{2}, \pi\right]$ and that the range of arccosecant is $\left[-\frac{\pi}{2}, 0\right) \cup \left(0, \frac{\pi}{2}\right]$ when finding the exact value.

119. $\text{arcsec}\left(\sec\left(\frac{\pi}{4}\right)\right)$
120. $\text{arcsec}\left(\sec\left(\frac{4\pi}{3}\right)\right)$
121. $\text{arcsec}\left(\sec\left(\frac{5\pi}{6}\right)\right)$
122. $\text{arcsec}\left(\sec\left(-\frac{\pi}{2}\right)\right)$
123. $\text{arcsec}\left(\sec\left(\frac{5\pi}{3}\right)\right)$
124. $\text{arccsc}\left(\csc\left(\frac{\pi}{6}\right)\right)$
125. $\text{arccsc}\left(\csc\left(\frac{5\pi}{4}\right)\right)$
126. $\text{arccsc}\left(\csc\left(\frac{2\pi}{3}\right)\right)$
127. $\text{arccsc}\left(\csc\left(-\frac{\pi}{2}\right)\right)$
128. $\text{arccsc}\left(\csc\left(\frac{11\pi}{6}\right)\right)$
129. $\text{arcsec}\left(\sec\left(\frac{11\pi}{12}\right)\right)$
130. $\text{arccsc}\left(\csc\left(\frac{9\pi}{8}\right)\right)$

In Exercises 131 - 154, find the exact value or state that it is undefined.

131. $\sin\left(\arccos\left(-\frac{1}{2}\right)\right)$

132. $\sin\left(\arccos\left(\frac{3}{5}\right)\right)$

133. $\sin(\arctan(-2))$

134. $\sin\left(\operatorname{arccot}\left(\sqrt{5}\right)\right)$

135. $\sin(\operatorname{arccsc}(-3))$

136. $\cos\left(\arcsin\left(-\frac{5}{13}\right)\right)$

137. $\cos\left(\arctan\left(\sqrt{7}\right)\right)$

138. $\cos(\operatorname{arccot}(3))$

139. $\cos(\operatorname{arcsec}(5))$

140. $\tan\left(\arcsin\left(-\frac{2\sqrt{5}}{5}\right)\right)$

141. $\tan\left(\arccos\left(-\frac{1}{2}\right)\right)$

142. $\tan\left(\operatorname{arcsec}\left(\frac{5}{3}\right)\right)$

143. $\tan(\operatorname{arccot}(12))$

144. $\cot\left(\arcsin\left(\frac{12}{13}\right)\right)$

145. $\cot\left(\arccos\left(\frac{\sqrt{3}}{2}\right)\right)$

146. $\cot\left(\operatorname{arccsc}\left(\sqrt{5}\right)\right)$

147. $\cot(\arctan(0.25))$

148. $\sec\left(\arccos\left(\frac{\sqrt{3}}{2}\right)\right)$

149. $\sec\left(\arcsin\left(-\frac{12}{13}\right)\right)$

150. $\sec(\arctan(10))$

151. $\sec\left(\operatorname{arccot}\left(-\frac{\sqrt{10}}{10}\right)\right)$

152. $\csc(\operatorname{arccot}(9))$

153. $\csc\left(\arcsin\left(\frac{3}{5}\right)\right)$

154. $\csc\left(\arctan\left(-\frac{2}{3}\right)\right)$

In Exercises 155 - 164, find the exact value or state that it is undefined.

155. $\sin\left(\arcsin\left(\frac{5}{13}\right)+\frac{\pi}{4}\right)$

156. $\cos(\operatorname{arcsec}(3)+\arctan(2))$

157. $\tan\left(\arctan(3)+\arccos\left(-\frac{3}{5}\right)\right)$

158. $\sin\left(2\arcsin\left(-\frac{4}{5}\right)\right)$

159. $\sin\left(2\operatorname{arccsc}\left(\frac{13}{5}\right)\right)$

160. $\sin(2\arctan(2))$

161. $\cos\left(2\arcsin\left(\frac{3}{5}\right)\right)$

162. $\cos\left(2\operatorname{arcsec}\left(\frac{25}{7}\right)\right)$

163. $\cos\left(2\operatorname{arccot}\left(-\sqrt{5}\right)\right)$

164. $\sin\left(\frac{\arctan(2)}{2}\right)$

In Exercises 165 - 184, rewrite the quantity as algebraic expressions of x and state the domain on which the equivalence is valid.

165. $\sin(\arccos(x))$

166. $\cos(\arctan(x))$

167. $\tan(\arcsin(x))$

168. $\sec(\arctan(x))$

169. $\csc(\arccos(x))$

170. $\sin(2\arctan(x))$

171. $\sin(2\arccos(x))$

172. $\cos(2\arctan(x))$

173. $\sin(\arccos(2x))$

174. $\sin\left(\arccos\left(\frac{x}{5}\right)\right)$

175. $\cos\left(\arcsin\left(\frac{x}{2}\right)\right)$

176. $\cos(\arctan(3x))$

177. $\sin(2\arcsin(7x))$

178. $\sin\left(2\arcsin\left(\frac{x\sqrt{3}}{3}\right)\right)$

179. $\cos(2\arcsin(4x))$

180. $\sec(\arctan(2x))\tan(\arctan(2x))$

181. $\sin(\arcsin(x) + \arccos(x))$

182. $\cos(\arcsin(x) + \arctan(x))$

183. $\tan(2\arcsin(x))$

184. $\sin\left(\frac{1}{2}\arctan(x)\right)$

185. If $\sin(\theta) = \frac{x}{2}$ for $-\frac{\pi}{2} < \theta < \frac{\pi}{2}$, find an expression for $\theta + \sin(2\theta)$ in terms of x.

186. If $\tan(\theta) = \frac{x}{7}$ for $-\frac{\pi}{2} < \theta < \frac{\pi}{2}$, find an expression for $\frac{1}{2}\theta - \frac{1}{2}\sin(2\theta)$ in terms of x.

187. If $\sec(\theta) = \frac{x}{4}$ for $0 < \theta < \frac{\pi}{2}$, find an expression for $4\tan(\theta) - 4\theta$ in terms of x.

In Exercises 188 - 207, solve the equation using the techniques discussed in Example 10.6.7 then approximate the solutions which lie in the interval $[0, 2\pi)$ to four decimal places.

188. $\sin(x) = \frac{7}{11}$

189. $\cos(x) = -\frac{2}{9}$

190. $\sin(x) = -0.569$

191. $\cos(x) = 0.117$

192. $\sin(x) = 0.008$

193. $\cos(x) = \frac{359}{360}$

194. $\tan(x) = 117$

195. $\cot(x) = -12$

196. $\sec(x) = \frac{3}{2}$

197. $\csc(x) = -\frac{90}{17}$

198. $\tan(x) = -\sqrt{10}$

199. $\sin(x) = \frac{3}{8}$

200. $\cos(x) = -\frac{7}{16}$

201. $\tan(x) = 0.03$

202. $\sin(x) = 0.3502$

203. $\sin(x) = -0.721$

204. $\cos(x) = 0.9824$

205. $\cos(x) = -0.5637$

206. $\cot(x) = \frac{1}{117}$

207. $\tan(x) = -0.6109$

In Exercises 208 - 210, find the two acute angles in the right triangle whose sides have the given lengths. Express your answers using degree measure rounded to two decimal places.

208. 3, 4 and 5

209. 5, 12 and 13

210. 336, 527 and 625

211. A guy wire 1000 feet long is attached to the top of a tower. When pulled taut it touches level ground 360 feet from the base of the tower. What angle does the wire make with the ground? Express your answer using degree measure rounded to one decimal place.

212. At Cliffs of Insanity Point, The Great Sasquatch Canyon is 7117 feet deep. From that point, a fire is seen at a location known to be 10 miles away from the base of the sheer canyon wall. What angle of depression is made by the line of sight from the canyon edge to the fire? Express your answer using degree measure rounded to one decimal place.

213. Shelving is being built at the Utility Muffin Research Library which is to be 14 inches deep. An 18-inch rod will be attached to the wall and the underside of the shelf at its edge away from the wall, forming a right triangle under the shelf to support it. What angle, to the nearest degree, will the rod make with the wall?

214. A parasailor is being pulled by a boat on Lake Ippizuti. The cable is 300 feet long and the parasailor is 100 feet above the surface of the water. What is the angle of elevation from the boat to the parasailor? Express your answer using degree measure rounded to one decimal place.

215. A tag-and-release program to study the Sasquatch population of the eponymous Sasquatch National Park is begun. From a 200 foot tall tower, a ranger spots a Sasquatch lumbering through the wilderness directly towards the tower. Let θ denote the angle of depression from the top of the tower to a point on the ground. If the range of the rifle with a tranquilizer dart is 300 feet, find the smallest value of θ for which the corresponding point on the ground is in range of the rifle. Round your answer to the nearest hundreth of a degree.

In Exercises 216 - 221, rewrite the given function as a sinusoid of the form $S(x) = A\sin(\omega x + \phi)$ using Exercises 35 and 36 in Section 10.5 for reference. Approximate the value of ϕ (which is in radians, of course) to four decimal places.

216. $f(x) = 5\sin(3x) + 12\cos(3x)$

217. $f(x) = 3\cos(2x) + 4\sin(2x)$

218. $f(x) = \cos(x) - 3\sin(x)$

219. $f(x) = 7\sin(10x) - 24\cos(10x)$

220. $f(x) = -\cos(x) - 2\sqrt{2}\sin(x)$

221. $f(x) = 2\sin(x) - \cos(x)$

In Exercises 222 - 233, find the domain of the given function. Write your answers in interval notation.

222. $f(x) = \arcsin(5x)$

223. $f(x) = \arccos\left(\dfrac{3x-1}{2}\right)$

224. $f(x) = \arcsin\left(2x^2\right)$

225. $f(x) = \arccos\left(\dfrac{1}{x^2-4}\right)$

226. $f(x) = \arctan(4x)$

227. $f(x) = \operatorname{arccot}\left(\dfrac{2x}{x^2-9}\right)$

228. $f(x) = \arctan(\ln(2x-1))$

229. $f(x) = \operatorname{arccot}(\sqrt{2x-1})$

230. $f(x) = \operatorname{arcsec}(12x)$

231. $f(x) = \operatorname{arccsc}(x+5)$

232. $f(x) = \operatorname{arcsec}\left(\dfrac{x^3}{8}\right)$

233. $f(x) = \operatorname{arccsc}\left(e^{2x}\right)$

234. Show that $\operatorname{arcsec}(x) = \arccos\left(\dfrac{1}{x}\right)$ for $|x| \geq 1$ as long as we use $\left[0, \dfrac{\pi}{2}\right) \cup \left(\dfrac{\pi}{2}, \pi\right]$ as the range of $f(x) = \operatorname{arcsec}(x)$.

235. Show that $\operatorname{arccsc}(x) = \arcsin\left(\dfrac{1}{x}\right)$ for $|x| \geq 1$ as long as we use $\left[-\dfrac{\pi}{2}, 0\right) \cup \left(0, \dfrac{\pi}{2}\right]$ as the range of $f(x) = \operatorname{arccsc}(x)$.

236. Show that $\arcsin(x) + \arccos(x) = \dfrac{\pi}{2}$ for $-1 \leq x \leq 1$.

237. Discuss with your classmates why $\arcsin\left(\dfrac{1}{2}\right) \neq 30^\circ$.

238. Use the following picture and the series of exercises on the next page to show that

$$\arctan(1) + \arctan(2) + \arctan(3) = \pi$$

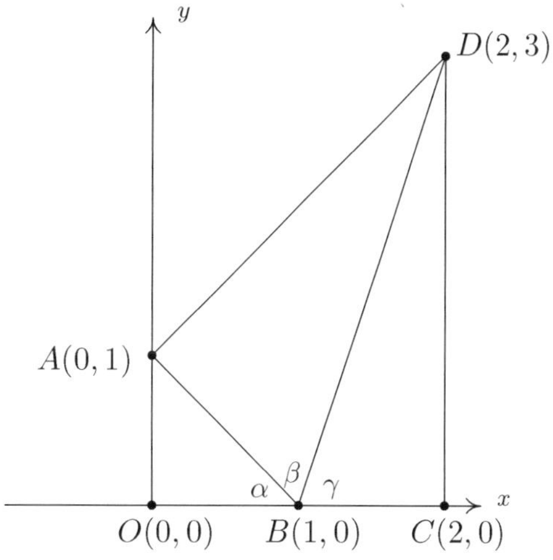

(a) Clearly $\triangle AOB$ and $\triangle BCD$ are right triangles because the line through O and A and the line through C and D are perpendicular to the x-axis. Use the distance formula to show that $\triangle BAD$ is also a right triangle (with $\angle BAD$ being the right angle) by showing that the sides of the triangle satisfy the Pythagorean Theorem.

(b) Use $\triangle AOB$ to show that $\alpha = \arctan(1)$

(c) Use $\triangle BAD$ to show that $\beta = \arctan(2)$

(d) Use $\triangle BCD$ to show that $\gamma = \arctan(3)$

(e) Use the fact that O, B and C all lie on the x-axis to conclude that $\alpha + \beta + \gamma = \pi$. Thus $\arctan(1) + \arctan(2) + \arctan(3) = \pi$.

10.6.6 Answers

1. $\arcsin(-1) = -\frac{\pi}{2}$
2. $\arcsin\left(-\frac{\sqrt{3}}{2}\right) = -\frac{\pi}{3}$
3. $\arcsin\left(-\frac{\sqrt{2}}{2}\right) = -\frac{\pi}{4}$
4. $\arcsin\left(-\frac{1}{2}\right) = -\frac{\pi}{6}$
5. $\arcsin(0) = 0$
6. $\arcsin\left(\frac{1}{2}\right) = \frac{\pi}{6}$
7. $\arcsin\left(\frac{\sqrt{2}}{2}\right) = \frac{\pi}{4}$
8. $\arcsin\left(\frac{\sqrt{3}}{2}\right) = \frac{\pi}{3}$
9. $\arcsin(1) = \frac{\pi}{2}$
10. $\arccos(-1) = \pi$
11. $\arccos\left(-\frac{\sqrt{3}}{2}\right) = \frac{5\pi}{6}$
12. $\arccos\left(-\frac{\sqrt{2}}{2}\right) = \frac{3\pi}{4}$
13. $\arccos\left(-\frac{1}{2}\right) = \frac{2\pi}{3}$
14. $\arccos(0) = \frac{\pi}{2}$
15. $\arccos\left(\frac{1}{2}\right) = \frac{\pi}{3}$
16. $\arccos\left(\frac{\sqrt{2}}{2}\right) = \frac{\pi}{4}$
17. $\arccos\left(\frac{\sqrt{3}}{2}\right) = \frac{\pi}{6}$
18. $\arccos(1) = 0$
19. $\arctan\left(-\sqrt{3}\right) = -\frac{\pi}{3}$
20. $\arctan(-1) = -\frac{\pi}{4}$
21. $\arctan\left(-\frac{\sqrt{3}}{3}\right) = -\frac{\pi}{6}$
22. $\arctan(0) = 0$
23. $\arctan\left(\frac{\sqrt{3}}{3}\right) = \frac{\pi}{6}$
24. $\arctan(1) = \frac{\pi}{4}$
25. $\arctan\left(\sqrt{3}\right) = \frac{\pi}{3}$
26. $\operatorname{arccot}\left(-\sqrt{3}\right) = \frac{5\pi}{6}$
27. $\operatorname{arccot}(-1) = \frac{3\pi}{4}$
28. $\operatorname{arccot}\left(-\frac{\sqrt{3}}{3}\right) = \frac{2\pi}{3}$
29. $\operatorname{arccot}(0) = \frac{\pi}{2}$
30. $\operatorname{arccot}\left(\frac{\sqrt{3}}{3}\right) = \frac{\pi}{3}$
31. $\operatorname{arccot}(1) = \frac{\pi}{4}$
32. $\operatorname{arccot}\left(\sqrt{3}\right) = \frac{\pi}{6}$
33. $\operatorname{arcsec}(2) = \frac{\pi}{3}$
34. $\operatorname{arccsc}(2) = \frac{\pi}{6}$
35. $\operatorname{arcsec}\left(\sqrt{2}\right) = \frac{\pi}{4}$
36. $\operatorname{arccsc}\left(\sqrt{2}\right) = \frac{\pi}{4}$
37. $\operatorname{arcsec}\left(\frac{2\sqrt{3}}{3}\right) = \frac{\pi}{6}$
38. $\operatorname{arccsc}\left(\frac{2\sqrt{3}}{3}\right) = \frac{\pi}{3}$
39. $\operatorname{arcsec}(1) = 0$
40. $\operatorname{arccsc}(1) = \frac{\pi}{2}$
41. $\operatorname{arcsec}(-2) = \frac{4\pi}{3}$
42. $\operatorname{arcsec}\left(-\sqrt{2}\right) = \frac{5\pi}{4}$

43. $\operatorname{arcsec}\left(-\frac{2\sqrt{3}}{3}\right) = \frac{7\pi}{6}$

44. $\operatorname{arcsec}(-1) = \pi$

45. $\operatorname{arccsc}(-2) = \frac{7\pi}{6}$

46. $\operatorname{arccsc}(-\sqrt{2}) = \frac{5\pi}{4}$

47. $\operatorname{arccsc}\left(-\frac{2\sqrt{3}}{3}\right) = \frac{4\pi}{3}$

48. $\operatorname{arccsc}(-1) = \frac{3\pi}{2}$

49. $\operatorname{arcsec}(-2) = \frac{2\pi}{3}$

50. $\operatorname{arcsec}(-\sqrt{2}) = \frac{3\pi}{4}$

51. $\operatorname{arcsec}\left(-\frac{2\sqrt{3}}{3}\right) = \frac{5\pi}{6}$

52. $\operatorname{arcsec}(-1) = \pi$

53. $\operatorname{arccsc}(-2) = -\frac{\pi}{6}$

54. $\operatorname{arccsc}(-\sqrt{2}) = -\frac{\pi}{4}$

55. $\operatorname{arccsc}\left(-\frac{2\sqrt{3}}{3}\right) = -\frac{\pi}{3}$

56. $\operatorname{arccsc}(-1) = -\frac{\pi}{2}$

57. $\sin\left(\arcsin\left(\frac{1}{2}\right)\right) = \frac{1}{2}$

58. $\sin\left(\arcsin\left(-\frac{\sqrt{2}}{2}\right)\right) = -\frac{\sqrt{2}}{2}$

59. $\sin\left(\arcsin\left(\frac{3}{5}\right)\right) = \frac{3}{5}$

60. $\sin(\arcsin(-0.42)) = -0.42$

61. $\sin\left(\arcsin\left(\frac{5}{4}\right)\right)$ is undefined.

62. $\cos\left(\arccos\left(\frac{\sqrt{2}}{2}\right)\right) = \frac{\sqrt{2}}{2}$

63. $\cos\left(\arccos\left(-\frac{1}{2}\right)\right) = -\frac{1}{2}$

64. $\cos\left(\arccos\left(\frac{5}{13}\right)\right) = \frac{5}{13}$

65. $\cos(\arccos(-0.998)) = -0.998$

66. $\cos(\arccos(\pi))$ is undefined.

67. $\tan(\arctan(-1)) = -1$

68. $\tan\left(\arctan(\sqrt{3})\right) = \sqrt{3}$

69. $\tan\left(\arctan\left(\frac{5}{12}\right)\right) = \frac{5}{12}$

70. $\tan(\arctan(0.965)) = 0.965$

71. $\tan(\arctan(3\pi)) = 3\pi$

72. $\cot(\operatorname{arccot}(1)) = 1$

73. $\cot\left(\operatorname{arccot}(-\sqrt{3})\right) = -\sqrt{3}$

74. $\cot\left(\operatorname{arccot}\left(-\frac{7}{24}\right)\right) = -\frac{7}{24}$

75. $\cot(\operatorname{arccot}(-0.001)) = -0.001$

76. $\cot\left(\operatorname{arccot}\left(\frac{17\pi}{4}\right)\right) = \frac{17\pi}{4}$

77. $\sec(\operatorname{arcsec}(2)) = 2$

78. $\sec(\operatorname{arcsec}(-1)) = -1$

79. $\sec\left(\text{arcsec}\left(\frac{1}{2}\right)\right)$ is undefined.

80. $\sec(\text{arcsec}(0.75))$ is undefined.

81. $\sec(\text{arcsec}(117\pi)) = 117\pi$

82. $\csc\left(\text{arccsc}\left(\sqrt{2}\right)\right) = \sqrt{2}$

83. $\csc\left(\text{arccsc}\left(-\frac{2\sqrt{3}}{3}\right)\right) = -\frac{2\sqrt{3}}{3}$

84. $\csc\left(\text{arccsc}\left(\frac{\sqrt{2}}{2}\right)\right)$ is undefined.

85. $\csc(\text{arccsc}(1.0001)) = 1.0001$

86. $\csc\left(\text{arccsc}\left(\frac{\pi}{4}\right)\right)$ is undefined.

87. $\arcsin\left(\sin\left(\frac{\pi}{6}\right)\right) = \frac{\pi}{6}$

88. $\arcsin\left(\sin\left(-\frac{\pi}{3}\right)\right) = -\frac{\pi}{3}$

89. $\arcsin\left(\sin\left(\frac{3\pi}{4}\right)\right) = \frac{\pi}{4}$

90. $\arcsin\left(\sin\left(\frac{11\pi}{6}\right)\right) = -\frac{\pi}{6}$

91. $\arcsin\left(\sin\left(\frac{4\pi}{3}\right)\right) = -\frac{\pi}{3}$

92. $\arccos\left(\cos\left(\frac{\pi}{4}\right)\right) = \frac{\pi}{4}$

93. $\arccos\left(\cos\left(\frac{2\pi}{3}\right)\right) = \frac{2\pi}{3}$

94. $\arccos\left(\cos\left(\frac{3\pi}{2}\right)\right) = \frac{\pi}{2}$

95. $\arccos\left(\cos\left(-\frac{\pi}{6}\right)\right) = \frac{\pi}{6}$

96. $\arccos\left(\cos\left(\frac{5\pi}{4}\right)\right) = \frac{3\pi}{4}$

97. $\arctan\left(\tan\left(\frac{\pi}{3}\right)\right) = \frac{\pi}{3}$

98. $\arctan\left(\tan\left(-\frac{\pi}{4}\right)\right) = -\frac{\pi}{4}$

99. $\arctan(\tan(\pi)) = 0$

100. $\arctan\left(\tan\left(\frac{\pi}{2}\right)\right)$ is undefined

101. $\arctan\left(\tan\left(\frac{2\pi}{3}\right)\right) = -\frac{\pi}{3}$

102. $\text{arccot}\left(\cot\left(\frac{\pi}{3}\right)\right) = \frac{\pi}{3}$

103. $\text{arccot}\left(\cot\left(-\frac{\pi}{4}\right)\right) = \frac{3\pi}{4}$

104. $\text{arccot}(\cot(\pi))$ is undefined

105. $\text{arccot}\left(\cot\left(\frac{3\pi}{2}\right)\right) = \frac{\pi}{2}$

106. $\text{arccot}\left(\cot\left(\frac{2\pi}{3}\right)\right) = \frac{2\pi}{3}$

107. $\text{arcsec}\left(\sec\left(\frac{\pi}{4}\right)\right) = \frac{\pi}{4}$

108. $\text{arcsec}\left(\sec\left(\frac{4\pi}{3}\right)\right) = \frac{4\pi}{3}$

109. $\text{arcsec}\left(\sec\left(\frac{5\pi}{6}\right)\right) = \frac{7\pi}{6}$

110. $\text{arcsec}\left(\sec\left(-\frac{\pi}{2}\right)\right)$ is undefined.

111. $\text{arcsec}\left(\sec\left(\frac{5\pi}{3}\right)\right) = \frac{\pi}{3}$

112. $\text{arccsc}\left(\csc\left(\frac{\pi}{6}\right)\right) = \frac{\pi}{6}$

113. $\operatorname{arccsc}\left(\csc\left(\frac{5\pi}{4}\right)\right) = \frac{5\pi}{4}$

114. $\operatorname{arccsc}\left(\csc\left(\frac{2\pi}{3}\right)\right) = \frac{\pi}{3}$

115. $\operatorname{arccsc}\left(\csc\left(-\frac{\pi}{2}\right)\right) = \frac{3\pi}{2}$

116. $\operatorname{arccsc}\left(\csc\left(\frac{11\pi}{6}\right)\right) = \frac{7\pi}{6}$

117. $\operatorname{arcsec}\left(\sec\left(\frac{11\pi}{12}\right)\right) = \frac{13\pi}{12}$

118. $\operatorname{arccsc}\left(\csc\left(\frac{9\pi}{8}\right)\right) = \frac{9\pi}{8}$

119. $\operatorname{arcsec}\left(\sec\left(\frac{\pi}{4}\right)\right) = \frac{\pi}{4}$

120. $\operatorname{arcsec}\left(\sec\left(\frac{4\pi}{3}\right)\right) = \frac{2\pi}{3}$

121. $\operatorname{arcsec}\left(\sec\left(\frac{5\pi}{6}\right)\right) = \frac{5\pi}{6}$

122. $\operatorname{arcsec}\left(\sec\left(-\frac{\pi}{2}\right)\right)$ is undefined.

123. $\operatorname{arcsec}\left(\sec\left(\frac{5\pi}{3}\right)\right) = \frac{\pi}{3}$

124. $\operatorname{arccsc}\left(\csc\left(\frac{\pi}{6}\right)\right) = \frac{\pi}{6}$

125. $\operatorname{arccsc}\left(\csc\left(\frac{5\pi}{4}\right)\right) = -\frac{\pi}{4}$

126. $\operatorname{arccsc}\left(\csc\left(\frac{2\pi}{3}\right)\right) = \frac{\pi}{3}$

127. $\operatorname{arccsc}\left(\csc\left(-\frac{\pi}{2}\right)\right) = -\frac{\pi}{2}$

128. $\operatorname{arccsc}\left(\csc\left(\frac{11\pi}{6}\right)\right) = -\frac{\pi}{6}$

129. $\operatorname{arcsec}\left(\sec\left(\frac{11\pi}{12}\right)\right) = \frac{11\pi}{12}$

130. $\operatorname{arccsc}\left(\csc\left(\frac{9\pi}{8}\right)\right) = -\frac{\pi}{8}$

131. $\sin\left(\arccos\left(-\frac{1}{2}\right)\right) = \frac{\sqrt{3}}{2}$

132. $\sin\left(\arccos\left(\frac{3}{5}\right)\right) = \frac{4}{5}$

133. $\sin(\arctan(-2)) = -\frac{2\sqrt{5}}{5}$

134. $\sin\left(\operatorname{arccot}\left(\sqrt{5}\right)\right) = \frac{\sqrt{6}}{6}$

135. $\sin(\operatorname{arccsc}(-3)) = -\frac{1}{3}$

136. $\cos\left(\arcsin\left(-\frac{5}{13}\right)\right) = \frac{12}{13}$

137. $\cos\left(\arctan\left(\sqrt{7}\right)\right) = \frac{\sqrt{2}}{4}$

138. $\cos(\operatorname{arccot}(3)) = \frac{3\sqrt{10}}{10}$

139. $\cos(\operatorname{arcsec}(5)) = \frac{1}{5}$

140. $\tan\left(\arcsin\left(-\frac{2\sqrt{5}}{5}\right)\right) = -2$

141. $\tan\left(\arccos\left(-\frac{1}{2}\right)\right) = -\sqrt{3}$

142. $\tan\left(\operatorname{arcsec}\left(\frac{5}{3}\right)\right) = \frac{4}{3}$

143. $\tan(\operatorname{arccot}(12)) = \frac{1}{12}$

144. $\cot\left(\arcsin\left(\frac{12}{13}\right)\right) = \frac{5}{12}$

145. $\cot\left(\arccos\left(\frac{\sqrt{3}}{2}\right)\right) = \sqrt{3}$

146. $\cot\left(\operatorname{arccsc}\left(\sqrt{5}\right)\right) = 2$

147. $\cot(\arctan(0.25)) = 4$

148. $\sec\left(\arccos\left(\frac{\sqrt{3}}{2}\right)\right) = \frac{2\sqrt{3}}{3}$

149. $\sec\left(\arcsin\left(-\frac{12}{13}\right)\right) = \frac{13}{5}$

150. $\sec(\arctan(10)) = \sqrt{101}$

151. $\sec\left(\operatorname{arccot}\left(-\frac{\sqrt{10}}{10}\right)\right) = -\sqrt{11}$

152. $\csc(\operatorname{arccot}(9)) = \sqrt{82}$

153. $\csc\left(\arcsin\left(\frac{3}{5}\right)\right) = \frac{5}{3}$

154. $\csc\left(\arctan\left(-\frac{2}{3}\right)\right) = -\frac{\sqrt{13}}{2}$

155. $\sin\left(\arcsin\left(\frac{5}{13}\right) + \frac{\pi}{4}\right) = \frac{17\sqrt{2}}{26}$

156. $\cos(\operatorname{arcsec}(3) + \arctan(2)) = \frac{\sqrt{5} - 4\sqrt{10}}{15}$

157. $\tan\left(\arctan(3) + \arccos\left(-\frac{3}{5}\right)\right) = \frac{1}{3}$

158. $\sin\left(2\arcsin\left(-\frac{4}{5}\right)\right) = -\frac{24}{25}$

159. $\sin\left(2\operatorname{arccsc}\left(\frac{13}{5}\right)\right) = \frac{120}{169}$

160. $\sin(2\arctan(2)) = \frac{4}{5}$

161. $\cos\left(2\arcsin\left(\frac{3}{5}\right)\right) = \frac{7}{25}$

162. $\cos\left(2\operatorname{arcsec}\left(\frac{25}{7}\right)\right) = -\frac{527}{625}$

163. $\cos\left(2\operatorname{arccot}\left(-\sqrt{5}\right)\right) = \frac{2}{3}$

164. $\sin\left(\frac{\arctan(2)}{2}\right) = \sqrt{\frac{5-\sqrt{5}}{10}}$

165. $\sin(\arccos(x)) = \sqrt{1-x^2}$ for $-1 \leq x \leq 1$

166. $\cos(\arctan(x)) = \frac{1}{\sqrt{1+x^2}}$ for all x

167. $\tan(\arcsin(x)) = \frac{x}{\sqrt{1-x^2}}$ for $-1 < x < 1$

168. $\sec(\arctan(x)) = \sqrt{1+x^2}$ for all x

169. $\csc(\arccos(x)) = \frac{1}{\sqrt{1-x^2}}$ for $-1 < x < 1$

170. $\sin(2\arctan(x)) = \frac{2x}{x^2+1}$ for all x

171. $\sin(2\arccos(x)) = 2x\sqrt{1-x^2}$ for $-1 \leq x \leq 1$

172. $\cos\left(2\arctan\left(x\right)\right) = \dfrac{1-x^2}{1+x^2}$ for all x

173. $\sin(\arccos(2x)) = \sqrt{1-4x^2}$ for $-\frac{1}{2} \leq x \leq \frac{1}{2}$

174. $\sin\left(\arccos\left(\dfrac{x}{5}\right)\right) = \dfrac{\sqrt{25-x^2}}{5}$ for $-5 \leq x \leq 5$

175. $\cos\left(\arcsin\left(\dfrac{x}{2}\right)\right) = \dfrac{\sqrt{4-x^2}}{2}$ for $-2 \leq x \leq 2$

176. $\cos\left(\arctan\left(3x\right)\right) = \dfrac{1}{\sqrt{1+9x^2}}$ for all x

177. $\sin(2\arcsin(7x)) = 14x\sqrt{1-49x^2}$ for $-\dfrac{1}{7} \leq x \leq \dfrac{1}{7}$

178. $\sin\left(2\arcsin\left(\dfrac{x\sqrt{3}}{3}\right)\right) = \dfrac{2x\sqrt{3-x^2}}{3}$ for $-\sqrt{3} \leq x \leq \sqrt{3}$

179. $\cos(2\arcsin(4x)) = 1-32x^2$ for $-\dfrac{1}{4} \leq x \leq \dfrac{1}{4}$

180. $\sec(\arctan(2x))\tan(\arctan(2x)) = 2x\sqrt{1+4x^2}$ for all x

181. $\sin\left(\arcsin(x) + \arccos(x)\right) = 1$ for $-1 \leq x \leq 1$

182. $\cos\left(\arcsin(x) + \arctan(x)\right) = \dfrac{\sqrt{1-x^2}-x^2}{\sqrt{1+x^2}}$ for $-1 \leq x \leq 1$

183. [10] $\tan\left(2\arcsin(x)\right) = \dfrac{2x\sqrt{1-x^2}}{1-2x^2}$ for x in $\left(-1, -\dfrac{\sqrt{2}}{2}\right) \cup \left(-\dfrac{\sqrt{2}}{2}, \dfrac{\sqrt{2}}{2}\right) \cup \left(\dfrac{\sqrt{2}}{2}, 1\right)$

184. $\sin\left(\dfrac{1}{2}\arctan(x)\right) = \begin{cases} \sqrt{\dfrac{\sqrt{x^2+1}-1}{2\sqrt{x^2+1}}} & \text{for } x \geq 0 \\ -\sqrt{\dfrac{\sqrt{x^2+1}-1}{2\sqrt{x^2+1}}} & \text{for } x < 0 \end{cases}$

185. If $\sin(\theta) = \dfrac{x}{2}$ for $-\dfrac{\pi}{2} < \theta < \dfrac{\pi}{2}$, then $\theta + \sin(2\theta) = \arcsin\left(\dfrac{x}{2}\right) + \dfrac{x\sqrt{4-x^2}}{2}$

186. If $\tan(\theta) = \dfrac{x}{7}$ for $-\dfrac{\pi}{2} < \theta < \dfrac{\pi}{2}$, then $\dfrac{1}{2}\theta - \dfrac{1}{2}\sin(2\theta) = \dfrac{1}{2}\arctan\left(\dfrac{x}{7}\right) - \dfrac{7x}{x^2+49}$

[10] The equivalence for $x = \pm 1$ can be verified independently of the derivation of the formula, but Calculus is required to fully understand what is happening at those x values. You'll see what we mean when you work through the details of the identity for $\tan(2t)$. For now, we exclude $x = \pm 1$ from our answer.

187. If $\sec(\theta) = \frac{x}{4}$ for $0 < \theta < \frac{\pi}{2}$, then $4\tan(\theta) - 4\theta = \sqrt{x^2 - 16} - 4\text{arcsec}\left(\frac{x}{4}\right)$

188. $x = \arcsin\left(\frac{7}{11}\right) + 2\pi k$ or $x = \pi - \arcsin\left(\frac{7}{11}\right) + 2\pi k$, in $[0, 2\pi)$, $x \approx 0.6898, 2.4518$

189. $x = \arccos\left(-\frac{2}{9}\right) + 2\pi k$ or $x = -\arccos\left(-\frac{2}{9}\right) + 2\pi k$, in $[0, 2\pi)$, $x \approx 1.7949, 4.4883$

190. $x = \pi + \arcsin(0.569) + 2\pi k$ or $x = 2\pi - \arcsin(0.569) + 2\pi k$, in $[0, 2\pi)$, $x \approx 3.7469, 5.6779$

191. $x = \arccos(0.117) + 2\pi k$ or $x = 2\pi - \arccos(0.117) + 2\pi k$, in $[0, 2\pi)$, $x \approx 1.4535, 4.8297$

192. $x = \arcsin(0.008) + 2\pi k$ or $x = \pi - \arcsin(0.008) + 2\pi k$, in $[0, 2\pi)$, $x \approx 0.0080, 3.1336$

193. $x = \arccos\left(\frac{359}{360}\right) + 2\pi k$ or $x = 2\pi - \arccos\left(\frac{359}{360}\right) + 2\pi k$, in $[0, 2\pi)$, $x \approx 0.0746, 6.2086$

194. $x = \arctan(117) + \pi k$, in $[0, 2\pi)$, $x \approx 1.56225, 4.70384$

195. $x = \arctan\left(-\frac{1}{12}\right) + \pi k$, in $[0, 2\pi)$, $x \approx 3.0585, 6.2000$

196. $x = \arccos\left(\frac{2}{3}\right) + 2\pi k$ or $x = 2\pi - \arccos\left(\frac{2}{3}\right) + 2\pi k$, in $[0, 2\pi)$, $x \approx 0.8411, 5.4422$

197. $x = \pi + \arcsin\left(\frac{17}{90}\right) + 2\pi k$ or $x = 2\pi - \arcsin\left(\frac{17}{90}\right) + 2\pi k$, in $[0, 2\pi)$, $x \approx 3.3316, 6.0932$

198. $x = \arctan\left(-\sqrt{10}\right) + \pi k$, in $[0, 2\pi)$, $x \approx 1.8771, 5.0187$

199. $x = \arcsin\left(\frac{3}{8}\right) + 2\pi k$ or $x = \pi - \arcsin\left(\frac{3}{8}\right) + 2\pi k$, in $[0, 2\pi)$, $x \approx 0.3844, 2.7572$

200. $x = \arccos\left(-\frac{7}{16}\right) + 2\pi k$ or $x = -\arccos\left(-\frac{7}{16}\right) + 2\pi k$, in $[0, 2\pi)$, $x \approx 2.0236, 4.2596$

201. $x = \arctan(0.03) + \pi k$, in $[0, 2\pi)$, $x \approx 0.0300, 3.1716$

202. $x = \arcsin(0.3502) + 2\pi k$ or $x = \pi - \arcsin(0.3502) + 2\pi k$, in $[0, 2\pi)$, $x \approx 0.3578, 2.784$

203. $x = \pi + \arcsin(0.721) + 2\pi k$ or $x = 2\pi - \arcsin(0.721) + 2\pi k$, in $[0, 2\pi)$, $x \approx 3.9468, 5.4780$

204. $x = \arccos(0.9824) + 2\pi k$ or $x = 2\pi - \arccos(0.9824) + 2\pi k$, in $[0, 2\pi)$, $x \approx 0.1879, 6.0953$

205. $x = \arccos(-0.5637) + 2\pi k$ or $x = -\arccos(-0.5637) + 2\pi k$, in $[0, 2\pi)$, $x \approx 2.1697, 4.1135$

206. $x = \arctan(117) + \pi k$, in $[0, 2\pi)$, $x \approx 1.5622, 4.7038$

207. $x = \arctan(-0.6109) + \pi k$, in $[0, 2\pi)$, $x \approx 2.5932, 5.7348$

208. 36.87° and 53.13°

209. 22.62° and 67.38°

210. 32.52° and 57.48°

211. 68.9°

212. 7.7°

213. 51°

214. 19.5°

215. 41.81°

216. $f(x) = 5\sin(3x) + 12\cos(3x) = 13\sin\left(3x + \arcsin\left(\frac{12}{13}\right)\right) \approx 13\sin(3x + 1.1760)$

217. $f(x) = 3\cos(2x) + 4\sin(2x) = 5\sin\left(2x + \arcsin\left(\frac{3}{5}\right)\right) \approx 5\sin(2x + 0.6435)$

218. $f(x) = \cos(x) - 3\sin(x) = \sqrt{10}\sin\left(x + \arccos\left(-\frac{3\sqrt{10}}{10}\right)\right) \approx \sqrt{10}\sin(x + 2.8198)$

219. $f(x) = 7\sin(10x) - 24\cos(10x) = 25\sin\left(10x + \arcsin\left(-\frac{24}{25}\right)\right) \approx 25\sin(10x - 1.2870)$

220. $f(x) = -\cos(x) - 2\sqrt{2}\sin(x) = 3\sin\left(x + \pi + \arcsin\left(\frac{1}{3}\right)\right) \approx 3\sin(x + 3.4814)$

221. $f(x) = 2\sin(x) - \cos(x) = \sqrt{5}\sin\left(x + \arcsin\left(-\frac{\sqrt{5}}{5}\right)\right) \approx \sqrt{5}\sin(x - 0.4636)$

222. $\left[-\frac{1}{5}, \frac{1}{5}\right]$

223. $\left[-\frac{1}{3}, 1\right]$

224. $\left[-\frac{\sqrt{2}}{2}, \frac{\sqrt{2}}{2}\right]$

225. $(-\infty, -\sqrt{5}] \cup [-\sqrt{3}, \sqrt{3}] \cup [\sqrt{5}, \infty)$

226. $(-\infty, \infty)$

227. $(-\infty, -3) \cup (-3, 3) \cup (3, \infty)$

228. $\left(\frac{1}{2}, \infty\right)$

229. $\left[\frac{1}{2}, \infty\right)$

230. $\left(-\infty, -\frac{1}{12}\right] \cup \left[\frac{1}{12}, \infty\right)$

231. $(-\infty, -6] \cup [-4, \infty)$

232. $(-\infty, -2] \cup [2, \infty)$

233. $[0, \infty)$

10.7 Trigonometric Equations and Inequalities

In Sections 10.2, 10.3 and most recently 10.6, we solved some basic equations involving the trigonometric functions. Below we summarize the techniques we've employed thus far. Note that we use the neutral letter 'u' as the argument[1] of each circular function for generality.

Strategies for Solving Basic Equations Involving Trigonometric Functions

- To solve $\cos(u) = c$ or $\sin(u) = c$ for $-1 \leq c \leq 1$, first solve for u in the interval $[0, 2\pi)$ and add integer multiples of the period 2π. If $c < -1$ or of $c > 1$, there are no real solutions.
- To solve $\sec(u) = c$ or $\csc(u) = c$ for $c \leq -1$ or $c \geq 1$, convert to cosine or sine, respectively, and solve as above. If $-1 < c < 1$, there are no real solutions.
- To solve $\tan(u) = c$ for any real number c, first solve for u in the interval $\left(-\frac{\pi}{2}, \frac{\pi}{2}\right)$ and add integer multiples of the period π.
- To solve $\cot(u) = c$ for $c \neq 0$, convert to tangent and solve as above. If $c = 0$, the solution to $\cot(u) = 0$ is $u = \frac{\pi}{2} + \pi k$ for integers k.

Using the above guidelines, we can comfortably solve $\sin(x) = \frac{1}{2}$ and find the solution $x = \frac{\pi}{6} + 2\pi k$ or $x = \frac{5\pi}{6} + 2\pi k$ for integers k. How do we solve something like $\sin(3x) = \frac{1}{2}$? Since this equation has the *form* $\sin(u) = \frac{1}{2}$, we know the solutions take the form $u = \frac{\pi}{6} + 2\pi k$ or $u = \frac{5\pi}{6} + 2\pi k$ for integers k. Since the argument of sine here is $3x$, we have $3x = \frac{\pi}{6} + 2\pi k$ or $3x = \frac{5\pi}{6} + 2\pi k$ for integers k. To solve for x, we divide both sides[2] of these equations by 3, and obtain $x = \frac{\pi}{18} + \frac{2\pi}{3}k$ or $x = \frac{5\pi}{18} + \frac{2\pi}{3}k$ for integers k. This is the technique employed in the example below.

Example 10.7.1. Solve the following equations and check your answers analytically. List the solutions which lie in the interval $[0, 2\pi)$ and verify them using a graphing utility.

1. $\cos(2x) = -\frac{\sqrt{3}}{2}$
2. $\csc\left(\frac{1}{3}x - \pi\right) = \sqrt{2}$
3. $\cot(3x) = 0$
4. $\sec^2(x) = 4$
5. $\tan\left(\frac{x}{2}\right) = -3$
6. $\sin(2x) = 0.87$

Solution.

1. The solutions to $\cos(u) = -\frac{\sqrt{3}}{2}$ are $u = \frac{5\pi}{6} + 2\pi k$ or $u = \frac{7\pi}{6} + 2\pi k$ for integers k. Since the argument of cosine here is $2x$, this means $2x = \frac{5\pi}{6} + 2\pi k$ or $2x = \frac{7\pi}{6} + 2\pi k$ for integers k. Solving for x gives $x = \frac{5\pi}{12} + \pi k$ or $x = \frac{7\pi}{12} + \pi k$ for integers k. To check these answers analytically, we substitute them into the original equation. For any integer k we have

$$\begin{aligned} \cos\left(2\left[\frac{5\pi}{12} + \pi k\right]\right) &= \cos\left(\frac{5\pi}{6} + 2\pi k\right) \\ &= \cos\left(\frac{5\pi}{6}\right) \qquad \text{(the period of cosine is } 2\pi\text{)} \\ &= -\frac{\sqrt{3}}{2} \end{aligned}$$

[1] See the comments at the beginning of Section 10.5 for a review of this concept.

[2] Don't forget to divide the $2\pi k$ by 3 as well!

Similarly, we find $\cos\left(2\left[\frac{7\pi}{12}+\pi k\right]\right) = \cos\left(\frac{7\pi}{6}+2\pi k\right) = \cos\left(\frac{7\pi}{6}\right) = -\frac{\sqrt{3}}{2}$. To determine which of our solutions lie in $[0, 2\pi)$, we substitute integer values for k. The solutions we keep come from the values of $k = 0$ and $k = 1$ and are $x = \frac{5\pi}{12}, \frac{7\pi}{12}, \frac{17\pi}{12}$ and $\frac{19\pi}{12}$. Using a calculator, we graph $y = \cos(2x)$ and $y = -\frac{\sqrt{3}}{2}$ over $[0, 2\pi)$ and examine where these two graphs intersect. We see that the x-coordinates of the intersection points correspond to the decimal representations of our exact answers.

2. Since this equation has the form $\csc(u) = \sqrt{2}$, we rewrite this as $\sin(u) = \frac{\sqrt{2}}{2}$ and find $u = \frac{\pi}{4} + 2\pi k$ or $u = \frac{3\pi}{4} + 2\pi k$ for integers k. Since the argument of cosecant here is $\left(\frac{1}{3}x - \pi\right)$,

$$\frac{1}{3}x - \pi = \frac{\pi}{4} + 2\pi k \quad \text{or} \quad \frac{1}{3}x - \pi = \frac{3\pi}{4} + 2\pi k$$

To solve $\frac{1}{3}x - \pi = \frac{\pi}{4} + 2\pi k$, we first add π to both sides

$$\frac{1}{3}x = \frac{\pi}{4} + 2\pi k + \pi$$

A common error is to treat the '$2\pi k$' and 'π' terms as 'like' terms and try to combine them when they are not.[3] We can, however, combine the 'π' and '$\frac{\pi}{4}$' terms to get

$$\frac{1}{3}x = \frac{5\pi}{4} + 2\pi k$$

We now finish by multiplying both sides by 3 to get

$$x = 3\left(\frac{5\pi}{4} + 2\pi k\right) = \frac{15\pi}{4} + 6\pi k$$

Solving the other equation, $\frac{1}{3}x - \pi = \frac{3\pi}{4} + 2\pi k$ produces $x = \frac{21\pi}{4} + 6\pi k$ for integers k. To check the first family of answers, we substitute, combine line terms, and simplify.

$$\begin{aligned}
\csc\left(\tfrac{1}{3}\left[\tfrac{15\pi}{4} + 6\pi k\right] - \pi\right) &= \csc\left(\tfrac{5\pi}{4} + 2\pi k - \pi\right) \\
&= \csc\left(\tfrac{\pi}{4} + 2\pi k\right) \\
&= \csc\left(\tfrac{\pi}{4}\right) \qquad \text{(the period of cosecant is } 2\pi) \\
&= \sqrt{2}
\end{aligned}$$

The family $x = \frac{21\pi}{4} + 6\pi k$ checks similarly. Despite having infinitely many solutions, we find that *none* of them lie in $[0, 2\pi)$. To verify this graphically, we use a reciprocal identity to rewrite the cosecant as a sine and we find that $y = \frac{1}{\sin\left(\frac{1}{3}x - \pi\right)}$ and $y = \sqrt{2}$ do not intersect at all over the interval $[0, 2\pi)$.

[3] Do you see why?

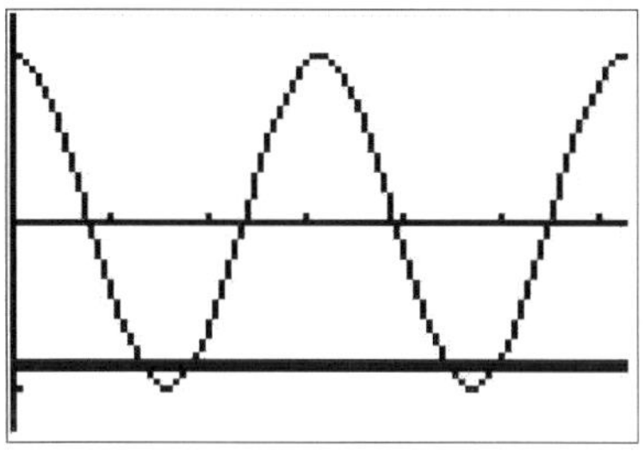

$y = \cos(2x)$ and $\boldsymbol{y = -\frac{\sqrt{3}}{2}}$

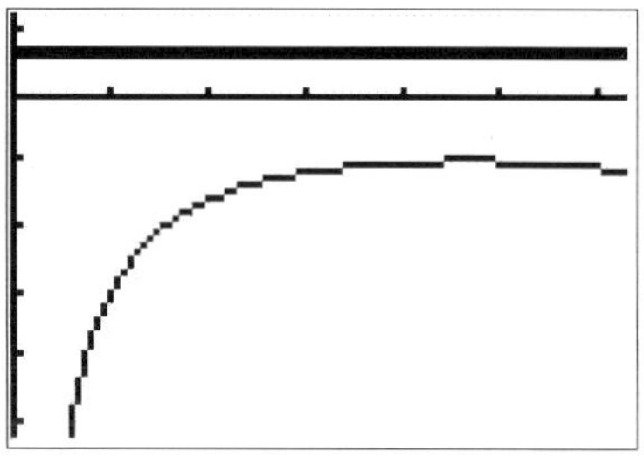

$y = \frac{1}{\sin\left(\frac{1}{3}x - \pi\right)}$ and $\boldsymbol{y = \sqrt{2}}$

3. Since $\cot(3x) = 0$ has the form $\cot(u) = 0$, we know $u = \frac{\pi}{2} + \pi k$, so, in this case, $3x = \frac{\pi}{2} + \pi k$ for integers k. Solving for x yields $x = \frac{\pi}{6} + \frac{\pi}{3}k$. Checking our answers, we get

$$\begin{aligned} \cot\left(3\left[\tfrac{\pi}{6} + \tfrac{\pi}{3}k\right]\right) &= \cot\left(\tfrac{\pi}{2} + \pi k\right) \\ &= \cot\left(\tfrac{\pi}{2}\right) \qquad \text{(the period of cotangent is } \pi) \\ &= 0 \end{aligned}$$

As k runs through the integers, we obtain six answers, corresponding to $k = 0$ through $k = 5$, which lie in $[0, 2\pi)$: $x = \frac{\pi}{6}, \frac{\pi}{2}, \frac{5\pi}{6}, \frac{7\pi}{6}, \frac{3\pi}{2}$ and $\frac{11\pi}{6}$. To confirm these graphically, we must be careful. On many calculators, there is no function button for cotangent. We choose[4] to use the quotient identity $\cot(3x) = \frac{\cos(3x)}{\sin(3x)}$. Graphing $y = \frac{\cos(3x)}{\sin(3x)}$ and $y = 0$ (the x-axis), we see that the x-coordinates of the intersection points approximately match our solutions.

4. The complication in solving an equation like $\sec^2(x) = 4$ comes not from the argument of secant, which is just x, but rather, the fact the secant is being squared. To get this equation to look like one of the forms listed on page 857, we extract square roots to get $\sec(x) = \pm 2$. Converting to cosines, we have $\cos(x) = \pm\frac{1}{2}$. For $\cos(x) = \frac{1}{2}$, we get $x = \frac{\pi}{3} + 2\pi k$ or $x = \frac{5\pi}{3} + 2\pi k$ for integers k. For $\cos(x) = -\frac{1}{2}$, we get $x = \frac{2\pi}{3} + 2\pi k$ or $x = \frac{4\pi}{3} + 2\pi k$ for integers k. If we take a step back and think of these families of solutions geometrically, we see we are finding the measures of all angles with a reference angle of $\frac{\pi}{3}$. As a result, these solutions can be combined and we may write our solutions as $x = \frac{\pi}{3} + \pi k$ and $x = \frac{2\pi}{3} + \pi k$ for integers k. To check the first family of solutions, we note that, depending on the integer k, $\sec\left(\frac{\pi}{3} + \pi k\right)$ doesn't always equal $\sec\left(\frac{\pi}{3}\right)$. However, it is true that for all integers k, $\sec\left(\frac{\pi}{3} + \pi k\right) = \pm\sec\left(\frac{\pi}{3}\right) = \pm 2$. (Can you show this?) As a result,

$$\begin{aligned} \sec^2\left(\tfrac{\pi}{3} + \pi k\right) &= \left(\pm\sec\left(\tfrac{\pi}{3}\right)\right)^2 \\ &= (\pm 2)^2 \\ &= 4 \end{aligned}$$

The same holds for the family $x = \frac{2\pi}{3} + \pi k$. The solutions which lie in $[0, 2\pi)$ come from the values $k = 0$ and $k = 1$, namely $x = \frac{\pi}{3}, \frac{2\pi}{3}, \frac{4\pi}{3}$ and $\frac{5\pi}{3}$. To confirm graphically, we use

[4] The reader is encouraged to see what happens if we had chosen the reciprocal identity $\cot(3x) = \frac{1}{\tan(3x)}$ instead. The graph on the calculator *appears* identical, but what happens when you try to find the intersection points?

a reciprocal identity to rewrite the secant as cosine. The x-coordinates of the intersection points of $y = \frac{1}{(\cos(x))^2}$ and $y = 4$ verify our answers.

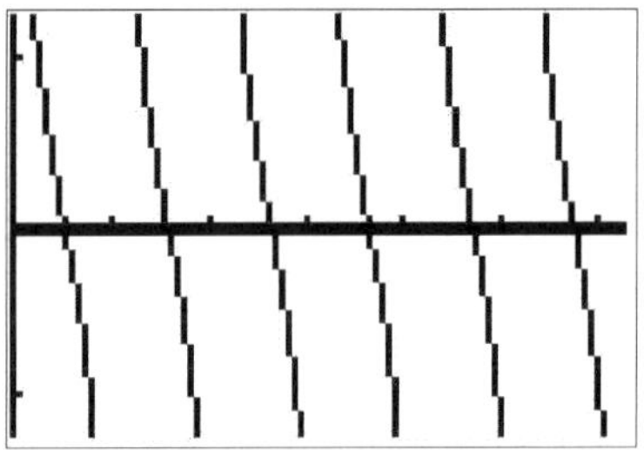

$y = \frac{\cos(3x)}{\sin(3x)}$ and $\boldsymbol{y = 0}$

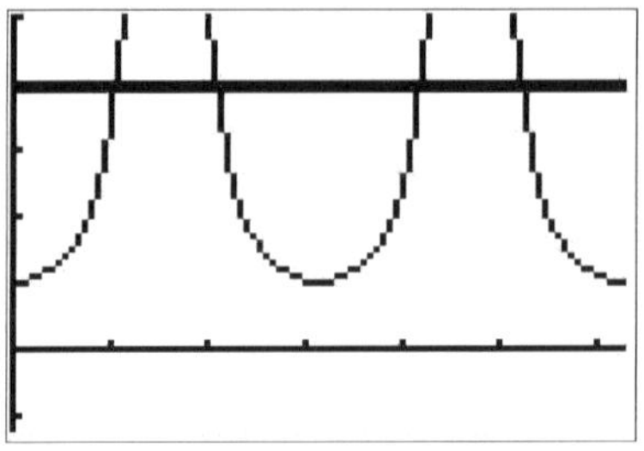

$y = \frac{1}{\cos^2(x)}$ and $\boldsymbol{y = 4}$

5. The equation $\tan\left(\frac{x}{2}\right) = -3$ has the form $\tan(u) = -3$, whose solution is $u = \arctan(-3) + \pi k$. Hence, $\frac{x}{2} = \arctan(-3) + \pi k$, so $x = 2\arctan(-3) + 2\pi k$ for integers k. To check, we note

$$\begin{aligned} \tan\left(\frac{2\arctan(-3)+2\pi k}{2}\right) &= \tan\left(\arctan(-3) + \pi k\right) \\ &= \tan\left(\arctan(-3)\right) && \text{(the period of tangent is } \pi\text{)} \\ &= -3 && \text{(See Theorem 10.27)} \end{aligned}$$

To determine which of our answers lie in the interval $[0, 2\pi)$, we first need to get an idea of the value of $2\arctan(-3)$. While we could easily find an approximation using a calculator,[5] we proceed analytically. Since $-3 < 0$, it follows that $-\frac{\pi}{2} < \arctan(-3) < 0$. Multiplying through by 2 gives $-\pi < 2\arctan(-3) < 0$. We are now in a position to argue which of the solutions $x = 2\arctan(-3) + 2\pi k$ lie in $[0, 2\pi)$. For $k = 0$, we get $x = 2\arctan(-3) < 0$, so we discard this answer and all answers $x = 2\arctan(-3) + 2\pi k$ where $k < 0$. Next, we turn our attention to $k = 1$ and get $x = 2\arctan(-3) + 2\pi$. Starting with the inequality $-\pi < 2\arctan(-3) < 0$, we add 2π and get $\pi < 2\arctan(-3) + 2\pi < 2\pi$. This means $x = 2\arctan(-3) + 2\pi$ lies in $[0, 2\pi)$. Advancing k to 2 produces $x = 2\arctan(-3) + 4\pi$. Once again, we get from $-\pi < 2\arctan(-3) < 0$ that $3\pi < 2\arctan(-3) + 4\pi < 4\pi$. Since this is outside the interval $[0, 2\pi)$, we discard $x = 2\arctan(-3) + 4\pi$ and all solutions of the form $x = 2\arctan(-3) + 2\pi k$ for $k > 2$. Graphically, we see $y = \tan\left(\frac{x}{2}\right)$ and $y = -3$ intersect only once on $[0, 2\pi)$ at $x = 2\arctan(-3) + 2\pi \approx 3.7851$.

6. To solve $\sin(2x) = 0.87$, we first note that it has the form $\sin(u) = 0.87$, which has the family of solutions $u = \arcsin(0.87) + 2\pi k$ or $u = \pi - \arcsin(0.87) + 2\pi k$ for integers k. Since the argument of sine here is $2x$, we get $2x = \arcsin(0.87) + 2\pi k$ or $2x = \pi - \arcsin(0.87) + 2\pi k$ which gives $x = \frac{1}{2}\arcsin(0.87) + \pi k$ or $x = \frac{\pi}{2} - \frac{1}{2}\arcsin(0.87) + \pi k$ for integers k. To check,

[5] Your instructor will let you know if you should abandon the analytic route at this point and use your calculator. But seriously, what fun would that be?

$$\begin{aligned} \sin\left(2\left[\tfrac{1}{2}\arcsin(0.87)+\pi k\right]\right) &= \sin\left(\arcsin(0.87)+2\pi k\right) \\ &= \sin\left(\arcsin(0.87)\right) && \text{(the period of sine is } 2\pi) \\ &= 0.87 && \text{(See Theorem 10.26)} \end{aligned}$$

For the family $x = \frac{\pi}{2} - \frac{1}{2}\arcsin(0.87) + \pi k$, we get

$$\begin{aligned} \sin\left(2\left[\tfrac{\pi}{2} - \tfrac{1}{2}\arcsin(0.87)+\pi k\right]\right) &= \sin\left(\pi - \arcsin(0.87)+2\pi k\right) \\ &= \sin\left(\pi - \arcsin(0.87)\right) && \text{(the period of sine is } 2\pi) \\ &= \sin\left(\arcsin(0.87)\right) && (\sin(\pi - t) = \sin(t)) \\ &= 0.87 && \text{(See Theorem 10.26)} \end{aligned}$$

To determine which of these solutions lie in $[0, 2\pi)$, we first need to get an idea of the value of $x = \frac{1}{2}\arcsin(0.87)$. Once again, we could use the calculator, but we adopt an analytic route here. By definition, $0 < \arcsin(0.87) < \frac{\pi}{2}$ so that multiplying through by $\frac{1}{2}$ gives us $0 < \frac{1}{2}\arcsin(0.87) < \frac{\pi}{4}$. Starting with the family of solutions $x = \frac{1}{2}\arcsin(0.87) + \pi k$, we use the same kind of arguments as in our solution to number 5 above and find only the solutions corresponding to $k = 0$ and $k = 1$ lie in $[0, 2\pi)$: $x = \frac{1}{2}\arcsin(0.87)$ and $x = \frac{1}{2}\arcsin(0.87) + \pi$. Next, we move to the family $x = \frac{\pi}{2} - \frac{1}{2}\arcsin(0.87) + \pi k$ for integers k. Here, we need to get a better estimate of $\frac{\pi}{2} - \frac{1}{2}\arcsin(0.87)$. From the inequality $0 < \frac{1}{2}\arcsin(0.87) < \frac{\pi}{4}$, we first multiply through by -1 and then add $\frac{\pi}{2}$ to get $\frac{\pi}{2} > \frac{\pi}{2} - \frac{1}{2}\arcsin(0.87) > \frac{\pi}{4}$, or $\frac{\pi}{4} < \frac{\pi}{2} - \frac{1}{2}\arcsin(0.87) < \frac{\pi}{2}$. Proceeding with the usual arguments, we find the only solutions which lie in $[0, 2\pi)$ correspond to $k = 0$ and $k = 1$, namely $x = \frac{\pi}{2} - \frac{1}{2}\arcsin(0.87)$ and $x = \frac{3\pi}{2} - \frac{1}{2}\arcsin(0.87)$. All told, we have found four solutions to $\sin(2x) = 0.87$ in $[0, 2\pi)$: $x = \frac{1}{2}\arcsin(0.87)$, $x = \frac{1}{2}\arcsin(0.87) + \pi$, $x = \frac{\pi}{2} - \frac{1}{2}\arcsin(0.87)$ and $x = \frac{3\pi}{2} - \frac{1}{2}\arcsin(0.87)$. By graphing $y = \sin(2x)$ and $y = 0.87$, we confirm our results.

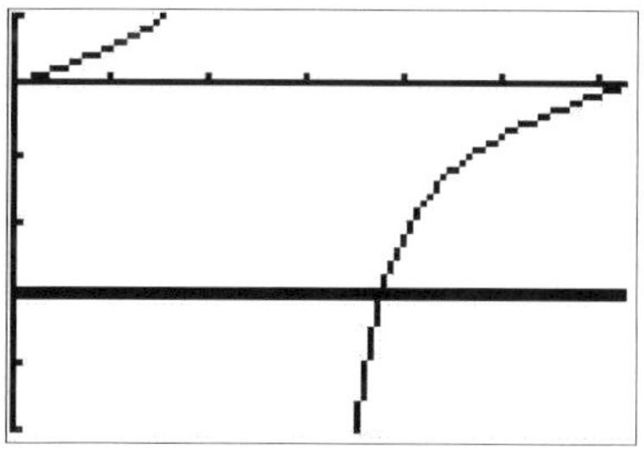

$y = \tan\left(\frac{x}{2}\right)$ and $\boldsymbol{y = -3}$

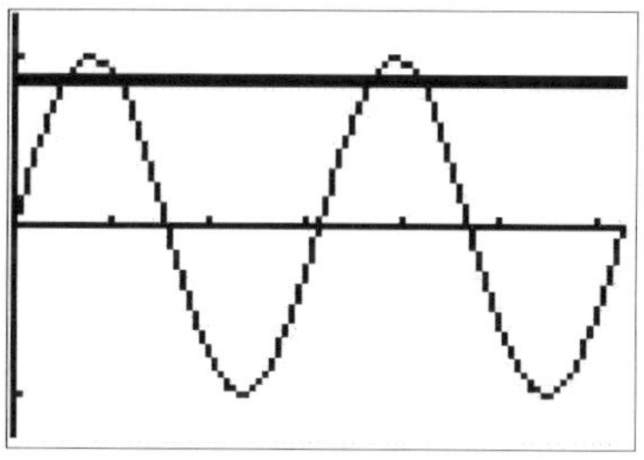

$y = \sin(2x)$ and $\boldsymbol{y = 0.87}$ □

Each of the problems in Example 10.7.1 featured one trigonometric function. If an equation involves two different trigonometric functions or if the equation contains the same trigonometric function but with different arguments, we will need to use identities and Algebra to reduce the equation to the same form as those given on page 857.

Example 10.7.2. Solve the following equations and list the solutions which lie in the interval $[0, 2\pi)$. Verify your solutions on $[0, 2\pi)$ graphically.

1. $3\sin^3(x) = \sin^2(x)$
2. $\sec^2(x) = \tan(x) + 3$
3. $\cos(2x) = 3\cos(x) - 2$
4. $\cos(3x) = 2 - \cos(x)$
5. $\cos(3x) = \cos(5x)$
6. $\sin(2x) = \sqrt{3}\cos(x)$
7. $\sin(x)\cos\left(\frac{x}{2}\right) + \cos(x)\sin\left(\frac{x}{2}\right) = 1$
8. $\cos(x) - \sqrt{3}\sin(x) = 2$

Solution.

1. We resist the temptation to divide both sides of $3\sin^3(x) = \sin^2(x)$ by $\sin^2(x)$ (What goes wrong if you do?) and instead gather all of the terms to one side of the equation and factor.

$$\begin{array}{rcll} 3\sin^3(x) & = & \sin^2(x) & \\ 3\sin^3(x) - \sin^2(x) & = & 0 & \\ \sin^2(x)(3\sin(x) - 1) & = & 0 & \text{Factor out } \sin^2(x) \text{ from both terms.} \end{array}$$

We get $\sin^2(x) = 0$ or $3\sin(x) - 1 = 0$. Solving for $\sin(x)$, we find $\sin(x) = 0$ or $\sin(x) = \frac{1}{3}$. The solution to the first equation is $x = \pi k$, with $x = 0$ and $x = \pi$ being the two solutions which lie in $[0, 2\pi)$. To solve $\sin(x) = \frac{1}{3}$, we use the arcsine function to get $x = \arcsin\left(\frac{1}{3}\right) + 2\pi k$ or $x = \pi - \arcsin\left(\frac{1}{3}\right) + 2\pi k$ for integers k. We find the two solutions here which lie in $[0, 2\pi)$ to be $x = \arcsin\left(\frac{1}{3}\right)$ and $x = \pi - \arcsin\left(\frac{1}{3}\right)$. To check graphically, we plot $y = 3(\sin(x))^3$ and $y = (\sin(x))^2$ and find the x-coordinates of the intersection points of these two curves. Some extra zooming is required near $x = 0$ and $x = \pi$ to verify that these two curves do in fact intersect four times.[6]

2. Analysis of $\sec^2(x) = \tan(x) + 3$ reveals two different trigonometric functions, so an identity is in order. Since $\sec^2(x) = 1 + \tan^2(x)$, we get

$$\begin{array}{rcll} \sec^2(x) & = & \tan(x) + 3 & \\ 1 + \tan^2(x) & = & \tan(x) + 3 & (\text{Since } \sec^2(x) = 1 + \tan^2(x).) \\ \tan^2(x) - \tan(x) - 2 & = & 0 & \\ u^2 - u - 2 & = & 0 & \text{Let } u = \tan(x). \\ (u + 1)(u - 2) & = & 0 & \end{array}$$

[6]Note that we are *not* counting the point $(2\pi, 0)$ in our solution set since $x = 2\pi$ is not in the interval $[0, 2\pi)$. In the forthcoming solutions, remember that while $x = 2\pi$ may be a solution to the equation, it isn't counted among the solutions in $[0, 2\pi)$.

This gives $u = -1$ or $u = 2$. Since $u = \tan(x)$, we have $\tan(x) = -1$ or $\tan(x) = 2$. From $\tan(x) = -1$, we get $x = -\frac{\pi}{4} + \pi k$ for integers k. To solve $\tan(x) = 2$, we employ the arctangent function and get $x = \arctan(2) + \pi k$ for integers k. From the first set of solutions, we get $x = \frac{3\pi}{4}$ and $x = \frac{7\pi}{4}$ as our answers which lie in $[0, 2\pi)$. Using the same sort of argument we saw in Example 10.7.1, we get $x = \arctan(2)$ and $x = \pi + \arctan(2)$ as answers from our second set of solutions which lie in $[0, 2\pi)$. Using a reciprocal identity, we rewrite the secant as a cosine and graph $y = \frac{1}{(\cos(x))^2}$ and $y = \tan(x) + 3$ to find the x-values of the points where they intersect.

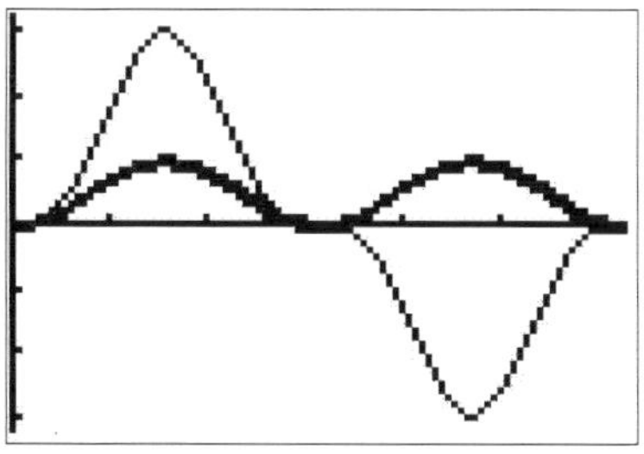

$y = 3(\sin(x))^3$ and $\boldsymbol{y = (\sin(x))^2}$

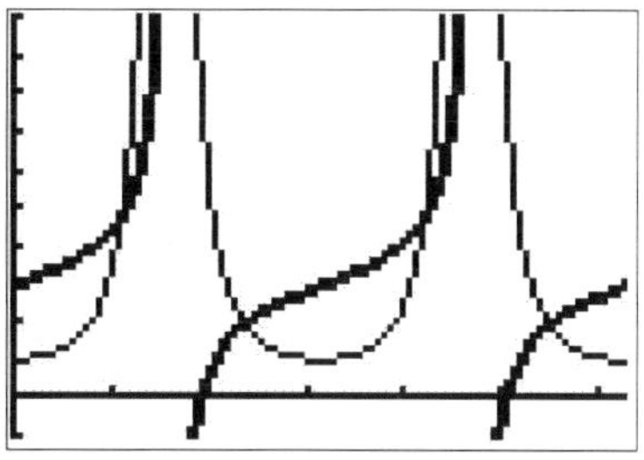

$y = \frac{1}{(\cos(x))^2}$ and $\boldsymbol{y = \tan(x) + 3}$

3. In the equation $\cos(2x) = 3\cos(x) - 2$, we have the same circular function, namely cosine, on both sides but the arguments differ. Using the identity $\cos(2x) = 2\cos^2(x) - 1$, we obtain a 'quadratic in disguise' and proceed as we have done in the past.

$$\begin{array}{rcll} \cos(2x) & = & 3\cos(x) - 2 & \\ 2\cos^2(x) - 1 & = & 3\cos(x) - 2 & (\text{Since } \cos(2x) = 2\cos^2(x) - 1.) \\ 2\cos^2(x) - 3\cos(x) + 1 & = & 0 & \\ 2u^2 - 3u + 1 & = & 0 & \text{Let } u = \cos(x). \\ (2u - 1)(u - 1) & = & 0 & \end{array}$$

This gives $u = \frac{1}{2}$ or $u = 1$. Since $u = \cos(x)$, we get $\cos(x) = \frac{1}{2}$ or $\cos(x) = 1$. Solving $\cos(x) = \frac{1}{2}$, we get $x = \frac{\pi}{3} + 2\pi k$ or $x = \frac{5\pi}{3} + 2\pi k$ for integers k. From $\cos(x) = 1$, we get $x = 2\pi k$ for integers k. The answers which lie in $[0, 2\pi)$ are $x = 0$, $\frac{\pi}{3}$, and $\frac{5\pi}{3}$. Graphing $y = \cos(2x)$ and $y = 3\cos(x) - 2$, we find, after a little extra effort, that the curves intersect in three places on $[0, 2\pi)$, and the x-coordinates of these points confirm our results.

4. To solve $\cos(3x) = 2 - \cos(x)$, we use the same technique as in the previous problem. From Example 10.4.3, number 4, we know that $\cos(3x) = 4\cos^3(x) - 3\cos(x)$. This transforms the equation into a polynomial in terms of $\cos(x)$.

$$\begin{array}{rcll} \cos(3x) & = & 2 - \cos(x) & \\ 4\cos^3(x) - 3\cos(x) & = & 2 - \cos(x) & \\ 2\cos^3(x) - 2\cos(x) - 2 & = & 0 & \\ 4u^3 - 2u - 2 & = & 0 & \text{Let } u = \cos(x). \end{array}$$

To solve $4u^3 - 2u - 2 = 0$, we need the techniques in Chapter 3 to factor $4u^3 - 2u - 2$ into $(u-1)\left(4u^2+4u+2\right)$. We get either $u-1=0$ or $4u^2+2u+2=0$, and since the discriminant of the latter is negative, the only real solution to $4u^3-2u-2=0$ is $u=1$. Since $u=\cos(x)$, we get $\cos(x)=1$, so $x=2\pi k$ for integers k. The only solution which lies in $[0,2\pi)$ is $x=0$. Graphing $y=\cos(3x)$ and $y=2-\cos(x)$ on the same set of axes over $[0,2\pi)$ shows that the graphs intersect at what appears to be $(0,1)$, as required.

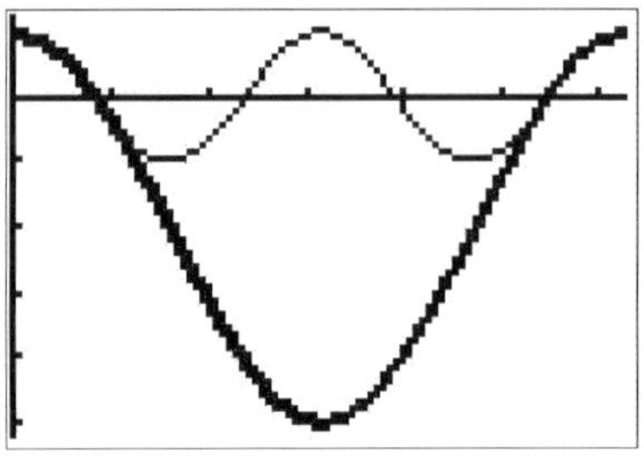

$y=\cos(2x)$ and $\boldsymbol{y=3\cos(x)-2}$

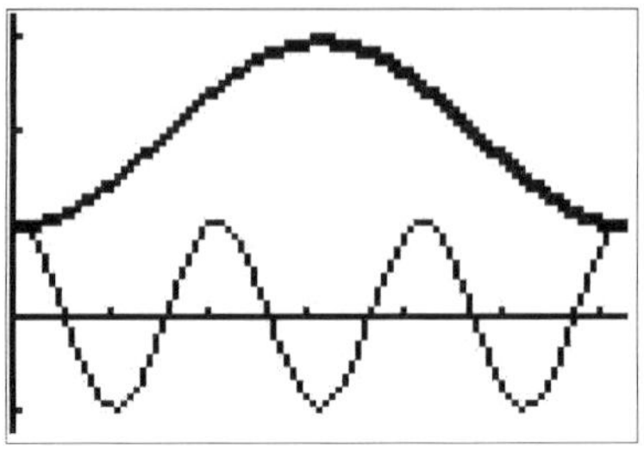

$y=\cos(3x)$ and $\boldsymbol{y=2-\cos(x)}$

5. While we could approach $\cos(3x)=\cos(5x)$ in the same manner as we did the previous two problems, we choose instead to showcase the utility of the Sum to Product Identities. From $\cos(3x)=\cos(5x)$, we get $\cos(5x)-\cos(3x)=0$, and it is the presence of 0 on the right hand side that indicates a switch to a product would be a good move.[7] Using Theorem 10.21, we have that $\cos(5x)-\cos(3x)=-2\sin\left(\frac{5x+3x}{2}\right)\sin\left(\frac{5x-3x}{2}\right)=-2\sin(4x)\sin(x)$. Hence, the equation $\cos(5x)=\cos(3x)$ is equivalent to $-2\sin(4x)\sin(x)=0$. From this, we get $\sin(4x)=0$ or $\sin(x)=0$. Solving $\sin(4x)=0$ gives $x=\frac{\pi}{4}k$ for integers k, and the solution to $\sin(x)=0$ is $x=\pi k$ for integers k. The second set of solutions is contained in the first set of solutions,[8] so our final solution to $\cos(5x)=\cos(3x)$ is $x=\frac{\pi}{4}k$ for integers k. There are eight of these answers which lie in $[0,2\pi)$: $x=0,\ \frac{\pi}{4},\ \frac{\pi}{2},\ \frac{3\pi}{4},\ \pi,\ \frac{5\pi}{4},\ \frac{3\pi}{2}$ and $\frac{7\pi}{4}$. Our plot of the graphs of $y=\cos(3x)$ and $y=\cos(5x)$ below (after some careful zooming) bears this out.

6. In examining the equation $\sin(2x)=\sqrt{3}\cos(x)$, not only do we have different circular functions involved, namely sine and cosine, we also have different arguments to contend with, namely $2x$ and x. Using the identity $\sin(2x)=2\sin(x)\cos(x)$ makes all of the arguments the same and we proceed as we would solving any nonlinear equation – gather all of the nonzero terms on one side of the equation and factor.

$$\begin{aligned}
\sin(2x) &= \sqrt{3}\cos(x) \\
2\sin(x)\cos(x) &= \sqrt{3}\cos(x) \quad (\text{Since } \sin(2x)=2\sin(x)\cos(x).) \\
2\sin(x)\cos(x)-\sqrt{3}\cos(x) &= 0 \\
\cos(x)(2\sin(x)-\sqrt{3}) &= 0
\end{aligned}$$

from which we get $\cos(x)=0$ or $\sin(x)=\frac{\sqrt{3}}{2}$. From $\cos(x)=0$, we obtain $x=\frac{\pi}{2}+\pi k$ for integers k. From $\sin(x)=\frac{\sqrt{3}}{2}$, we get $x=\frac{\pi}{3}+2\pi k$ or $x=\frac{2\pi}{3}+2\pi k$ for integers k. The answers

[7] As always, experience is the greatest teacher here!

[8] As always, when in doubt, write it out!

which lie in $[0, 2\pi)$ are $x = \frac{\pi}{2}, \frac{3\pi}{2}, \frac{\pi}{3}$ and $\frac{2\pi}{3}$. We graph $y = \sin(2x)$ and $y = \sqrt{3}\cos(x)$ and, after some careful zooming, verify our answers.

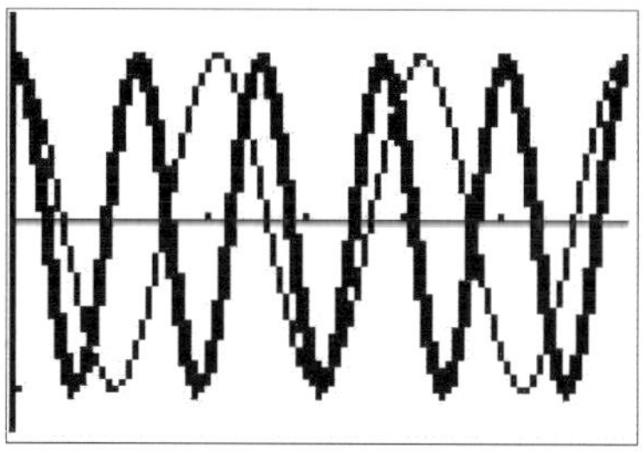

$y = \cos(3x)$ and $\boldsymbol{y = \cos(5x)}$

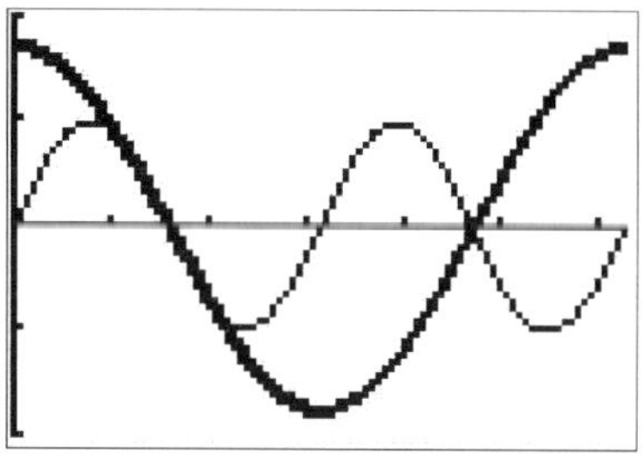

$y = \sin(2x)$ and $\boldsymbol{y = \sqrt{3}\cos(x)}$

7. Unlike the previous problem, there seems to be no quick way to get the circular functions or their arguments to match in the equation $\sin(x)\cos\left(\frac{x}{2}\right) + \cos(x)\sin\left(\frac{x}{2}\right) = 1$. If we stare at it long enough, however, we realize that the left hand side is the expanded form of the sum formula for $\sin\left(x + \frac{x}{2}\right)$. Hence, our original equation is equivalent to $\sin\left(\frac{3}{2}x\right) = 1$. Solving, we find $x = \frac{\pi}{3} + \frac{4\pi}{3}k$ for integers k. Two of these solutions lie in $[0, 2\pi)$: $x = \frac{\pi}{3}$ and $x = \frac{5\pi}{3}$. Graphing $y = \sin(x)\cos\left(\frac{x}{2}\right) + \cos(x)\sin\left(\frac{x}{2}\right)$ and $y = 1$ validates our solutions.

8. With the absence of double angles or squares, there doesn't seem to be much we can do. However, since the arguments of the cosine and sine are the same, we can rewrite the left hand side of this equation as a sinusoid.[9] To fit $f(x) = \cos(x) - \sqrt{3}\sin(x)$ to the form $A\sin(\omega t + \phi) + B$, we use what we learned in Example 10.5.3 and find $A = 2$, $B = 0$, $\omega = 1$ and $\phi = \frac{5\pi}{6}$. Hence, we can rewrite the equation $\cos(x) - \sqrt{3}\sin(x) = 2$ as $2\sin\left(x + \frac{5\pi}{6}\right) = 2$, or $\sin\left(x + \frac{5\pi}{6}\right) = 1$. Solving the latter, we get $x = -\frac{\pi}{3} + 2\pi k$ for integers k. Only one of these solutions, $x = \frac{5\pi}{3}$, which corresponds to $k = 1$, lies in $[0, 2\pi)$. Geometrically, we see that $y = \cos(x) - \sqrt{3}\sin(x)$ and $y = 2$ intersect just once, supporting our answer.

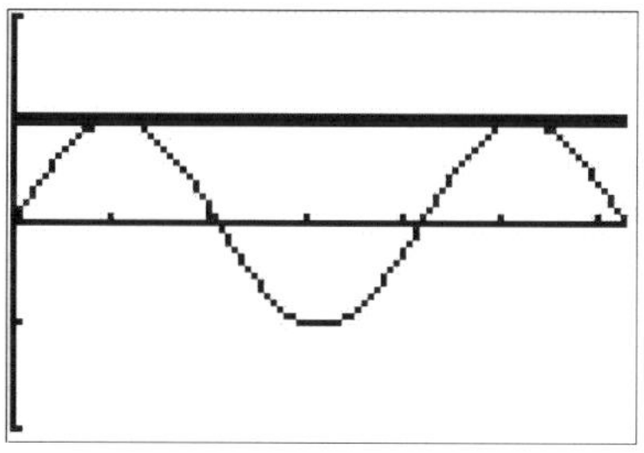

$y = \sin(x)\cos\left(\frac{x}{2}\right) + \cos(x)\sin\left(\frac{x}{2}\right)$ and $\boldsymbol{y = 1}$

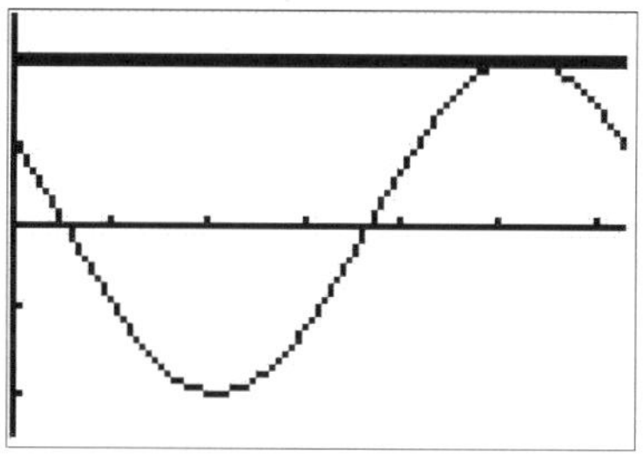

$y = \cos(x) - \sqrt{3}\sin(x)$ and $\boldsymbol{y = 2}$

□

We repeat here the advice given when solving systems of nonlinear equations in section 8.7 – when it comes to solving equations involving the trigonometric functions, it helps to just try something.

[9] We are essentially 'undoing' the sum / difference formula for cosine or sine, depending on which form we use, so this problem is actually closely related to the previous one!

Next, we focus on solving inequalities involving the trigonometric functions. Since these functions are continuous on their domains, we may use the sign diagram technique we've used in the past to solve the inequalities.[10]

Example 10.7.3. Solve the following inequalities on $[0, 2\pi)$. Express your answers using interval notation and verify your answers graphically.

1. $2\sin(x) \leq 1$
2. $\sin(2x) > \cos(x)$
3. $\tan(x) \geq 3$

Solution.

1. We begin solving $2\sin(x) \leq 1$ by collecting all of the terms on one side of the equation and zero on the other to get $2\sin(x) - 1 \leq 0$. Next, we let $f(x) = 2\sin(x) - 1$ and note that our original inequality is equivalent to solving $f(x) \leq 0$. We now look to see where, if ever, f is undefined and where $f(x) = 0$. Since the domain of f is all real numbers, we can immediately set about finding the zeros of f. Solving $f(x) = 0$, we have $2\sin(x) - 1 = 0$ or $\sin(x) = \frac{1}{2}$. The solutions here are $x = \frac{\pi}{6} + 2\pi k$ and $x = \frac{5\pi}{6} + 2\pi k$ for integers k. Since we are restricting our attention to $[0, 2\pi)$, only $x = \frac{\pi}{6}$ and $x = \frac{5\pi}{6}$ are of concern to us. Next, we choose test values in $[0, 2\pi)$ other than the zeros and determine if f is positive or negative there. For $x = 0$ we have $f(0) = -1$, for $x = \frac{\pi}{2}$ we get $f\left(\frac{\pi}{2}\right) = 1$ and for $x = \pi$ we get $f(\pi) = -1$. Since our original inequality is equivalent to $f(x) \leq 0$, we are looking for where the function is negative $(-)$ or 0, and we get the intervals $\left[0, \frac{\pi}{6}\right] \cup \left[\frac{5\pi}{6}, 2\pi\right)$. We can confirm our answer graphically by seeing where the graph of $y = 2\sin(x)$ crosses or is below the graph of $y = 1$.

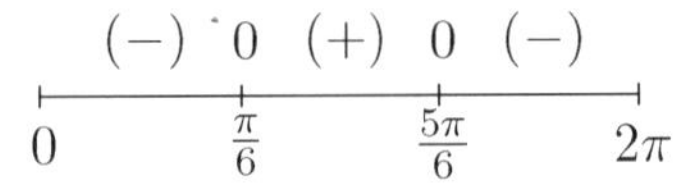

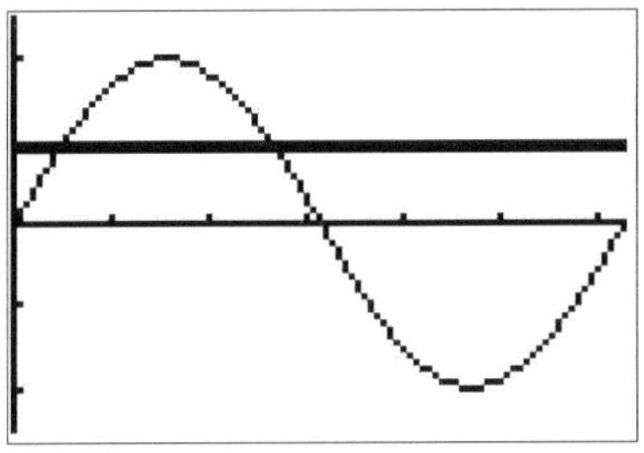

$y = 2\sin(x)$ and $\boldsymbol{y = 1}$

2. We first rewrite $\sin(2x) > \cos(x)$ as $\sin(2x) - \cos(x) > 0$ and let $f(x) = \sin(2x) - \cos(x)$. Our original inequality is thus equivalent to $f(x) > 0$. The domain of f is all real numbers, so we can advance to finding the zeros of f. Setting $f(x) = 0$ yields $\sin(2x) - \cos(x) = 0$, which, by way of the double angle identity for sine, becomes $2\sin(x)\cos(x) - \cos(x) = 0$ or $\cos(x)(2\sin(x) - 1) = 0$. From $\cos(x) = 0$, we get $x = \frac{\pi}{2} + \pi k$ for integers k of which only $x = \frac{\pi}{2}$ and $x = \frac{3\pi}{2}$ lie in $[0, 2\pi)$. For $2\sin(x) - 1 = 0$, we get $\sin(x) = \frac{1}{2}$ which gives $x = \frac{\pi}{6} + 2\pi k$ or $x = \frac{5\pi}{6} + 2\pi k$ for integers k. Of those, only $x = \frac{\pi}{6}$ and $x = \frac{5\pi}{6}$ lie in $[0, 2\pi)$. Next, we choose

[10] See page 214, Example 3.1.5, page 321, page 399, Example 6.3.2 and Example 6.4.2 for discussion of this technique.

our test values. For $x = 0$ we find $f(0) = -1$; when $x = \frac{\pi}{4}$ we get $f\left(\frac{\pi}{4}\right) = 1 - \frac{\sqrt{2}}{2} = \frac{2-\sqrt{2}}{2}$; for $x = \frac{3\pi}{4}$ we get $f\left(\frac{3\pi}{4}\right) = -1 + \frac{\sqrt{2}}{2} = \frac{\sqrt{2}-2}{2}$; when $x = \pi$ we have $f(\pi) = 1$, and lastly, for $x = \frac{7\pi}{4}$ we get $f\left(\frac{7\pi}{4}\right) = -1 - \frac{\sqrt{2}}{2} = \frac{-2-\sqrt{2}}{2}$. We see $f(x) > 0$ on $\left(\frac{\pi}{6}, \frac{\pi}{2}\right) \cup \left(\frac{5\pi}{6}, \frac{3\pi}{2}\right)$, so this is our answer. We can use the calculator to check that the graph of $y = \sin(2x)$ is indeed above the graph of $y = \cos(x)$ on those intervals.

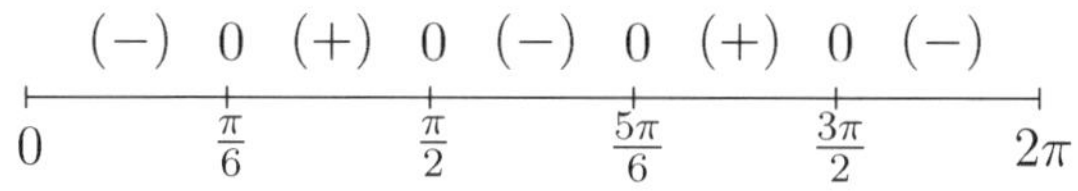

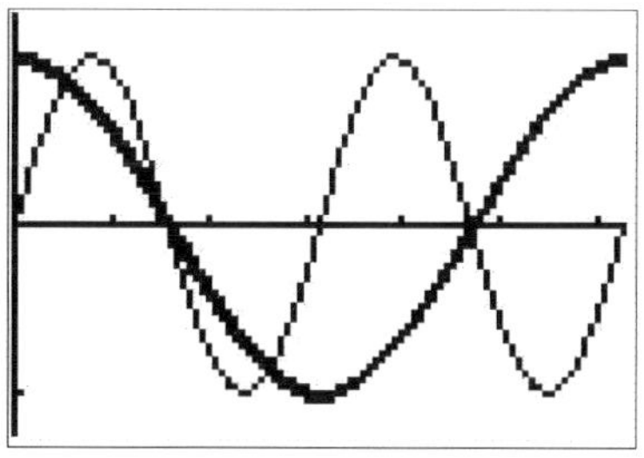

$y = \sin(2x)$ and $\boldsymbol{y = \cos(x)}$

3. Proceeding as in the last two problems, we rewrite $\tan(x) \geq 3$ as $\tan(x) - 3 \geq 0$ and let $f(x) = \tan(x) - 3$. We note that on $[0, 2\pi)$, f is undefined at $x = \frac{\pi}{2}$ and $\frac{3\pi}{2}$, so those values will need the usual disclaimer on the sign diagram.[11] Moving along to zeros, solving $f(x) = \tan(x) - 3 = 0$ requires the arctangent function. We find $x = \arctan(3) + \pi k$ for integers k and of these, only $x = \arctan(3)$ and $x = \arctan(3) + \pi$ lie in $[0, 2\pi)$. Since $3 > 0$, we know $0 < \arctan(3) < \frac{\pi}{2}$ which allows us to position these zeros correctly on the sign diagram. To choose test values, we begin with $x = 0$ and find $f(0) = -3$. Finding a convenient test value in the interval $\left(\arctan(3), \frac{\pi}{2}\right)$ is a bit more challenging. Keep in mind that the arctangent function is increasing and is bounded above by $\frac{\pi}{2}$. This means that the number $x = \arctan(117)$ is guaranteed[12] to lie between $\arctan(3)$ and $\frac{\pi}{2}$. We see that $f(\arctan(117)) = \tan(\arctan(117)) - 3 = 114$. For our next test value, we take $x = \pi$ and find $f(\pi) = -3$. To find our next test value, we note that since $\arctan(3) < \arctan(117) < \frac{\pi}{2}$, it follows[13] that $\arctan(3) + \pi < \arctan(117) + \pi < \frac{3\pi}{2}$. Evaluating f at $x = \arctan(117) + \pi$ yields $f(\arctan(117) + \pi) = \tan(\arctan(117) + \pi) - 3 = \tan(\arctan(117)) - 3 = 114$. We choose our last test value to be $x = \frac{7\pi}{4}$ and find $f\left(\frac{7\pi}{4}\right) = -4$. Since we want $f(x) \geq 0$, we see that our answer is $\left[\arctan(3), \frac{\pi}{2}\right) \cup \left[\arctan(3) + \pi, \frac{3\pi}{2}\right)$. Using the graphs of $y = \tan(x)$ and $y = 3$, we see when the graph of the former is above (or meets) the graph of the latter.

[11] See page 321 for a discussion of the non-standard character known as the interrobang.

[12] We could have chosen any value $\arctan(t)$ where $t > 3$.

[13] ... by adding π through the inequality ...

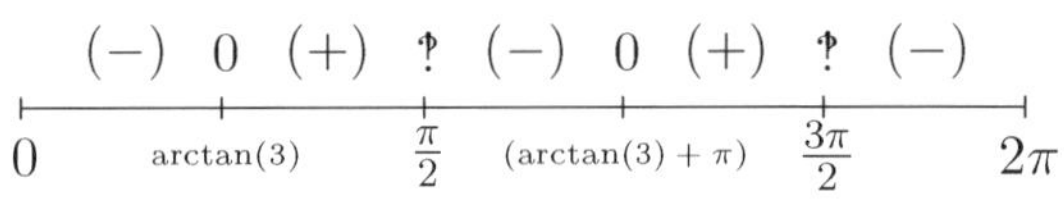

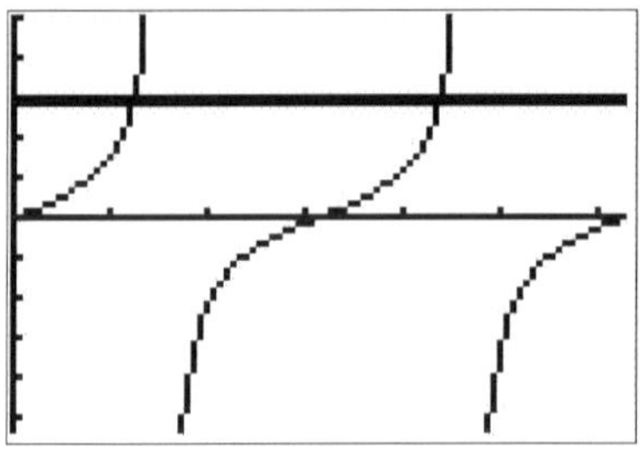

$y = \tan(x)$ and $\boldsymbol{y = 3}$

□

Our next example puts solving equations and inequalities to good use – finding domains of functions.

Example 10.7.4. Express the domain of the following functions using extended interval notation.[14]

1. $f(x) = \csc\left(2x + \frac{\pi}{3}\right)$
2. $f(x) = \dfrac{\sin(x)}{2\cos(x) - 1}$
3. $f(x) = \sqrt{1 - \cot(x)}$

Solution.

1. To find the domain of $f(x) = \csc\left(2x + \frac{\pi}{3}\right)$, we rewrite f in terms of sine as $f(x) = \frac{1}{\sin\left(2x+\frac{\pi}{3}\right)}$. Since the sine function is defined everywhere, our only concern comes from zeros in the denominator. Solving $\sin\left(2x + \frac{\pi}{3}\right) = 0$, we get $x = -\frac{\pi}{6} + \frac{\pi}{2}k$ for integers k. In set-builder notation, our domain is $\left\{x : x \neq -\frac{\pi}{6} + \frac{\pi}{2}k \text{ for integers } k\right\}$. To help visualize the domain, we follow the old mantra 'When in doubt, write it out!' We get $\left\{x : x \neq -\frac{\pi}{6}, \frac{2\pi}{6}, -\frac{4\pi}{6}, \frac{5\pi}{6}, -\frac{7\pi}{6}, \frac{8\pi}{6}, \ldots\right\}$, where we have kept the denominators 6 throughout to help see the pattern. Graphing the situation on a numberline, we have

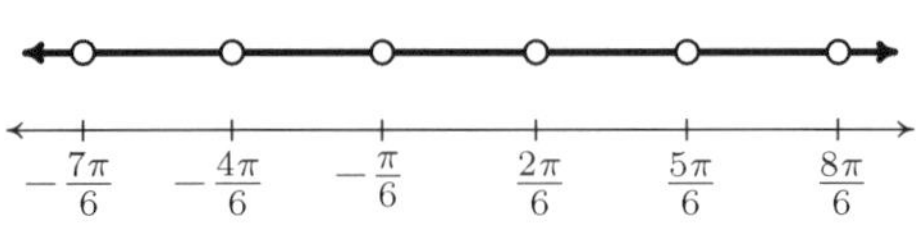

Proceeding as we did on page 756 in Section 10.3.1, we let x_k denote the kth number excluded from the domain and we have $x_k = -\frac{\pi}{6} + \frac{\pi}{2}k = \frac{(3k-1)\pi}{6}$ for integers k. The intervals which comprise the domain are of the form $(x_k, x_{k+1}) = \left(\frac{(3k-1)\pi}{6}, \frac{(3k+2)\pi}{6}\right)$ as k runs through the integers. Using extended interval notation, we have that the domain is

$$\bigcup_{k=-\infty}^{\infty} \left(\frac{(3k-1)\pi}{6}, \frac{(3k+2)\pi}{6}\right)$$

We can check our answer by substituting in values of k to see that it matches our diagram.

[14]See page 756 for details about this notation.

2. Since the domains of $\sin(x)$ and $\cos(x)$ are all real numbers, the only concern when finding the domain of $f(x) = \frac{\sin(x)}{2\cos(x)-1}$ is division by zero so we set the denominator equal to zero and solve. From $2\cos(x) - 1 = 0$ we get $\cos(x) = \frac{1}{2}$ so that $x = \frac{\pi}{3} + 2\pi k$ or $x = \frac{5\pi}{3} + 2\pi k$ for integers k. Using set-builder notation, the domain is $\left\{x : x \neq \frac{\pi}{3} + 2\pi k \text{ and } x \neq \frac{5\pi}{3} + 2\pi k \text{ for integers } k\right\}$, or $\left\{x : x \neq \pm\frac{\pi}{3}, \pm\frac{5\pi}{3}, \pm\frac{7\pi}{3}, \pm\frac{11\pi}{3}, \ldots\right\}$, so we have

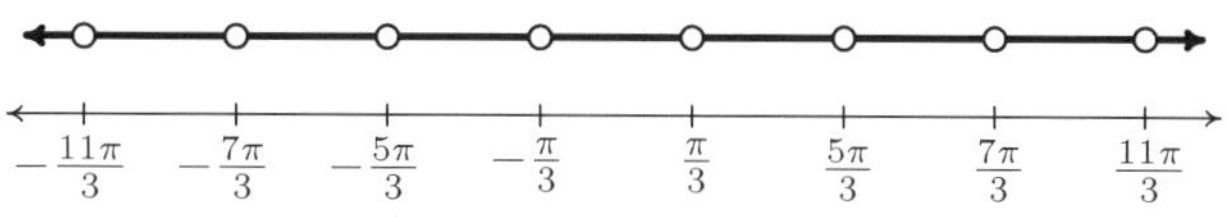

Unlike the previous example, we have *two* different families of points to consider, and we present two ways of dealing with this kind of situation. One way is to generalize what we did in the previous example and use the formulas we found in our domain work to describe the intervals. To that end, we let $a_k = \frac{\pi}{3} + 2\pi k = \frac{(6k+1)\pi}{3}$ and $b_k = \frac{5\pi}{3} + 2\pi k = \frac{(6k+5)\pi}{3}$ for integers k. The goal now is to write the domain in terms of the a's an b's. We find $a_0 = \frac{\pi}{3}$, $a_1 = \frac{7\pi}{3}$, $a_{-1} = -\frac{5\pi}{3}$, $a_2 = \frac{13\pi}{3}$, $a_{-2} = -\frac{11\pi}{3}$, $b_0 = \frac{5\pi}{3}$, $b_1 = \frac{11\pi}{3}$, $b_{-1} = -\frac{\pi}{3}$, $b_2 = \frac{17\pi}{3}$ and $b_{-2} = -\frac{7\pi}{3}$. Hence, in terms of the a's and b's, our domain is

$$\ldots (a_{-2}, b_{-2}) \cup (b_{-2}, a_{-1}) \cup (a_{-1}, b_{-1}) \cup (b_{-1}, a_0) \cup (a_0, b_0) \cup (b_0, a_1) \cup (a_1, b_1) \cup \ldots$$

If we group these intervals in pairs, $(a_{-2}, b_{-2}) \cup (b_{-2}, a_{-1})$, $(a_{-1}, b_{-1}) \cup (b_{-1}, a_0)$, $(a_0, b_0) \cup (b_0, a_1)$ and so forth, we see a pattern emerge of the form $(a_k, b_k) \cup (b_k, a_{k+1})$ for integers k so that our domain can be written as

$$\bigcup_{k=-\infty}^{\infty} (a_k, b_k) \cup (b_k, a_{k+1}) = \bigcup_{k=-\infty}^{\infty} \left(\frac{(6k+1)\pi}{3}, \frac{(6k+5)\pi}{3}\right) \cup \left(\frac{(6k+5)\pi}{3}, \frac{(6k+7)\pi}{3}\right)$$

A second approach to the problem exploits the periodic nature of f. Since $\cos(x)$ and $\sin(x)$ have period 2π, it's not too difficult to show the function f repeats itself every 2π units.[15] This means if we can find a formula for the domain on an interval of length 2π, we can express the entire domain by translating our answer left and right on the x-axis by adding integer multiples of 2π. One such interval that arises from our domain work is $\left[\frac{\pi}{3}, \frac{7\pi}{3}\right]$. The portion of the domain here is $\left(\frac{\pi}{3}, \frac{5\pi}{3}\right) \cup \left(\frac{5\pi}{3}, \frac{7\pi}{3}\right)$. Adding integer multiples of 2π, we get the family of intervals $\left(\frac{\pi}{3} + 2\pi k, \frac{5\pi}{3} + 2\pi k\right) \cup \left(\frac{5\pi}{3} + 2\pi k, \frac{7\pi}{3} + 2\pi k\right)$ for integers k. We leave it to the reader to show that getting common denominators leads to our previous answer.

[15]This doesn't necessarily mean the period of f is 2π. The tangent function is comprised of $\cos(x)$ and $\sin(x)$, but its period is half theirs. The reader is invited to investigate the period of f.

3. To find the domain of $f(x) = \sqrt{1 - \cot(x)}$, we first note that, due to the presence of the $\cot(x)$ term, $x \neq \pi k$ for integers k. Next, we recall that for the square root to be defined, we need $1 - \cot(x) \geq 0$. Unlike the inequalities we solved in Example 10.7.3, we are not restricted here to a given interval. Our strategy is to solve this inequality over $(0, \pi)$ (the same interval which generates a fundamental cycle of cotangent) and then add integer multiples of the period, in this case, π. We let $g(x) = 1 - \cot(x)$ and set about making a sign diagram for g over the interval $(0, \pi)$ to find where $g(x) \geq 0$. We note that g is undefined for $x = \pi k$ for integers k, in particular, at the endpoints of our interval $x = 0$ and $x = \pi$. Next, we look for the zeros of g. Solving $g(x) = 0$, we get $\cot(x) = 1$ or $x = \frac{\pi}{4} + \pi k$ for integers k and only one of these, $x = \frac{\pi}{4}$, lies in $(0, \pi)$. Choosing the test values $x = \frac{\pi}{6}$ and $x = \frac{\pi}{2}$, we get $g\left(\frac{\pi}{6}\right) = 1 - \sqrt{3}$, and $g\left(\frac{\pi}{2}\right) = 1$.

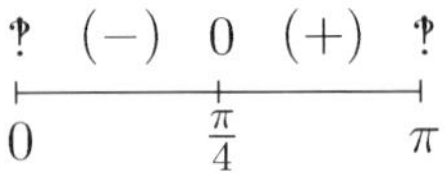

We find $g(x) \geq 0$ on $\left[\frac{\pi}{4}, \pi\right)$. Adding multiples of the period we get our solution to consist of the intervals $\left[\frac{\pi}{4} + \pi k, \pi + \pi k\right) = \left[\frac{(4k+1)\pi}{4}, (k+1)\pi\right)$. Using extended interval notation, we express our final answer as

$$\bigcup_{k=-\infty}^{\infty} \left[\frac{(4k+1)\pi}{4}, (k+1)\pi\right)$$

□

We close this section with an example which demonstrates how to solve equations and inequalities involving the inverse trigonometric functions.

Example 10.7.5. Solve the following equations and inequalities analytically. Check your answers using a graphing utility.

1. $\arcsin(2x) = \frac{\pi}{3}$
2. $4\arccos(x) - 3\pi = 0$
3. $3\,\mathrm{arcsec}(2x - 1) + \pi = 2\pi$
4. $4\arctan^2(x) - 3\pi\arctan(x) - \pi^2 = 0$
5. $\pi^2 - 4\arccos^2(x) < 0$
6. $4\,\mathrm{arccot}(3x) > \pi$

Solution.

1. To solve $\arcsin(2x) = \frac{\pi}{3}$, we first note that $\frac{\pi}{3}$ is in the range of the arcsine function (so a solution exists!) Next, we exploit the inverse property of sine and arcsine from Theorem 10.26

$$\begin{aligned} \arcsin(2x) &= \frac{\pi}{3} \\ \sin(\arcsin(2x)) &= \sin\left(\frac{\pi}{3}\right) \\ 2x &= \frac{\sqrt{3}}{2} \qquad \text{Since } \sin(\arcsin(u)) = u \\ x &= \frac{\sqrt{3}}{4} \end{aligned}$$

Graphing $y = \arcsin(2x)$ and the horizontal line $y = \frac{\pi}{3}$, we see they intersect at $\frac{\sqrt{3}}{4} \approx 0.4430$.

2. Our first step in solving $4\arccos(x) - 3\pi = 0$ is to isolate the arccosine. Doing so, we get $\arccos(x) = \frac{3\pi}{4}$. Since $\frac{3\pi}{4}$ is in the range of arccosine, we may apply Theorem 10.26

$$\begin{aligned} \arccos(x) &= \frac{3\pi}{4} \\ \cos(\arccos(x)) &= \cos\left(\frac{3\pi}{4}\right) \\ x &= -\frac{\sqrt{2}}{2} \qquad \text{Since } \cos(\arccos(u)) = u \end{aligned}$$

The calculator confirms $y = 4\arccos(x) - 3\pi$ crosses $y = 0$ (the x-axis) at $-\frac{\sqrt{2}}{2} \approx -0.7071$.

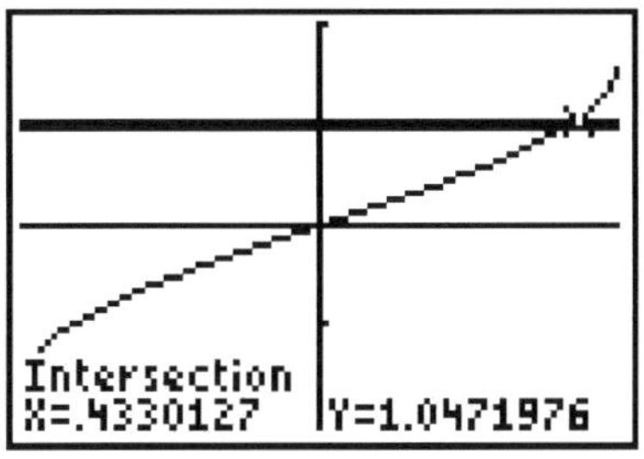

$y = \arcsin(2x)$ and $y = \frac{\pi}{3}$

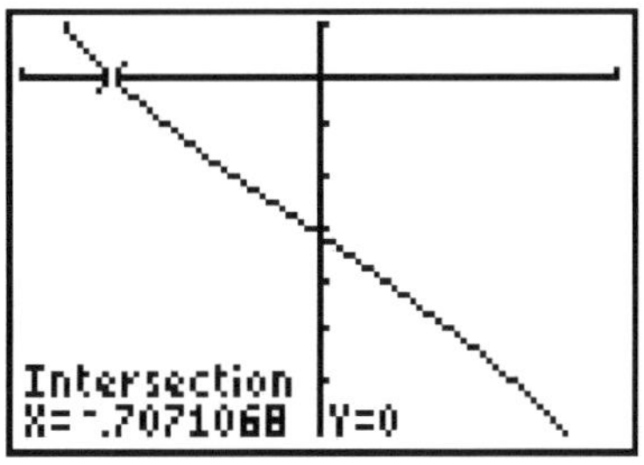

$y = 4\arccos(x) - 3\pi$

3. From $3\,\text{arcsec}(2x-1) + \pi = 2\pi$, we get $\text{arcsec}(2x-1) = \frac{\pi}{3}$. As we saw in Section 10.6, there are two possible ranges for the arcsecant function. Fortunately, both ranges contain $\frac{\pi}{3}$. Applying Theorem 10.28 / 10.29, we get

$$\begin{aligned} \text{arcsec}(2x-1) &= \frac{\pi}{3} \\ \sec(\text{arcsec}(2x-1)) &= \sec\left(\frac{\pi}{3}\right) \\ 2x-1 &= 2 \qquad \text{Since } \sec(\text{arcsec}(u)) = u \\ x &= \frac{3}{2} \end{aligned}$$

To check using our calculator, we need to graph $y = 3\,\text{arcsec}(2x-1) + \pi$. To do so, we make use of the identity $\text{arcsec}(u) = \arccos\left(\frac{1}{u}\right)$ from Theorems 10.28 and 10.29.[16] We see the graph of $y = 3\arccos\left(\frac{1}{2x-1}\right) + \pi$ and the horizontal line $y = 2\pi$ intersect at $\frac{3}{2} = 1.5$.

[16] Since we are checking for solutions where arcsecant is positive, we know $u = 2x - 1 \geq 1$, and so the identity applies in both cases.

4. With the presence of both $\arctan^2(x)$ ($= (\arctan(x))^2$) and $\arctan(x)$, we substitute $u = \arctan(x)$. The equation $4\arctan^2(x) - 3\pi\arctan(x) - \pi^2 = 0$ becomes $4u^2 - 3\pi u - \pi^2 = 0$. Factoring,[17] we get $(4u + \pi)(u - \pi) = 0$, so $u = \arctan(x) = -\frac{\pi}{4}$ or $u = \arctan(x) = \pi$. Since $-\frac{\pi}{4}$ is in the range of arctangent, but π is not, we only get solutions from the first equation. Using Theorem 10.27, we get

$$\begin{aligned} \arctan(x) &= -\tfrac{\pi}{4} \\ \tan(\arctan(x)) &= \tan\left(-\tfrac{\pi}{4}\right) \\ x &= -1 \qquad \text{Since } \tan(\arctan(u)) = u. \end{aligned}$$

The calculator verifies our result.

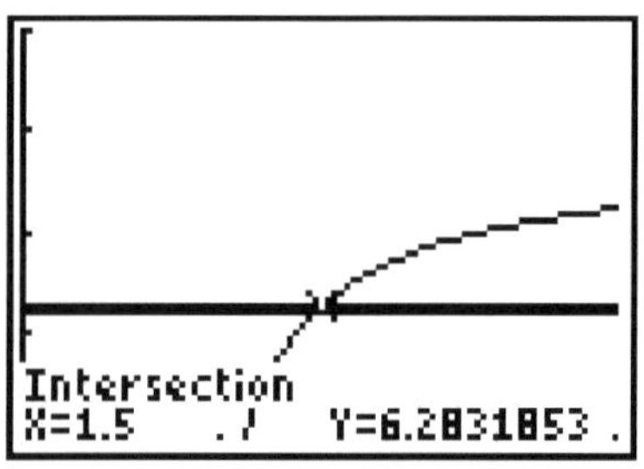

$y = 3\,\text{arcsec}(2x - 1) + \pi$ and $\boldsymbol{y = 2\pi}$

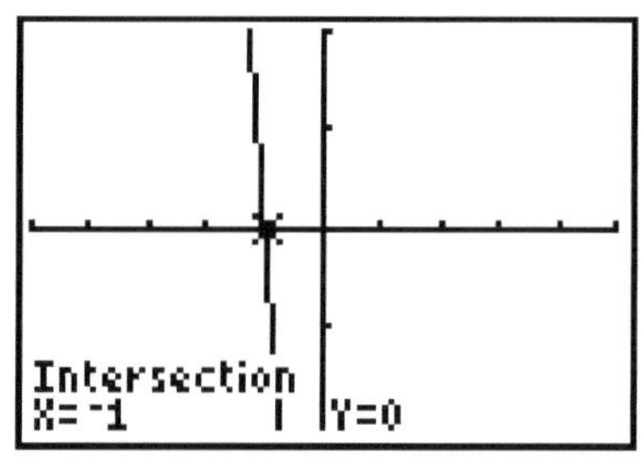

$y = 4\arctan^2(x) - 3\pi\arctan(x) - \pi^2$

5. Since the inverse trigonometric functions are continuous on their domains, we can solve inequalities featuring these functions using sign diagrams. Since all of the nonzero terms of $\pi^2 - 4\arccos^2(x) < 0$ are on one side of the inequality, we let $f(x) = \pi^2 - 4\arccos^2(x)$ and note the domain of f is limited by the $\arccos(x)$ to $[-1, 1]$. Next, we find the zeros of f by setting $f(x) = \pi^2 - 4\arccos^2(x) = 0$. We get $\arccos(x) = \pm\frac{\pi}{2}$, and since the range of arccosine is $[0, \pi]$, we focus our attention on $\arccos(x) = \frac{\pi}{2}$. Using Theorem 10.26, we get $x = \cos\left(\frac{\pi}{2}\right) = 0$ as our only zero. Hence, we have two test intervals, $[-1, 0)$ and $(0, 1]$. Choosing test values $x = \pm 1$, we get $f(-1) = -3\pi^2 < 0$ and $f(1) = \pi^2 > 0$. Since we are looking for where $f(x) = \pi^2 - 4\arccos^2(x) < 0$, our answer is $[-1, 0)$. The calculator confirms that for these values of x, the graph of $y = \pi^2 - 4\arccos^2(x)$ is below $y = 0$ (the x-axis.)

[17] It's not as bad as it looks... don't let the π throw you!

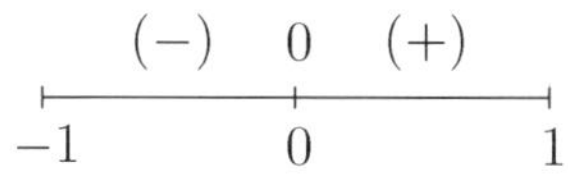

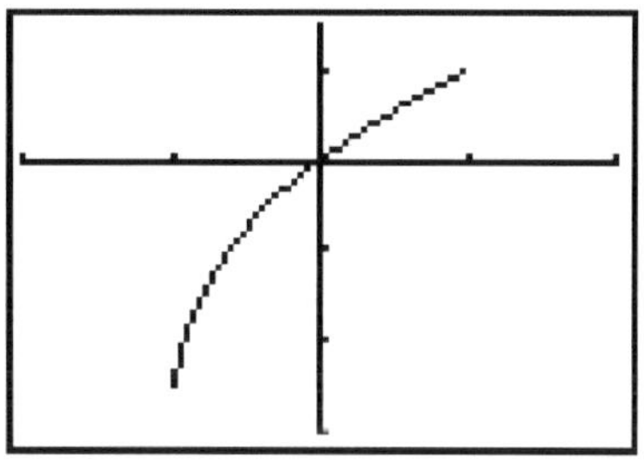

$y = \pi^2 - 4\arccos^2(x)$

6. To begin, we rewrite $4\,\text{arccot}(3x) > \pi$ as $4\,\text{arccot}(3x) - \pi > 0$. We let $f(x) = 4\,\text{arccot}(3x) - \pi$, and note the domain of f is all real numbers, $(-\infty, \infty)$. To find the zeros of f, we set $f(x) = 4\,\text{arccot}(3x) - \pi = 0$ and solve. We get $\text{arccot}(3x) = \frac{\pi}{4}$, and since $\frac{\pi}{4}$ is in the range of arccotangent, we may apply Theorem 10.27 and solve

$$\begin{aligned} \text{arccot}(3x) &= \tfrac{\pi}{4} \\ \cot(\text{arccot}(3x)) &= \cot\left(\tfrac{\pi}{4}\right) \\ 3x &= 1 \qquad \text{Since } \cot(\text{arccot}(u)) = u. \\ x &= \tfrac{1}{3} \end{aligned}$$

Next, we make a sign diagram for f. Since the domain of f is all real numbers, and there is only one zero of f, $x = \frac{1}{3}$, we have two test intervals, $\left(-\infty, \frac{1}{3}\right)$ and $\left(\frac{1}{3}, \infty\right)$. Ideally, we wish to find test values x in these intervals so that $\text{arccot}(4x)$ corresponds to one of our oft-used 'common' angles. After a bit of computation,[18] we choose $x = 0$ for $x < \frac{1}{3}$ and for $x > \frac{1}{3}$, we choose $x = \frac{\sqrt{3}}{3}$. We find $f(0) = \pi > 0$ and $f\left(\frac{\sqrt{3}}{3}\right) = -\frac{\pi}{3} < 0$. Since we are looking for where $f(x) = 4\,\text{arccot}(3x) - \pi > 0$, we get our answer $\left(-\infty, \frac{1}{3}\right)$. To check graphically, we use the technique in number 2c of Example 10.6.5 in Section 10.6 to graph $y = 4\,\text{arccot}(3x)$ and we see it is above the horizontal line $y = \pi$ on $\left(-\infty, \frac{1}{3}\right) = \left(-\infty, 0.\overline{3}\right)$.

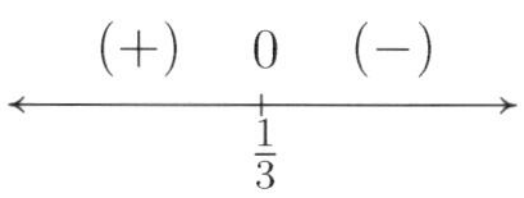

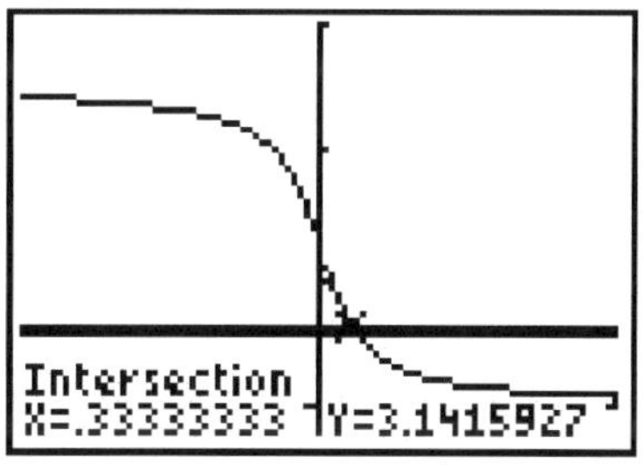

$y = 4\,\text{arccot}(3x)$ and $\boldsymbol{y = \pi}$

□

[18] Set $3x$ equal to the cotangents of the 'common angles' and choose accordingly.

10.7.1 Exercises

In Exercises 1 - 18, find all of the exact solutions of the equation and then list those solutions which are in the interval $[0, 2\pi)$.

1. $\sin(5x) = 0$
2. $\cos(3x) = \frac{1}{2}$
3. $\sin(-2x) = \frac{\sqrt{3}}{2}$
4. $\tan(6x) = 1$
5. $\csc(4x) = -1$
6. $\sec(3x) = \sqrt{2}$
7. $\cot(2x) = -\frac{\sqrt{3}}{3}$
8. $\cos(9x) = 9$
9. $\sin\left(\frac{x}{3}\right) = \frac{\sqrt{2}}{2}$
10. $\cos\left(x + \frac{5\pi}{6}\right) = 0$
11. $\sin\left(2x - \frac{\pi}{3}\right) = -\frac{1}{2}$
12. $2\cos\left(x + \frac{7\pi}{4}\right) = \sqrt{3}$
13. $\csc(x) = 0$
14. $\tan(2x - \pi) = 1$
15. $\tan^2(x) = 3$
16. $\sec^2(x) = \frac{4}{3}$
17. $\cos^2(x) = \frac{1}{2}$
18. $\sin^2(x) = \frac{3}{4}$

In Exercises 19 - 42, solve the equation, giving the exact solutions which lie in $[0, 2\pi)$

19. $\sin(x) = \cos(x)$
20. $\sin(2x) = \sin(x)$
21. $\sin(2x) = \cos(x)$
22. $\cos(2x) = \sin(x)$
23. $\cos(2x) = \cos(x)$
24. $\cos(2x) = 2 - 5\cos(x)$
25. $3\cos(2x) + \cos(x) + 2 = 0$
26. $\cos(2x) = 5\sin(x) - 2$
27. $3\cos(2x) = \sin(x) + 2$
28. $2\sec^2(x) = 3 - \tan(x)$
29. $\tan^2(x) = 1 - \sec(x)$
30. $\cot^2(x) = 3\csc(x) - 3$
31. $\sec(x) = 2\csc(x)$
32. $\cos(x)\csc(x)\cot(x) = 6 - \cot^2(x)$
33. $\sin(2x) = \tan(x)$
34. $\cot^4(x) = 4\csc^2(x) - 7$
35. $\cos(2x) + \csc^2(x) = 0$
36. $\tan^3(x) = 3\tan(x)$
37. $\tan^2(x) = \frac{3}{2}\sec(x)$
38. $\cos^3(x) = -\cos(x)$
39. $\tan(2x) - 2\cos(x) = 0$
40. $\csc^3(x) + \csc^2(x) = 4\csc(x) + 4$
41. $2\tan(x) = 1 - \tan^2(x)$
42. $\tan(x) = \sec(x)$

In Exercises 43 - 58, solve the equation, giving the exact solutions which lie in $[0, 2\pi)$

43. $\sin(6x)\cos(x) = -\cos(6x)\sin(x)$
44. $\sin(3x)\cos(x) = \cos(3x)\sin(x)$
45. $\cos(2x)\cos(x) + \sin(2x)\sin(x) = 1$
46. $\cos(5x)\cos(3x) - \sin(5x)\sin(3x) = \frac{\sqrt{3}}{2}$
47. $\sin(x) + \cos(x) = 1$
48. $\sin(x) + \sqrt{3}\cos(x) = 1$
49. $\sqrt{2}\cos(x) - \sqrt{2}\sin(x) = 1$
50. $\sqrt{3}\sin(2x) + \cos(2x) = 1$
51. $\cos(2x) - \sqrt{3}\sin(2x) = \sqrt{2}$
52. $3\sqrt{3}\sin(3x) - 3\cos(3x) = 3\sqrt{3}$
53. $\cos(3x) = \cos(5x)$
54. $\cos(4x) = \cos(2x)$
55. $\sin(5x) = \sin(3x)$
56. $\cos(5x) = -\cos(2x)$
57. $\sin(6x) + \sin(x) = 0$
58. $\tan(x) = \cos(x)$

In Exercises 59 - 68, solve the equation.

59. $\arccos(2x) = \pi$
60. $\pi - 2\arcsin(x) = 2\pi$
61. $4\arctan(3x - 1) - \pi = 0$
62. $6\,\text{arccot}(2x) - 5\pi = 0$
63. $4\,\text{arcsec}\left(\frac{x}{2}\right) = \pi$
64. $12\,\text{arccsc}\left(\frac{x}{3}\right) = 2\pi$
65. $9\arcsin^2(x) - \pi^2 = 0$
66. $9\arccos^2(x) - \pi^2 = 0$
67. $8\,\text{arccot}^2(x) + 3\pi^2 = 10\pi\,\text{arccot}(x)$
68. $6\arctan(x)^2 = \pi\arctan(x) + \pi^2$

In Exercises 69 - 80, solve the inequality. Express the exact answer in interval notation, restricting your attention to $0 \leq x \leq 2\pi$.

69. $\sin(x) \leq 0$
70. $\tan(x) \geq \sqrt{3}$
71. $\sec^2(x) \leq 4$
72. $\cos^2(x) > \frac{1}{2}$
73. $\cos(2x) \leq 0$
74. $\sin\left(x + \frac{\pi}{3}\right) > \frac{1}{2}$
75. $\cot^2(x) \geq \frac{1}{3}$
76. $2\cos(x) \geq 1$
77. $\sin(5x) \geq 5$
78. $\cos(3x) \leq 1$
79. $\sec(x) \leq \sqrt{2}$
80. $\cot(x) \leq 4$

In Exercises 81 - 86, solve the inequality. Express the exact answer in interval notation, restricting your attention to $-\pi \leq x \leq \pi$.

81. $\cos(x) > \frac{\sqrt{3}}{2}$

82. $\sin(x) > \frac{1}{3}$

83. $\sec(x) \leq 2$

84. $\sin^2(x) < \frac{3}{4}$

85. $\cot(x) \geq -1$

86. $\cos(x) \geq \sin(x)$

In Exercises 87 - 92, solve the inequality. Express the exact answer in interval notation, restricting your attention to $-2\pi \leq x \leq 2\pi$.

87. $\csc(x) > 1$

88. $\cos(x) \leq \frac{5}{3}$

89. $\cot(x) \geq 5$

90. $\tan^2(x) \geq 1$

91. $\sin(2x) \geq \sin(x)$

92. $\cos(2x) \leq \sin(x)$

In Exercises 93 - 98, solve the given inequality.

93. $\arcsin(2x) > 0$

94. $3\arccos(x) \leq \pi$

95. $6\,\text{arccot}(7x) \geq \pi$

96. $\pi > 2\arctan(x)$

97. $2\arcsin(x)^2 > \pi \arcsin(x)$

98. $12\arccos(x)^2 + 2\pi^2 > 11\pi \arccos(x)$

In Exercises 99 - 107, express the domain of the function using the extended interval notation. (See page 756 in Section 10.3.1 for details.)

99. $f(x) = \dfrac{1}{\cos(x) - 1}$

100. $f(x) = \dfrac{\cos(x)}{\sin(x) + 1}$

101. $f(x) = \sqrt{\tan^2(x) - 1}$

102. $f(x) = \sqrt{2 - \sec(x)}$

103. $f(x) = \csc(2x)$

104. $f(x) = \dfrac{\sin(x)}{2 + \cos(x)}$

105. $f(x) = 3\csc(x) + 4\sec(x)$

106. $f(x) = \ln(|\cos(x)|)$

107. $f(x) = \arcsin(\tan(x))$

108. With the help of your classmates, determine the number of solutions to $\sin(x) = \frac{1}{2}$ in $[0, 2\pi)$. Then find the number of solutions to $\sin(2x) = \frac{1}{2}$, $\sin(3x) = \frac{1}{2}$ and $\sin(4x) = \frac{1}{2}$ in $[0, 2\pi)$. A pattern should emerge. Explain how this pattern would help you solve equations like $\sin(11x) = \frac{1}{2}$. Now consider $\sin\left(\frac{x}{2}\right) = \frac{1}{2}$, $\sin\left(\frac{3x}{2}\right) = \frac{1}{2}$ and $\sin\left(\frac{5x}{2}\right) = \frac{1}{2}$. What do you find? Replace $\frac{1}{2}$ with -1 and repeat the whole exploration.

10.7.2 Answers

1. $x = \frac{\pi k}{5}$; $x = 0, \frac{\pi}{5}, \frac{2\pi}{5}, \frac{3\pi}{5}, \frac{4\pi}{5}, \pi, \frac{6\pi}{5}, \frac{7\pi}{5}, \frac{8\pi}{5}, \frac{9\pi}{5}$

2. $x = \frac{\pi}{9} + \frac{2\pi k}{3}$ or $x = \frac{5\pi}{9} + \frac{2\pi k}{3}$; $x = \frac{\pi}{9}, \frac{5\pi}{9}, \frac{7\pi}{9}, \frac{11\pi}{9}, \frac{13\pi}{9}, \frac{17\pi}{9}$

3. $x = \frac{2\pi}{3} + \pi k$ or $x = \frac{5\pi}{6} + \pi k$; $x = \frac{2\pi}{3}, \frac{5\pi}{6}, \frac{5\pi}{3}, \frac{11\pi}{6}$

4. $x = \frac{\pi}{24} + \frac{\pi k}{6}$; $x = \frac{\pi}{24}, \frac{5\pi}{24}, \frac{3\pi}{8}, \frac{13\pi}{24}, \frac{17\pi}{24}, \frac{7\pi}{8}, \frac{25\pi}{24}, \frac{29\pi}{24}, \frac{11\pi}{8}, \frac{37\pi}{24}, \frac{41\pi}{24}, \frac{15\pi}{8}$

5. $x = \frac{3\pi}{8} + \frac{\pi k}{2}$; $x = \frac{3\pi}{8}, \frac{7\pi}{8}, \frac{11\pi}{8}, \frac{15\pi}{8}$

6. $x = \frac{\pi}{12} + \frac{2\pi k}{3}$ or $x = \frac{7\pi}{12} + \frac{2\pi k}{3}$; $x = \frac{\pi}{12}, \frac{7\pi}{12}, \frac{3\pi}{4}, \frac{5\pi}{4}, \frac{17\pi}{12}, \frac{23\pi}{12}$

7. $x = \frac{\pi}{3} + \frac{\pi k}{2}$; $x = \frac{\pi}{3}, \frac{5\pi}{6}, \frac{4\pi}{3}, \frac{11\pi}{6}$

8. No solution

9. $x = \frac{3\pi}{4} + 6\pi k$ or $x = \frac{9\pi}{4} + 6\pi k$; $x = \frac{3\pi}{4}$

10. $x = -\frac{\pi}{3} + \pi k$; $x = \frac{2\pi}{3}, \frac{5\pi}{3}$

11. $x = \frac{3\pi}{4} + \pi k$ or $x = \frac{13\pi}{12} + \pi k$; $x = \frac{\pi}{12}, \frac{3\pi}{4}, \frac{13\pi}{12}, \frac{7\pi}{4}$

12. $x = -\frac{19\pi}{12} + 2\pi k$ or $x = \frac{\pi}{12} + 2\pi k$; $x = \frac{\pi}{12}, \frac{5\pi}{12}$

13. No solution

14. $x = \frac{5\pi}{8} + \frac{\pi k}{2}$; $x = \frac{\pi}{8}, \frac{5\pi}{8}, \frac{9\pi}{8}, \frac{13\pi}{8}$

15. $x = \frac{\pi}{3} + \pi k$ or $x = \frac{2\pi}{3} + \pi k$; $x = \frac{\pi}{3}, \frac{2\pi}{3}, \frac{4\pi}{3}, \frac{5\pi}{3}$

16. $x = \frac{\pi}{6} + \pi k$ or $x = \frac{5\pi}{6} + \pi k$; $x = \frac{\pi}{6}, \frac{5\pi}{6}, \frac{7\pi}{6}, \frac{11\pi}{6}$

17. $x = \frac{\pi}{4} + \frac{\pi k}{2}$; $x = \frac{\pi}{4}, \frac{3\pi}{4}, \frac{5\pi}{4}, \frac{7\pi}{4}$

18. $x = \frac{\pi}{3} + \pi k$ or $x = \frac{2\pi}{3} + \pi k$; $x = \frac{\pi}{3}, \frac{2\pi}{3}, \frac{4\pi}{3}, \frac{5\pi}{3}$

19. $x = \frac{\pi}{4}, \frac{5\pi}{4}$

20. $x = 0, \frac{\pi}{3}, \pi, \frac{5\pi}{3}$

21. $x = \frac{\pi}{6}, \frac{\pi}{2}, \frac{5\pi}{6}, \frac{3\pi}{2}$

22. $x = \frac{\pi}{6}, \frac{5\pi}{6}, \frac{3\pi}{2}$

23. $x = 0, \frac{2\pi}{3}, \frac{4\pi}{3}$

24. $x = \frac{\pi}{3}, \frac{5\pi}{3}$

25. $x = \frac{2\pi}{3}, \frac{4\pi}{3}, \arccos\left(\frac{1}{3}\right), 2\pi - \arccos\left(\frac{1}{3}\right)$

26. $x = \frac{\pi}{6}, \frac{5\pi}{6}$

27. $x = \frac{7\pi}{6}, \frac{11\pi}{6}, \arcsin\left(\frac{1}{3}\right), \pi - \arcsin\left(\frac{1}{3}\right)$

28. $x = \frac{3\pi}{4}, \frac{7\pi}{4}, \arctan\left(\frac{1}{2}\right), \pi + \arctan\left(\frac{1}{2}\right)$

29. $x = 0, \frac{2\pi}{3}, \frac{4\pi}{3}$

30. $x = \frac{\pi}{6}, \frac{5\pi}{6}, \frac{\pi}{2}$

31. $x = \arctan(2), \pi + \arctan(2)$

32. $x = \frac{\pi}{6}, \frac{7\pi}{6}, \frac{5\pi}{6}, \frac{11\pi}{6}$

33. $x = 0, \pi, \frac{\pi}{4}, \frac{3\pi}{4}, \frac{5\pi}{4}, \frac{7\pi}{4}$

34. $x = \frac{\pi}{6}, \frac{\pi}{4}, \frac{3\pi}{4}, \frac{5\pi}{6}, \frac{7\pi}{6}, \frac{5\pi}{4}, \frac{7\pi}{4}, \frac{11\pi}{6}$

35. $x = \frac{\pi}{2}, \frac{3\pi}{2}$

36. $x = 0, \frac{\pi}{3}, \frac{2\pi}{3}, \pi, \frac{4\pi}{3}, \frac{5\pi}{3}$

37. $x = \frac{\pi}{3}, \frac{5\pi}{3}$

38. $x = \frac{\pi}{2}, \frac{3\pi}{2}$

39. $x = \frac{\pi}{6}, \frac{\pi}{2}, \frac{5\pi}{6}, \frac{3\pi}{2}$

40. $x = \frac{\pi}{6}, \frac{5\pi}{6}, \frac{7\pi}{6}, \frac{3\pi}{2}, \frac{11\pi}{6}$

41. $x = \frac{\pi}{8}, \frac{5\pi}{8}, \frac{9\pi}{8}, \frac{13\pi}{8}$

42. No solution

43. $x = 0, \frac{\pi}{7}, \frac{2\pi}{7}, \frac{3\pi}{7}, \frac{4\pi}{7}, \frac{5\pi}{7}, \frac{6\pi}{7}, \pi, \frac{8\pi}{7}, \frac{9\pi}{7}, \frac{10\pi}{7}, \frac{11\pi}{7}, \frac{12\pi}{7}, \frac{13\pi}{7}$

44. $x = 0, \frac{\pi}{2}, \pi, \frac{3\pi}{2}$

45. $x = 0$

46. $x = \frac{\pi}{48}, \frac{11\pi}{48}, \frac{13\pi}{48}, \frac{23\pi}{48}, \frac{25\pi}{48}, \frac{35\pi}{48}, \frac{37\pi}{48}, \frac{47\pi}{48}, \frac{49\pi}{48}, \frac{59\pi}{48}, \frac{61\pi}{48}, \frac{71\pi}{48}, \frac{73\pi}{48}, \frac{83\pi}{48}, \frac{85\pi}{48}, \frac{95\pi}{48}$

47. $x = 0, \frac{\pi}{2}$

48. $x = \frac{\pi}{2}, \frac{11\pi}{6}$

49. $x = \frac{\pi}{12}, \frac{17\pi}{12}$

50. $x = 0, \pi, \frac{\pi}{3}, \frac{4\pi}{3}$

51. $x = \frac{17\pi}{24}, \frac{41\pi}{24}, \frac{23\pi}{24}, \frac{47\pi}{24}$

52. $x = \frac{\pi}{6}, \frac{5\pi}{18}, \frac{5\pi}{6}, \frac{17\pi}{18}, \frac{3\pi}{2}, \frac{29\pi}{18}$

53. $x = 0, \frac{\pi}{4}, \frac{\pi}{2}, \frac{3\pi}{4}, \pi, \frac{5\pi}{4}, \frac{3\pi}{2}, \frac{7\pi}{4}$

54. $x = 0, \frac{\pi}{3}, \frac{2\pi}{3}, \pi, \frac{4\pi}{3}, \frac{5\pi}{3}$

55. $x = 0, \frac{\pi}{8}, \frac{3\pi}{8}, \frac{5\pi}{8}, \frac{7\pi}{8}, \pi, \frac{9\pi}{8}, \frac{11\pi}{8}, \frac{13\pi}{8}, \frac{15\pi}{8}$

56. $x = \frac{\pi}{7}, \frac{\pi}{3}, \frac{3\pi}{7}, \frac{5\pi}{7}, \pi, \frac{9\pi}{7}, \frac{11\pi}{7}, \frac{5\pi}{3}, \frac{13\pi}{7}$

57. $x = 0, \frac{2\pi}{7}, \frac{4\pi}{7}, \frac{6\pi}{7}, \frac{8\pi}{7}, \frac{10\pi}{7}, \frac{12\pi}{7}, \frac{\pi}{5}, \frac{3\pi}{5}, \pi, \frac{7\pi}{5}, \frac{9\pi}{5}$

58. $x = \arcsin\left(\frac{-1+\sqrt{5}}{2}\right) \approx 0.6662, \pi - \arcsin\left(\frac{-1+\sqrt{5}}{2}\right) \approx 2.4754$

59. $x = -\frac{1}{2}$

60. $x = -1$

61. $x = \frac{2}{3}$

62. $x = -\frac{\sqrt{3}}{2}$

63. $x = 2\sqrt{2}$

64. $x = 6$

65. $x = \pm\frac{\sqrt{3}}{2}$

66. $x = \frac{1}{2}$

67. $x = -1, 0$

68. $x = -\sqrt{3}$

69. $[\pi, 2\pi]$

70. $\left[\frac{\pi}{3}, \frac{\pi}{2}\right) \cup \left[\frac{4\pi}{3}, \frac{3\pi}{2}\right)$

71. $\left[0, \frac{\pi}{3}\right] \cup \left[\frac{2\pi}{3}, \frac{4\pi}{3}\right] \cup \left[\frac{5\pi}{3}, 2\pi\right]$

72. $\left[0, \frac{\pi}{4}\right) \cup \left(\frac{3\pi}{4}, \frac{5\pi}{4}\right) \cup \left(\frac{7\pi}{4}, 2\pi\right]$

73. $\left[\frac{\pi}{4}, \frac{3\pi}{4}\right] \cup \left[\frac{5\pi}{4}, \frac{7\pi}{4}\right]$

74. $\left[0, \frac{\pi}{2}\right) \cup \left(\frac{11\pi}{6}, 2\pi\right]$

75. $\left(0, \frac{\pi}{3}\right] \cup \left[\frac{2\pi}{3}, \pi\right) \cup \left(\pi, \frac{4\pi}{3}\right] \cup \left[\frac{5\pi}{3}, 2\pi\right)$

76. $\left[0, \frac{\pi}{3}\right] \cup \left[\frac{5\pi}{3}, 2\pi\right]$

77. No solution

78. $[0, 2\pi]$

79. $\left[0, \frac{\pi}{4}\right] \cup \left(\frac{\pi}{2}, \frac{3\pi}{2}\right) \cup \left[\frac{7\pi}{4}, 2\pi\right]$

80. $[\operatorname{arccot}(4), \pi) \cup [\pi + \operatorname{arccot}(4), 2\pi)$

81. $\left(-\frac{\pi}{6}, \frac{\pi}{6}\right)$

82. $\left(\arcsin\left(\frac{1}{3}\right), \pi - \arcsin\left(\frac{1}{3}\right)\right)$

83. $\left[-\pi, -\frac{\pi}{2}\right) \cup \left[-\frac{\pi}{3}, \frac{\pi}{3}\right] \cup \left(\frac{\pi}{2}, \pi\right]$

84. $\left(-\frac{2\pi}{3}, -\frac{\pi}{3}\right) \cup \left(\frac{\pi}{3}, \frac{2\pi}{3}\right)$

85. $\left(-\pi,-\frac{\pi}{4}\right]\cup\left(0,\frac{3\pi}{4}\right]$

86. $\left[-\frac{3\pi}{4},\frac{\pi}{4}\right]$

87. $\left(-2\pi,-\frac{3\pi}{2}\right)\cup\left(-\frac{3\pi}{2},-\pi\right)\cup\left(0,\frac{\pi}{2}\right)\cup\left(\frac{\pi}{2},\pi\right)$

88. $[-2\pi,2\pi]$

89. $(-2\pi,\operatorname{arccot}(5)-2\pi]\cup(-\pi,\operatorname{arccot}(5)-\pi]\cup(0,\operatorname{arccot}(5)]\cup(\pi,\pi+\operatorname{arccot}(5)]$

90. $\left[-\frac{7\pi}{4},-\frac{3\pi}{2}\right)\cup\left(-\frac{3\pi}{2},-\frac{5\pi}{4}\right]\cup\left[-\frac{3\pi}{4},-\frac{\pi}{2}\right)\cup\left(-\frac{\pi}{2},-\frac{\pi}{4}\right]\cup\left[\frac{\pi}{4},\frac{\pi}{2}\right)\cup\left(\frac{\pi}{2},\frac{3\pi}{4}\right]\cup\left[\frac{5\pi}{4},\frac{3\pi}{2}\right)\cup\left(\frac{3\pi}{2},\frac{7\pi}{4}\right]$

91. $\left[-2\pi,-\frac{5\pi}{3}\right]\cup\left[-\pi,-\frac{\pi}{3}\right]\cup\left[0,\frac{\pi}{3}\right]\cup\left[\pi,\frac{5\pi}{3}\right]$

92. $\left[-\frac{11\pi}{6},-\frac{7\pi}{6}\right]\cup\left[\frac{\pi}{6},\frac{5\pi}{6}\right]\cup,\left\{-\frac{\pi}{2},\frac{3\pi}{2}\right\}$

93. $\left(0,\frac{1}{2}\right]$

94. $\left[\frac{1}{2},1\right]$

95. $\left(-\infty,\frac{\sqrt{3}}{7}\right]$

96. $(-\infty,\infty)$

97. $[-1,0)$

98. $\left[-1,-\frac{1}{2}\right)\cup\left(\frac{\sqrt{2}}{2},1\right]$

99. $\displaystyle\bigcup_{k=-\infty}^{\infty}(2k\pi,(2k+2)\pi)$

100. $\displaystyle\bigcup_{k=-\infty}^{\infty}\left(\frac{(4k-1)\pi}{2},\frac{(4k+3)\pi}{2}\right)$

101. $\displaystyle\bigcup_{k=-\infty}^{\infty}\left\{\left[\frac{(4k+1)\pi}{4},\frac{(2k+1)\pi}{2}\right)\cup\left(\frac{(2k+1)\pi}{2},\frac{(4k+3)\pi}{4}\right]\right\}$

102. $\displaystyle\bigcup_{k=-\infty}^{\infty}\left\{\left[\frac{(6k-1)\pi}{3},\frac{(6k+1)\pi}{3}\right]\cup\left(\frac{(4k+1)\pi}{2},\frac{(4k+3)\pi}{2}\right)\right\}$

103. $\displaystyle\bigcup_{k=-\infty}^{\infty}\left(\frac{k\pi}{2},\frac{(k+1)\pi}{2}\right)$

104. $(-\infty,\infty)$

105. $\displaystyle\bigcup_{k=-\infty}^{\infty}\left(\frac{k\pi}{2},\frac{(k+1)\pi}{2}\right)$

106. $\displaystyle\bigcup_{k=-\infty}^{\infty}\left(\frac{(2k-1)\pi}{2},\frac{(2k+1)\pi}{2}\right)$

107. $\displaystyle\bigcup_{k=-\infty}^{\infty}\left[\frac{(4k-1)\pi}{4},\frac{(4k+1)\pi}{4}\right]$

Chapter 11

Applications of Trigonometry

11.1 Applications of Sinusoids

In the same way exponential functions can be used to model a wide variety of phenomena in nature,[1] the cosine and sine functions can be used to model their fair share of natural behaviors. In section 10.5, we introduced the concept of a sinusoid as a function which can be written either in the form $C(x) = A\cos(\omega x+\phi)+B$ for $\omega > 0$ or equivalently, in the form $S(x) = A\sin(\omega x+\phi)+B$ for $\omega > 0$. At the time, we remained undecided as to which form we preferred, but the time for such indecision is over. For clarity of exposition we focus on the sine function[2] in this section and switch to the independent variable t, since the applications in this section are time-dependent. We reintroduce and summarize all of the important facts and definitions about this form of the sinusoid below.

Properties of the Sinusoid $S(t) = A\sin(\omega t + \phi) + B$

- The **amplitude** is $|A|$
- The **angular frequency** is ω and the **ordinary frequency** is $f = \dfrac{\omega}{2\pi}$
- The **period** is $T = \dfrac{1}{f} = \dfrac{2\pi}{\omega}$
- The **phase** is ϕ and the **phase shift** is $-\dfrac{\phi}{\omega}$
- The **vertical shift** or **baseline** is B

Along with knowing these formulas, it is helpful to remember what these quantities mean in context. The amplitude measures the maximum displacement of the sine wave from its baseline (determined by the vertical shift), the period is the length of time it takes to complete one cycle of the sinusoid, the angular frequency tells how many cycles are completed over an interval of length 2π, and the ordinary frequency measures how many cycles occur per unit of time. The phase indicates what

[1] See Section 6.5.

[2] Sine haters can use the co-function identity $\cos\left(\frac{\pi}{2} - \theta\right) = \sin(\theta)$ to turn all of the sines into cosines.

angle ϕ corresponds to $t = 0$, and the phase shift represents how much of a 'head start' the sinusoid has over the un-shifted sine function. The figure below is repeated from Section 10.5.

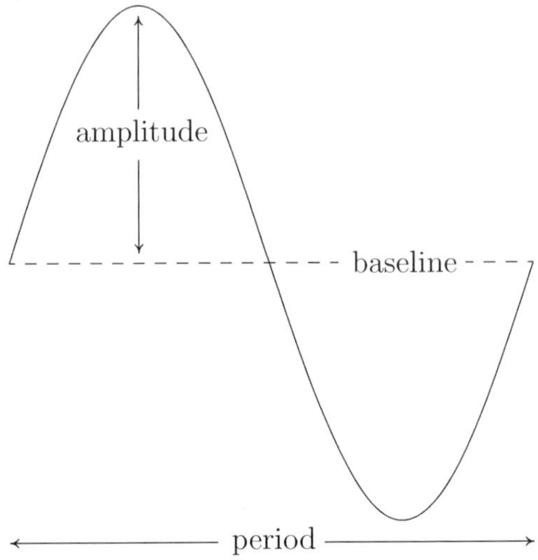

In Section 10.1.1, we introduced the concept of circular motion and in Section 10.2.1, we developed formulas for circular motion. Our first foray into sinusoidal motion puts these notions to good use.

Example 11.1.1. Recall from Exercise 55 in Section 10.1 that The Giant Wheel at Cedar Point is a circle with diameter 128 feet which sits on an 8 foot tall platform making its overall height 136 feet. It completes two revolutions in 2 minutes and 7 seconds. Assuming that the riders are at the edge of the circle, find a sinusoid which describes the height of the passengers above the ground t seconds after they pass the point on the wheel closest to the ground.

Solution. We sketch the problem situation below and assume a counter-clockwise rotation.[3]

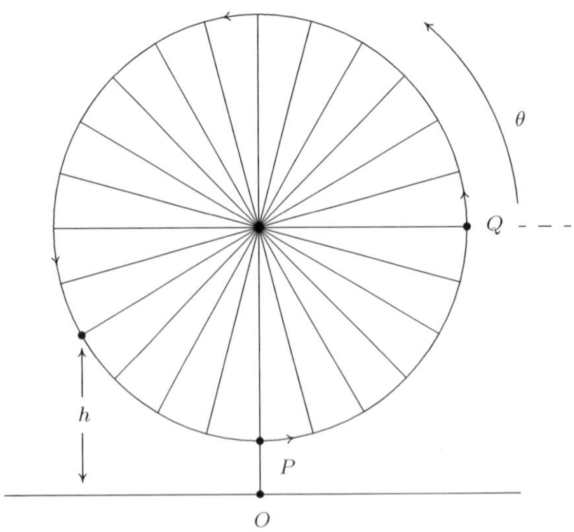

[3]Otherwise, we could just observe the motion of the wheel from the other side.

We know from the equations given on page 732 in Section 10.2.1 that the y-coordinate for counter-clockwise motion on a circle of radius r centered at the origin with constant angular velocity (frequency) ω is given by $y = r\sin(\omega t)$. Here, $t = 0$ corresponds to the point $(r, 0)$ so that θ, the angle measuring the amount of rotation, is in standard position. In our case, the diameter of the wheel is 128 feet, so the radius is $r = 64$ feet. Since the wheel completes two revolutions in 2 minutes and 7 seconds (which is 127 seconds) the period $T = \frac{1}{2}(127) = \frac{127}{2}$ seconds. Hence, the angular frequency is $\omega = \frac{2\pi}{T} = \frac{4\pi}{127}$ radians per second. Putting these two pieces of information together, we have that $y = 64\sin\left(\frac{4\pi}{127}t\right)$ describes the y-coordinate on the Giant Wheel after t seconds, assuming it is centered at $(0,0)$ with $t = 0$ corresponding to the point Q. In order to find an expression for h, we take the point O in the figure as the origin. Since the base of the Giant Wheel ride is 8 feet above the ground and the Giant Wheel itself has a radius of 64 feet, its center is 72 feet above the ground. To account for this vertical shift upward,[4] we add 72 to our formula for y to obtain the new formula $h = y + 72 = 64\sin\left(\frac{4\pi}{127}t\right) + 72$. Next, we need to adjust things so that $t = 0$ corresponds to the point P instead of the point Q. This is where the phase comes into play. Geometrically, we need to shift the angle θ in the figure back $\frac{\pi}{2}$ radians. From Section 10.2.1, we know $\theta = \omega t = \frac{4\pi}{127}t$, so we (temporarily) write the height in terms of θ as $h = 64\sin(\theta) + 72$. Subtracting $\frac{\pi}{2}$ from θ gives the final answer $h(t) = 64\sin\left(\theta - \frac{\pi}{2}\right) + 72 = 64\sin\left(\frac{4\pi}{127}t - \frac{\pi}{2}\right) + 72$. We can check the reasonableness of our answer by graphing $y = h(t)$ over the interval $\left[0, \frac{127}{2}\right]$.

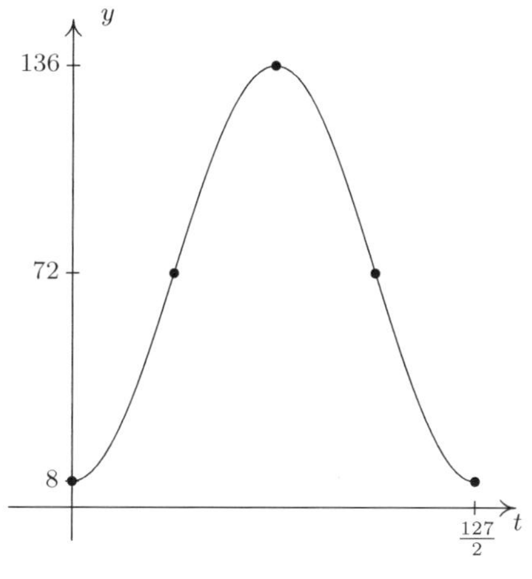

□

A few remarks about Example 11.1.1 are in order. First, note that the amplitude of 64 in our answer corresponds to the radius of the Giant Wheel. This means that passengers on the Giant Wheel never stray more than 64 feet vertically from the center of the Wheel, which makes sense. Second, the phase shift of our answer works out to be $\frac{\pi/2}{4\pi/127} = \frac{127}{8} = 15.875$. This represents the 'time delay' (in seconds) we introduce by starting the motion at the point P as opposed to the point Q. Said differently, passengers which 'start' at P take 15.875 seconds to 'catch up' to the point Q.

Our next example revisits the daylight data first introduced in Section 2.5, Exercise 6b.

[4]We are readjusting our 'baseline' from $y = 0$ to $y = 72$.

Example 11.1.2. According to the U.S. Naval Observatory website, the number of hours H of daylight that Fairbanks, Alaska received on the 21st day of the nth month of 2009 is given below. Here $t = 1$ represents January 21, 2009, $t = 2$ represents February 21, 2009, and so on.

Month Number	1	2	3	4	5	6	7	8	9	10	11	12
Hours of Daylight	5.8	9.3	12.4	15.9	19.4	21.8	19.4	15.6	12.4	9.1	5.6	3.3

1. Find a sinusoid which models these data and use a graphing utility to graph your answer along with the data.

2. Compare your answer to part 1 to one obtained using the regression feature of a calculator.

Solution.

1. To get a feel for the data, we plot it below.

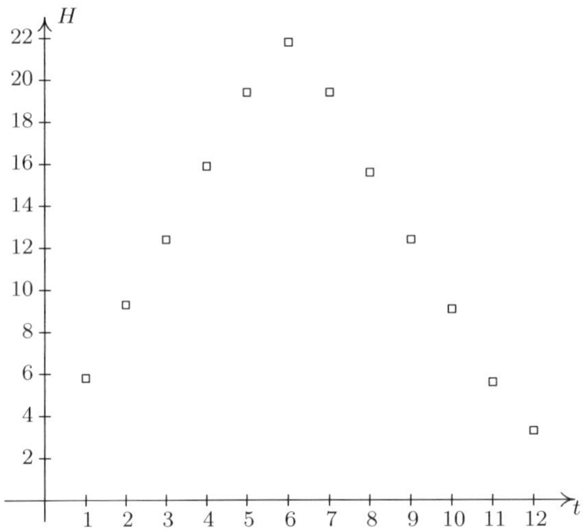

The data certainly appear sinusoidal,[5] but when it comes down to it, fitting a sinusoid to data manually is not an exact science. We do our best to find the constants A, ω, ϕ and B so that the function $H(t) = A\sin(\omega t + \phi) + B$ closely matches the data. We first go after the vertical shift B whose value determines the baseline. In a typical sinusoid, the value of B is the average of the maximum and minimum values. So here we take $B = \frac{3.3+21.8}{2} = 12.55$. Next is the amplitude A which is the displacement from the baseline to the maximum (and minimum) values. We find $A = 21.8 - 12.55 = 12.55 - 3.3 = 9.25$. At this point, we have $H(t) = 9.25\sin(\omega t + \phi) + 12.55$. Next, we go after the angular frequency ω. Since the data collected is over the span of a year (12 months), we take the period $T = 12$ months.[6] This

[5]Okay, it appears to be the '∧' shape we saw in some of the graphs in Section 2.2. Just humor us.

[6]Even though the data collected lies in the interval $[1, 12]$, which has a length of 11, we need to think of the data point at $t = 1$ as a representative sample of the amount of daylight for every day in January. That is, it represents $H(t)$ over the interval $[0, 1]$. Similarly, $t = 2$ is a sample of $H(t)$ over $[1, 2]$, and so forth.

means $\omega = \frac{2\pi}{T} = \frac{2\pi}{12} = \frac{\pi}{6}$. The last quantity to find is the phase ϕ. Unlike the previous example, it is easier in this case to find the phase shift $-\frac{\phi}{\omega}$. Since we picked $A > 0$, the phase shift corresponds to the first value of t with $H(t) = 12.55$ (the baseline value).[7] Here, we choose $t = 3$, since its corresponding H value of 12.4 is closer to 12.55 than the next value, 15.9, which corresponds to $t = 4$. Hence, $-\frac{\phi}{\omega} = 3$, so $\phi = -3\omega = -3\left(\frac{\pi}{6}\right) = -\frac{\pi}{2}$. We have $H(t) = 9.25\sin\left(\frac{\pi}{6}t - \frac{\pi}{2}\right) + 12.55$. Below is a graph of our data with the curve $y = H(t)$.

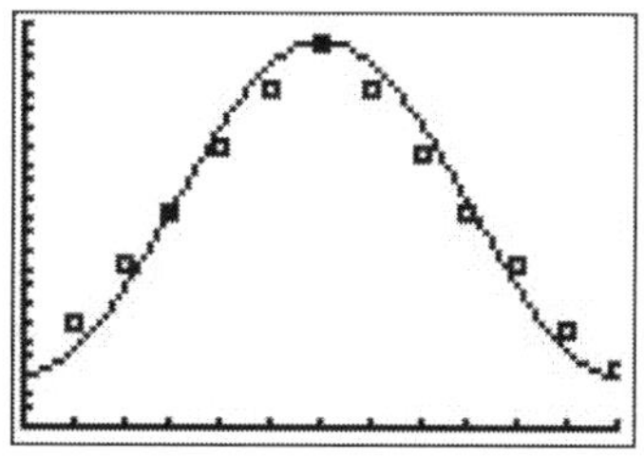

2. Using the 'SinReg' command, we graph the calculator's regression below.

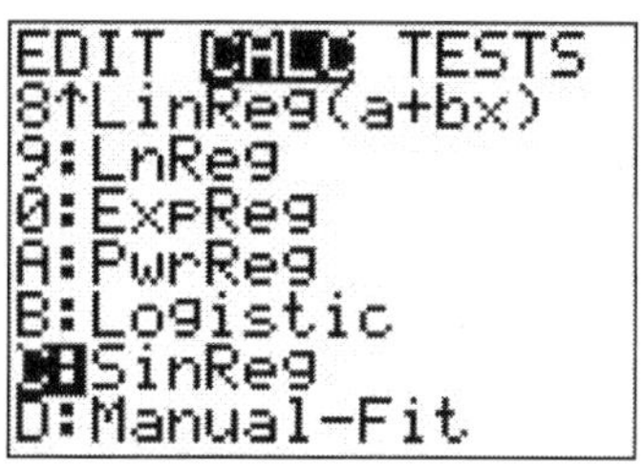

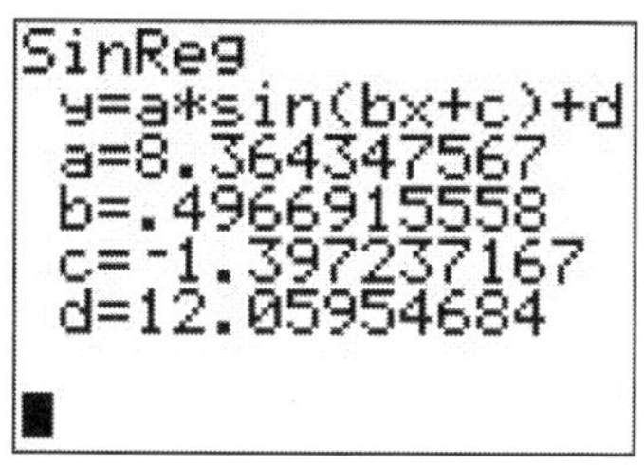

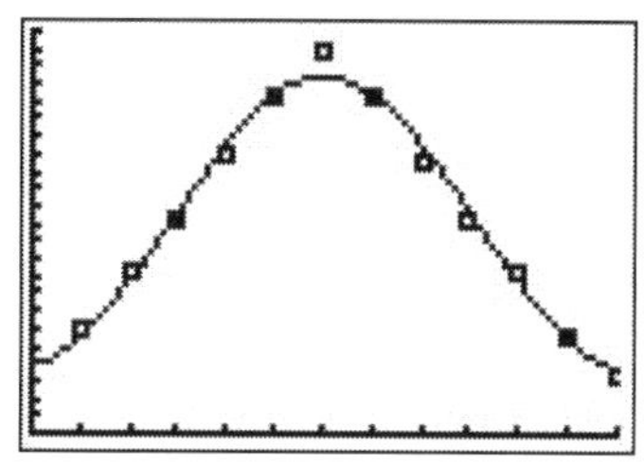

While both models seem to be reasonable fits to the data, the calculator model is possibly the better fit. The calculator does not give us an r^2 value like it did for linear regressions in Section 2.5, nor does it give us an R^2 value like it did for quadratic, cubic and quartic regressions as in Section 3.1. The reason for this, much like the reason for the absence of R^2 for the logistic model in Section 6.5, is beyond the scope of this course. We'll just have to use our own good judgment when choosing the best sinusoid model. □

11.1.1 Harmonic Motion

One of the major applications of sinusoids in Science and Engineering is the study of **harmonic motion**. The equations for harmonic motion can be used to describe a wide range of phenomena, from the motion of an object on a spring, to the response of an electronic circuit. In this subsection, we restrict our attention to modeling a simple spring system. Before we jump into the Mathematics, there are some Physics terms and concepts we need to discuss. In Physics, 'mass' is defined as a measure of an object's resistance to straight-line motion whereas 'weight' is the amount of force (pull) gravity exerts on an object. An object's mass cannot change,[8] while its weight could change.

[7] See the figure on page 882.

[8] Well, assuming the object isn't subjected to relativistic speeds ...

An object which weighs 6 pounds on the surface of the Earth would weigh 1 pound on the surface of the Moon, but its mass is the same in both places. In the English system of units, 'pounds' (lbs.) is a measure of force (weight), and the corresponding unit of mass is the 'slug'. In the SI system, the unit of force is 'Newtons' (N) and the associated unit of mass is the 'kilogram' (kg). We convert between mass and weight using the formula[9] $w = mg$. Here, w is the weight of the object, m is the mass and g is the acceleration due to gravity. In the English system, $g = 32 \frac{\text{feet}}{\text{second}^2}$, and in the SI system, $g = 9.8 \frac{\text{meters}}{\text{second}^2}$. Hence, on Earth a *mass* of 1 slug *weighs* 32 lbs. and a *mass* of 1 kg *weighs* 9.8 N.[10] Suppose we attach an object with mass m to a spring as depicted below. The weight of the object will stretch the spring. The system is said to be in 'equilibrium' when the weight of the object is perfectly balanced with the restorative force of the spring. How far the spring stretches to reach equilibrium depends on the spring's 'spring constant'. Usually denoted by the letter k, the spring constant relates the force F applied to the spring to the amount d the spring stretches in accordance with Hooke's Law[11] $F = kd$. If the object is released above or below the equilibrium position, or if the object is released with an upward or downward velocity, the object will bounce up and down on the end of the spring until some external force stops it. If we let $x(t)$ denote the object's displacement from the equilibrium position at time t, then $x(t) = 0$ means the object is at the equilibrium position, $x(t) < 0$ means the object is *above* the equilibrium position, and $x(t) > 0$ means the object is *below* the equilibrium position. The function $x(t)$ is called the 'equation of motion' of the object.[12]

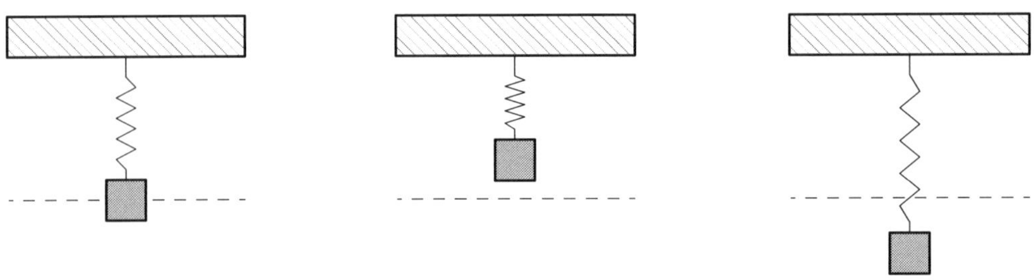

$x(t) = 0$ at the equilibrium position

$x(t) < 0$ above the equilibrium position

$x(t) > 0$ below the equilibrium position

If we ignore all other influences on the system except gravity and the spring force, then Physics tells us that gravity and the spring force will battle each other forever and the object will oscillate indefinitely. In this case, we describe the motion as 'free' (meaning there is no external force causing the motion) and 'undamped' (meaning we ignore friction caused by surrounding medium, which in our case is air). The following theorem, which comes from Differential Equations, gives $x(t)$ as a function of the mass m of the object, the spring constant k, the initial displacement x_0 of the

[9]This is a consequence of Newton's Second Law of Motion $F = ma$ where F is force, m is mass and a is acceleration. In our present setting, the force involved is weight which is caused by the acceleration due to gravity.

[10]Note that 1 pound $= 1 \frac{\text{slug foot}}{\text{second}^2}$ and 1 Newton $= 1 \frac{\text{kg meter}}{\text{second}^2}$.

[11]Look familiar? We saw Hooke's Law in Section 4.3.1.

[12]To keep units compatible, if we are using the English system, we use feet (ft.) to measure displacement. If we are in the SI system, we measure displacement in meters (m). Time is always measured in seconds (s).

object and initial velocity v_0 of the object. As with $x(t)$, $x_0 = 0$ means the object is released from the equilibrium position, $x_0 < 0$ means the object is released *above* the equilibrium position and $x_0 > 0$ means the object is released *below* the equilibrium position. As far as the initial velocity v_0 is concerned, $v_0 = 0$ means the object is released 'from rest,' $v_0 < 0$ means the object is heading *upwards* and $v_0 > 0$ means the object is heading *downwards*.[13]

Theorem 11.1. Equation for Free Undamped Harmonic Motion: Suppose an object of mass m is suspended from a spring with spring constant k. If the initial displacement from the equilibrium position is x_0 and the initial velocity of the object is v_0, then the displacement x from the equilibrium position at time t is given by $x(t) = A\sin(\omega t + \phi)$ where

- $\omega = \sqrt{\dfrac{k}{m}}$ and $A = \sqrt{x_0^2 + \left(\dfrac{v_0}{\omega}\right)^2}$
- $A\sin(\phi) = x_0$ and $A\omega\cos(\phi) = v_0$.

It is a great exercise in 'dimensional analysis' to verify that the formulas given in Theorem 11.1 work out so that ω has units $\frac{1}{s}$ and A has units ft. or m, depending on which system we choose.

Example 11.1.3. Suppose an object weighing 64 pounds stretches a spring 8 feet.

1. If the object is attached to the spring and released 3 feet below the equilibrium position from rest, find the equation of motion of the object, $x(t)$. When does the object first pass through the equilibrium position? Is the object heading upwards or downwards at this instant?

2. If the object is attached to the spring and released 3 feet below the equilibrium position with an upward velocity of 8 feet per second, find the equation of motion of the object, $x(t)$. What is the longest distance the object travels *above* the equilibrium position? When does this first happen? Confirm your result using a graphing utility.

Solution. In order to use the formulas in Theorem 11.1, we first need to determine the spring constant k and the mass of the object m. To find k, we use Hooke's Law $F = kd$. We know the object weighs 64 lbs. and stretches the spring 8 ft.. Using $F = 64$ and $d = 8$, we get $64 = k \cdot 8$, or $k = 8\frac{\text{lbs.}}{\text{ft.}}$. To find m, we use $w = mg$ with $w = 64$ lbs. and $g = 32\frac{\text{ft.}}{s^2}$. We get $m = 2$ slugs. We can now proceed to apply Theorem 11.1.

1. With $k = 8$ and $m = 2$, we get $\omega = \sqrt{\frac{k}{m}} = \sqrt{\frac{8}{2}} = 2$. We are told that the object is released 3 feet *below* the equilibrium position 'from rest.' This means $x_0 = 3$ and $v_0 = 0$. Therefore, $A = \sqrt{x_0^2 + \left(\frac{v_0}{\omega}\right)^2} = \sqrt{3^2 + 0^2} = 3$. To determine the phase ϕ, we have $A\sin(\phi) = x_0$, which in this case gives $3\sin(\phi) = 3$ so $\sin(\phi) = 1$. Only $\phi = \frac{\pi}{2}$ and angles coterminal to it

[13]The sign conventions here are carried over from Physics. If not for the spring, the object would fall towards the ground, which is the 'natural' or 'positive' direction. Since the spring force acts in direct opposition to gravity, any movement upwards is considered 'negative'.

satisfy this condition, so we pick[14] the phase to be $\phi = \frac{\pi}{2}$. Hence, the equation of motion is $x(t) = 3\sin\left(2t + \frac{\pi}{2}\right)$. To find when the object passes through the equilibrium position we solve $x(t) = 3\sin\left(2t + \frac{\pi}{2}\right) = 0$. Going through the usual analysis we find $t = -\frac{\pi}{4} + \frac{\pi}{2}k$ for integers k. Since we are interested in the first time the object passes through the equilibrium position, we look for the smallest positive t value which in this case is $t = \frac{\pi}{4} \approx 0.78$ seconds after the start of the motion. Common sense suggests that if we release the object below the equilibrium position, the object should be traveling upwards when it first passes through it. To check this answer, we graph one cycle of $x(t)$. Since our applied domain in this situation is $t \geq 0$, and the period of $x(t)$ is $T = \frac{2\pi}{\omega} = \frac{2\pi}{2} = \pi$, we graph $x(t)$ over the interval $[0, \pi]$. Remembering that $x(t) > 0$ means the object is below the equilibrium position and $x(t) < 0$ means the object is above the equilibrium position, the fact our graph is crossing through the t-axis from positive x to negative x at $t = \frac{\pi}{4}$ confirms our answer.

2. The only difference between this problem and the previous problem is that we now release the object with an upward velocity of $8\,\frac{\text{ft}}{s}$. We still have $\omega = 2$ and $x_0 = 3$, but now we have $v_0 = -8$, the negative indicating the velocity is directed upwards. Here, we get $A = \sqrt{x_0^2 + \left(\frac{v_0}{\omega}\right)^2} = \sqrt{3^2 + (-4)^2} = 5$. From $A\sin(\phi) = x_0$, we get $5\sin(\phi) = 3$ which gives $\sin(\phi) = \frac{3}{5}$. From $A\omega\cos(\phi) = v_0$, we get $10\cos(\phi) = -8$, or $\cos(\phi) = -\frac{4}{5}$. This means that ϕ is a Quadrant II angle which we can describe in terms of either arcsine or arccosine. Since $x(t)$ is expressed in terms of sine, we choose to express $\phi = \pi - \arcsin\left(\frac{3}{5}\right)$. Hence, $x(t) = 5\sin\left(2t + \left[\pi - \arcsin\left(\frac{3}{5}\right)\right]\right)$. Since the amplitude of $x(t)$ is 5, the object will travel at most 5 feet above the equilibrium position. To find when this happens, we solve the equation $x(t) = 5\sin\left(2t + \left[\pi - \arcsin\left(\frac{3}{5}\right)\right]\right) = -5$, the negative once again signifying that the object is *above* the equilibrium position. Going through the usual machinations, we get $t = \frac{1}{2}\arcsin\left(\frac{3}{5}\right) + \frac{\pi}{4} + \pi k$ for integers k. The smallest of these values occurs when $k = 0$, that is, $t = \frac{1}{2}\arcsin\left(\frac{3}{5}\right) + \frac{\pi}{4} \approx 1.107$ seconds after the start of the motion. To check our answer using the calculator, we graph $y = 5\sin\left(2x + \left[\pi - \arcsin\left(\frac{3}{5}\right)\right]\right)$ on a graphing utility and confirm the coordinates of the first relative minimum to be approximately $(1.107, -5)$.

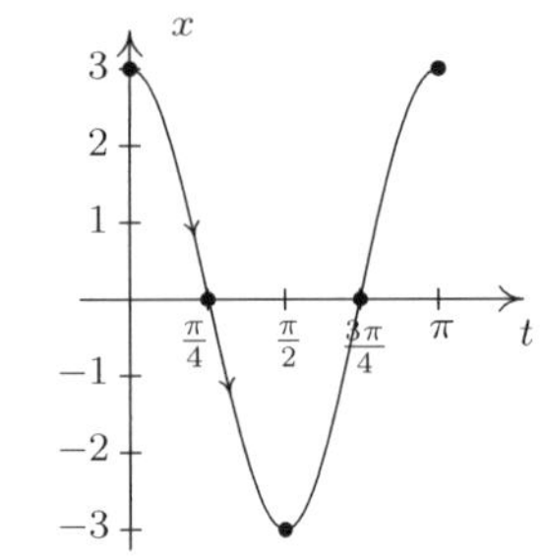

$x(t) = 3\sin\left(2t + \frac{\pi}{2}\right)$

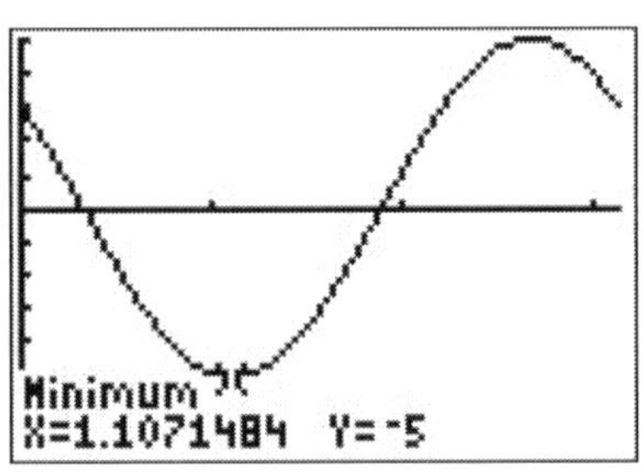

$y = 5\sin\left(2x + \left[\pi - \arcsin\left(\frac{3}{5}\right)\right]\right)$

□

It is possible, though beyond the scope of this course, to model the effects of friction and other external forces acting on the system.[15] While we may not have the Physics and Calculus background

[14] For confirmation, we note that $A\omega\cos(\phi) = v_0$, which in this case reduces to $6\cos(\phi) = 0$.

[15] Take a good Differential Equations class to see this!

to *derive* equations of motion for these scenarios, we can certainly analyze them. We examine three cases in the following example.

Example 11.1.4.

1. Write $x(t) = 5e^{-t/5}\cos(t) + 5e^{-t/5}\sqrt{3}\sin(t)$ in the form $x(t) = A(t)\sin(\omega t + \phi)$. Graph $x(t)$ using a graphing utility.

2. Write $x(t) = (t+3)\sqrt{2}\cos(2t) + (t+3)\sqrt{2}\sin(2t)$ in the form $x(t) = A(t)\sin(\omega t + \phi)$. Graph $x(t)$ using a graphing utility.

3. Find the period of $x(t) = 5\sin(6t) - 5\sin(8t)$. Graph $x(t)$ using a graphing utility.

Solution.

1. We start rewriting $x(t) = 5e^{-t/5}\cos(t) + 5e^{-t/5}\sqrt{3}\sin(t)$ by factoring out $5e^{-t/5}$ from both terms to get $x(t) = 5e^{-t/5}\left(\cos(t) + \sqrt{3}\sin(t)\right)$. We convert what's left in parentheses to the required form using the formulas introduced in Exercise 36 from Section 10.5. We find $\left(\cos(t) + \sqrt{3}\sin(t)\right) = 2\sin\left(t + \frac{\pi}{3}\right)$ so that $x(t) = 10e^{-t/5}\sin\left(t + \frac{\pi}{3}\right)$. Graphing this on the calculator as $y = 10e^{-x/5}\sin\left(x + \frac{\pi}{3}\right)$ reveals some interesting behavior. The sinusoidal nature continues indefinitely, but it is being attenuated. In the sinusoid $A\sin(\omega x + \phi)$, the coefficient A of the sine function is the amplitude. In the case of $y = 10e^{-x/5}\sin\left(x + \frac{\pi}{3}\right)$, we can think of the *function* $A(x) = 10e^{-x/5}$ as the amplitude. As $x \to \infty$, $10e^{-x/5} \to 0$ which means the amplitude continues to shrink towards zero. Indeed, if we graph $y = \pm 10e^{-x/5}$ along with $y = 10e^{-x/5}\sin\left(x + \frac{\pi}{3}\right)$, we see this attenuation taking place. This equation corresponds to the motion of an object on a spring where there is a slight force which acts to 'damp', or slow the motion. An example of this kind of force would be the friction of the object against the air. In this model, the object oscillates forever, but with smaller and smaller amplitude.

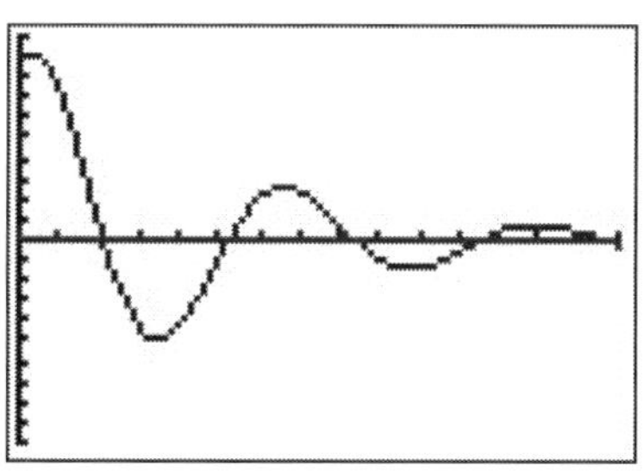

$y = 10e^{-x/5}\sin\left(x + \frac{\pi}{3}\right)$

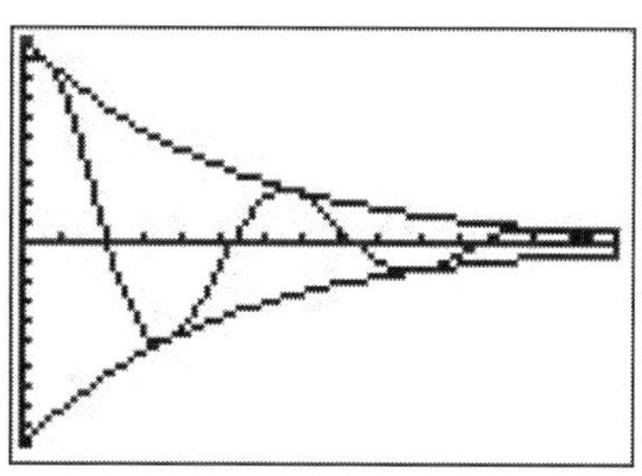

$y = 10e^{-x/5}\sin\left(x + \frac{\pi}{3}\right)$, $y = \pm 10e^{-x/5}$

2. Proceeding as in the first example, we factor out $(t+3)\sqrt{2}$ from each term in the function $x(t) = (t+3)\sqrt{2}\cos(2t) + (t+3)\sqrt{2}\sin(2t)$ to get $x(t) = (t+3)\sqrt{2}(\cos(2t) + \sin(2t))$. We find $(\cos(2t) + \sin(2t)) = \sqrt{2}\sin\left(2t + \frac{\pi}{4}\right)$, so $x(t) = 2(t+3)\sin\left(2t + \frac{\pi}{4}\right)$. Graphing this on the calculator as $y = 2(x+3)\sin\left(2x + \frac{\pi}{4}\right)$, we find the sinusoid's amplitude growing. Since our amplitude function here is $A(x) = 2(x+3) = 2x+6$, which continues to grow without bound

as $x \to \infty$, this is hardly surprising. The phenomenon illustrated here is 'forced' motion. That is, we imagine that the entire apparatus on which the spring is attached is oscillating as well. In this case, we are witnessing a 'resonance' effect – the frequency of the external oscillation matches the frequency of the motion of the object on the spring.[16]

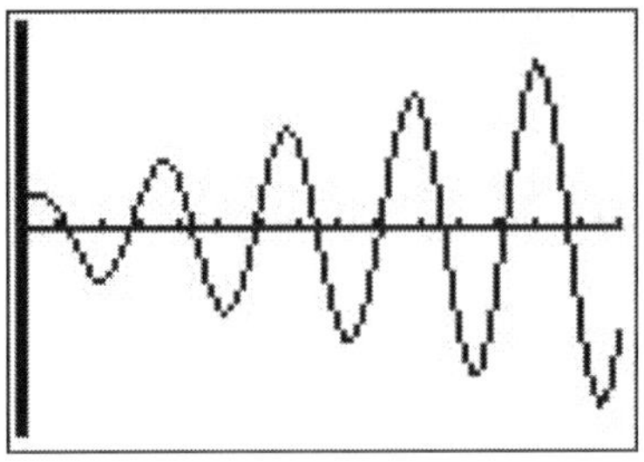

$y = 2(x+3)\sin\left(2x + \frac{\pi}{4}\right)$

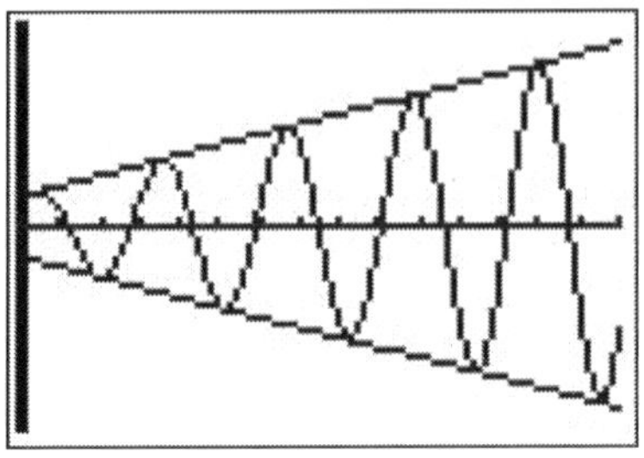

$y = 2(x+3)\sin\left(2x + \frac{\pi}{4}\right)$
$y = \pm 2(x+3)$

3. Last, but not least, we come to $x(t) = 5\sin(6t) - 5\sin(8t)$. To find the period of this function, we need to determine the length of the smallest interval on which both $f(t) = 5\sin(6t)$ and $g(t) = 5\sin(8t)$ complete a whole number of cycles. To do this, we take the ratio of their frequencies and reduce to lowest terms: $\frac{6}{8} = \frac{3}{4}$. This tells us that for every 3 cycles f makes, g makes 4. In other words, the period of $x(t)$ is three times the period of $f(t)$ (which is four times the period of $g(t)$), or π. We graph $y = 5\sin(6x) - 5\sin(8x)$ over $[0, \pi]$ on the calculator to check this. This equation of motion also results from 'forced' motion, but here the frequency of the external oscillation is different than that of the object on the spring. Since the sinusoids here have different frequencies, they are 'out of sync' and do not amplify each other as in the previous example. Taking things a step further, we can use a sum to product identity to rewrite $x(t) = 5\sin(6t) - 5\sin(8t)$ as $x(t) = -10\sin(t)\cos(7t)$. The lower frequency factor in this expression, $-10\sin(t)$, plays an interesting role in the graph of $x(t)$. Below we graph $y = 5\sin(6x) - 5\sin(8x)$ and $y = \pm 10\sin(x)$ over $[0, 2\pi]$. This is an example of the 'beat' phenomena, and the curious reader is invited to explore this concept as well.[17]

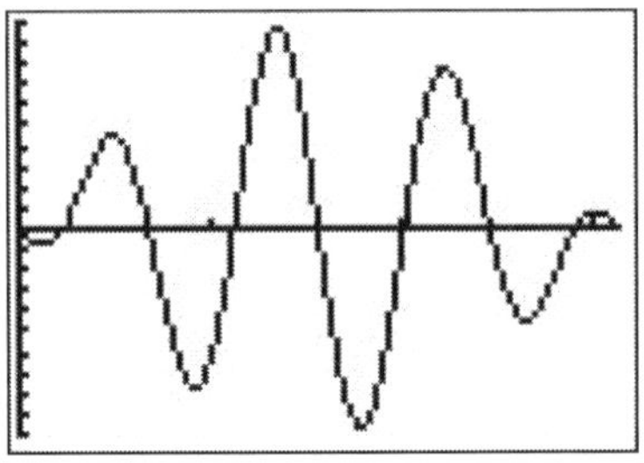

$y = 5\sin(6x) - 5\sin(8x)$ over $[0, \pi]$

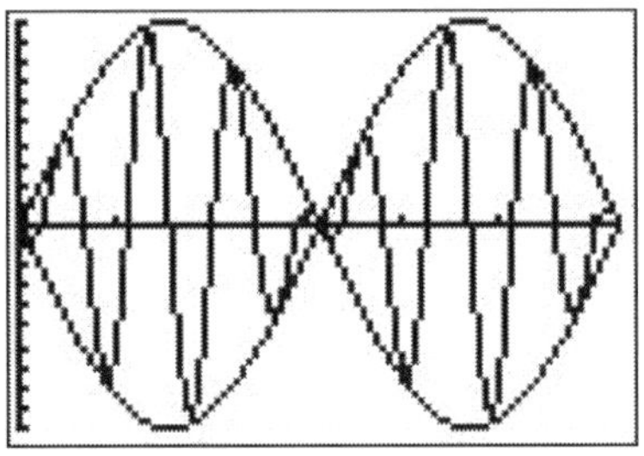

$y = 5\sin(6x) - 5\sin(8x)$ and
$y = \pm 10\sin(x)$ over $[0, 2\pi]$ □

[16] The reader is invited to investigate the destructive implications of resonance.
[17] A good place to start is this article on beats.

11.1.2 Exercises

1. The sounds we hear are made up of mechanical waves. The note 'A' above the note 'middle C' is a sound wave with ordinary frequency $f = 440$ Hertz $= 440\frac{\text{cycles}}{\text{second}}$. Find a sinusoid which models this note, assuming that the amplitude is 1 and the phase shift is 0.

2. The voltage V in an alternating current source has amplitude $220\sqrt{2}$ and ordinary frequency $f = 60$ Hertz. Find a sinusoid which models this voltage. Assume that the phase is 0.

3. The London Eye is a popular tourist attraction in London, England and is one of the largest Ferris Wheels in the world. It has a diameter of 135 meters and makes one revolution (counter-clockwise) every 30 minutes. It is constructed so that the lowest part of the Eye reaches ground level, enabling passengers to simply walk on to, and off of, the ride. Find a sinsuoid which models the height h of the passenger above the ground in meters t minutes after they board the Eye at ground level.

4. On page 732 in Section 10.2.1, we found the x-coordinate of counter-clockwise motion on a circle of radius r with angular frequency ω to be $x = r\cos(\omega t)$, where $t = 0$ corresponds to the point $(r, 0)$. Suppose we are in the situation of Exercise 3 above. Find a sinsusoid which models the horizontal *displacement* x of the passenger from the center of the Eye in meters t minutes after they board the Eye. Here we take $x(t) > 0$ to mean the passenger is to the *right* of the center, while $x(t) < 0$ means the passenger is to the *left* of the center.

5. In Exercise 52 in Section 10.1, we introduced the yo-yo trick 'Around the World' in which a yo-yo is thrown so it sweeps out a vertical circle. As in that exercise, suppose the yo-yo string is 28 inches and it completes one revolution in 3 seconds. If the closest the yo-yo ever gets to the ground is 2 inches, find a sinsuoid which models the height h of the yo-yo above the ground in inches t seconds after it leaves its lowest point.

6. Suppose an object weighing 10 pounds is suspended from the ceiling by a spring which stretches 2 feet to its equilibrium position when the object is attached.

   (a) Find the spring constant k in $\frac{\text{lbs.}}{\text{ft.}}$ and the mass of the object in slugs.

   (b) Find the equation of motion of the object if it is released from 1 foot *below* the equilibrium position from rest. When is the first time the object passes through the equilibrium position? In which direction is it heading?

   (c) Find the equation of motion of the object if it is released from 6 inches *above* the equilibrium position with a *downward* velocity of 2 feet per second. Find when the object passes through the equilibrium position heading downwards for the third time.

7. Consider the pendulum below. Ignoring air resistance, the angular displacement of the pendulum from the vertical position, θ, can be modeled as a sinusoid.[18]

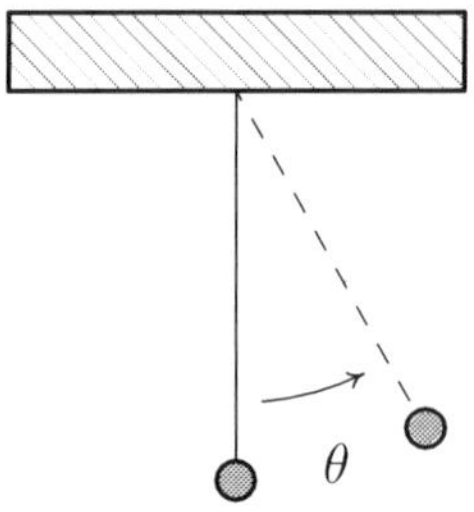

The amplitude of the sinusoid is the same as the initial angular displacement, θ_0, of the pendulum and the period of the motion is given by

$$T = 2\pi\sqrt{\frac{l}{g}}$$

where l is the length of the pendulum and g is the acceleration due to gravity.

(a) Find a sinusoid which gives the angular displacement θ as a function of time, t. Arrange things so $\theta(0) = \theta_0$.

(b) In Exercise 40 section 5.3, you found the length of the pendulum needed in Jeff's antique Seth-Thomas clock to ensure the period of the pendulum is $\frac{1}{2}$ of a second. Assuming the initial displacement of the pendulum is 15°, find a sinusoid which models the displacement of the pendulum θ as a function of time, t, in seconds.

8. The table below lists the average temperature of Lake Erie as measured in Cleveland, Ohio on the first of the month for each month during the years 1971 – 2000.[19] For example, $t = 3$ represents the average of the temperatures recorded for Lake Erie on every March 1 for the years 1971 through 2000.

Month Number, t	1	2	3	4	5	6	7	8	9	10	11	12
Temperature ($^\circ$ F), T	36	33	34	38	47	57	67	74	73	67	56	46

(a) Using the techniques discussed in Example 11.1.2, fit a sinusoid to these data.

(b) Using a graphing utility, graph your model along with the data set to judge the reasonableness of the fit.

[18] Provided θ is kept 'small.' Carl remembers the 'Rule of Thumb' as being 20° or less. Check with your friendly neighborhood physicist to make sure.

[19] See this website: http://www.erh.noaa.gov/cle/climate/cle/normals/laketempcle.html.

(c) Use the model you found in part 8a to predict the average temperature recorded for Lake Erie on April 15th and September 15th during the years 1971–2000.[20]

(d) Compare your results to those obtained using a graphing utility.

9. The fraction of the moon illuminated at midnight Eastern Standard Time on the tth day of June, 2009 is given in the table below.[21]

Day of June, t	3	6	9	12	15	18	21	24	27	30
Fraction Illuminated, F	0.81	0.98	0.98	0.83	0.57	0.27	0.04	0.03	0.26	0.58

(a) Using the techniques discussed in Example 11.1.2, fit a sinusoid to these data.[22]

(b) Using a graphing utility, graph your model along with the data set to judge the reasonableness of the fit.

(c) Use the model you found in part 9a to predict the fraction of the moon illuminated on June 1, 2009. [23]

(d) Compare your results to those obtained using a graphing utility.

10. With the help of your classmates, research the phenomena mentioned in Example 11.1.4, namely resonance and beats.

11. With the help of your classmates, research Amplitude Modulation and Frequency Modulation.

12. What other things in the world might be roughly sinusoidal? Look to see what models you can find for them and share your results with your class.

[20] The computed average is 41°F for April 15th and 71°F for September 15th.

[21] See this website: http://www.usno.navy.mil/USNO/astronomical-applications/data-services/frac-moon-ill.

[22] You may want to plot the data before you find the phase shift.

[23] The listed fraction is 0.62.

11.1.3 Answers

1. $S(t) = \sin(880\pi t)$
2. $V(t) = 220\sqrt{2}\sin(120\pi t)$
3. $h(t) = 67.5\sin\left(\frac{\pi}{15}t - \frac{\pi}{2}\right) + 67.5$
4. $x(t) = 67.5\cos\left(\frac{\pi}{15}t - \frac{\pi}{2}\right) = 67.5\sin\left(\frac{\pi}{15}t\right)$
5. $h(t) = 28\sin\left(\frac{2\pi}{3}t - \frac{\pi}{2}\right) + 30$
6. (a) $k = 5\,\frac{\text{lbs.}}{\text{ft.}}$ and $m = \frac{5}{16}$ slugs

   (b) $x(t) = \sin\left(4t + \frac{\pi}{2}\right)$. The object first passes through the equilibrium point when $t = \frac{\pi}{8} \approx 0.39$ seconds after the motion starts. At this time, the object is heading upwards.

   (c) $x(t) = \frac{\sqrt{2}}{2}\sin\left(4t + \frac{7\pi}{4}\right)$. The object passes through the equilibrium point heading downwards for the third time when $t = \frac{17\pi}{16} \approx 3.34$ seconds.
7. (a) $\theta(t) = \theta_0 \sin\left(\sqrt{\frac{g}{l}}\,t + \frac{\pi}{2}\right)$ (b) $\theta(t) = \frac{\pi}{12}\sin\left(4\pi t + \frac{\pi}{2}\right)$
8. (a) $T(t) = 20.5\sin\left(\frac{\pi}{6}t - \pi\right) + 53.5$

   (b) Our function and the data set are graphed below. The sinusoid seems to be shifted to the right of our data.

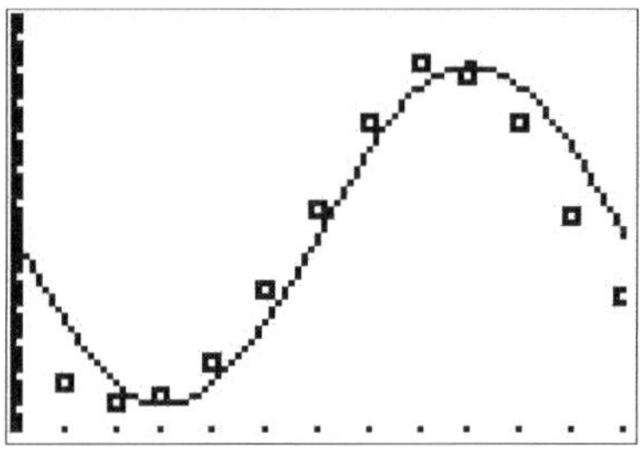

   (c) The average temperature on April 15th is approximately $T(4.5) \approx 39.00$°F and the average temperature on September 15th is approximately $T(9.5) \approx 73.38$°F.

   (d) Using a graphing calculator, we get the following

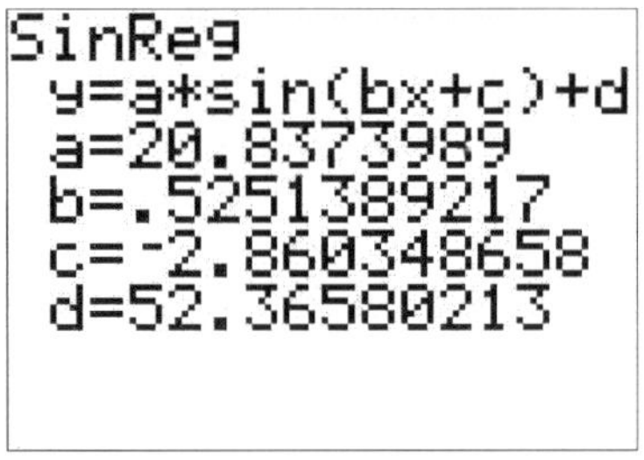

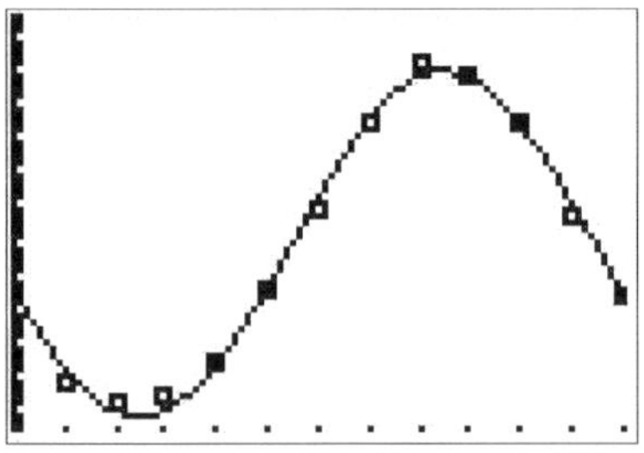

   This model predicts the average temperature for April 15th to be approximately 42.43°F and the average temperature on September 15th to be approximately 70.05°F. This model appears to be more accurate.

9. (a) Based on the shape of the data, we either choose $A < 0$ or we find the *second* value of t which closely approximates the 'baseline' value, $F = 0.505$. We choose the latter to obtain $F(t) = 0.475 \sin\left(\frac{\pi}{15}t - 2\pi\right) + 0.505 = 0.475 \sin\left(\frac{\pi}{15}t\right) + 0.505$

   (b) Our function and the data set are graphed below. It's a pretty good fit.

   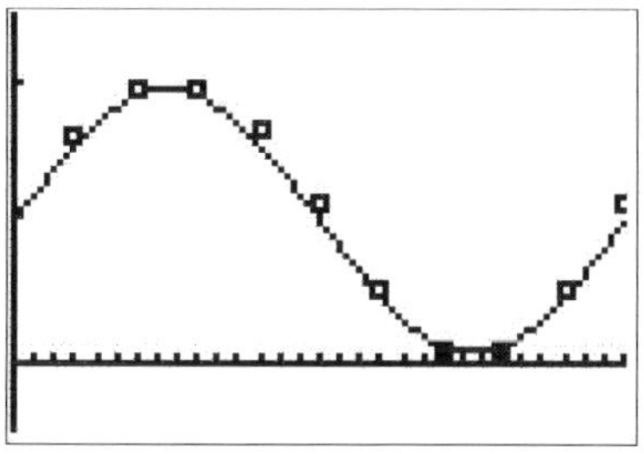

   (c) The fraction of the moon illuminated on June 1st, 2009 is approximately $F(1) \approx 0.60$

   (d) Using a graphing calculator, we get the following.

   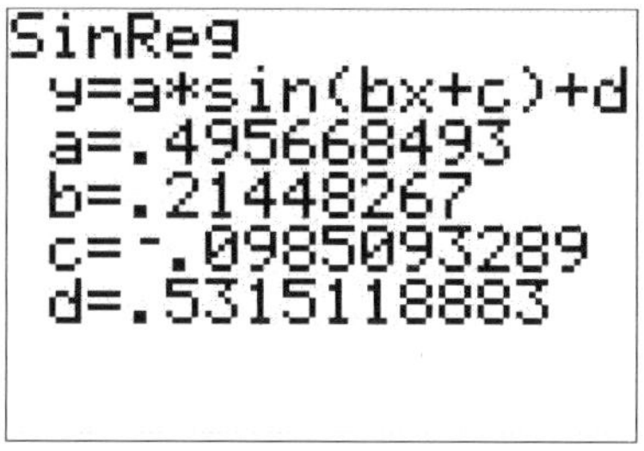

   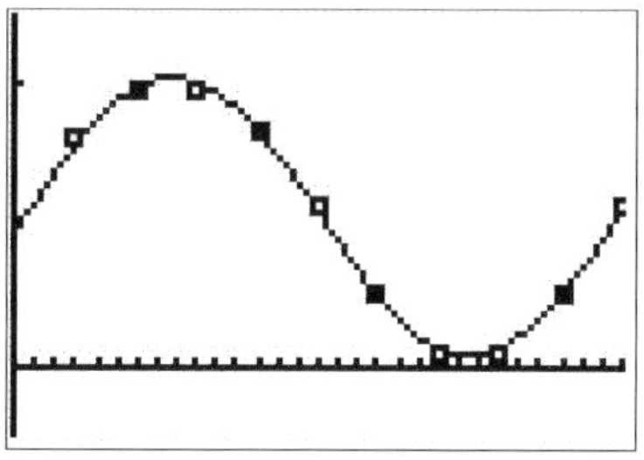

   This model predicts that the fraction of the moon illuminated on June 1st, 2009 is approximately 0.59. This appears to be a better fit to the data than our first model.

11.2 The Law of Sines

Trigonometry literally means 'measuring triangles' and with Chapter 10 under our belts, we are more than prepared to do just that. The main goal of this section and the next is to develop theorems which allow us to 'solve' triangles – that is, find the length of each side of a triangle and the measure of each of its angles. In Sections 10.2, 10.3 and 10.6, we've had some experience solving right triangles. The following example reviews what we know.

Example 11.2.1. Given a right triangle with a hypotenuse of length 7 units and one leg of length 4 units, find the length of the remaining side and the measures of the remaining angles. Express the angles in decimal degrees, rounded to the nearest hundredth of a degree.

Solution. For definitiveness, we label the triangle below.

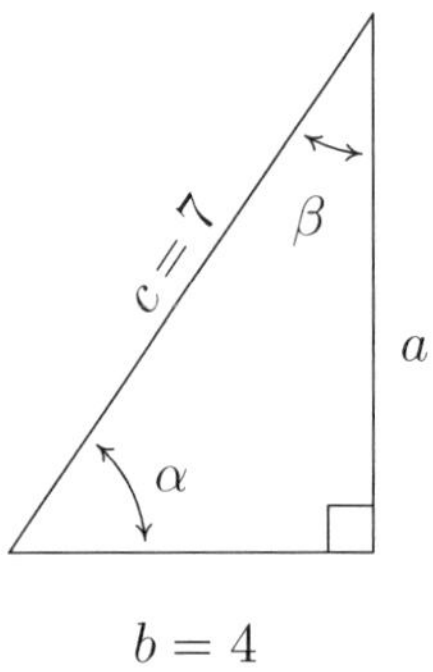

To find the length of the missing side a, we use the Pythagorean Theorem to get $a^2 + 4^2 = 7^2$ which then yields $a = \sqrt{33}$ units. Now that all three sides of the triangle are known, there are several ways we can find α using the inverse trigonometric functions. To decrease the chances of propagating error, however, we stick to using the data given to us in the problem. In this case, the lengths 4 and 7 were given, so we want to relate these to α. According to Theorem 10.4, $\cos(\alpha) = \frac{4}{7}$. Since α is an acute angle, $\alpha = \arccos\left(\frac{4}{7}\right)$ radians. Converting to degrees, we find $\alpha \approx 55.15^\circ$. Now that we have the measure of angle α, we could find the measure of angle β using the fact that α and β are complements so $\alpha + \beta = 90^\circ$. Once again, we opt to use the data given to us in the problem. According to Theorem 10.4, we have that $\sin(\beta) = \frac{4}{7}$ so $\beta = \arcsin\left(\frac{4}{7}\right)$ radians and we have $\beta \approx 34.85^\circ$. □

A few remarks about Example 11.2.1 are in order. First, we adhere to the convention that a lower case Greek letter denotes an angle[1] and the corresponding lowercase English letter represents the side[2] opposite that angle. Thus, a is the side opposite α, b is the side opposite β and c is the side opposite γ. Taken together, the pairs (α, a), (β, b) and (γ, c) are called *angle-side opposite pairs.* Second, as mentioned earlier, we will strive to solve for quantities using the original data given in the problem whenever possible. While this is not always the easiest or fastest way to proceed, it

[1] as well as the measure of said angle

[2] as well as the length of said side

minimizes the chances of propagated error.[3] Third, since many of the applications which require solving triangles 'in the wild' rely on degree measure, we shall adopt this convention for the time being.[4] The Pythagorean Theorem along with Theorems 10.4 and 10.10 allow us to easily handle any given right triangle problem, but what if the triangle isn't a right triangle? In certain cases, we can use the **Law of Sines** to help.

Theorem 11.2. The Law of Sines: Given a triangle with angle-side opposite pairs (α, a), (β, b) and (γ, c), the following ratios hold

$$\frac{\sin(\alpha)}{a} = \frac{\sin(\beta)}{b} = \frac{\sin(\gamma)}{c}$$

or, equivalently,

$$\frac{a}{\sin(\alpha)} = \frac{b}{\sin(\beta)} = \frac{c}{\sin(\gamma)}$$

The proof of the Law of Sines can be broken into three cases. For our first case, consider the triangle $\triangle ABC$ below, all of whose angles are acute, with angle-side opposite pairs (α, a), (β, b) and (γ, c). If we drop an altitude from vertex B, we divide the triangle into two right triangles: $\triangle ABQ$ and $\triangle BCQ$. If we call the length of the altitude h (for height), we get from Theorem 10.4 that $\sin(\alpha) = \frac{h}{c}$ and $\sin(\gamma) = \frac{h}{a}$ so that $h = c\sin(\alpha) = a\sin(\gamma)$. After some rearrangement of the last equation, we get $\frac{\sin(\alpha)}{a} = \frac{\sin(\gamma)}{c}$. If we drop an altitude from vertex A, we can proceed as above using the triangles $\triangle ABQ$ and $\triangle ACQ$ to get $\frac{\sin(\beta)}{b} = \frac{\sin(\gamma)}{c}$, completing the proof for this case.

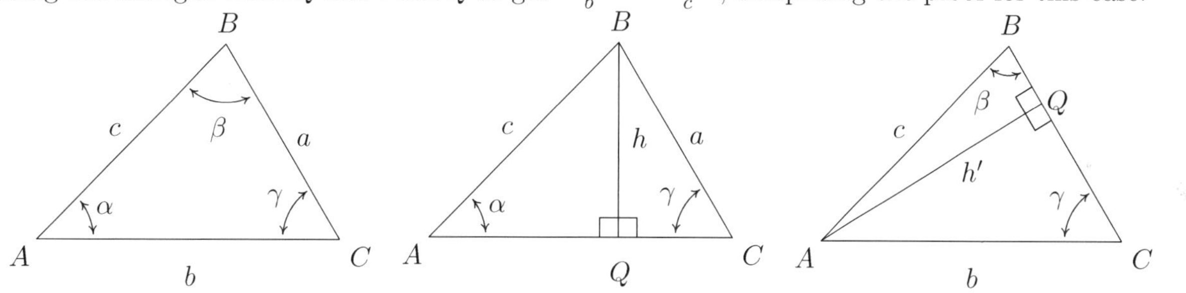

For our next case consider the triangle $\triangle ABC$ below with <u>obtuse</u> angle α. Extending an altitude from vertex A gives two right triangles, as in the previous case: $\triangle ABQ$ and $\triangle ACQ$. Proceeding as before, we get $h = b\sin(\gamma)$ and $h = c\sin(\beta)$ so that $\frac{\sin(\beta)}{b} = \frac{\sin(\gamma)}{c}$.

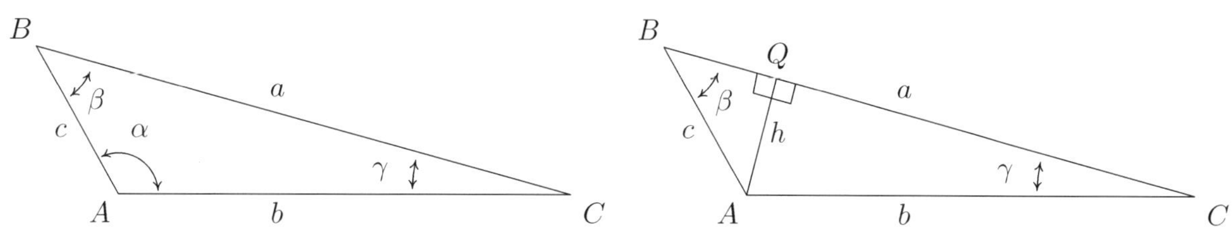

[3]Your Science teachers should thank us for this.

[4]Don't worry! Radians will be back before you know it!

Dropping an altitude from vertex B also generates two right triangles, $\triangle ABQ$ and $\triangle BCQ$. We know that $\sin(\alpha') = \frac{h'}{c}$ so that $h' = c\sin(\alpha')$. Since $\alpha' = 180^\circ - \alpha$, $\sin(\alpha') = \sin(\alpha)$, so in fact, we have $h' = c\sin(\alpha)$. Proceeding to $\triangle BCQ$, we get $\sin(\gamma) = \frac{h'}{a}$ so $h' = a\sin(\gamma)$. Putting this together with the previous equation, we get $\frac{\sin(\gamma)}{c} = \frac{\sin(\alpha)}{a}$, and we are finished with this case.

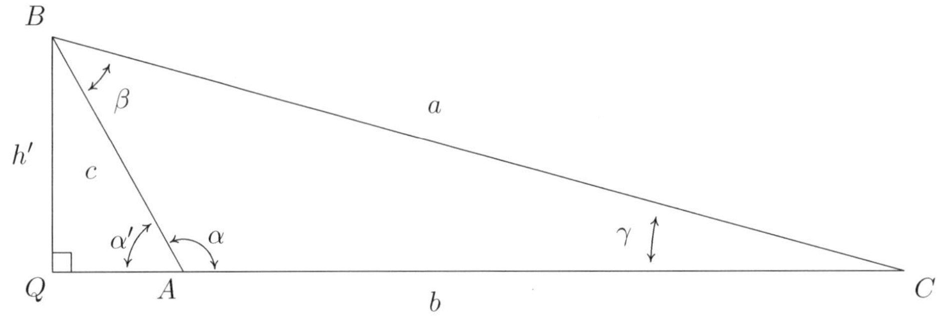

The remaining case is when $\triangle ABC$ is a right triangle. In this case, the Law of Sines reduces to the formulas given in Theorem 10.4 and is left to the reader. In order to use the Law of Sines to solve a triangle, we need at least one angle-side opposite pair. The next example showcases some of the power, and the pitfalls, of the Law of Sines.

Example 11.2.2. Solve the following triangles. Give exact answers and decimal approximations (rounded to hundredths) and sketch the triangle.

1. $\alpha = 120^\circ$, $a = 7$ units, $\beta = 45^\circ$
2. $\alpha = 85^\circ$, $\beta = 30^\circ$, $c = 5.25$ units
3. $\alpha = 30^\circ$, $a = 1$ units, $c = 4$ units
4. $\alpha = 30^\circ$, $a = 2$ units, $c = 4$ units
5. $\alpha = 30^\circ$, $a = 3$ units, $c = 4$ units
6. $\alpha = 30^\circ$, $a = 4$ units, $c = 4$ units

Solution.

1. Knowing an angle-side opposite pair, namely α and a, we may proceed in using the Law of Sines. Since $\beta = 45^\circ$, we use $\frac{b}{\sin(45^\circ)} = \frac{7}{\sin(120^\circ)}$ so $b = \frac{7\sin(45^\circ)}{\sin(120^\circ)} = \frac{7\sqrt{6}}{3} \approx 5.72$ units. Now that we have two angle-side pairs, it is time to find the third. To find γ, we use the fact that the sum of the measures of the angles in a triangle is 180°. Hence, $\gamma = 180^\circ - 120^\circ - 45^\circ = 15^\circ$. To find c, we have no choice but to used the derived value $\gamma = 15^\circ$, yet we can minimize the propagation of error here by using the given angle-side opposite pair (α, a). The Law of Sines gives us $\frac{c}{\sin(15^\circ)} = \frac{7}{\sin(120^\circ)}$ so that $c = \frac{7\sin(15^\circ)}{\sin(120^\circ)} \approx 2.09$ units.[5]

2. In this example, we are not immediately given an angle-side opposite pair, but as we have the measures of α and β, we can solve for γ since $\gamma = 180^\circ - 85^\circ - 30^\circ = 65^\circ$. As in the previous example, we are forced to use a derived value in our computations since the only

[5]The exact value of $\sin(15^\circ)$ could be found using the difference identity for sine or a half-angle formula, but that becomes unnecessarily messy for the discussion at hand. Thus "exact" here means $\frac{7\sin(15^\circ)}{\sin(120^\circ)}$.

angle-side pair available is (γ, c). The Law of Sines gives $\frac{a}{\sin(85^\circ)} = \frac{5.25}{\sin(65^\circ)}$. After the usual rearrangement, we get $a = \frac{5.25\sin(85^\circ)}{\sin(65^\circ)} \approx 5.77$ units. To find b we use the angle-side pair (γ, c) which yields $\frac{b}{\sin(30^\circ)} = \frac{5.25}{\sin(65^\circ)}$ hence $b = \frac{5.25\sin(30^\circ)}{\sin(65^\circ)} \approx 2.90$ units.

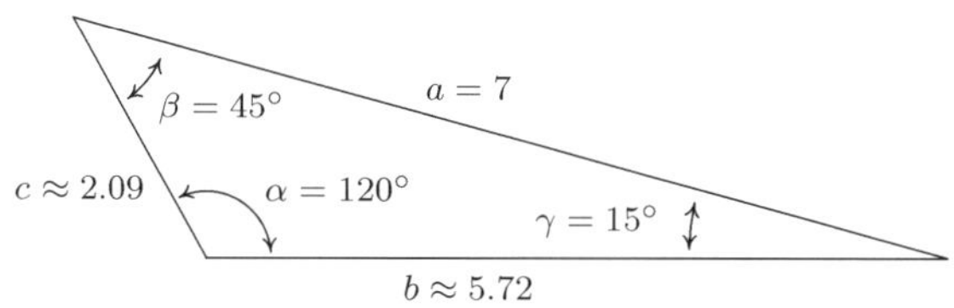

Triangle for number 1

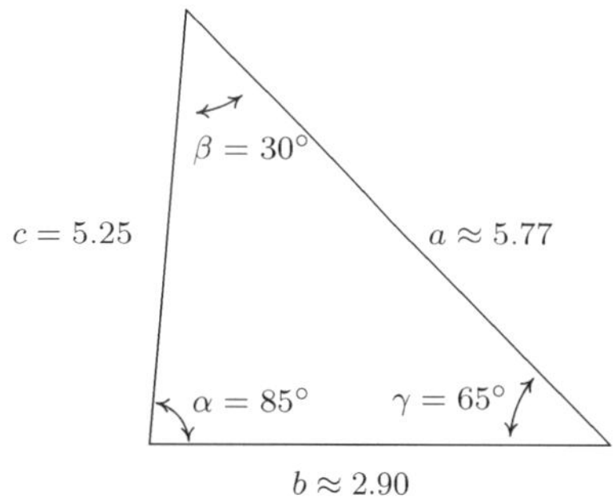

Triangle for number 2

3. Since we are given (α, a) and c, we use the Law of Sines to find the measure of γ. We start with $\frac{\sin(\gamma)}{4} = \frac{\sin(30^\circ)}{1}$ and get $\sin(\gamma) = 4\sin(30^\circ) = 2$. Since the range of the sine function is $[-1, 1]$, there is no real number with $\sin(\gamma) = 2$. Geometrically, we see that side a is just too short to make a triangle. The next three examples keep the same values for the measure of α and the length of c while varying the length of a. We will discuss this case in more detail after we see what happens in those examples.

4. In this case, we have the measure of $\alpha = 30^\circ$, $a = 2$ and $c = 4$. Using the Law of Sines, we get $\frac{\sin(\gamma)}{4} = \frac{\sin(30^\circ)}{2}$ so $\sin(\gamma) = 2\sin(30^\circ) = 1$. Now γ is an angle in a triangle which also contains $\alpha = 30^\circ$. This means that γ must measure between 0° and 150° in order to fit inside the triangle with α. The only angle that satisfies this requirement and has $\sin(\gamma) = 1$ is $\gamma = 90^\circ$. In other words, we have a right triangle. We find the measure of β to be $\beta = 180^\circ - 30^\circ - 90^\circ = 60^\circ$ and then determine b using the Law of Sines. We find $b = \frac{2\sin(60^\circ)}{\sin(30^\circ)} = 2\sqrt{3} \approx 3.46$ units. In this case, the side a is precisely long enough to form a unique right triangle.

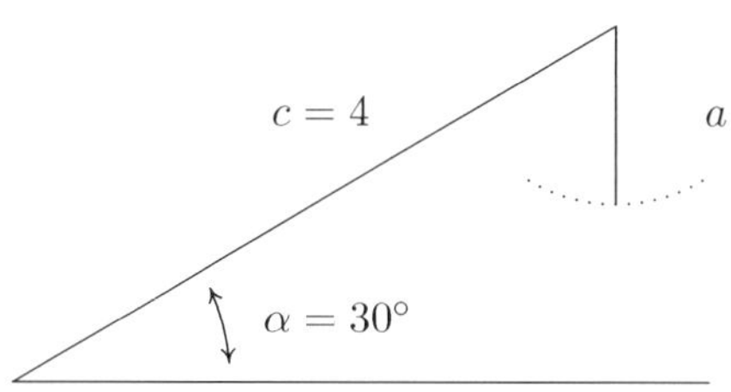

Diagram for number 3

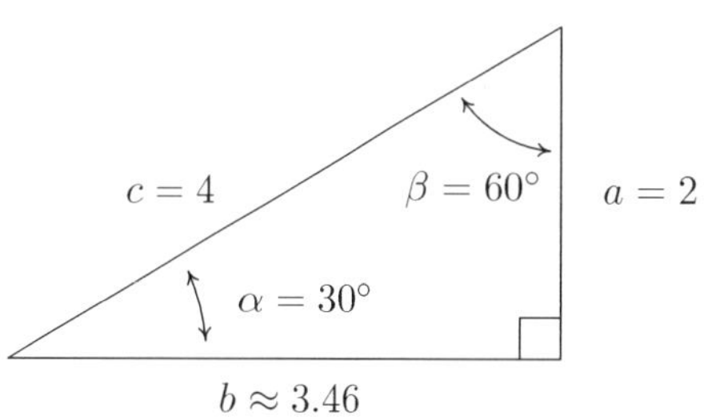

Triangle for number 4

5. Proceeding as we have in the previous two examples, we use the Law of Sines to find γ. In this case, we have $\frac{\sin(\gamma)}{4} = \frac{\sin(30^\circ)}{3}$ or $\sin(\gamma) = \frac{4\sin(30^\circ)}{3} = \frac{2}{3}$. Since γ lies in a triangle with $\alpha = 30^\circ$,

we must have that $0^\circ < \gamma < 150^\circ$. There are two angles γ that fall in this range and have $\sin(\gamma) = \frac{2}{3}$: $\gamma = \arcsin\left(\frac{2}{3}\right)$ radians $\approx 41.81^\circ$ and $\gamma = \pi - \arcsin\left(\frac{2}{3}\right)$ radians $\approx 138.19^\circ$. At this point, we pause to see if it makes sense that we actually have two viable cases to consider. As we have discussed, both candidates for γ are 'compatible' with the given angle-side pair $(\alpha, a) = (30^\circ, 3)$ in that both choices for γ can fit in a triangle with α and both have a sine of $\frac{2}{3}$. The only other given piece of information is that $c = 4$ units. Since $c > a$, it must be true that γ, which is opposite c, has greater measure than α which is opposite a. In both cases, $\gamma > \alpha$, so both candidates for γ are compatible with this last piece of given information as well. Thus have two triangles on our hands. In the case $\gamma = \arcsin\left(\frac{2}{3}\right)$ radians $\approx 41.81^\circ$, we find[6] $\beta \approx 180^\circ - 30^\circ - 41.81^\circ = 108.19^\circ$. Using the Law of Sines with the angle-side opposite pair (α, a) and β, we find $b \approx \frac{3\sin(108.19^\circ)}{\sin(30^\circ)} \approx 5.70$ units. In the case $\gamma = \pi - \arcsin\left(\frac{2}{3}\right)$ radians $\approx 138.19^\circ$, we repeat the exact same steps and find $\beta \approx 11.81^\circ$ and $b \approx 1.23$ units.[7] Both triangles are drawn below.

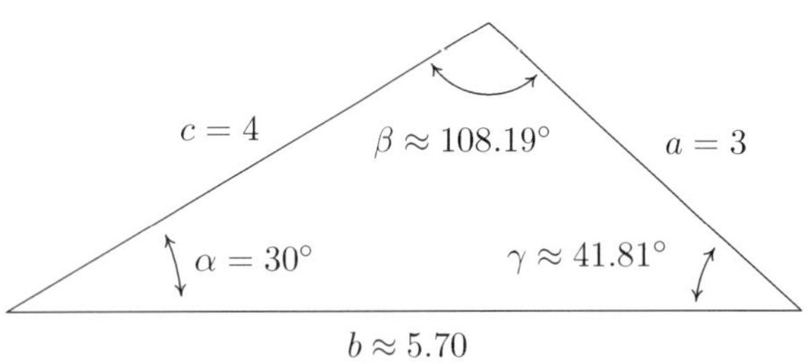

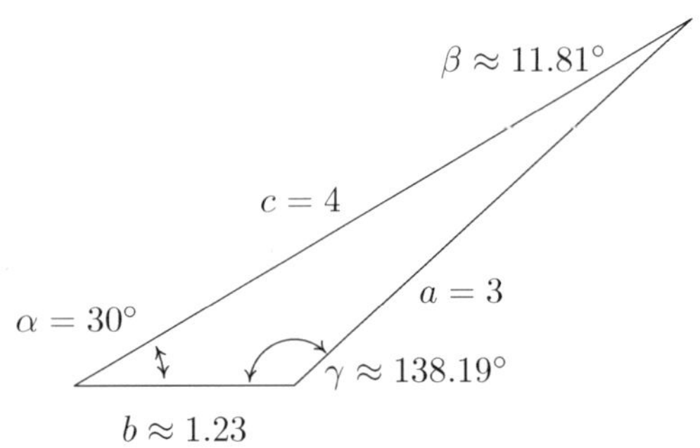

6. For this last problem, we repeat the usual Law of Sines routine to find that $\frac{\sin(\gamma)}{4} = \frac{\sin(30^\circ)}{4}$ so that $\sin(\gamma) = \frac{1}{2}$. Since γ must inhabit a triangle with $\alpha = 30^\circ$, we must have $0^\circ < \gamma < 150^\circ$. Since the measure of γ must be *strictly* less than 150°, there is just one angle which satisfies both required conditions, namely $\gamma = 30^\circ$. So $\beta = 180^\circ - 30^\circ - 30^\circ = 120^\circ$ and, using the Law of Sines one last time, $b = \frac{4\sin(120^\circ)}{\sin(30^\circ)} = 4\sqrt{3} \approx 6.93$ units.

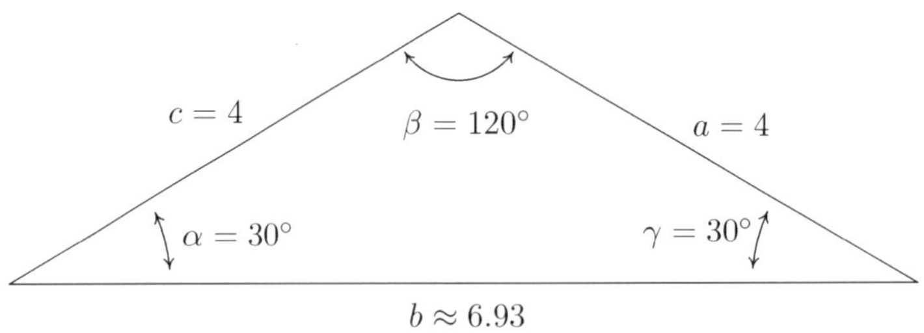

□

Some remarks about Example 11.2.2 are in order. We first note that if we are given the measures of two of the angles in a triangle, say α and β, the measure of the third angle γ is uniquely

[6] To find an exact expression for β, we convert everything back to radians: $\alpha = 30^\circ = \frac{\pi}{6}$ radians, $\gamma = \arcsin\left(\frac{2}{3}\right)$ radians and $180^\circ = \pi$ radians. Hence, $\beta = \pi - \frac{\pi}{6} - \arcsin\left(\frac{2}{3}\right) = \frac{5\pi}{6} - \arcsin\left(\frac{2}{3}\right)$ radians $\approx 108.19^\circ$.

[7] An exact answer for β in this case is $\beta = \arcsin\left(\frac{2}{3}\right) - \frac{\pi}{6}$ radians $\approx 11.81^\circ$.

determined using the equation $\gamma = 180^\circ - \alpha - \beta$. Knowing the measures of all three angles of a triangle completely determines its *shape*. If in addition we are given the length of one of the sides of the triangle, we can then use the Law of Sines to find the lengths of the remaining two sides to determine the *size* of the triangle. Such is the case in numbers 1 and 2 above. In number 1, the given side is adjacent to just one of the angles – this is called the 'Angle-Angle-Side' (AAS) case.[8] In number 2, the given side is adjacent to both angles which means we are in the so-called 'Angle-Side-Angle' (ASA) case. If, on the other hand, we are given the measure of just one of the angles in the triangle along with the length of two sides, only one of which is adjacent to the given angle, we are in the 'Angle-Side-Side' (ASS) case.[9] In number 3, the length of the one given side a was too short to even form a triangle; in number 4, the length of a was just long enough to form a right triangle; in 5, a was long enough, but not too long, so that two triangles were possible; and in number 6, side a was long enough to form a triangle but too long to swing back and form two. These four cases exemplify all of the possibilities in the Angle-Side-Side case which are summarized in the following theorem.

Theorem 11.3. Suppose (α, a) and (γ, c) are intended to be angle-side pairs in a triangle where α, a and c are given. Let $h = c\sin(\alpha)$

- If $a < h$, then no triangle exists which satisfies the given criteria.
- If $a = h$, then $\gamma = 90^\circ$ so exactly one (right) triangle exists which satisfies the criteria.
- If $h < a < c$, then two distinct triangles exist which satisfy the given criteria.
- If $a \geq c$, then γ is acute and exactly one triangle exists which satisfies the given criteria

Theorem 11.3 is proved on a case-by-case basis. If $a < h$, then $a < c\sin(\alpha)$. If a triangle were to exist, the Law of Sines would have $\frac{\sin(\gamma)}{c} = \frac{\sin(\alpha)}{a}$ so that $\sin(\gamma) = \frac{c\sin(\alpha)}{a} > \frac{a}{a} = 1$, which is impossible. In the figure below, we see geometrically why this is the case.

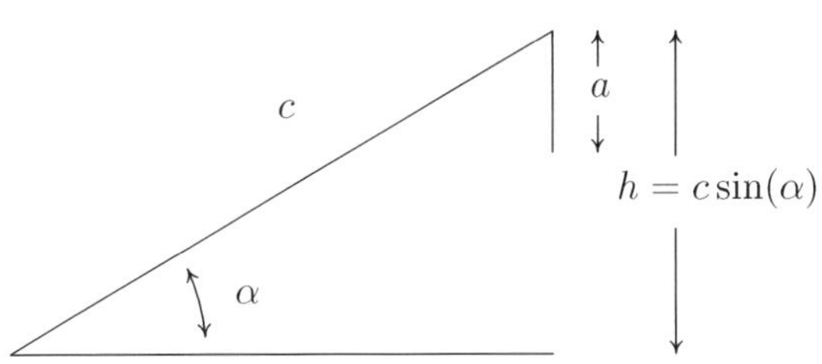

$a < h$, no triangle

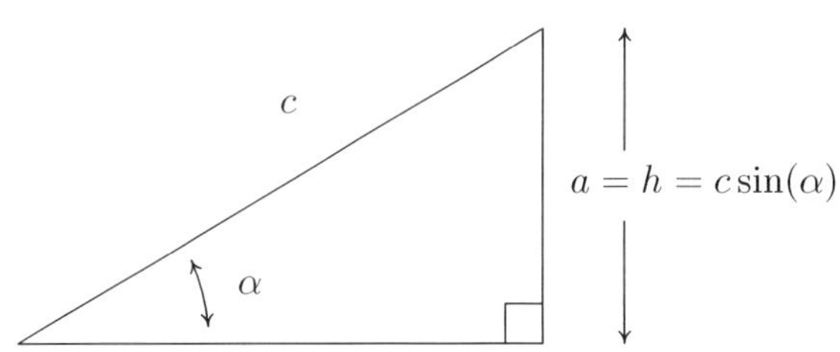

$a = h$, $\gamma = 90^\circ$

Simply put, if $a < h$ the side a is too short to connect to form a triangle. This means if $a \geq h$, we are always guaranteed to have at least one triangle, and the remaining parts of the theorem

[8] If this sounds familiar, it should. From high school Geometry, we know there are four congruence conditions for triangles: Angle-Angle-Side (AAS), Angle-Side-Angle (ASA), Side-Angle-Side (SAS) and Side-Side-Side (SSS). If we are given information about a triangle that meets one of these four criteria, then we are guaranteed that exactly one triangle exists which satisfies the given criteria.

[9] In more reputable books, this is called the 'Side-Side-Angle' or SSA case.

tell us what kind and how many triangles to expect in each case. If $a = h$, then $a = c\sin(\alpha)$ and the Law of Sines gives $\frac{\sin(\alpha)}{a} = \frac{\sin(\gamma)}{c}$ so that $\sin(\gamma) = \frac{c\sin(\alpha)}{a} = \frac{a}{a} = 1$. Here, $\gamma = 90^\circ$ as required. Moving along, now suppose $h < a < c$. As before, the Law of Sines[10] gives $\sin(\gamma) = \frac{c\sin(\alpha)}{a}$. Since $h < a$, $c\sin(\alpha) < a$ or $\frac{c\sin(\alpha)}{a} < 1$ which means there are two solutions to $\sin(\gamma) = \frac{c\sin(\alpha)}{a}$: an acute angle which we'll call γ_0, and its supplement, $180^\circ - \gamma_0$. We need to argue that each of these angles 'fit' into a triangle with α. Since (α, a) and (γ_0, c) are angle-side opposite pairs, the assumption $c > a$ in this case gives us $\gamma_0 > \alpha$. Since γ_0 is acute, we must have that α is acute as well. This means one triangle can contain both α and γ_0, giving us one of the triangles promised in the theorem. If we manipulate the inequality $\gamma_0 > \alpha$ a bit, we have $180^\circ - \gamma_0 < 180^\circ - \alpha$ which gives $(180^\circ - \gamma_0) + \alpha < 180^\circ$. This proves a triangle can contain both of the angles α and $(180^\circ - \gamma_0)$, giving us the second triangle predicted in the theorem. To prove the last case in the theorem, we assume $a \geq c$. Then $\alpha \geq \gamma$, which forces γ to be an acute angle. Hence, we get only one triangle in this case, completing the proof.

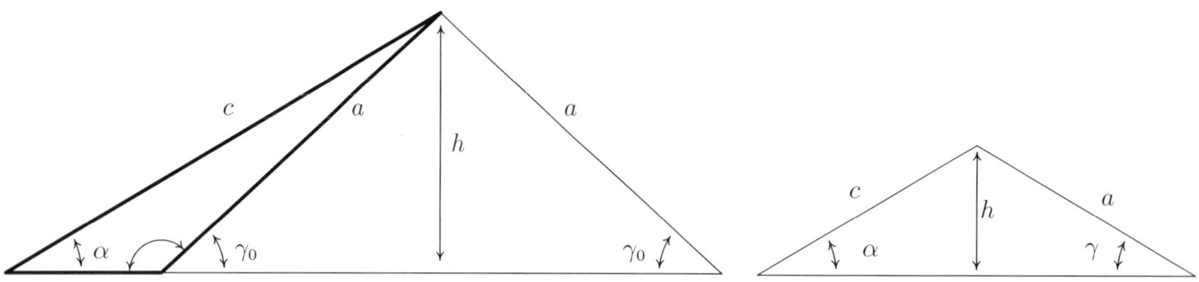

$h < a < c$, two triangles $\qquad$ $a \geq c$, one triangle

One last comment before we use the Law of Sines to solve an application problem. In the Angle-Side-Side case, if you are given an obtuse angle to begin with then it is impossible to have the two triangle case. Think about this before reading further.

Example 11.2.3. Sasquatch Island lies off the coast of Ippizuti Lake. Two sightings, taken 5 miles apart, are made to the island. The angle between the shore and the island at the first observation point is 30° and at the second point the angle is 45°. Assuming a straight coastline, find the distance from the second observation point to the island. What point on the shore is closest to the island? How far is the island from this point?

Solution. We sketch the problem below with the first observation point labeled as P and the second as Q. In order to use the Law of Sines to find the distance d from Q to the island, we first need to find the measure of β which is the angle opposite the side of length 5 miles. To that end, we note that the angles γ and 45° are supplemental, so that $\gamma = 180^\circ - 45^\circ = 135^\circ$. We can now find $\beta = 180^\circ - 30^\circ - \gamma = 180^\circ - 30^\circ - 135^\circ = 15^\circ$. By the Law of Sines, we have $\frac{d}{\sin(30^\circ)} = \frac{5}{\sin(15^\circ)}$ which gives $d = \frac{5\sin(30^\circ)}{\sin(15^\circ)} \approx 9.66$ miles. Next, to find the point on the coast closest to the island, which we've labeled as C, we need to find the perpendicular distance from the island to the coast.[11]

[10] Remember, we have already argued that a triangle exists in this case!

[11] Do you see why C must lie to the right of Q?

Let x denote the distance from the second observation point Q to the point C and let y denote the distance from C to the island. Using Theorem 10.4, we get $\sin(45^\circ) = \frac{y}{d}$. After some rearranging, we find $y = d\sin(45^\circ) \approx 9.66\left(\frac{\sqrt{2}}{2}\right) \approx 6.83$ miles. Hence, the island is approximately 6.83 miles from the coast. To find the distance from Q to C, we note that $\beta = 180^\circ - 90^\circ - 45^\circ = 45^\circ$ so by symmetry,[12] we get $x = y \approx 6.83$ miles. Hence, the point on the shore closest to the island is approximately 6.83 miles down the coast from the second observation point.

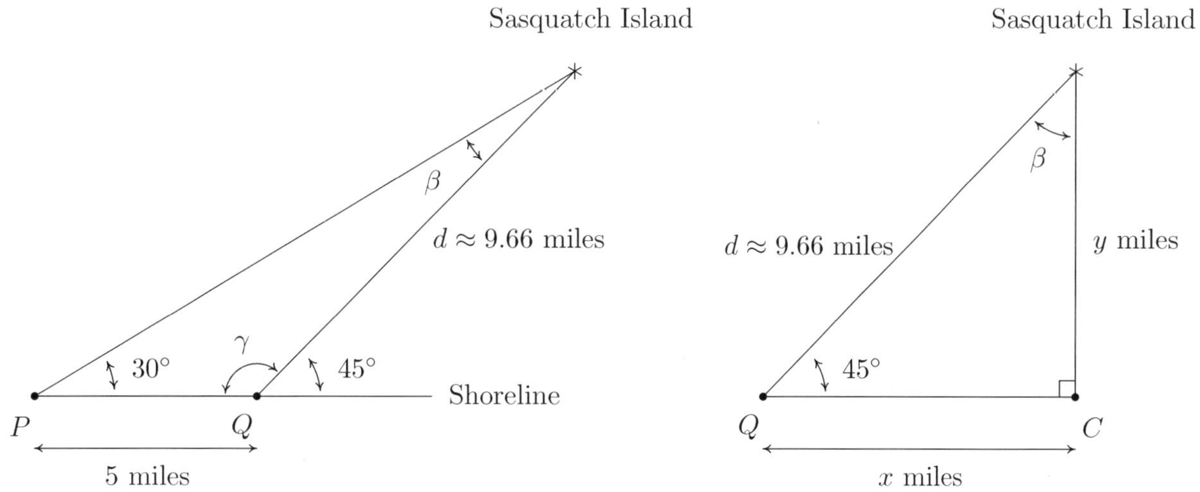

□

We close this section with a new formula to compute the area enclosed by a triangle. Its proof uses the same cases and diagrams as the proof of the Law of Sines and is left as an exercise.

Theorem 11.4. Suppose (α, a), (β, b) and (γ, c) are the angle-side opposite pairs of a triangle. Then the area A enclosed by the triangle is given by

$$A = \frac{1}{2}bc\sin(\alpha) = \frac{1}{2}ac\sin(\beta) = \frac{1}{2}ab\sin(\gamma)$$

Example 11.2.4. Find the area of the triangle in Example 11.2.2 number 1.

Solution. From our work in Example 11.2.2 number 1, we have all three angles and all three sides to work with. However, to minimize propagated error, we choose $A = \frac{1}{2}ac\sin(\beta)$ from Theorem 11.4 because it uses the most pieces of given information. We are given $a = 7$ and $\beta = 45^\circ$, and we calculated $c = \frac{7\sin(15^\circ)}{\sin(120^\circ)}$. Using these values, we find $A = \frac{1}{2}(7)\left(\frac{7\sin(15^\circ)}{\sin(120^\circ)}\right)\sin(45^\circ) =\approx 5.18$ square units. The reader is encouraged to check this answer against the results obtained using the other formulas in Theorem 11.4. □

[12] Or by Theorem 10.4 again ...

11.2.1 EXERCISES

In Exercises 1 - 20, solve for the remaining side(s) and angle(s) if possible. As in the text, (α, a), (β, b) and (γ, c) are angle-side opposite pairs.

1. $\alpha = 13^\circ,\ \beta = 17^\circ,\ a = 5$
2. $\alpha = 73.2^\circ,\ \beta = 54.1^\circ,\ a = 117$
3. $\alpha = 95^\circ,\ \beta = 85^\circ,\ a = 33.33$
4. $\alpha = 95^\circ,\ \beta = 62^\circ,\ a = 33.33$
5. $\alpha = 117^\circ,\ a = 35,\ b = 42$
6. $\alpha = 117^\circ,\ a = 45,\ b = 42$
7. $\alpha = 68.7^\circ,\ a = 88,\ b = 92$
8. $\alpha = 42^\circ,\ a = 17,\ b = 23.5$
9. $\alpha = 68.7^\circ,\ a = 70,\ b = 90$
10. $\alpha = 30^\circ,\ a = 7,\ b = 14$
11. $\alpha = 42^\circ,\ a = 39,\ b = 23.5$
12. $\gamma = 53^\circ,\ \alpha = 53^\circ,\ c = 28.01$
13. $\alpha = 6^\circ,\ a = 57,\ b = 100$
14. $\gamma = 74.6^\circ,\ c = 3,\ a = 3.05$
15. $\beta = 102^\circ,\ b = 16.75,\ c = 13$
16. $\beta = 102^\circ,\ b = 16.75,\ c = 18$
17. $\beta = 102^\circ,\ \gamma = 35^\circ,\ b = 16.75$
18. $\beta = 29.13^\circ,\ \gamma = 83.95^\circ,\ b = 314.15$
19. $\gamma = 120^\circ,\ \beta = 61^\circ,\ c = 4$
20. $\alpha = 50^\circ,\ a = 25,\ b = 12.5$

21. Find the area of the triangles given in Exercises 1, 12 and 20 above.

(Another Classic Application: Grade of a Road) The grade of a road is much like the pitch of a roof (See Example 10.6.6) in that it expresses the ratio of rise/run. In the case of a road, this ratio is always positive because it is measured going uphill and it is usually given as a percentage. For example, a road which rises 7 feet for every 100 feet of (horizontal) forward progress is said to have a 7% grade. However, if we want to apply any Trigonometry to a story problem involving roads going uphill or downhill, we need to view the grade as an angle with respect to the horizontal. In Exercises 22 - 24, we first have you change road grades into angles and then use the Law of Sines in an application.

22. Using a right triangle with a horizontal leg of length 100 and vertical leg with length 7, show that a 7% grade means that the road (hypotenuse) makes about a 4° angle with the horizontal. (It will not be exactly 4°, but it's pretty close.)
23. What grade is given by a 9.65° angle made by the road and the horizontal?[13]

[13]I have friends who live in Pacifica, CA and their road is actually this steep. It's not a nice road to drive.

24. Along a long, straight stretch of mountain road with a 7% grade, you see a tall tree standing perfectly plumb alongside the road.[14] From a point 500 feet downhill from the tree, the angle of inclination from the road to the top of the tree is 6°. Use the Law of Sines to find the height of the tree. (Hint: First show that the tree makes a 94° angle with the road.)

(Another Classic Application: Bearings) In the next several exercises we introduce and work with the navigation tool known as bearings. Simply put, a bearing is the direction you are heading according to a compass. The classic nomenclature for bearings, however, is not given as an angle in standard position, so we must first understand the notation. A bearing is given as an acute angle of rotation (to the east or to the west) away from the north-south (up and down) line of a compass rose. For example, N40°E (read "40° east of north") is a bearing which is rotated clockwise 40° from due north. If we imagine standing at the origin in the Cartesian Plane, this bearing would have us heading into Quadrant I along the terminal side of $\theta = 50^\circ$. Similarly, S50°W would point into Quadrant III along the terminal side of $\theta = 220^\circ$ because we started out pointing due south (along $\theta = 270^\circ$) and rotated clockwise 50° back to 220°. Counter-clockwise rotations would be found in the bearings N60°W (which is on the terminal side of $\theta = 150^\circ$) and S27°E (which lies along the terminal side of $\theta = 297^\circ$). These four bearings are drawn in the plane below.

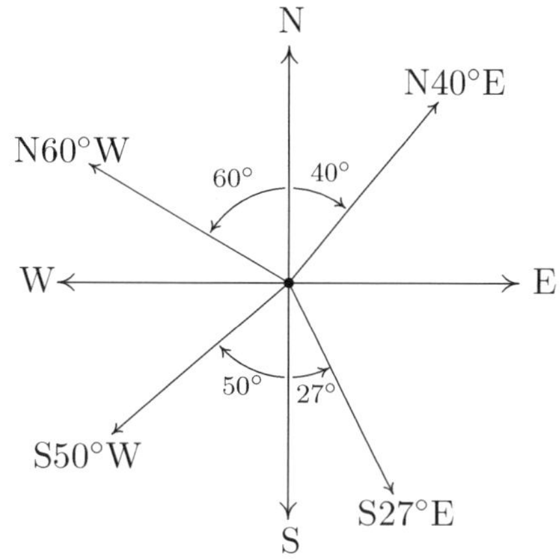

The cardinal directions north, south, east and west are usually not given as bearings in the fashion described above, but rather, one just refers to them as 'due north', 'due south', 'due east' and 'due west', respectively, and it is assumed that you know which quadrantal angle goes with each cardinal direction. (Hint: Look at the diagram above.)

25. Find the angle θ in standard position with $0^\circ \leq \theta < 360^\circ$ which corresponds to each of the bearings given below.

(a) due west (b) S83°E (c) N5.5°E (d) due south

[14]The word 'plumb' here means that the tree is perpendicular to the horizontal.

(e) N31.25°W (f) S72°41′12″W[15] (g) N45°E (h) S45°W

26. The Colonel spots a campfire at a of bearing N42°E from his current position. Sarge, who is positioned 3000 feet due east of the Colonel, reckons the bearing to the fire to be N20°W from his current position. Determine the distance from the campfire to each man, rounded to the nearest foot.

27. A hiker starts walking due west from Sasquatch Point and gets to the Chupacabra Trailhead before she realizes that she hasn't reset her pedometer. From the Chupacabra Trailhead she hikes for 5 miles along a bearing of N53°W which brings her to the Muffin Ridge Observatory. From there, she knows a bearing of S65°E will take her straight back to Sasquatch Point. How far will she have to walk to get from the Muffin Ridge Observatory to Sasquach Point? What is the distance between Sasquatch Point and the Chupacabra Trailhead?

28. The captain of the SS Bigfoot sees a signal flare at a bearing of N15°E from her current location. From his position, the captain of the HMS Sasquatch finds the signal flare to be at a bearing of N75°W. If the SS Bigfoot is 5 miles from the HMS Sasquatch and the bearing from the SS Bigfoot to the HMS Sasquatch is N50°E, find the distances from the flare to each vessel, rounded to the nearest tenth of a mile.

29. Carl spies a potential Sasquatch nest at a bearing of N10°E and radios Jeff, who is at a bearing of N50°E from Carl's position. From Jeff's position, the nest is at a bearing of S70°W. If Jeff and Carl are 500 feet apart, how far is Jeff from the Sasquatch nest? Round your answer to the nearest foot.

30. A hiker determines the bearing to a lodge from her current position is S40°W. She proceeds to hike 2 miles at a bearing of S20°E at which point she determines the bearing to the lodge is S75°W. How far is she from the lodge at this point? Round your answer to the nearest hundredth of a mile.

31. A watchtower spots a ship off shore at a bearing of N70°E. A second tower, which is 50 miles from the first at a bearing of S80°E from the first tower, determines the bearing to the ship to be N25°W. How far is the boat from the second tower? Round your answer to the nearest tenth of a mile.

32. Skippy and Sally decide to hunt UFOs. One night, they position themselves 2 miles apart on an abandoned stretch of desert runway. An hour into their investigation, Skippy spies a UFO hovering over a spot on the runway directly between him and Sally. He records the angle of inclination from the ground to the craft to be 75° and radios Sally immediately to find the angle of inclination from her position to the craft is 50°. How high off the ground is the UFO at this point? Round your answer to the nearest foot. (Recall: 1 mile is 5280 feet.)

[15]See Example 10.1.1 in Section 10.1 for a review of the DMS system.

33. The angle of depression from an observer in an apartment complex to a gargoyle on the building next door is 55°. From a point five stories below the original observer, the angle of inclination to the gargoyle is 20°. Find the distance from each observer to the gargoyle and the distance from the gargoyle to the apartment complex. Round your answers to the nearest foot. (Use the rule of thumb that one story of a building is 9 feet.)

34. Prove that the Law of Sines holds when $\triangle ABC$ is a right triangle.

35. Discuss with your classmates why knowing only the three angles of a triangle is not enough to determine any of the sides.

36. Discuss with your classmates why the Law of Sines cannot be used to find the angles in the triangle when only the three sides are given. Also discuss what happens if only two sides and the angle between them are given. (Said another way, explain why the Law of Sines cannot be used in the SSS and SAS cases.)

37. Given $\alpha = 30^\circ$ and $b = 10$, choose four different values for a so that

    (a) the information yields no triangle

    (b) the information yields exactly one right triangle

    (c) the information yields two distinct triangles

    (d) the information yields exactly one obtuse triangle

    Explain why you cannot choose a in such a way as to have $\alpha = 30^\circ$, $b = 10$ and your choice of a yield only one triangle where that unique triangle has three acute angles.

38. Use the cases and diagrams in the proof of the Law of Sines (Theorem 11.2) to prove the area formulas given in Theorem 11.4. Why do those formulas yield square units when four quantities are being multiplied together?

11.2.2 Answers

1. $\alpha = 13^\circ \quad \beta = 17^\circ \quad \gamma = 150^\circ$
   $a = 5 \quad b \approx 6.50 \quad c \approx 11.11$

2. $\alpha = 73.2^\circ \quad \beta = 54.1^\circ \quad \gamma = 52.7^\circ$
   $a = 117 \quad b \approx 99.00 \quad c \approx 97.22$

3. Information does not produce a triangle

4. $\alpha = 95^\circ \quad \beta = 62^\circ \quad \gamma = 23^\circ$
   $a = 33.33 \quad b \approx 29.54 \quad c \approx 13.07$

5. Information does not produce a triangle

6. $\alpha = 117^\circ \quad \beta \approx 56.3^\circ \quad \gamma \approx 6.7^\circ$
   $a = 45 \quad b = 42 \quad c \approx 5.89$

7. $\alpha = 68.7^\circ \quad \beta \approx 76.9^\circ \quad \gamma \approx 34.4^\circ$
   $a = 88 \quad b = 92 \quad c \approx 53.36$

   $\alpha = 68.7^\circ \quad \beta \approx 103.1^\circ \quad \gamma \approx 8.2^\circ$
   $a = 88 \quad b = 92 \quad c \approx 13.47$

8. $\alpha = 42^\circ \quad \beta \approx 67.66^\circ \quad \gamma \approx 70.34^\circ$
   $a = 17 \quad b = 23.5 \quad c \approx 23.93$

   $\alpha = 42^\circ \quad \beta \approx 112.34^\circ \quad \gamma \approx 25.66^\circ$
   $a = 17 \quad b = 23.5 \quad c \approx 11.00$

9. Information does not produce a triangle

10. $\alpha = 30^\circ \quad \beta = 90^\circ \quad \gamma = 60^\circ$
    $a = 7 \quad b = 14 \quad c = 7\sqrt{3}$

11. $\alpha = 42^\circ \quad \beta \approx 23.78^\circ \quad \gamma \approx 114.22^\circ$
    $a = 39 \quad b = 23.5 \quad c \approx 53.15$

12. $\alpha = 53^\circ \quad \beta = 74^\circ \quad \gamma = 53^\circ$
    $a = 28.01 \quad b \approx 33.71 \quad c = 28.01$

13. $\alpha = 6^\circ \quad \beta \approx 169.43^\circ \quad \gamma \approx 4.57^\circ$
    $a = 57 \quad b = 100 \quad c \approx 43.45$

    $\alpha = 6^\circ \quad \beta \approx 10.57^\circ \quad \gamma \approx 163.43^\circ$
    $a = 57 \quad b = 100 \quad c \approx 155.51$

14. $\alpha \approx 78.59^\circ \quad \beta \approx 26.81^\circ \quad \gamma = 74.6^\circ$
    $a = 3.05 \quad b \approx 1.40 \quad c = 3$

    $\alpha \approx 101.41^\circ \quad \beta \approx 3.99^\circ \quad \gamma = 74.6^\circ$
    $a = 3.05 \quad b \approx 0.217 \quad c = 3$

15. $\alpha \approx 28.61^\circ \quad \beta = 102^\circ \quad \gamma \approx 49.39^\circ$
    $a \approx 8.20 \quad b = 16.75 \quad c = 13$

16. Information does not produce a triangle

17. $\alpha = 43^\circ \quad \beta = 102^\circ \quad \gamma = 35^\circ$
    $a \approx 11.68 \quad b = 16.75 \quad c \approx 9.82$

18. $\alpha = 66.92^\circ \quad \beta = 29.13^\circ \quad \gamma = 83.95^\circ$
    $a \approx 593.69 \quad b = 314.15 \quad c \approx 641.75$

19. Information does not produce a triangle

20. $\alpha = 50^\circ \quad \beta \approx 22.52^\circ \quad \gamma \approx 107.48^\circ$
    $a = 25 \quad b = 12.5 \quad c \approx 31.13$

21. The area of the triangle from Exercise 1 is about 8.1 square units.
    The area of the triangle from Exercise 12 is about 377.1 square units.
    The area of the triangle from Exercise 20 is about 149 square units.

22. $\arctan\left(\frac{7}{100}\right) \approx 0.699$ radians, which is equivalent to 4.004°

23. About 17%

24. About 53 feet

25. (a) $\theta = 180^\circ$ (b) $\theta = 353^\circ$ (c) $\theta = 84.5^\circ$ (d) $\theta = 270^\circ$

    (e) $\theta = 121.25^\circ$ (f) $\theta = 197^\circ 18' 48''$ (g) $\theta = 45^\circ$ (h) $\theta = 225^\circ$

26. The Colonel is about 3193 feet from the campfire.
    Sarge is about 2525 feet to the campfire.

27. The distance from the Muffin Ridge Observatory to Sasquach Point is about 7.12 miles.
    The distance from Sasquatch Point to the Chupacabra Trailhead is about 2.46 miles.

28. The SS Bigfoot is about 4.1 miles from the flare.
    The HMS Sasquatch is about 2.9 miles from the flare.

29. Jeff is about 371 feet from the nest.

30. She is about 3.02 miles from the lodge

31. The boat is about 25.1 miles from the second tower.

32. The UFO is hovering about 9539 feet above the ground.

33. The gargoyle is about 44 feet from the observer on the upper floor.
    The gargoyle is about 27 feet from the observer on the lower floor.
    The gargoyle is about 25 feet from the other building.

11.3 THE LAW OF COSINES

In Section 11.2, we developed the Law of Sines (Theorem 11.2) to enable us to solve triangles in the 'Angle-Angle-Side' (AAS), the 'Angle-Side-Angle' (ASA) and the ambiguous 'Angle-Side-Side' (ASS) cases. In this section, we develop the Law of Cosines which handles solving triangles in the 'Side-Angle-Side' (SAS) and 'Side-Side-Side' (SSS) cases.[1] We state and prove the theorem below.

Theorem 11.5. Law of Cosines: Given a triangle with angle-side opposite pairs (α, a), (β, b) and (γ, c), the following equations hold

$$a^2 = b^2 + c^2 - 2bc\cos(\alpha) \qquad b^2 = a^2 + c^2 - 2ac\cos(\beta) \qquad c^2 = a^2 + b^2 - 2ab\cos(\gamma)$$

or, solving for the cosine in each equation, we have

$$\cos(\alpha) = \frac{b^2 + c^2 - a^2}{2bc} \qquad \cos(\beta) = \frac{a^2 + c^2 - b^2}{2ac} \qquad \cos(\gamma) = \frac{a^2 + b^2 - c^2}{2ab}$$

To prove the theorem, we consider a generic triangle with the vertex of angle α at the origin with side b positioned along the positive x-axis.

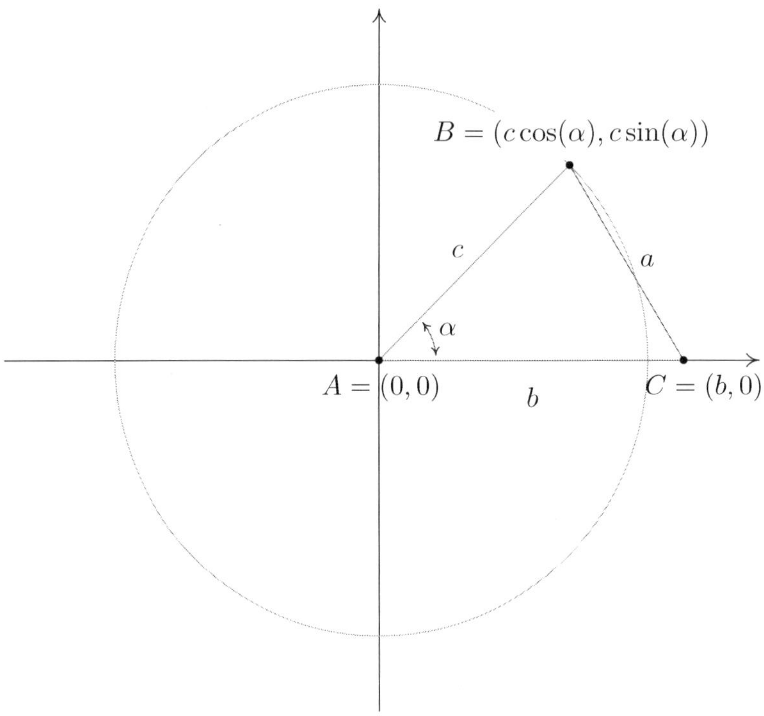

From this set-up, we immediately find that the coordinates of A and C are $A(0,0)$ and $C(b,0)$. From Theorem 10.3, we know that since the point $B(x,y)$ lies on a circle of radius c, the coordinates

[1] Here, 'Side-Angle-Side' means that we are given two sides and the 'included' angle - that is, the given angle is adjacent to both of the given sides.

of B are $B(x, y) = B(c\cos(\alpha), c\sin(\alpha))$. (This would be true even if α were an obtuse or right angle so although we have drawn the case when α is acute, the following computations hold for any angle α drawn in standard position where $0 < \alpha < 180°$.) We note that the distance between the points B and C is none other than the length of side a. Using the distance formula, Equation 1.1, we get

$$\begin{aligned}
a &= \sqrt{(c\cos(\alpha) - b)^2 + (c\sin(\alpha) - 0)^2} \\
a^2 &= \left(\sqrt{(c\cos(\alpha) - b)^2 + c^2\sin^2(\alpha)}\right)^2 \\
a^2 &= (c\cos(\alpha) - b)^2 + c^2\sin^2(\alpha) \\
a^2 &= c^2\cos^2(\alpha) - 2bc\cos(\alpha) + b^2 + c^2\sin^2(\alpha) \\
a^2 &= c^2\left(\cos^2(\alpha) + \sin^2(\alpha)\right) + b^2 - 2bc\cos(\alpha) \\
a^2 &= c^2(1) + b^2 - 2bc\cos(\alpha) \qquad \text{Since } \cos^2(\alpha) + \sin^2(\alpha) = 1 \\
a^2 &= c^2 + b^2 - 2bc\cos(\alpha)
\end{aligned}$$

The remaining formulas given in Theorem 11.5 can be shown by simply reorienting the triangle to place a different vertex at the origin. We leave these details to the reader. What's important about a and α in the above proof is that (α, a) is an angle-side opposite pair and b and c are the sides adjacent to α – the same can be said of any other angle-side opposite pair in the triangle. Notice that the proof of the Law of Cosines relies on the distance formula which has its roots in the Pythagorean Theorem. That being said, the Law of Cosines can be thought of as a generalization of the Pythagorean Theorem. If we have a triangle in which $\gamma = 90°$, then $\cos(\gamma) = \cos(90°) = 0$ so we get the familiar relationship $c^2 = a^2 + b^2$. What this means is that in the larger mathematical sense, the Law of Cosines and the Pythagorean Theorem amount to pretty much the same thing.[2]

Example 11.3.1. Solve the following triangles. Give exact answers and decimal approximations (rounded to hundredths) and sketch the triangle.

1. $\beta = 50°$, $a = 7$ units, $c = 2$ units
2. $a = 4$ units, $b = 7$ units, $c = 5$ units

Solution.

1. We are given the lengths of two sides, $a = 7$ and $c = 2$, and the measure of the included angle, $\beta = 50°$. With no angle-side opposite pair to use, we apply the Law of Cosines. We get $b^2 = 7^2 + 2^2 - 2(7)(2)\cos(50°)$ which yields $b = \sqrt{53 - 28\cos(50°)} \approx 5.92$ units. In order to determine the measures of the remaining angles α and γ, we are forced to used the derived value for b. There are two ways to proceed at this point. We could use the Law of Cosines again, or, since we have the angle-side opposite pair (β, b) we could use the Law of Sines. The advantage to using the Law of Cosines over the Law of Sines in cases like this is that unlike the sine function, the cosine function distinguishes between acute and obtuse angles. The cosine of an acute is positive, whereas the cosine of an obtuse angle is negative. Since the sine of both acute and obtuse angles are positive, the sine of an angle alone is not

[2]This shouldn't come as too much of a shock. All of the theorems in Trigonometry can ultimately be traced back to the definition of the circular functions along with the distance formula and hence, the Pythagorean Theorem.

enough to determine if the angle in question is acute or obtuse. Since both authors of the textbook prefer the Law of Cosines, we proceed with this method first. When using the Law of Cosines, it's always best to find the measure of the largest unknown angle first, since this will give us the obtuse angle of the triangle if there is one. Since the largest angle is opposite the longest side, we choose to find α first. To that end, we use the formula $\cos(\alpha) = \frac{b^2+c^2-a^2}{2bc}$ and substitute $a = 7$, $b = \sqrt{53 - 28\cos(50^\circ)}$ and $c = 2$. We get[3]

$$\cos(\alpha) = \frac{2 - 7\cos(50^\circ)}{\sqrt{53 - 28\cos(50^\circ)}}$$

Since α is an angle in a triangle, we know the radian measure of α must lie between 0 and π radians. This matches the range of the arccosine function, so we have

$$\alpha = \arccos\left(\frac{2 - 7\cos(50^\circ)}{\sqrt{53 - 28\cos(50^\circ)}}\right) \text{ radians} \approx 114.99^\circ$$

At this point, we could find γ using $\gamma = 180^\circ - \alpha - \beta \approx 180^\circ - 114.99^\circ - 50^\circ = 15.01^\circ$, that is if we trust our approximation for α. To minimize propagation of error, however, we could use the Law of Cosines again,[4] in this case using $\cos(\gamma) = \frac{a^2+b^2-c^2}{2ab}$. Plugging in $a = 7$, $b = \sqrt{53 - 28\cos(50^\circ)}$ and $c = 2$, we get $\gamma = \arccos\left(\frac{7-2\cos(50^\circ)}{\sqrt{53-28\cos(50^\circ)}}\right)$ radians $\approx 15.01^\circ$. We sketch the triangle below.

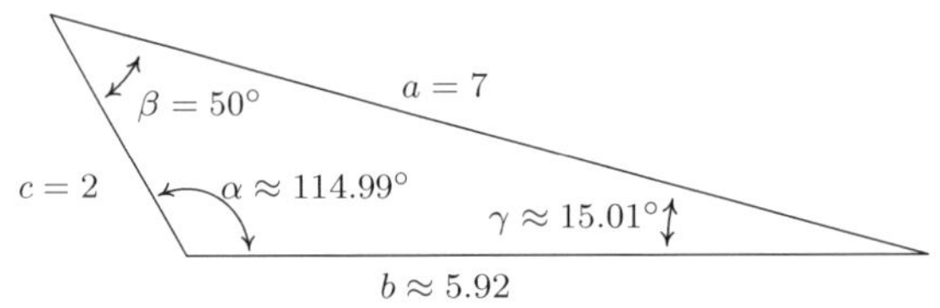

As we mentioned earlier, once we've determined b it is possible to use the Law of Sines to find the remaining angles. Here, however, we must proceed with caution as we are in the ambiguous (ASS) case. It is advisable to first find the *smallest* of the unknown angles, since we are guaranteed it will be acute.[5] In this case, we would find γ since the side opposite γ is smaller than the side opposite the other unknown angle, α. Using the angle-side opposite pair (β, b), we get $\frac{\sin(\gamma)}{2} = \frac{\sin(50^\circ)}{\sqrt{53-28\cos(50^\circ)}}$. The usual calculations produces $\gamma \approx 15.01^\circ$ and $\alpha = 180^\circ - \beta - \gamma \approx 180^\circ - 50^\circ - 15.01^\circ = 114.99^\circ$.

2. Since all three sides and no angles are given, we are forced to use the Law of Cosines. Following our discussion in the previous problem, we find β first, since it is opposite the longest side, b. We get $\cos(\beta) = \frac{a^2+c^2-b^2}{2ac} = -\frac{1}{5}$, so we get $\beta = \arccos\left(-\frac{1}{5}\right)$ radians $\approx 101.54^\circ$. As in

[3]after simplifying ...

[4]Your instructor will let you know which procedure to use. It all boils down to how much you trust your calculator.

[5]There can only be one *obtuse* angle in the triangle, and if there is one, it must be the largest.

the previous problem, now that we have obtained an angle-side opposite pair (β, b), we could proceed using the Law of Sines. The Law of Cosines, however, offers us a rare opportunity to find the remaining angles using *only* the data given to us in the statement of the problem. Using this, we get $\gamma = \arccos\left(\frac{5}{7}\right)$ radians $\approx 44.42^\circ$ and $\alpha = \arccos\left(\frac{29}{35}\right)$ radians $\approx 34.05^\circ$.

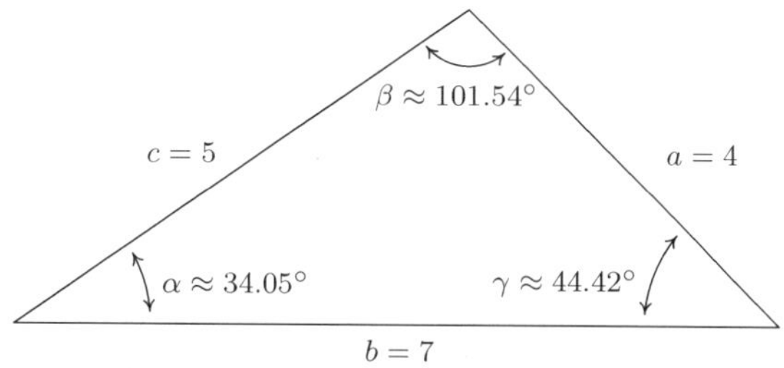

□

We note that, depending on how many decimal places are carried through successive calculations, and depending on which approach is used to solve the problem, the approximate answers you obtain may differ slightly from those the authors obtain in the Examples and the Exercises. A great example of this is number 2 in Example 11.3.1, where the *approximate* values we record for the measures of the angles sum to 180.01°, which is geometrically impossible. Next, we have an application of the Law of Cosines.

Example 11.3.2. A researcher wishes to determine the width of a vernal pond as drawn below. From a point P, he finds the distance to the eastern-most point of the pond to be 950 feet, while the distance to the western-most point of the pond from P is 1000 feet. If the angle between the two lines of sight is 60°, find the width of the pond.

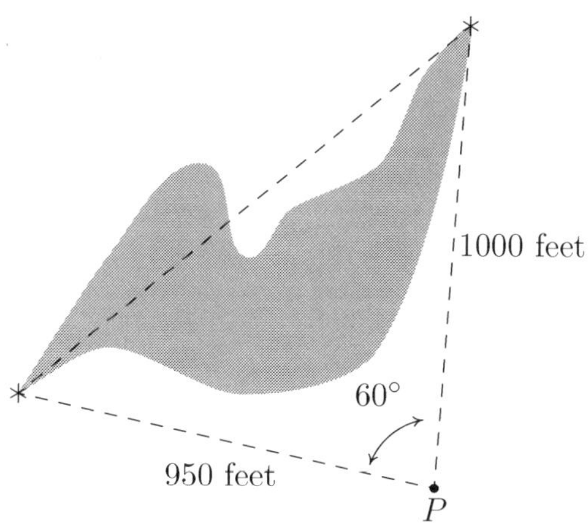

Solution. We are given the lengths of two sides and the measure of an included angle, so we may apply the Law of Cosines to find the length of the missing side opposite the given angle. Calling this length w (for *width*), we get $w^2 = 950^2 + 1000^2 - 2(950)(1000)\cos(60^\circ) = 952500$ from which we get $w = \sqrt{952500} \approx 976$ feet. □

In Section 11.2, we used the proof of the Law of Sines to develop Theorem 11.4 as an alternate formula for the area enclosed by a triangle. In this section, we use the Law of Cosines to derive another such formula - Heron's Formula.

Theorem 11.6. Heron's Formula: Suppose a, b and c denote the lengths of the three sides of a triangle. Let s be the semiperimeter of the triangle, that is, let $s = \frac{1}{2}(a+b+c)$. Then the area A enclosed by the triangle is given by

$$A = \sqrt{s(s-a)(s-b)(s-c)}$$

We prove Theorem 11.6 using Theorem 11.4. Using the convention that the angle γ is opposite the side c, we have $A = \frac{1}{2}ab\sin(\gamma)$ from Theorem 11.4. In order to simplify computations, we start by manipulating the expression for A^2.

$$\begin{aligned} A^2 &= \left(\frac{1}{2}ab\sin(\gamma)\right)^2 \\ &= \frac{1}{4}a^2b^2\sin^2(\gamma) \\ &= \frac{a^2b^2}{4}\left(1-\cos^2(\gamma)\right) \quad \text{since } \sin^2(\gamma) = 1-\cos^2(\gamma). \end{aligned}$$

The Law of Cosines tells us $\cos(\gamma) = \frac{a^2+b^2-c^2}{2ab}$, so substituting this into our equation for A^2 gives

$$\begin{aligned} A^2 &= \frac{a^2b^2}{4}\left(1-\cos^2(\gamma)\right) \\ &= \frac{a^2b^2}{4}\left[1-\left(\frac{a^2+b^2-c^2}{2ab}\right)^2\right] \\ &= \frac{a^2b^2}{4}\left[1-\frac{\left(a^2+b^2-c^2\right)^2}{4a^2b^2}\right] \\ &= \frac{a^2b^2}{4}\left[\frac{4a^2b^2-\left(a^2+b^2-c^2\right)^2}{4a^2b^2}\right] \\ &= \frac{4a^2b^2-\left(a^2+b^2-c^2\right)^2}{16} \\ &= \frac{(2ab)^2-\left(a^2+b^2-c^2\right)^2}{16} \\ &= \frac{\left(2ab-\left[a^2+b^2-c^2\right]\right)\left(2ab+\left[a^2+b^2-c^2\right]\right)}{16} \quad \text{difference of squares.} \\ &= \frac{\left(c^2-a^2+2ab-b^2\right)\left(a^2+2ab+b^2-c^2\right)}{16} \end{aligned}$$

$$
\begin{aligned}
A^2 &= \frac{\left(c^2 - \left[a^2 - 2ab + b^2\right]\right)\left(\left[a^2 + 2ab + b^2\right] - c^2\right)}{16} \\
&= \frac{\left(c^2 - (a-b)^2\right)\left((a+b)^2 - c^2\right)}{16} && \text{perfect square trinomials.} \\
&= \frac{(c-(a-b))(c+(a-b))((a+b)-c)((a+b)+c)}{16} && \text{difference of squares.} \\
&= \frac{(b+c-a)(a+c-b)(a+b-c)(a+b+c)}{16} \\
&= \frac{(b+c-a)}{2} \cdot \frac{(a+c-b)}{2} \cdot \frac{(a+b-c)}{2} \cdot \frac{(a+b+c)}{2}
\end{aligned}
$$

At this stage, we recognize the last factor as the semiperimeter, $s = \frac{1}{2}(a+b+c) = \frac{a+b+c}{2}$. To complete the proof, we note that

$$(s-a) = \frac{a+b+c}{2} - a = \frac{a+b+c-2a}{2} = \frac{b+c-a}{2}$$

Similarly, we find $(s-b) = \frac{a+c-b}{2}$ and $(s-c) = \frac{a+b-c}{2}$. Hence, we get

$$
\begin{aligned}
A^2 &= \frac{(b+c-a)}{2} \cdot \frac{(a+c-b)}{2} \cdot \frac{(a+b-c)}{2} \cdot \frac{(a+b+c)}{2} \\
&= (s-a)(s-b)(s-c)s
\end{aligned}
$$

so that $A = \sqrt{s(s-a)(s-b)(s-c)}$ as required.

We close with an example of Heron's Formula.

Example 11.3.3. Find the area enclosed of the triangle in Example 11.3.1 number 2.

Solution. We are given $a = 4$, $b = 7$ and $c = 5$. Using these values, we find $s = \frac{1}{2}(4+7+5) = 8$, $(s-a) = 8-4 = 4$, $(s-b) = 8-7 = 1$ and $(s-c) = 8-5 = 3$. Using Heron's Formula, we get $A = \sqrt{s(s-a)(s-b)(s-c)} = \sqrt{(8)(4)(1)(3)} = \sqrt{96} = 4\sqrt{6} \approx 9.80$ square units. □

11.3.1 Exercises

In Exercises 1 - 10, use the Law of Cosines to find the remaining side(s) and angle(s) if possible.

1. $a = 7,\ b = 12,\ \gamma = 59.3^\circ$
2. $\alpha = 104^\circ,\ b = 25,\ c = 37$
3. $a = 153,\ \beta = 8.2^\circ,\ c = 153$
4. $a = 3,\ b = 4,\ \gamma = 90^\circ$
5. $\alpha = 120^\circ,\ b = 3,\ c = 4$
6. $a = 7,\ b = 10,\ c = 13$
7. $a = 1,\ b = 2,\ c = 5$
8. $a = 300,\ b = 302,\ c = 48$
9. $a = 5,\ b = 5,\ c = 5$
10. $a = 5,\ b = 12, ; c = 13$

In Exercises 11 - 16, solve for the remaining side(s) and angle(s), if possible, using any appropriate technique.

11. $a = 18,\ \alpha = 63^\circ,\ b = 20$
12. $a = 37,\ b = 45,\ c = 26$
13. $a = 16,\ \alpha = 63^\circ,\ b = 20$
14. $a = 22,\ \alpha = 63^\circ,\ b = 20$
15. $\alpha = 42^\circ,\ b = 117,\ c = 88$
16. $\beta = 7^\circ,\ \gamma = 170^\circ,\ c = 98.6$
17. Find the area of the triangles given in Exercises 6, 8 and 10 above.
18. The hour hand on my antique Seth Thomas schoolhouse clock in 4 inches long and the minute hand is 5.5 inches long. Find the distance between the ends of the hands when the clock reads four o'clock. Round your answer to the nearest hundredth of an inch.
19. A geologist wants to measure the diameter of a crater. From her camp, it is 4 miles to the northern-most point of the crater and 2 miles to the southern-most point. If the angle between the two lines of sight is 117°, what is the diameter of the crater? Round your answer to the nearest hundredth of a mile.
20. From the Pedimaxus International Airport a tour helicopter can fly to Cliffs of Insanity Point by following a bearing of N8.2°E for 192 miles and it can fly to Bigfoot Falls by following a bearing of S68.5°E for 207 miles.[6] Find the distance between Cliffs of Insanity Point and Bigfoot Falls. Round your answer to the nearest mile.
21. Cliffs of Insanity Point and Bigfoot Falls from Exericse 20 above both lie on a straight stretch of the Great Sasquatch Canyon. What bearing would the tour helicopter need to follow to go directly from Bigfoot Falls to Cliffs of Insanity Point? Round your angle to the nearest tenth of a degree.

[6]Please refer to Page 905 in Section 11.2 for an introduction to bearings.

22. A naturalist sets off on a hike from a lodge on a bearing of S80°W. After 1.5 miles, she changes her bearing to S17°W and continues hiking for 3 miles. Find her distance from the lodge at this point. Round your answer to the nearest hundredth of a mile. What bearing should she follow to return to the lodge? Round your angle to the nearest degree.

23. The HMS Sasquatch leaves port on a bearing of N23°E and travels for 5 miles. It then changes course and follows a heading of S41°E for 2 miles. How far is it from port? Round your answer to the nearest hundredth of a mile. What is its bearing to port? Round your angle to the nearest degree.

24. The SS Bigfoot leaves a harbor bound for Nessie Island which is 300 miles away at a bearing of N32°E. A storm moves in and after 100 miles, the captain of the Bigfoot finds he has drifted off course. If his bearing to the harbor is now S70°W, how far is the SS Bigfoot from Nessie Island? Round your answer to the nearest hundredth of a mile. What course should the captain set to head to the island? Round your angle to the nearest tenth of a degree.

25. From a point 300 feet above level ground in a firetower, a ranger spots two fires in the Yeti National Forest. The angle of depression[7] made by the line of sight from the ranger to the first fire is 2.5° and the angle of depression made by line of sight from the ranger to the second fire is 1.3°. The angle formed by the two lines of sight is 117°. Find the distance between the two fires. Round your answer to the nearest foot. (Hint: In order to use the 117° angle between the lines of sight, you will first need to use right angle Trigonometry to find the lengths of the lines of sight. This will give you a Side-Angle-Side case in which to apply the Law of Cosines.)

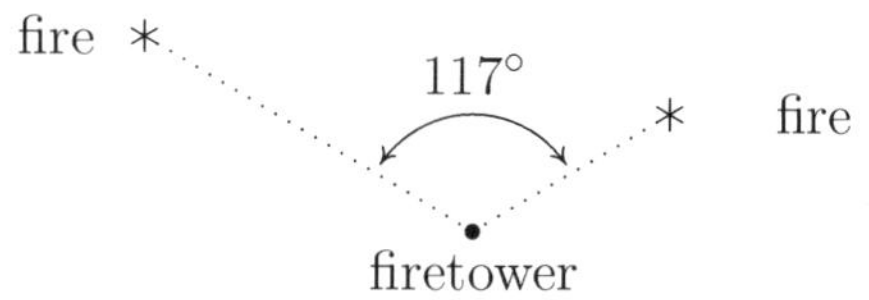

26. If you apply the Law of Cosines to the ambiguous Angle-Side-Side (ASS) case, the result is a quadratic equation whose variable is that of the missing side. If the equation has no positive real zeros then the information given does not yield a triangle. If the equation has only one positive real zero then exactly one triangle is formed and if the equation has two distinct positive real zeros then two distinct triangles are formed. Apply the Law of Cosines to Exercises 11, 13 and 14 above in order to demonstrate this result.

27. Discuss with your classmates why Heron's Formula yields an area in square units even though four lengths are being multiplied together.

[7]See Exercise 78 in Section 10.3 for the definition of this angle.

11.3.2 Answers

1. $\alpha \approx 35.54^\circ$ $\beta \approx 85.16^\circ$ $\gamma = 59.3^\circ$
   $a = 7$ $b = 12$ $c \approx 10.36$

2. $\alpha = 104^\circ$ $\beta \approx 29.40^\circ$ $\gamma \approx 46.60^\circ$
   $a \approx 49.41$ $b = 25$ $c = 37$

3. $\alpha \approx 85.90^\circ$ $\beta = 8.2^\circ$ $\gamma \approx 85.90^\circ$
   $a = 153$ $b \approx 21.88$ $c = 153$

4. $\alpha \approx 36.87^\circ$ $\beta \approx 53.13^\circ$ $\gamma = 90^\circ$
   $a = 3$ $b = 4$ $c = 5$

5. $\alpha = 120^\circ$ $\beta \approx 25.28^\circ$ $\gamma \approx 34.72^\circ$
   $a = \sqrt{37}$ $b = 3$ $c = 4$

6. $\alpha \approx 32.31^\circ$ $\beta \approx 49.58^\circ$ $\gamma \approx 98.21^\circ$
   $a = 7$ $b = 10$ $c = 13$

7. Information does not produce a triangle

8. $\alpha \approx 83.05^\circ$ $\beta \approx 87.81^\circ$ $\gamma \approx 9.14^\circ$
   $a = 300$ $b = 302$ $c = 48$

9. $\alpha = 60^\circ$ $\beta = 60^\circ$ $\gamma = 60^\circ$
   $a = 5$ $b = 5$ $c = 5$

10. $\alpha \approx 22.62^\circ$ $\beta \approx 67.38^\circ$ $\gamma = 90^\circ$
    $a = 5$ $b = 12$ $c = 13$

11. $\alpha = 63^\circ$ $\beta \approx 98.11^\circ$ $\gamma \approx 18.89^\circ$
    $a = 18$ $b = 20$ $c \approx 6.54$

    $\alpha = 63^\circ$ $\beta \approx 81.89^\circ$ $\gamma \approx 35.11^\circ$
    $a = 18$ $b = 20$ $c \approx 11.62$

12. $\alpha \approx 55.30^\circ$ $\beta \approx 89.40^\circ$ $\gamma \approx 35.30^\circ$
    $a = 37$ $b = 45$ $c = 26$

13. Information does not produce a triangle

14. $\alpha = 63^\circ$ $\beta \approx 54.1^\circ$ $\gamma \approx 62.9^\circ$
    $a = 22$ $b = 20$ $c \approx 21.98$

15. $\alpha = 42^\circ$ $\beta \approx 89.23^\circ$ $\gamma \approx 48.77^\circ$
    $a \approx 78.30$ $b = 117$ $c = 88$

16. $\alpha \approx 3^\circ$ $\beta = 7^\circ$ $\gamma = 170^\circ$
    $a \approx 29.72$ $b \approx 69.2$ $c = 98.6$

17. The area of the triangle given in Exercise 6 is $\sqrt{1200} = 20\sqrt{3} \approx 34.64$ square units.
    The area of the triangle given in Exercise 8 is $\sqrt{51764375} \approx 7194.75$ square units.
    The area of the triangle given in Exercise 10 is exactly 30 square units.

18. The distance between the ends of the hands at four o'clock is about 8.26 inches.

19. The diameter of the crater is about 5.22 miles.

20. About 313 miles

21. N31.8°W

22. She is about 3.92 miles from the lodge and her bearing to the lodge is N37°E.

23. It is about 4.50 miles from port and its heading to port is S47°W.

24. It is about 229.61 miles from the island and the captain should set a course of N16.4°E to reach the island.

25. The fires are about 17456 feet apart. (Try to avoid rounding errors.)

11.4 POLAR COORDINATES

In Section 1.1, we introduced the Cartesian coordinates of a point in the plane as a means of assigning ordered pairs of numbers to points in the plane. We defined the Cartesian coordinate plane using two number lines – one horizontal and one vertical – which intersect at right angles at a point we called the 'origin'. To plot a point, say $P(-3,4)$, we start at the origin, travel horizontally to the left 3 units, then up 4 units. Alternatively, we could start at the origin, travel up 4 units, then to the left 3 units and arrive at the same location. For the most part, the 'motions' of the Cartesian system (over and up) describe a rectangle, and most points can be thought of as the corner diagonally across the rectangle from the origin.[1] For this reason, the Cartesian coordinates of a point are often called 'rectangular' coordinates. In this section, we introduce a new system for assigning coordinates to points in the plane – **polar coordinates**. We start with an origin point, called the **pole**, and a ray called the **polar axis**. We then locate a point P using two coordinates, (r,θ), where r represents a *directed* distance from the pole[2] and θ is a measure of rotation from the polar axis. Roughly speaking, the polar coordinates (r,θ) of a point measure 'how far out' the point is from the pole (that's r), and 'how far to rotate' from the polar axis, (that's θ).

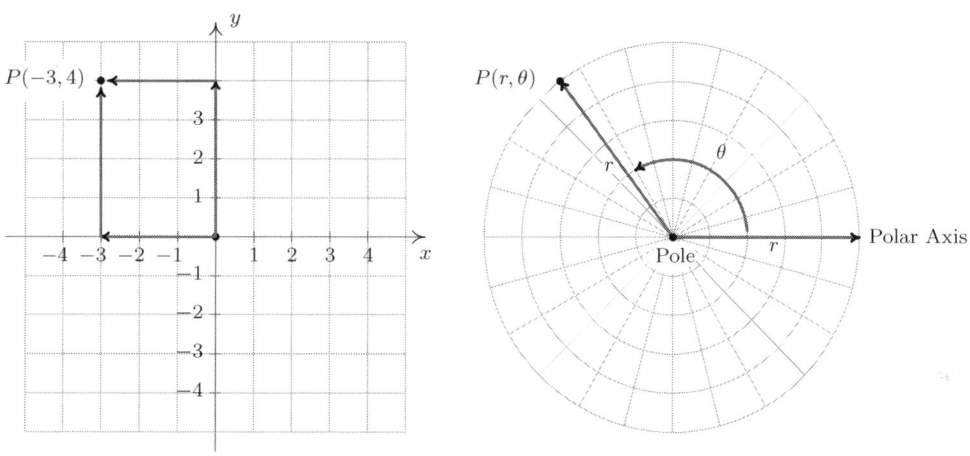

For example, if we wished to plot the point P with polar coordinates $\left(4, \frac{5\pi}{6}\right)$, we'd start at the pole, move out along the polar axis 4 units, then rotate $\frac{5\pi}{6}$ radians counter-clockwise.

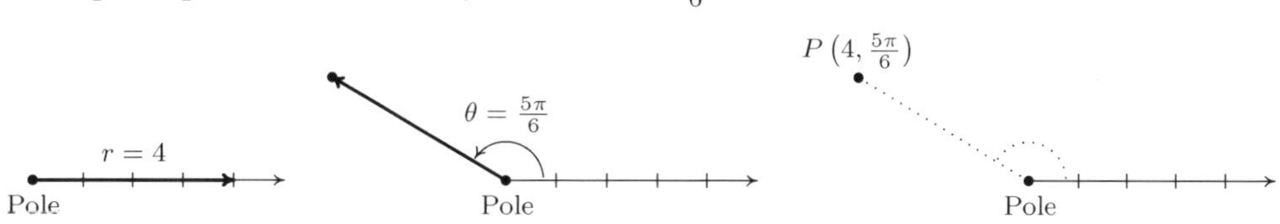

We may also visualize this process by thinking of the rotation first.[3] To plot $P\left(4, \frac{5\pi}{6}\right)$ this way, we rotate $\frac{5\pi}{6}$ counter-clockwise from the polar axis, then move outwards from the pole 4 units.

[1] Excluding, of course, the points in which one or both coordinates are 0.

[2] We will explain more about this momentarily.

[3] As with anything in Mathematics, the more ways you have to look at something, the better. The authors encourage the reader to take time to think about both approaches to plotting points given in polar coordinates.

Essentially we are locating a point on the terminal side of $\frac{5\pi}{6}$ which is 4 units away from the pole.

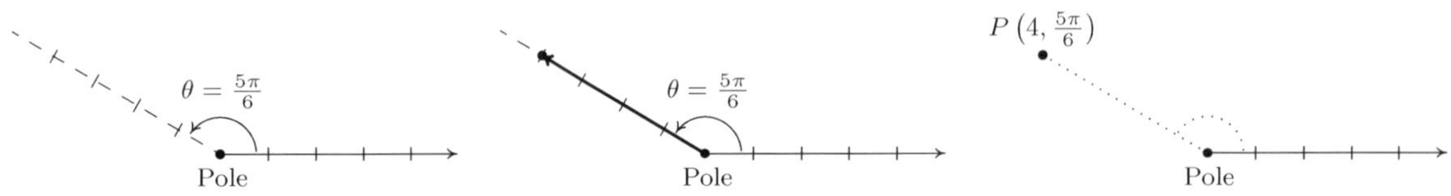

If $r < 0$, we begin by moving in the opposite direction on the polar axis from the pole. For example, to plot $Q\left(-3.5, \frac{\pi}{4}\right)$ we have

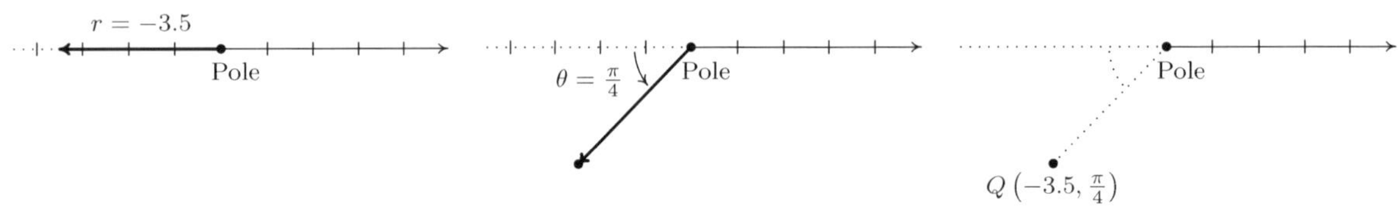

If we interpret the angle first, we rotate $\frac{\pi}{4}$ radians, then move back through the pole 3.5 units. Here we are locating a point 3.5 units away from the pole on the terminal side of $\frac{5\pi}{4}$, not $\frac{\pi}{4}$.

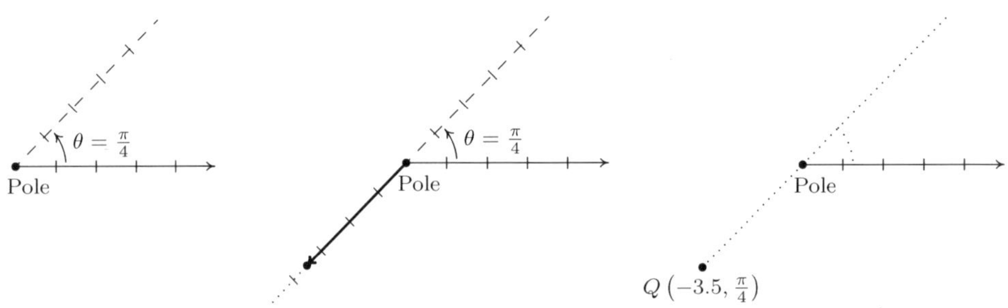

As you may have guessed, $\theta < 0$ means the rotation away from the polar axis is clockwise instead of counter-clockwise. Hence, to plot $R\left(3.5, -\frac{3\pi}{4}\right)$ we have the following.

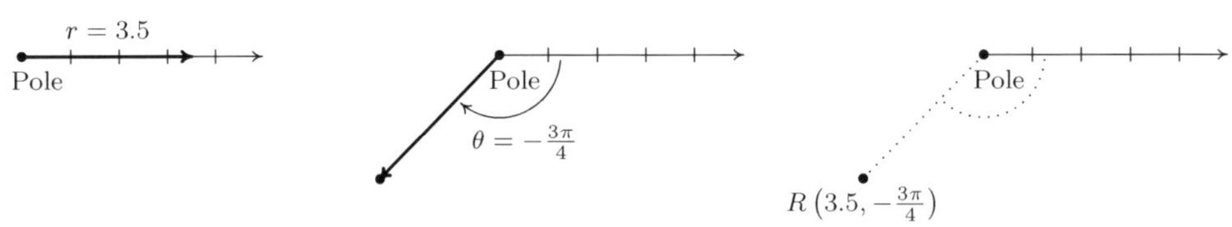

From an 'angles first' approach, we rotate $-\frac{3\pi}{4}$ then move out 3.5 units from the pole. We see that R is the point on the terminal side of $\theta = -\frac{3\pi}{4}$ which is 3.5 units from the pole.

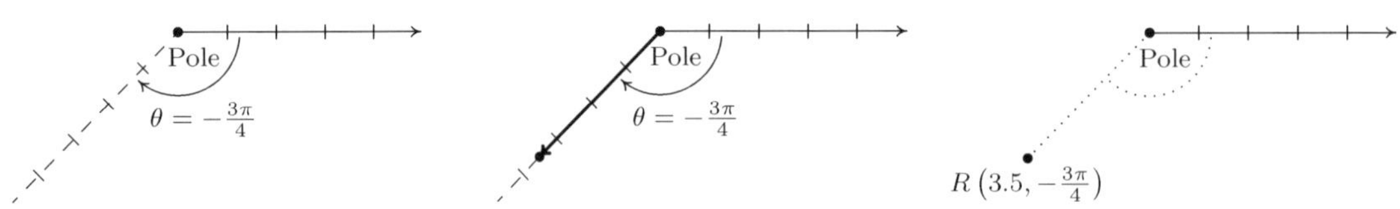

The points Q and R above are, in fact, the same point despite the fact that their polar coordinate representations are different. Unlike Cartesian coordinates where (a,b) and (c,d) represent the same point if and only if $a=c$ and $b=d$, a point can be represented by infinitely many polar coordinate pairs. We explore this notion more in the following example.

Example 11.4.1. For each point in polar coordinates given below plot the point and then give two additional expressions for the point, one of which has $r>0$ and the other with $r<0$.

1. $P(2,240^\circ)$
2. $P\left(-4,\frac{7\pi}{6}\right)$
3. $P\left(117,-\frac{5\pi}{2}\right)$
4. $P\left(-3,-\frac{\pi}{4}\right)$

Solution.

1. Whether we move 2 units along the polar axis and then rotate 240° or rotate 240° then move out 2 units from the pole, we plot $P(2,240^\circ)$ below.

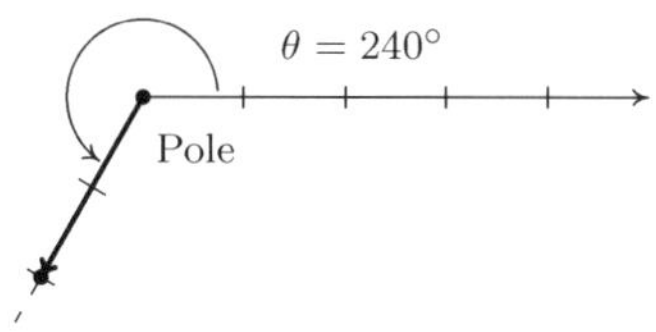

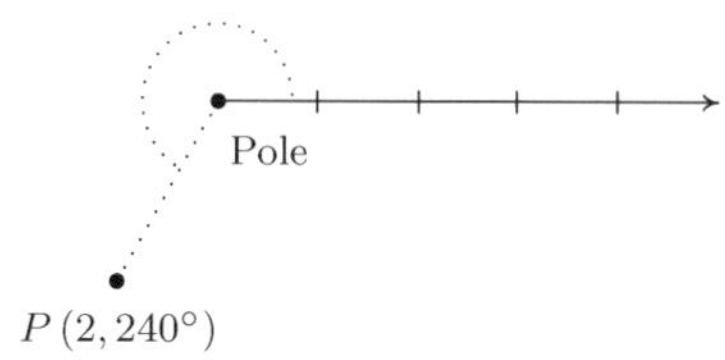

We now set about finding alternate descriptions (r,θ) for the point P. Since P is 2 units from the pole, $r=\pm 2$. Next, we choose angles θ for each of the r values. The given representation for P is $(2,240^\circ)$ so the angle θ we choose for the $r=2$ case must be coterminal with 240°. (Can you see why?) One such angle is $\theta=-120^\circ$ so one answer for this case is $(2,-120^\circ)$. For the case $r=-2$, we visualize our rotation starting 2 units to the left of the pole. From this position, we need only to rotate $\theta=60^\circ$ to arrive at location coterminal with 240°. Hence, our answer here is $(-2,60^\circ)$. We check our answers by plotting them.

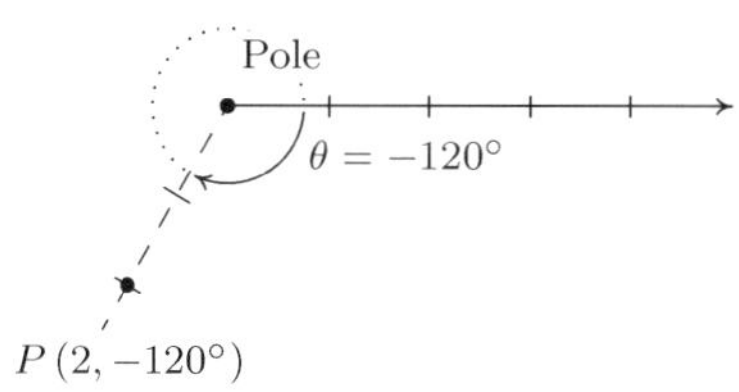

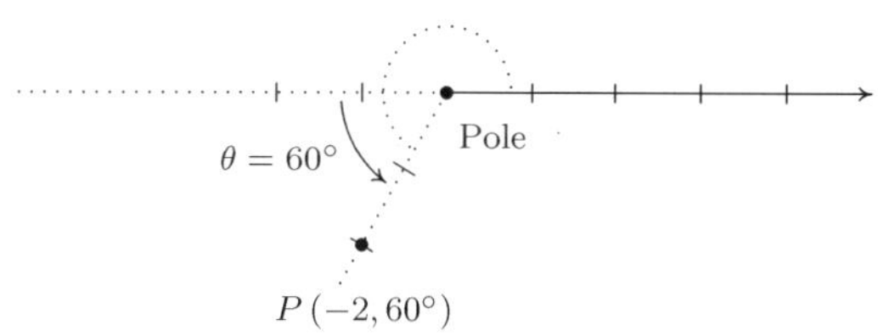

2. We plot $\left(-4,\frac{7\pi}{6}\right)$ by first moving 4 units to the left of the pole and then rotating $\frac{7\pi}{6}$ radians. Since $r=-4<0$, we find our point lies 4 units from the pole on the terminal side of $\frac{\pi}{6}$.

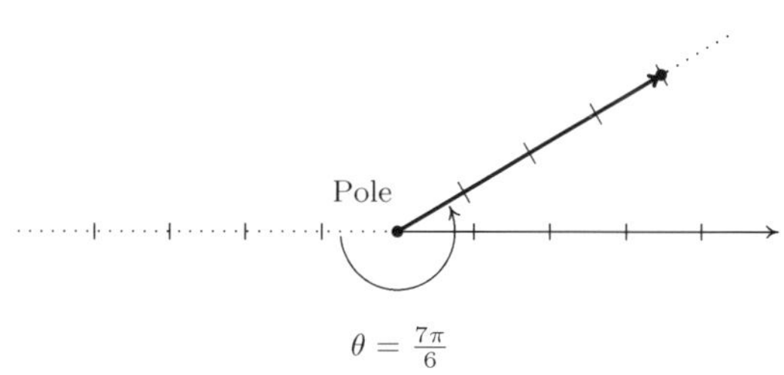

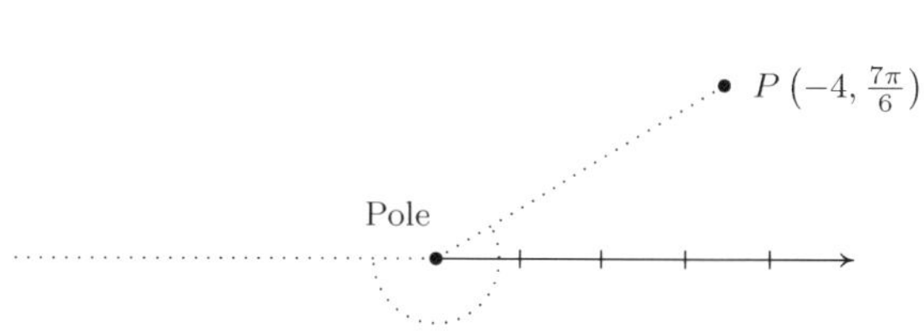

To find alternate descriptions for P, we note that the distance from P to the pole is 4 units, so any representation (r, θ) for P must have $r = \pm 4$. As we noted above, P lies on the terminal side of $\frac{\pi}{6}$, so this, coupled with $r = 4$, gives us $\left(4, \frac{\pi}{6}\right)$ as one of our answers. To find a different representation for P with $r = -4$, we may choose any angle coterminal with the angle in the original representation of P $\left(-4, \frac{7\pi}{6}\right)$. We pick $-\frac{5\pi}{6}$ and get $\left(-4, -\frac{5\pi}{6}\right)$ as our second answer.

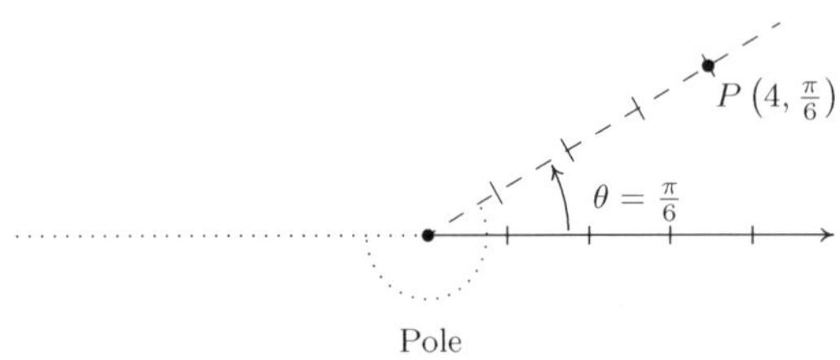

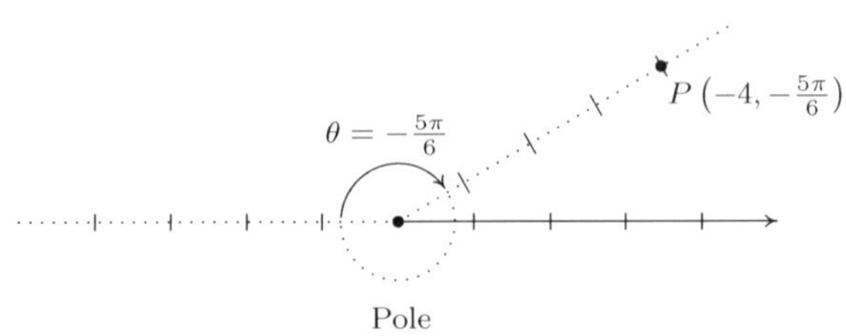

3. To plot $P\left(117, -\frac{5\pi}{2}\right)$, we move along the polar axis 117 units from the pole and rotate *clockwise* $\frac{5\pi}{2}$ radians as illustrated below.

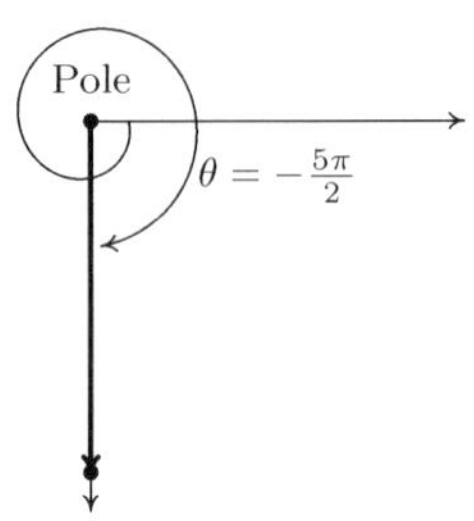

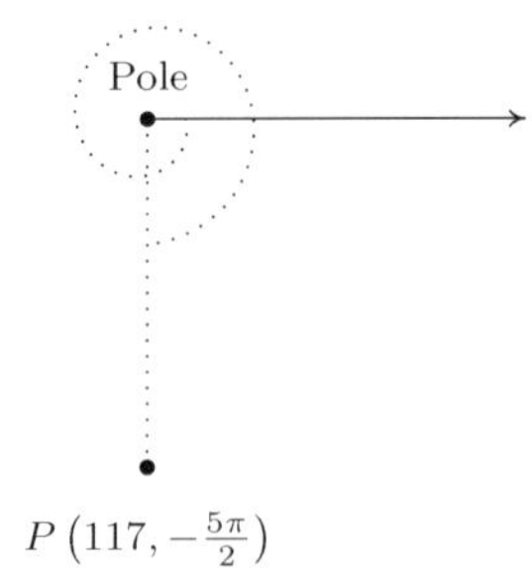

Since P is 117 units from the pole, any representation (r, θ) for P satisfies $r = \pm 117$. For the $r = 117$ case, we can take θ to be any angle coterminal with $-\frac{5\pi}{2}$. In this case, we choose $\theta = \frac{3\pi}{2}$, and get $\left(117, \frac{3\pi}{2}\right)$ as one answer. For the $r = -117$ case, we visualize moving left 117 units from the pole and then rotating through an angle θ to reach P. We find that $\theta = \frac{\pi}{2}$ satisfies this requirement, so our second answer is $\left(-117, \frac{\pi}{2}\right)$.

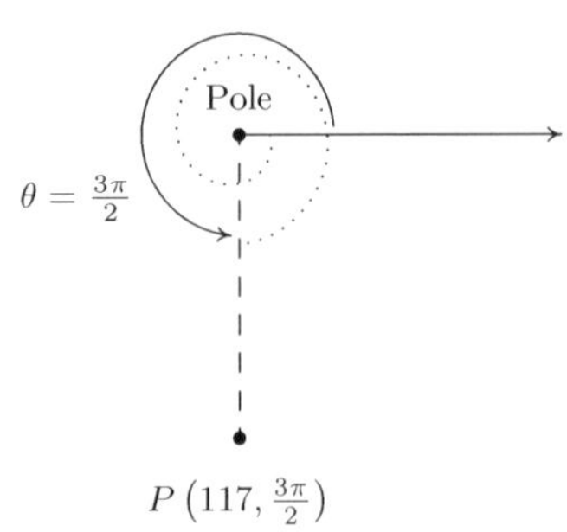

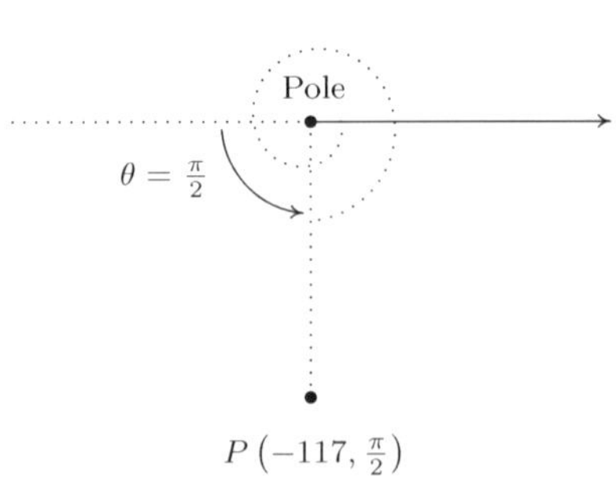

4. We move three units to the left of the pole and follow up with a clockwise rotation of $\frac{\pi}{4}$ radians to plot $P\left(-3, -\frac{\pi}{4}\right)$. We see that P lies on the terminal side of $\frac{3\pi}{4}$.

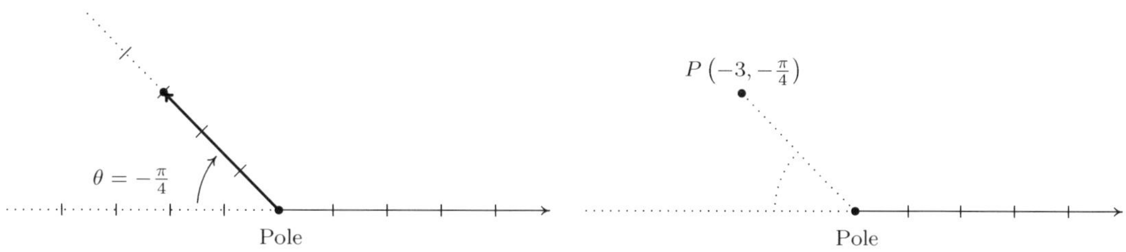

Since P lies on the terminal side of $\frac{3\pi}{4}$, one alternative representation for P is $\left(3, \frac{3\pi}{4}\right)$. To find a different representation for P with $r = -3$, we may choose any angle coterminal with $-\frac{\pi}{4}$. We choose $\theta = \frac{7\pi}{4}$ for our final answer $\left(-3, \frac{7\pi}{4}\right)$.

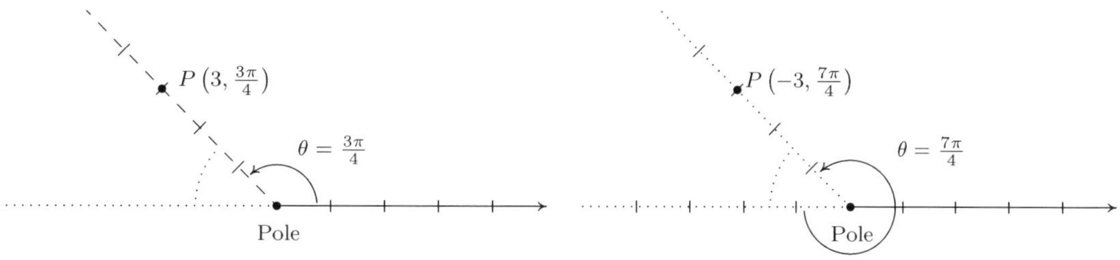

□

Now that we have had some practice with plotting points in polar coordinates, it should come as no surprise that any given point expressed in polar coordinates has infinitely many other representations in polar coordinates. The following result characterizes when two sets of polar coordinates determine the same point in the plane. It could be considered as a definition or a theorem, depending on your point of view. We state it as a property of the polar coordinate system.

Equivalent Representations of Points in Polar Coordinates

Suppose (r, θ) and (r', θ') are polar coordinates where $r \neq 0$, $r' \neq 0$ and the angles are measured in radians. Then (r, θ) and (r', θ') determine the same point P if and only if one of the following is true:

- $r' = r$ and $\theta' = \theta + 2\pi k$ for some integer k
- $r' = -r$ and $\theta' = \theta + (2k+1)\pi$ for some integer k

All polar coordinates of the form $(0, \theta)$ represent the pole regardless of the value of θ.

The key to understanding this result, and indeed the whole polar coordinate system, is to keep in mind that (r, θ) means (directed distance from pole, angle of rotation). If $r = 0$, then no matter how much rotation is performed, the point never leaves the pole. Thus $(0, \theta)$ is the pole for all values of θ. Now let's assume that neither r nor r' is zero. If (r, θ) and (r', θ') determine the same point P then the (non-zero) distance from P to the pole in each case must be the same. Since this distance is controlled by the first coordinate, we have that either $r' = r$ or $r' = -r$. If $r' = r$, then when plotting (r, θ) and (r', θ'), the angles θ and θ' have the same initial side. Hence, if (r, θ) and (r', θ') determine the same point, we must have that θ' is coterminal with θ. We know that this means $\theta' = \theta + 2\pi k$ for some integer k, as required. If, on the other hand, $r' = -r$, then when plotting (r, θ) and (r', θ'), the initial side of θ' is rotated π radians away from the initial side of θ. In this case, θ' must be coterminal with $\pi + \theta$. Hence, $\theta' = \pi + \theta + 2\pi k$ which we rewrite as $\theta' = \theta + (2k+1)\pi$ for some integer k. Conversely, if $r' = r$ and $\theta' = \theta + 2\pi k$ for some integer k, then the points $P(r, \theta)$ and $P'(r', \theta')$ lie the same (directed) distance from the pole on the terminal sides of coterminal angles, and hence are the same point. Now suppose $r' = -r$ and $\theta' = \theta + (2k+1)\pi$ for some integer k. To plot P, we first move a directed distance r from the pole; to plot P', our first step is to move the same distance from the pole as P, but in the opposite direction. At this intermediate stage, we have two points equidistant from the pole rotated exactly π radians apart. Since $\theta' = \theta + (2k+1)\pi = (\theta + \pi) + 2\pi k$ for some integer k, we see that θ' is coterminal to $(\theta + \pi)$ and it is this extra π radians of rotation which aligns the points P and P'.

Next, we marry the polar coordinate system with the Cartesian (rectangular) coordinate system. To do so, we identify the pole and polar axis in the polar system to the origin and positive x-axis, respectively, in the rectangular system. We get the following result.

Theorem 11.7. Conversion Between Rectangular and Polar Coordinates: Suppose P is represented in rectangular coordinates as (x, y) and in polar coordinates as (r, θ). Then

- $x = r\cos(\theta)$ and $y = r\sin(\theta)$
- $x^2 + y^2 = r^2$ and $\tan(\theta) = \dfrac{y}{x}$ (provided $x \neq 0$)

In the case $r > 0$, Theorem 11.7 is an immediate consequence of Theorem 10.3 along with the quotient identity $\tan(\theta) = \frac{\sin(\theta)}{\cos(\theta)}$. If $r < 0$, then we know an alternate representation for (r, θ) is $(-r, \theta + \pi)$. Since $\cos(\theta + \pi) = -\cos(\theta)$ and $\sin(\theta + \pi) = -\sin(\theta)$, applying the theorem to $(-r, \theta + \pi)$ gives $x = (-r)\cos(\theta + \pi) = (-r)(-\cos(\theta)) = r\cos(\theta)$ and $y = (-r)\sin(\theta + \pi) = (-r)(-\sin(\theta)) = r\sin(\theta)$. Moreover, $x^2 + y^2 = (-r)^2 = r^2$, and $\frac{y}{x} = \tan(\theta + \pi) = \tan(\theta)$, so the theorem is true in this case, too. The remaining case is $r = 0$, in which case $(r, \theta) = (0, \theta)$ is the pole. Since the pole is identified with the origin $(0, 0)$ in rectangular coordinates, the theorem in this case amounts to checking '$0 = 0$.' The following example puts Theorem 11.7 to good use.

Example 11.4.2. Convert each point in rectangular coordinates given below into polar coordinates with $r \geq 0$ and $0 \leq \theta < 2\pi$. Use exact values if possible and round any approximate values to two decimal places. Check your answer by converting them back to rectangular coordinates.

1. $P\left(2, -2\sqrt{3}\right)$
2. $Q(-3, -3)$
3. $R(0, -3)$
4. $S(-3, 4)$

Solution.

1. Even though we are not explicitly told to do so, we can avoid many common mistakes by taking the time to plot the points before we do any calculations. Plotting $P\left(2, -2\sqrt{3}\right)$ shows that it lies in Quadrant IV. With $x = 2$ and $y = -2\sqrt{3}$, we get $r^2 = x^2 + y^2 = (2)^2 + \left(-2\sqrt{3}\right)^2 = 4 + 12 = 16$ so $r = \pm 4$. Since we are asked for $r \geq 0$, we choose $r = 4$. To find θ, we have that $\tan(\theta) = \frac{y}{x} = \frac{-2\sqrt{3}}{2} = -\sqrt{3}$. This tells us θ has a reference angle of $\frac{\pi}{3}$, and since P lies in Quadrant IV, we know θ is a Quadrant IV angle. We are asked to have $0 \leq \theta < 2\pi$, so we choose $\theta = \frac{5\pi}{3}$. Hence, our answer is $\left(4, \frac{5\pi}{3}\right)$. To check, we convert $(r, \theta) = \left(4, \frac{5\pi}{3}\right)$ back to rectangular coordinates and we find $x = r\cos(\theta) = 4\cos\left(\frac{5\pi}{3}\right) = 4\left(\frac{1}{2}\right) = 2$ and $y = r\sin(\theta) = 4\sin\left(\frac{5\pi}{3}\right) = 4\left(-\frac{\sqrt{3}}{2}\right) = -2\sqrt{3}$, as required.

2. The point $Q(-3, -3)$ lies in Quadrant III. Using $x = y = -3$, we get $r^2 = (-3)^2 + (-3)^2 = 18$ so $r = \pm\sqrt{18} = \pm 3\sqrt{2}$. Since we are asked for $r \geq 0$, we choose $r = 3\sqrt{2}$. We find $\tan(\theta) = \frac{-3}{-3} = 1$, which means θ has a reference angle of $\frac{\pi}{4}$. Since Q lies in Quadrant III, we choose $\theta = \frac{5\pi}{4}$, which satisfies the requirement that $0 \leq \theta < 2\pi$. Our final answer is $(r, \theta) = \left(3\sqrt{2}, \frac{5\pi}{4}\right)$. To check, we find $x = r\cos(\theta) = (3\sqrt{2})\cos\left(\frac{5\pi}{4}\right) = (3\sqrt{2})\left(-\frac{\sqrt{2}}{2}\right) = -3$ and $y = r\sin(\theta) = (3\sqrt{2})\sin\left(\frac{5\pi}{4}\right) = (3\sqrt{2})\left(-\frac{\sqrt{2}}{2}\right) = -3$, so we are done.

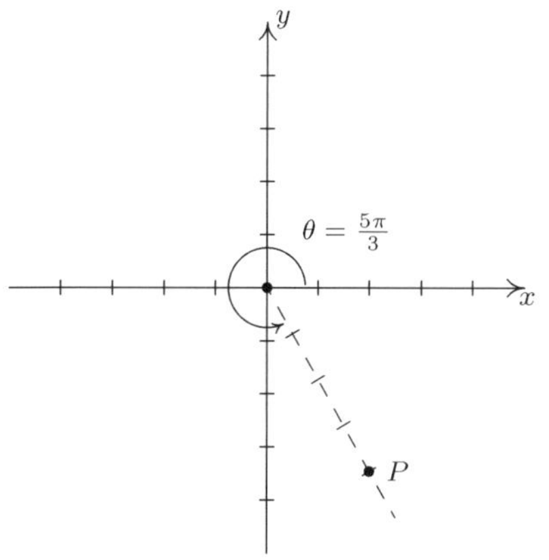

P has rectangular coordinates $(2, -2\sqrt{3})$
P has polar coordinates $\left(4, \frac{5\pi}{3}\right)$

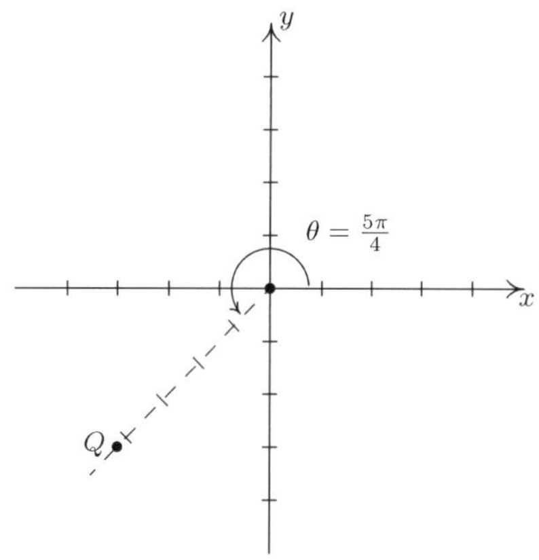

Q has rectangular coordinates $(-3, -3)$
Q has polar coordinates $\left(3\sqrt{2}, \frac{5\pi}{4}\right)$

3. The point $R(0, -3)$ lies along the negative y-axis. While we could go through the usual computations[4] to find the polar form of R, in this case we can find the polar coordinates of R using the definition. Since the pole is identified with the origin, we can easily tell the point R is 3 units from the pole, which means in the polar representation (r, θ) of R we know $r = \pm 3$. Since we require $r \geq 0$, we choose $r = 3$. Concerning θ, the angle $\theta = \frac{3\pi}{2}$ satisfies $0 \leq \theta < 2\pi$

[4]Since $x = 0$, we would have to determine θ geometrically.

with its terminal side along the negative y-axis, so our answer is $\left(3, \frac{3\pi}{2}\right)$. To check, we note $x = r\cos(\theta) = 3\cos\left(\frac{3\pi}{2}\right) = (3)(0) = 0$ and $y = r\sin(\theta) = 3\sin\left(\frac{3\pi}{2}\right) = 3(-1) = -3$.

4. The point $S(-3,4)$ lies in Quadrant II. With $x = -3$ and $y = 4$, we get $r^2 = (-3)^2 + (4)^2 = 25$ so $r = \pm 5$. As usual, we choose $r = 5 \geq 0$ and proceed to determine θ. We have $\tan(\theta) = \frac{y}{x} = \frac{4}{-3} = -\frac{4}{3}$, and since this isn't the tangent of one the common angles, we resort to using the arctangent function. Since θ lies in Quadrant II and must satisfy $0 \leq \theta < 2\pi$, we choose $\theta = \pi - \arctan\left(\frac{4}{3}\right)$ radians. Hence, our answer is $(r,\theta) = \left(5, \pi - \arctan\left(\frac{4}{3}\right)\right) \approx (5, 2.21)$. To check our answers requires a bit of tenacity since we need to simplify expressions of the form: $\cos\left(\pi - \arctan\left(\frac{4}{3}\right)\right)$ and $\sin\left(\pi - \arctan\left(\frac{4}{3}\right)\right)$. These are good review exercises and are hence left to the reader. We find $\cos\left(\pi - \arctan\left(\frac{4}{3}\right)\right) = -\frac{3}{5}$ and $\sin\left(\pi - \arctan\left(\frac{4}{3}\right)\right) = \frac{4}{5}$, so that $x = r\cos(\theta) = (5)\left(-\frac{3}{5}\right) = -3$ and $y = r\sin(\theta) = (5)\left(\frac{4}{5}\right) = 4$ which confirms our answer.

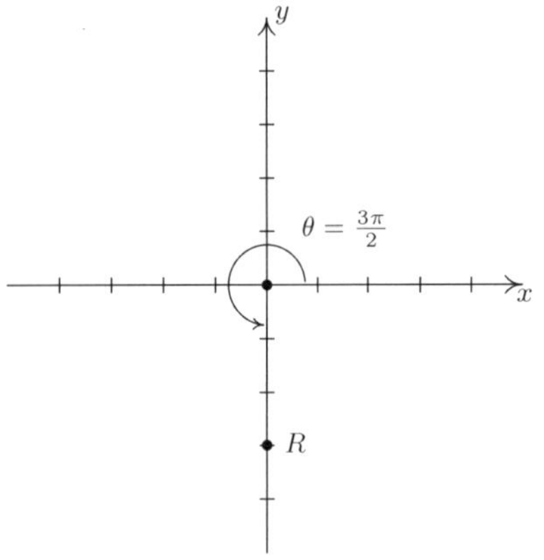

R has rectangular coordinates $(0,-3)$
R has polar coordinates $\left(3, \frac{3\pi}{2}\right)$

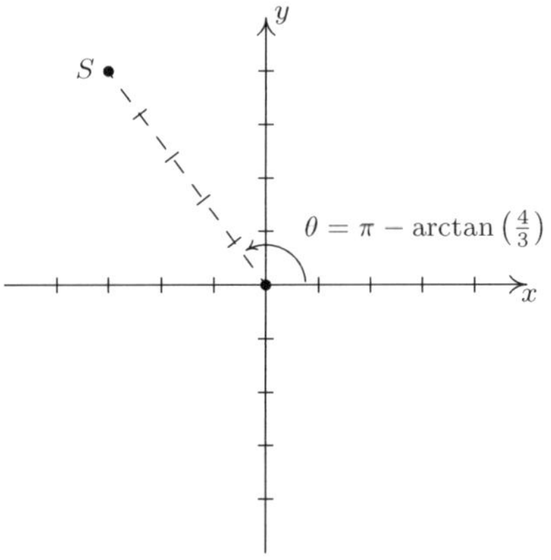

S has rectangular coordinates $(-3,4)$
S has polar coordinates $\left(5, \pi - \arctan\left(\frac{4}{3}\right)\right)$

□

Now that we've had practice converting representations of *points* between the rectangular and polar coordinate systems, we now set about converting *equations* from one system to another. Just as we've used equations in x and y to represent relations in rectangular coordinates, equations in the variables r and θ represent relations in polar coordinates. We convert equations between the two systems using Theorem 11.7 as the next example illustrates.

Example 11.4.3.

1. Convert each equation in rectangular coordinates into an equation in polar coordinates.

   (a) $(x-3)^2 + y^2 = 9$ (b) $y = -x$ (c) $y = x^2$

2. Convert each equation in polar coordinates into an equation in rectangular coordinates.

   (a) $r = -3$ (b) $\theta = \frac{4\pi}{3}$ (c) $r = 1 - \cos(\theta)$

Solution.

1. One strategy to convert an equation from rectangular to polar coordinates is to replace every occurrence of x with $r\cos(\theta)$ and every occurrence of y with $r\sin(\theta)$ and use identities to simplify. This is the technique we employ below.

   (a) We start by substituting $x = r\cos(\theta)$ and $y = \sin(\theta)$ into $(x-3)^2+y^2 = 9$ and simplifying. With no real direction in which to proceed, we follow our mathematical instincts and see where they take us.[5]

$$\begin{array}{rcll} (r\cos(\theta)-3)^2 + (r\sin(\theta))^2 & = & 9 & \\ r^2\cos^2(\theta) - 6r\cos(\theta) + 9 + r^2\sin^2(\theta) & = & 9 & \\ r^2\left(\cos^2(\theta)+\sin^2(\theta)\right) - 6r\cos(\theta) & = & 0 & \text{Subtract 9 from both sides.} \\ r^2 - 6r\cos(\theta) & = & 0 & \text{Since } \cos^2(\theta)+\sin^2(\theta) = 1 \\ r(r - 6\cos(\theta)) & = & 0 & \text{Factor.} \end{array}$$

   We get $r = 0$ or $r = 6\cos(\theta)$. From Section 7.2 we know the equation $(x-3)^2+y^2 = 9$ describes a circle, and since $r = 0$ describes just a point (namely the pole/origin), we choose $r = 6\cos(\theta)$ for our final answer.[6]

   (b) Substituting $x = r\cos(\theta)$ and $y = r\sin(\theta)$ into $y = -x$ gives $r\sin(\theta) = -r\cos(\theta)$. Rearranging, we get $r\cos(\theta)+r\sin(\theta) = 0$ or $r(\cos(\theta)+\sin(\theta)) = 0$. This gives $r = 0$ or $\cos(\theta)+\sin(\theta) = 0$. Solving the latter equation for θ, we get $\theta = -\frac{\pi}{4} + \pi k$ for integers k. As we did in the previous example, we take a step back and think geometrically. We know $y = -x$ describes a line through the origin. As before, $r = 0$ describes the origin, but nothing else. Consider the equation $\theta = -\frac{\pi}{4}$. In this equation, the variable r is free,[7] meaning it can assume any and all values including $r = 0$. If we imagine plotting points $(r, -\frac{\pi}{4})$ for all conceivable values of r (positive, negative and zero), we are essentially drawing the line containing the terminal side of $\theta = -\frac{\pi}{4}$ which is none other than $y = -x$. Hence, we can take as our final answer $\theta = -\frac{\pi}{4}$ here.[8]

   (c) We substitute $x = r\cos(\theta)$ and $y = r\sin(\theta)$ into $y = x^2$ and get $r\sin(\theta) = (r\cos(\theta))^2$, or $r^2\cos^2(\theta) - r\sin(\theta) = 0$. Factoring, we get $r(r\cos^2(\theta) - \sin(\theta)) = 0$ so that either $r = 0$ or $r\cos^2(\theta) = \sin(\theta)$. We can solve the latter equation for r by dividing both sides of the equation by $\cos^2(\theta)$, but as a general rule, we never divide through by a quantity that may be 0. In this particular case, we are safe since if $\cos^2(\theta) = 0$, then $\cos(\theta) = 0$, and for the equation $r\cos^2(\theta) = \sin(\theta)$ to hold, then $\sin(\theta)$ would also have to be 0. Since there are no angles with both $\cos(\theta) = 0$ and $\sin(\theta) = 0$, we are not losing any

[5]Experience is the mother of all instinct, and necessity is the mother of invention. Study this example and see what techniques are employed, then try your best to get your answers in the homework to match Jeff's.

[6]Note that when we substitute $\theta = \frac{\pi}{2}$ into $r = 6\cos(\theta)$, we recover the point $r = 0$, so we aren't losing anything by disregarding $r = 0$.

[7]See Section 8.1.

[8]We could take it to be *any* of $\theta = -\frac{\pi}{4} + \pi k$ for integers k.

information by dividing both sides of $r\cos^2(\theta) = \sin(\theta)$ by $\cos^2(\theta)$. Doing so, we get $r = \frac{\sin(\theta)}{\cos^2(\theta)}$, or $r = \sec(\theta)\tan(\theta)$. As before, the $r = 0$ case is recovered in the solution $r = \sec(\theta)\tan(\theta)$ (let $\theta = 0$), so we state the latter as our final answer.

2. As a general rule, converting equations from polar to rectangular coordinates isn't as straight forward as the reverse process. We could solve $r^2 = x^2 + y^2$ for r to get $r = \pm\sqrt{x^2+y^2}$ and solving $\tan(\theta) = \frac{y}{x}$ requires the arctangent function to get $\theta = \arctan\left(\frac{y}{x}\right) + \pi k$ for integers k. Neither of these expressions for r and θ are especially user-friendly, so we opt for a second strategy – rearrange the given polar equation so that the expressions $r^2 = x^2 + y^2$, $r\cos(\theta) = x$, $r\sin(\theta) = y$ and/or $\tan(\theta) = \frac{y}{x}$ present themselves.

   (a) Starting with $r = -3$, we can square both sides to get $r^2 = (-3)^2$ or $r^2 = 9$. We may now substitute $r^2 = x^2 + y^2$ to get the equation $x^2 + y^2 = 9$. As we have seen,[9] squaring an equation does not, in general, produce an equivalent equation. The concern here is that the equation $r^2 = 9$ might be satisfied by more points than $r = -3$. On the surface, this appears to be the case since $r^2 = 9$ is equivalent to $r = \pm 3$, not just $r = -3$. However, any point with polar coordinates $(3, \theta)$ can be represented as $(-3, \theta + \pi)$, which means any point (r, θ) whose polar coordinates satisfy the relation $r = \pm 3$ has an equivalent[10] representation which satisfies $r = -3$.

   (b) We take the tangent of both sides the equation $\theta = \frac{4\pi}{3}$ to get $\tan(\theta) = \tan\left(\frac{4\pi}{3}\right) = \sqrt{3}$. Since $\tan(\theta) = \frac{y}{x}$, we get $\frac{y}{x} = \sqrt{3}$ or $y = x\sqrt{3}$. Of course, we pause a moment to wonder if, geometrically, the equations $\theta = \frac{4\pi}{3}$ and $y = x\sqrt{3}$ generate the same set of points.[11] The same argument presented in number 1b applies equally well here so we are done.

   (c) Once again, we need to manipulate $r = 1 - \cos(\theta)$ a bit before using the conversion formulas given in Theorem 11.7. We could square both sides of this equation like we did in part 2a above to obtain an r^2 on the left hand side, but that does nothing helpful for the right hand side. Instead, we multiply both sides by r to obtain $r^2 = r - r\cos(\theta)$. We now have an r^2 and an $r\cos(\theta)$ in the equation, which we can easily handle, but we also have another r to deal with. Rewriting the equation as $r = r^2 + r\cos(\theta)$ and squaring both sides yields $r^2 = \left(r^2 + r\cos(\theta)\right)^2$. Substituting $r^2 = x^2 + y^2$ and $r\cos(\theta) = x$ gives $x^2 + y^2 = \left(x^2 + y^2 + x\right)^2$. Once again, we have performed some

[9] Exercise 5.3.1 in Section 5.3, for instance ...

[10] Here, 'equivalent' means they represent the same point in the plane. As ordered pairs, $(3, 0)$ and $(-3, \pi)$ are different, but when interpreted as polar coordinates, they correspond to the same point in the plane. Mathematically speaking, relations are sets of ordered pairs, so the equations $r^2 = 9$ and $r = -3$ represent different relations since they correspond to different sets of ordered pairs. Since polar coordinates were defined geometrically to describe the location of points in the plane, however, we concern ourselves only with ensuring that the sets of *points* in the plane generated by two equations are the same. This was not an issue, by the way, when we first defined relations as sets of points in the plane in Section 1.2. Back then, a point in the plane was identified with a unique ordered pair given by its Cartesian coordinates.

[11] In addition to taking the tangent of both sides of an equation (There are infinitely many solutions to $\tan(\theta) = \sqrt{3}$, and $\theta = \frac{4\pi}{3}$ is only one of them!), we also went from $\frac{y}{x} = \sqrt{3}$, in which x cannot be 0, to $y = x\sqrt{3}$ in which we assume x can be 0.

algebraic maneuvers which may have altered the set of points described by the original equation. First, we multiplied both sides by r. This means that now $r = 0$ is a viable solution to the equation. In the original equation, $r = 1 - \cos(\theta)$, we see that $\theta = 0$ gives $r = 0$, so the multiplication by r doesn't introduce any new points. The squaring of both sides of this equation is also a reason to pause. Are there points with coordinates (r, θ) which satisfy $r^2 = \left(r^2 + r\cos(\theta)\right)^2$ but do not satisfy $r = r^2 + r\cos(\theta)$? Suppose (r', θ') satisfies $r^2 = \left(r^2 + r\cos(\theta)\right)^2$. Then $r' = \pm\left((r')^2 + r'\cos(\theta')\right)$. If we have that $r' = (r')^2 + r'\cos(\theta')$, we are done. What if $r' = -\left((r')^2 + r'\cos(\theta')\right) = -(r')^2 - r'\cos(\theta')$? We claim that the coordinates $(-r', \theta' + \pi)$, which determine the same point as (r', θ'), satisfy $r = r^2 + r\cos(\theta)$. We substitute $r = -r'$ and $\theta = \theta' + \pi$ into $r = r^2 + r\cos(\theta)$ to see if we get a true statement.

$$\begin{array}{rcll}
-r' & \stackrel{?}{=} & (-r')^2 + (-r'\cos(\theta' + \pi)) & \\
-\left(-(r')^2 - r'\cos(\theta')\right) & \stackrel{?}{=} & (r')^2 - r'\cos(\theta' + \pi) & \text{Since } r' = -(r')^2 - r'\cos(\theta') \\
(r')^2 + r'\cos(\theta') & \stackrel{?}{=} & (r')^2 - r'(-\cos(\theta')) & \text{Since } \cos(\theta' + \pi) = -\cos(\theta') \\
(r')^2 + r'\cos(\theta') & \stackrel{\checkmark}{=} & (r')^2 + r'\cos(\theta') &
\end{array}$$

Since both sides worked out to be equal, $(-r', \theta' + \pi)$ satisfies $r = r^2 + r\cos(\theta)$ which means that any point (r, θ) which satisfies $r^2 = \left(r^2 + r\cos(\theta)\right)^2$ has a representation which satisfies $r = r^2 + r\cos(\theta)$, and we are done. □

In practice, much of the pedantic verification of the equivalence of equations in Example 11.4.3 is left unsaid. Indeed, in most textbooks, squaring equations like $r = -3$ to arrive at $r^2 = 9$ happens without a second thought. Your instructor will ultimately decide how much, if any, justification is warranted. If you take anything away from Example 11.4.3, it should be that relatively nice things in rectangular coordinates, such as $y = x^2$, can turn ugly in polar coordinates, and vice-versa. In the next section, we devote our attention to graphing equations like the ones given in Example 11.4.3 number 2 on the Cartesian coordinate plane without converting back to rectangular coordinates. If nothing else, number 2c above shows the price we pay if we insist on always converting to back to the more familiar rectangular coordinate system.

11.4.1 EXERCISES

In Exercises 1 - 16, plot the point given in polar coordinates and then give three different expressions for the point such that (a) $r < 0$ and $0 \leq \theta \leq 2\pi$, (b) $r > 0$ and $\theta \leq 0$ (c) $r > 0$ and $\theta \geq 2\pi$

1. $\left(2, \frac{\pi}{3}\right)$
2. $\left(5, \frac{7\pi}{4}\right)$
3. $\left(\frac{1}{3}, \frac{3\pi}{2}\right)$
4. $\left(\frac{5}{2}, \frac{5\pi}{6}\right)$
5. $\left(12, -\frac{7\pi}{6}\right)$
6. $\left(3, -\frac{5\pi}{4}\right)$
7. $(2\sqrt{2}, -\pi)$
8. $\left(\frac{7}{2}, -\frac{13\pi}{6}\right)$
9. $(-20, 3\pi)$
10. $\left(-4, \frac{5\pi}{4}\right)$
11. $\left(-1, \frac{2\pi}{3}\right)$
12. $\left(-3, \frac{\pi}{2}\right)$
13. $\left(-3, -\frac{11\pi}{6}\right)$
14. $\left(-2.5, -\frac{\pi}{4}\right)$
15. $\left(-\sqrt{5}, -\frac{4\pi}{3}\right)$
16. $(-\pi, -\pi)$

In Exercises 17 - 36, convert the point from polar coordinates into rectangular coordinates.

17. $\left(5, \frac{7\pi}{4}\right)$
18. $\left(2, \frac{\pi}{3}\right)$
19. $\left(11, -\frac{7\pi}{6}\right)$
20. $(-20, 3\pi)$
21. $\left(\frac{3}{5}, \frac{\pi}{2}\right)$
22. $\left(-4, \frac{5\pi}{6}\right)$
23. $\left(9, \frac{7\pi}{2}\right)$
24. $\left(-5, -\frac{9\pi}{4}\right)$
25. $\left(42, \frac{13\pi}{6}\right)$
26. $(-117, 117\pi)$
27. $(6, \arctan(2))$
28. $(10, \arctan(3))$
29. $\left(-3, \arctan\left(\frac{4}{3}\right)\right)$
30. $\left(5, \arctan\left(-\frac{4}{3}\right)\right)$
31. $\left(2, \pi - \arctan\left(\frac{1}{2}\right)\right)$
32. $\left(-\frac{1}{2}, \pi - \arctan(5)\right)$
33. $\left(-1, \pi + \arctan\left(\frac{3}{4}\right)\right)$
34. $\left(\frac{2}{3}, \pi + \arctan(2\sqrt{2})\right)$
35. $(\pi, \arctan(\pi))$
36. $\left(13, \arctan\left(\frac{12}{5}\right)\right)$

In Exercises 37 - 56, convert the point from rectangular coordinates into polar coordinates with $r \geq 0$ and $0 \leq \theta < 2\pi$.

37. $(0, 5)$
38. $(3, \sqrt{3})$
39. $(7, -7)$
40. $(-3, -\sqrt{3})$
41. $(-3, 0)$
42. $(-\sqrt{2}, \sqrt{2})$
43. $(-4, -4\sqrt{3})$
44. $\left(\frac{\sqrt{3}}{4}, -\frac{1}{4}\right)$

45. $\left(-\frac{3}{10}, -\frac{3\sqrt{3}}{10}\right)$ 46. $(-\sqrt{5}, -\sqrt{5})$ 47. $(6, 8)$ 48. $(\sqrt{5}, 2\sqrt{5})$

49. $(-8, 1)$ 50. $(-2\sqrt{10}, 6\sqrt{10})$ 51. $(-5, -12)$ 52. $\left(-\frac{\sqrt{5}}{15}, -\frac{2\sqrt{5}}{15}\right)$

53. $(24, -7)$ 54. $(12, -9)$ 55. $\left(\frac{\sqrt{2}}{4}, \frac{\sqrt{6}}{4}\right)$ 56. $\left(-\frac{\sqrt{65}}{5}, \frac{2\sqrt{65}}{5}\right)$

In Exercises 57 - 76, convert the equation from rectangular coordinates into polar coordinates. Solve for r in all but #60 through #63. In Exercises 60 - 63, you need to solve for θ

57. $x = 6$ 58. $x = -3$ 59. $y = 7$ 60. $y = 0$

61. $y = -x$ 62. $y = x\sqrt{3}$ 63. $y = 2x$ 64. $x^2 + y^2 = 25$

65. $x^2 + y^2 = 117$ 66. $y = 4x - 19$ 67. $x = 3y + 1$ 68. $y = -3x^2$

69. $4x = y^2$ 70. $x^2 + y^2 - 2y = 0$ 71. $x^2 - 4x + y^2 = 0$ 72. $x^2 + y^2 = x$

73. $y^2 = 7y - x^2$ 74. $(x+2)^2 + y^2 = 4$

75. $x^2 + (y-3)^2 = 9$ 76. $4x^2 + 4\left(y - \frac{1}{2}\right)^2 = 1$

In Exercises 77 - 96, convert the equation from polar coordinates into rectangular coordinates.

77. $r = 7$ 78. $r = -3$ 79. $r = \sqrt{2}$ 80. $\theta = \frac{\pi}{4}$

81. $\theta = \frac{2\pi}{3}$ 82. $\theta = \pi$ 83. $\theta = \frac{3\pi}{2}$ 84. $r = 4\cos(\theta)$

85. $5r = \cos(\theta)$ 86. $r = 3\sin(\theta)$ 87. $r = -2\sin(\theta)$ 88. $r = 7\sec(\theta)$

89. $12r = \csc(\theta)$ 90. $r = -2\sec(\theta)$ 91. $r = -\sqrt{5}\csc(\theta)$ 92. $r = 2\sec(\theta)\tan(\theta)$

93. $r = -\csc(\theta)\cot(\theta)$ 94. $r^2 = \sin(2\theta)$ 95. $r = 1 - 2\cos(\theta)$ 96. $r = 1 + \sin(\theta)$

97. Convert the origin $(0, 0)$ into polar coordinates in four different ways.

98. With the help of your classmates, use the Law of Cosines to develop a formula for the distance between two points in polar coordinates.

11.4.2 Answers

1. $\left(2, \frac{\pi}{3}\right)$, $\left(-2, \frac{4\pi}{3}\right)$
$\left(2, -\frac{5\pi}{3}\right)$, $\left(2, \frac{7\pi}{3}\right)$

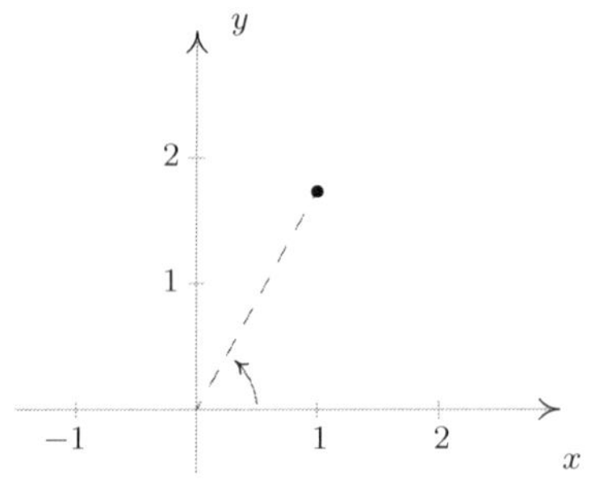

2. $\left(5, \frac{7\pi}{4}\right)$, $\left(-5, \frac{3\pi}{4}\right)$
$\left(5, -\frac{\pi}{4}\right)$, $\left(5, \frac{15\pi}{4}\right)$

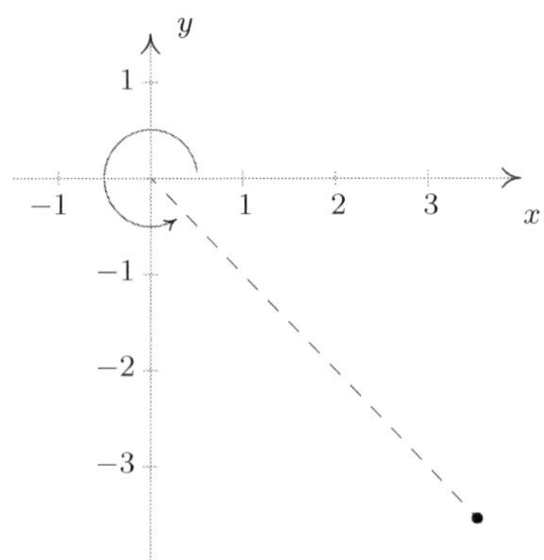

3. $\left(\frac{1}{3}, \frac{3\pi}{2}\right)$, $\left(-\frac{1}{3}, \frac{\pi}{2}\right)$
$\left(\frac{1}{3}, -\frac{\pi}{2}\right)$, $\left(\frac{1}{3}, \frac{7\pi}{2}\right)$

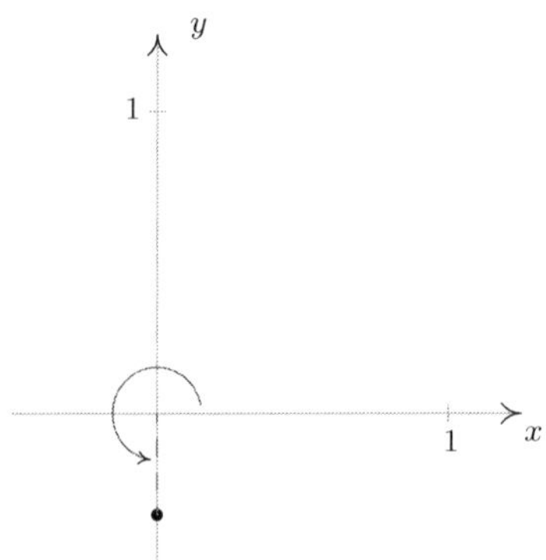

4. $\left(\frac{5}{2}, \frac{5\pi}{6}\right)$, $\left(-\frac{5}{2}, \frac{11\pi}{6}\right)$
$\left(\frac{5}{2}, -\frac{7\pi}{6}\right)$, $\left(\frac{5}{2}, \frac{17\pi}{6}\right)$

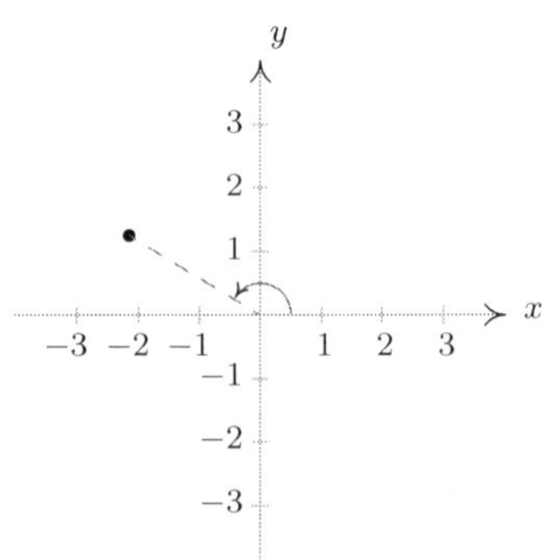

5. $\left(12, -\frac{7\pi}{6}\right)$, $\left(-12, \frac{11\pi}{6}\right)$
$\left(12, -\frac{19\pi}{6}\right)$, $\left(12, \frac{17\pi}{6}\right)$

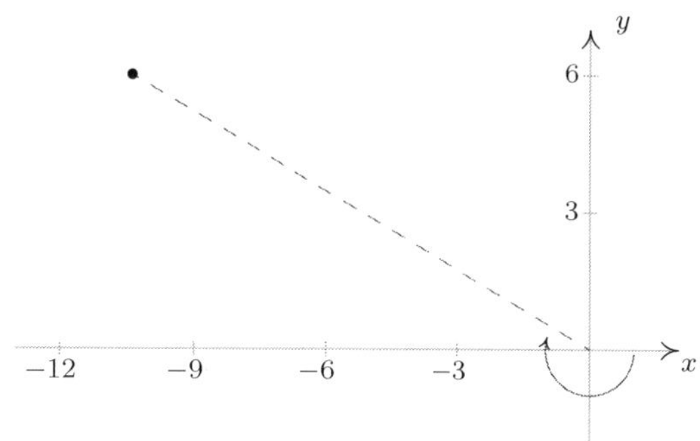

6. $\left(3, -\frac{5\pi}{4}\right)$, $\left(-3, \frac{7\pi}{4}\right)$
$\left(3, -\frac{13\pi}{4}\right)$, $\left(3, \frac{11\pi}{4}\right)$

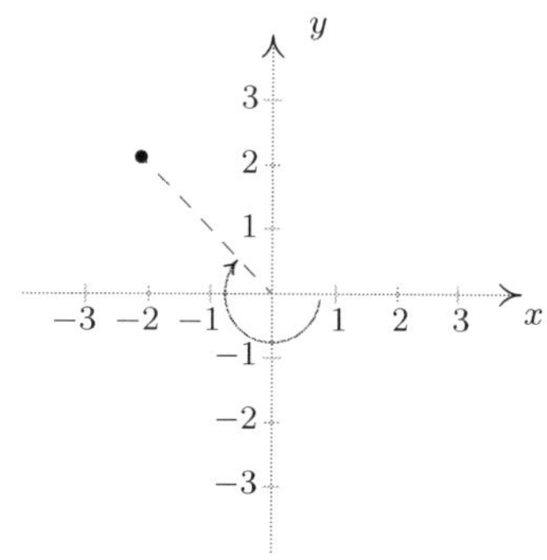

7. $\left(2\sqrt{2}, -\pi\right)$, $\left(-2\sqrt{2}, 0\right)$
$\left(2\sqrt{2}, -3\pi\right)$, $\left(2\sqrt{2}, 3\pi\right)$

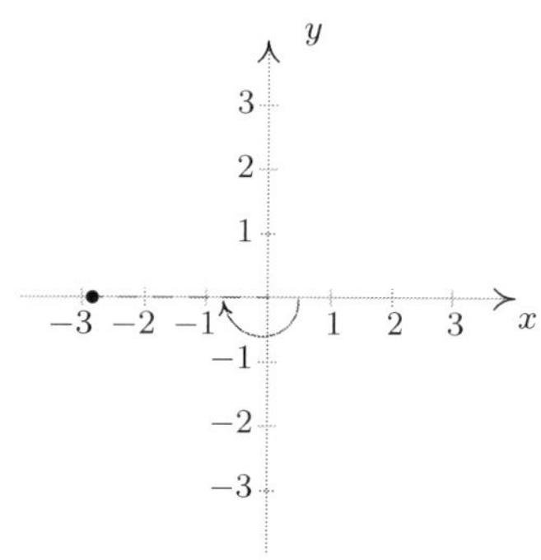

8. $\left(\frac{7}{2}, -\frac{13\pi}{6}\right)$, $\left(-\frac{7}{2}, \frac{5\pi}{6}\right)$
$\left(\frac{7}{2}, -\frac{\pi}{6}\right)$, $\left(\frac{7}{2}, \frac{23\pi}{6}\right)$

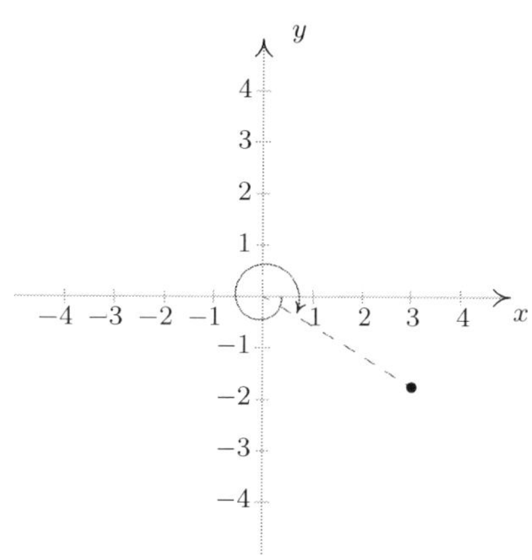

9. $(-20, 3\pi)$, $(-20, \pi)$
$(20, -2\pi)$, $(20, 4\pi)$

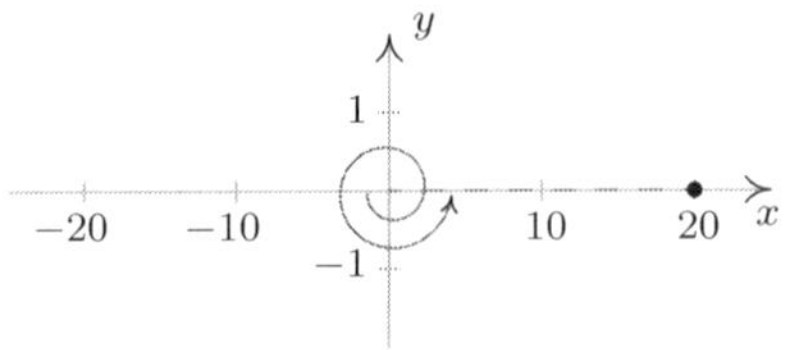

10. $\left(-4, \dfrac{5\pi}{4}\right)$, $\left(-4, \dfrac{13\pi}{4}\right)$
$\left(4, -\dfrac{7\pi}{4}\right)$, $\left(4, \dfrac{9\pi}{4}\right)$

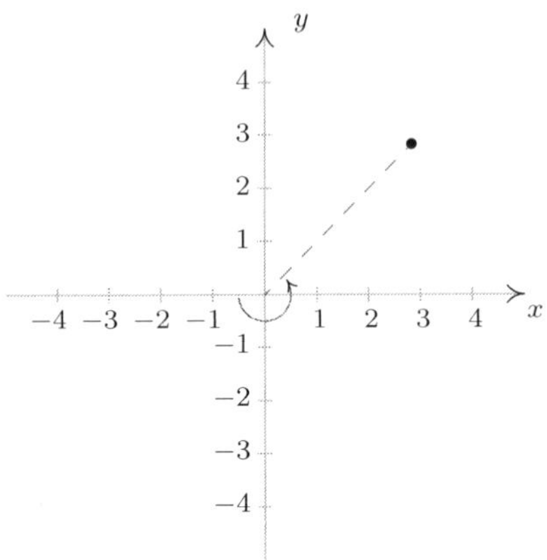

11. $\left(-1, \dfrac{2\pi}{3}\right)$, $\left(-1, \dfrac{8\pi}{3}\right)$
$\left(1, -\dfrac{\pi}{3}\right)$, $\left(1, \dfrac{11\pi}{3}\right)$

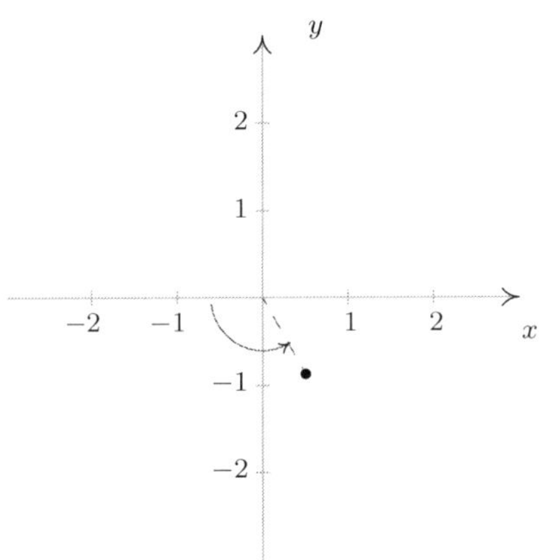

12. $\left(-3, \dfrac{\pi}{2}\right)$, $\left(-3, \dfrac{5\pi}{2}\right)$
$\left(3, -\dfrac{\pi}{2}\right)$, $\left(3, \dfrac{7\pi}{2}\right)$

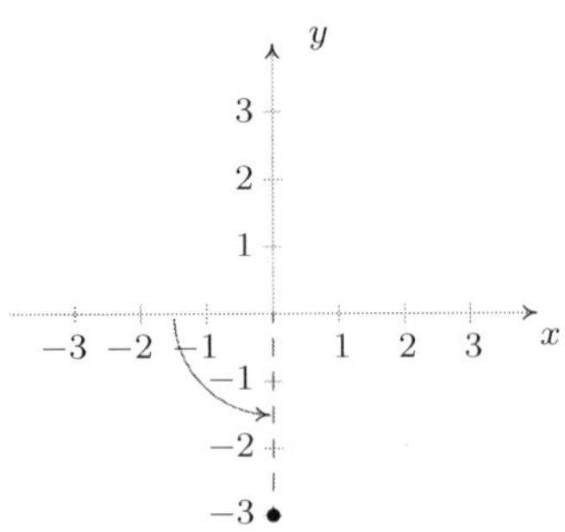

13. $\left(-3, -\frac{11\pi}{6}\right), \left(-3, \frac{\pi}{6}\right)$
$\left(3, -\frac{5\pi}{6}\right), \left(3, \frac{19\pi}{6}\right)$

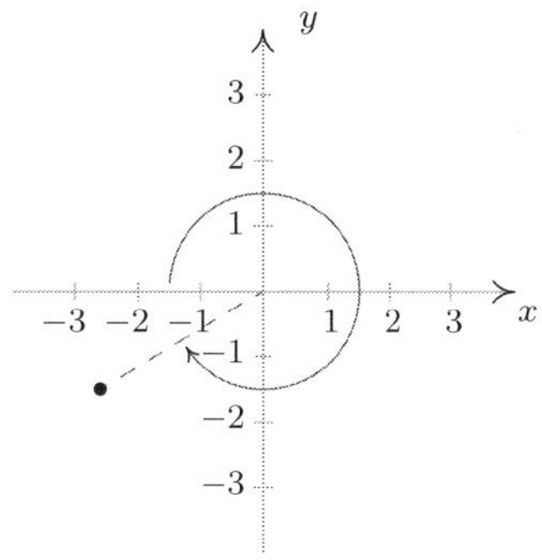

14. $\left(-2.5, -\frac{\pi}{4}\right), \left(-2.5, \frac{7\pi}{4}\right)$
$\left(2.5, -\frac{5\pi}{4}\right), \left(2.5, \frac{11\pi}{4}\right)$

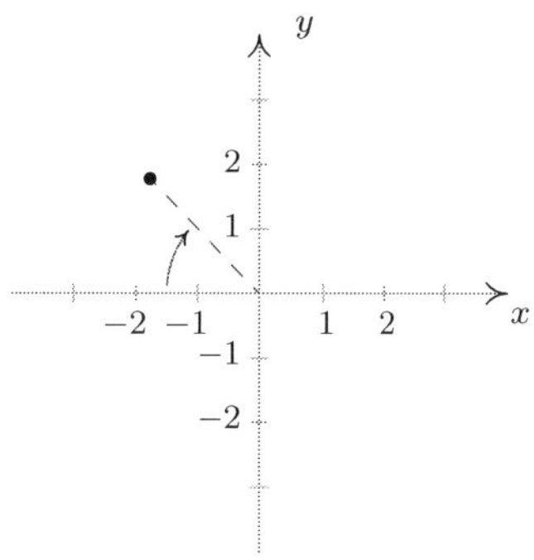

15. $\left(-\sqrt{5}, -\frac{4\pi}{3}\right), \left(-\sqrt{5}, \frac{2\pi}{3}\right)$
$\left(\sqrt{5}, -\frac{\pi}{3}\right), \left(\sqrt{5}, \frac{11\pi}{3}\right)$

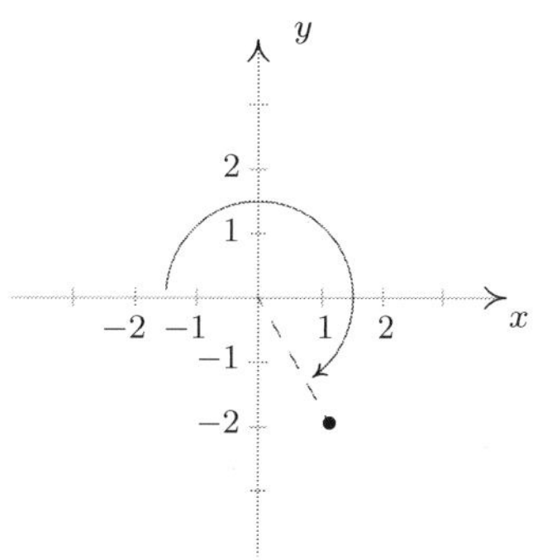

16. $(-\pi, -\pi), (-\pi, \pi)$
$(\pi, -2\pi), (\pi, 2\pi)$

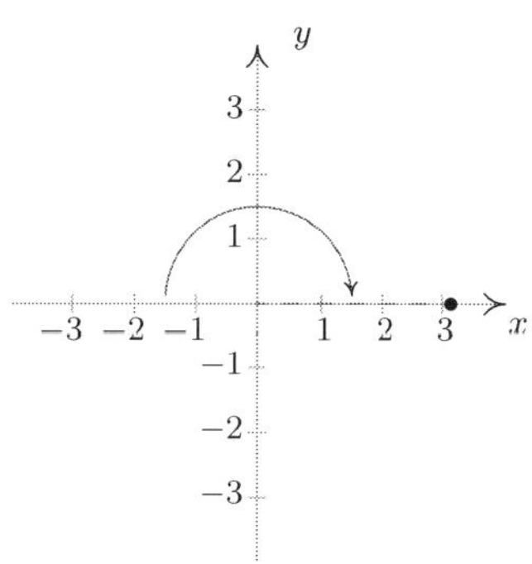

17. $\left(\frac{5\sqrt{2}}{2}, -\frac{5\sqrt{2}}{2}\right)$
18. $(1, \sqrt{3})$
19. $\left(-\frac{11\sqrt{3}}{2}, \frac{11}{2}\right)$
20. $(20, 0)$
21. $\left(0, \frac{3}{5}\right)$
22. $(2\sqrt{3}, -2)$
23. $(0, -9)$
24. $\left(-\frac{5\sqrt{2}}{2}, \frac{5\sqrt{2}}{2}\right)$
25. $(21\sqrt{3}, 21)$
26. $(117, 0)$
27. $\left(\frac{6\sqrt{5}}{5}, \frac{12\sqrt{5}}{5}\right)$
28. $(\sqrt{10}, 3\sqrt{10})$
29. $\left(-\frac{9}{5}, -\frac{12}{5}\right)$
30. $(3, -4)$
31. $\left(-\frac{4\sqrt{5}}{5}, \frac{2\sqrt{5}}{5}\right)$
32. $\left(\frac{\sqrt{26}}{52}, -\frac{5\sqrt{26}}{52}\right)$
33. $\left(\frac{4}{5}, \frac{3}{5}\right)$
34. $\left(-\frac{2}{9}, -\frac{4\sqrt{2}}{9}\right)$
35. $\left(\frac{\pi}{\sqrt{1+\pi^2}}, \frac{\pi^2}{\sqrt{1+\pi^2}}\right)$
36. $(5, 12)$
37. $\left(5, \frac{\pi}{2}\right)$
38. $\left(2\sqrt{3}, \frac{\pi}{6}\right)$
39. $\left(7\sqrt{2}, \frac{7\pi}{4}\right)$
40. $\left(2\sqrt{3}, \frac{7\pi}{6}\right)$
41. $(3, \pi)$
42. $\left(2, \frac{3\pi}{4}\right)$
43. $\left(8, \frac{4\pi}{3}\right)$
44. $\left(\frac{1}{2}, \frac{11\pi}{6}\right)$
45. $\left(\frac{3}{5}, \frac{4\pi}{3}\right)$
46. $\left(\sqrt{10}, \frac{5\pi}{4}\right)$
47. $\left(10, \arctan\left(\frac{4}{3}\right)\right)$
48. $(5, \arctan(2))$
49. $\left(\sqrt{65}, \pi - \arctan\left(\frac{1}{8}\right)\right)$
50. $(20, \pi - \arctan(3))$
51. $\left(13, \pi + \arctan\left(\frac{12}{5}\right)\right)$
52. $\left(\frac{1}{3}, \pi + \arctan(2)\right)$
53. $\left(25, 2\pi - \arctan\left(\frac{7}{24}\right)\right)$
54. $\left(15, 2\pi - \arctan\left(\frac{3}{4}\right)\right)$
55. $\left(\frac{\sqrt{2}}{2}, \frac{\pi}{3}\right)$
56. $(\sqrt{13}, \pi - \arctan(2))$
57. $r = 6\sec(\theta)$
58. $r = -3\sec(\theta)$
59. $r = 7\csc(\theta)$
60. $\theta = 0$
61. $\theta = \frac{3\pi}{4}$
62. $\theta = \frac{\pi}{3}$
63. $\theta = \arctan(2)$
64. $r = 5$
65. $r = \sqrt{117}$
66. $r = \frac{19}{4\cos(\theta)-\sin(\theta)}$
67. $x = \frac{1}{\cos(\theta)-3\sin(\theta)}$
68. $r = \frac{-\sec(\theta)\tan(\theta)}{3}$
69. $r = 4\csc(\theta)\cot(\theta)$
70. $r = 2\sin(\theta)$
71. $r = 4\cos(\theta)$
72. $r = \cos(\theta)$

73. $r = 7\sin(\theta)$

74. $r = -4\cos(\theta)$

75. $r = 6\sin(\theta)$

76. $r = \sin(\theta)$

77. $x^2 + y^2 = 49$

78. $x^2 + y^2 = 9$

79. $x^2 + y^2 = 2$

80. $y = x$

81. $y = -\sqrt{3}x$

82. $y = 0$

83. $x = 0$

84. $x^2 + y^2 = 4x$ or $(x - 2)^2 + y^2 = 4$

85. $5x^2 + 5y^2 = x$ or $\left(x - \frac{1}{10}\right)^2 + y^2 = \frac{1}{100}$

86. $x^2 + y^2 = 3y$ or $x^2 + \left(y - \frac{3}{2}\right)^2 = \frac{9}{4}$

87. $x^2 + y^2 = -2y$ or $x^2 + (y + 1)^2 = 1$

88. $x = 7$

89. $y = \frac{1}{12}$

90. $x = -2$

91. $y = -\sqrt{5}$

92. $x^2 = 2y$

93. $y^2 = -x$

94. $\left(x^2 + y^2\right)^2 = 2xy$

95. $\left(x^2 + 2x + y^2\right)^2 = x^2 + y^2$

96. $\left(x^2 + y^2 + y\right)^2 = x^2 + y^2$

97. Any point of the form $(0, \theta)$ will work, e.g. $(0, \pi)$, $(0, -117)$, $\left(0, \frac{23\pi}{4}\right)$ and $(0, 0)$.

11.5 GRAPHS OF POLAR EQUATIONS

In this section, we discuss how to graph equations in polar coordinates on the rectangular coordinate plane. Since any given point in the plane has infinitely many different representations in polar coordinates, our 'Fundamental Graphing Principle' in this section is not as clean as it was for graphs of rectangular equations on page 23. We state it below for completeness.

The Fundamental Graphing Principle for Polar Equations
The graph of an equation in polar coordinates is the set of points which satisfy the equation. That is, a point $P(r,\theta)$ is on the graph of an equation if and only if there is a representation of P, say (r',θ'), such that r' and θ' satisfy the equation.

Our first example focuses on some of the more structurally simple polar equations.

Example 11.5.1. Graph the following polar equations.

1. $r = 4$
2. $r = -3\sqrt{2}$
3. $\theta = \frac{5\pi}{4}$
4. $\theta = -\frac{3\pi}{2}$

Solution. In each of these equations, only one of the variables r and θ is present making the other variable free.[1] This makes these graphs easier to visualize than others.

1. In the equation $r = 4$, θ is free. The graph of this equation is, therefore, all points which have a polar coordinate representation $(4,\theta)$, for any choice of θ. Graphically this translates into tracing out all of the points 4 units away from the origin. This is exactly the definition of circle, centered at the origin, with a radius of 4.

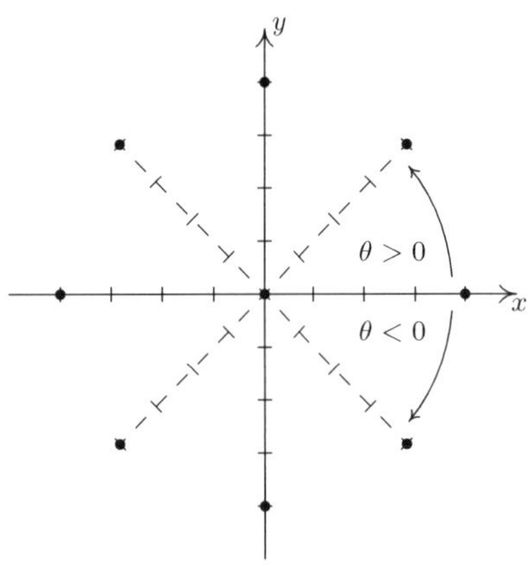

In $r = 4$, θ is free

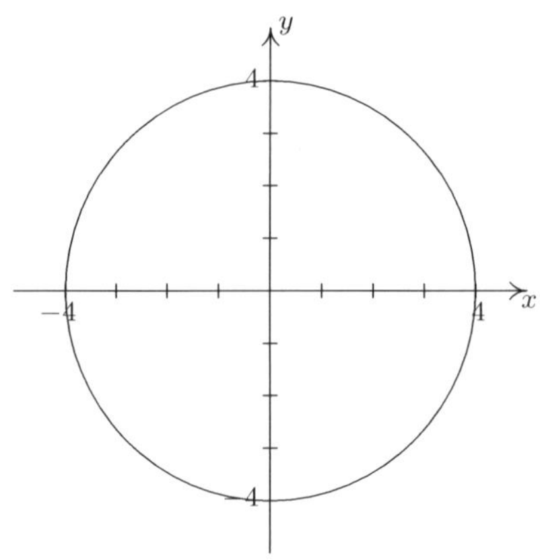

The graph of $r = 4$

2. Once again we have θ being free in the equation $r = -3\sqrt{2}$. Plotting all of the points of the form $(-3\sqrt{2},\theta)$ gives us a circle of radius $3\sqrt{2}$ centered at the origin.

[1]See the discussion in Example 11.4.3 number 2a.

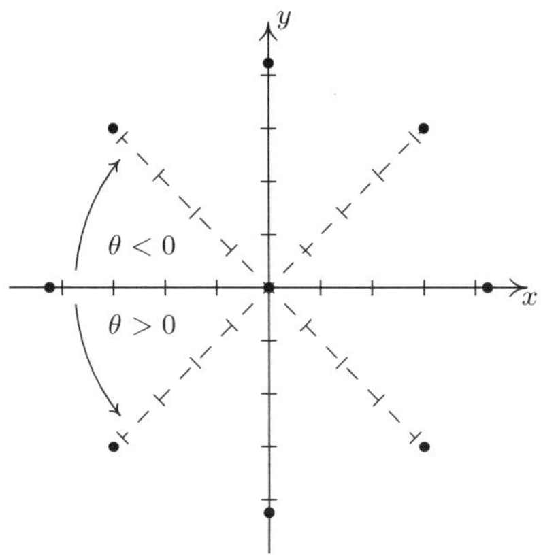

In $r = -3\sqrt{2}$, θ is free

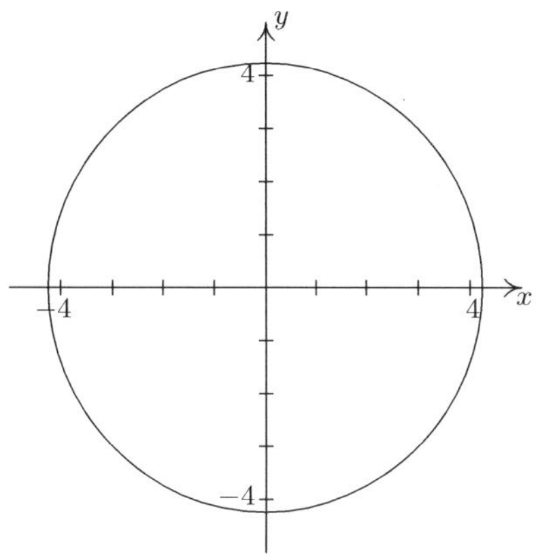

The graph of $r = -3\sqrt{2}$

3. In the equation $\theta = \frac{5\pi}{4}$, r is free, so we plot all of the points with polar representation $\left(r, \frac{5\pi}{4}\right)$. What we find is that we are tracing out the line which contains the terminal side of $\theta = \frac{5\pi}{4}$ when plotted in standard position.

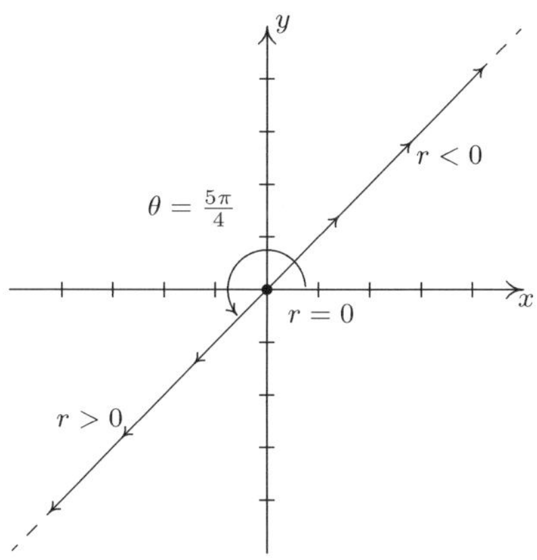

In $\theta = \frac{5\pi}{4}$, r is free

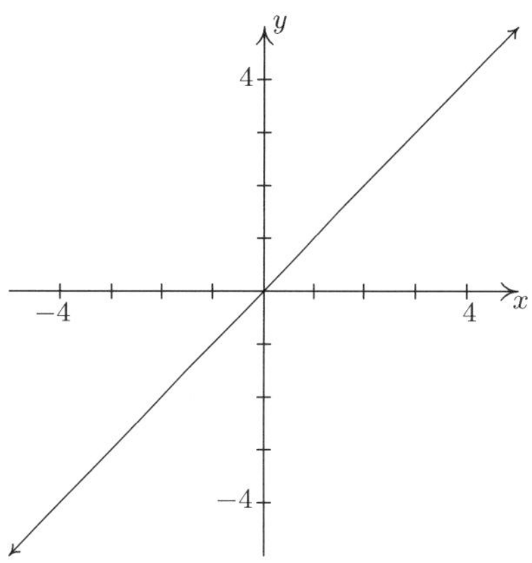

The graph of $\theta = \frac{5\pi}{4}$

4. As in the previous example, the variable r is free in the equation $\theta = -\frac{3\pi}{2}$. Plotting $\left(r, -\frac{3\pi}{2}\right)$ for various values of r shows us that we are tracing out the y-axis.

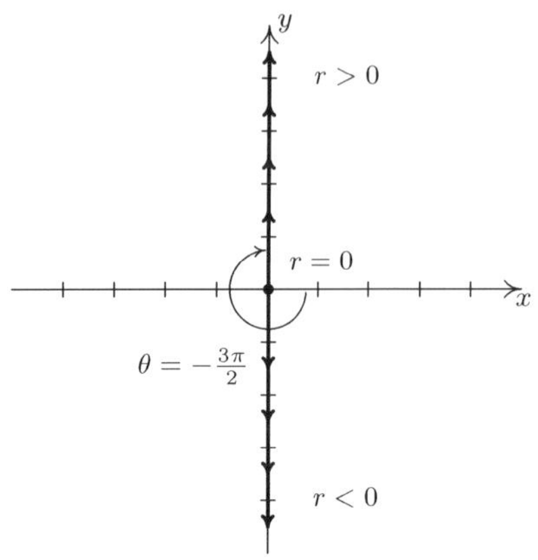

In $\theta = -\frac{3\pi}{2}$, r is free

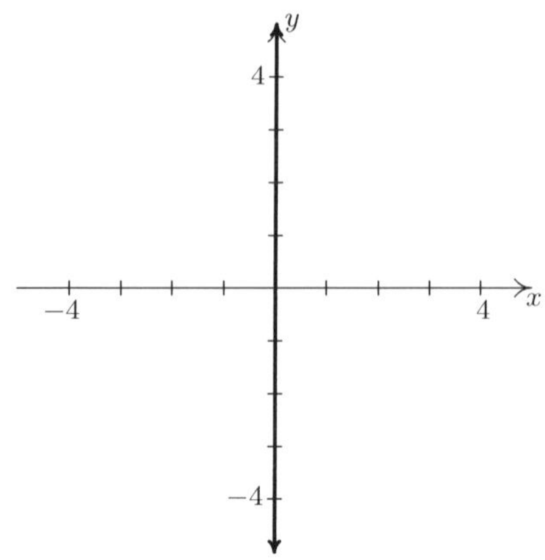

The graph of $\theta = -\frac{3\pi}{2}$ □

Hopefully, our experience in Example 11.5.1 makes the following result clear.

Theorem 11.8. Graphs of Constant r and θ: Suppose a and α are constants, $a \neq 0$.

- The graph of the polar equation $r = a$ on the Cartesian plane is a circle centered at the origin of radius $|a|$.
- The graph of the polar equation $\theta = \alpha$ on the Cartesian plane is the line containing the terminal side of α when plotted in standard position.

Suppose we wish to graph $r = 6\cos(\theta)$. A reasonable way to start is to treat θ as the independent variable, r as the dependent variable, evaluate $r = f(\theta)$ at some 'friendly' values of θ and plot the resulting points.[2] We generate the table below.

θ	$r = 6\cos(\theta)$	(r, θ)
0	6	$(6, 0)$
$\frac{\pi}{4}$	$3\sqrt{2}$	$\left(3\sqrt{2}, \frac{\pi}{4}\right)$
$\frac{\pi}{2}$	0	$\left(0, \frac{\pi}{2}\right)$
$\frac{3\pi}{4}$	$-3\sqrt{2}$	$\left(-3\sqrt{2}, \frac{3\pi}{4}\right)$
π	-6	$(-6, \pi)$
$\frac{5\pi}{4}$	$-3\sqrt{2}$	$\left(-3\sqrt{2}, \frac{5\pi}{4}\right)$
$\frac{3\pi}{2}$	0	$\left(0, \frac{3\pi}{2}\right)$
$\frac{7\pi}{4}$	$3\sqrt{2}$	$\left(3\sqrt{2}, \frac{7\pi}{4}\right)$
2π	6	$(6, 2\pi)$

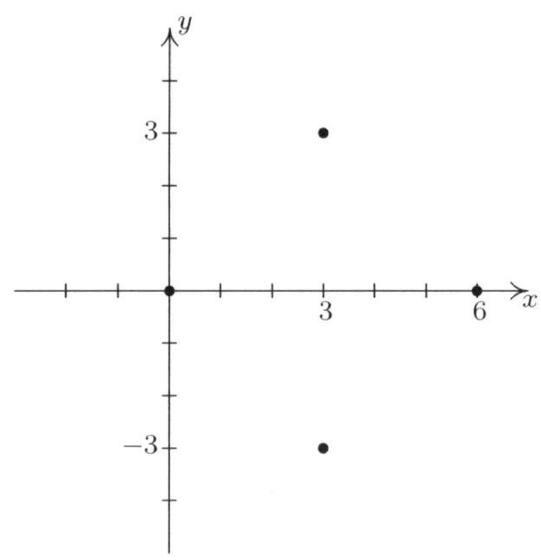

[2]For a review of these concepts and this process, see Sections 1.4 and 1.6.

Despite having nine ordered pairs, we get only four distinct points on the graph. For this reason, we employ a slightly different strategy. We graph one cycle of $r = 6\cos(\theta)$ on the θr-plane[3] and use it to help graph the equation on the xy-plane. We see that as θ ranges from 0 to $\frac{\pi}{2}$, r ranges from 6 to 0. In the xy-plane, this means that the curve starts 6 units from the origin on the positive x-axis ($\theta = 0$) and gradually returns to the origin by the time the curve reaches the y-axis ($\theta = \frac{\pi}{2}$). The arrows drawn in the figure below are meant to help you visualize this process. In the θr-plane, the arrows are drawn from the θ-axis to the curve $r = 6\cos(\theta)$. In the xy-plane, each of these arrows starts at the origin and is rotated through the corresponding angle θ, in accordance with how we plot polar coordinates. It is a less-precise way to generate the graph than computing the actual function values, but it is markedly faster.

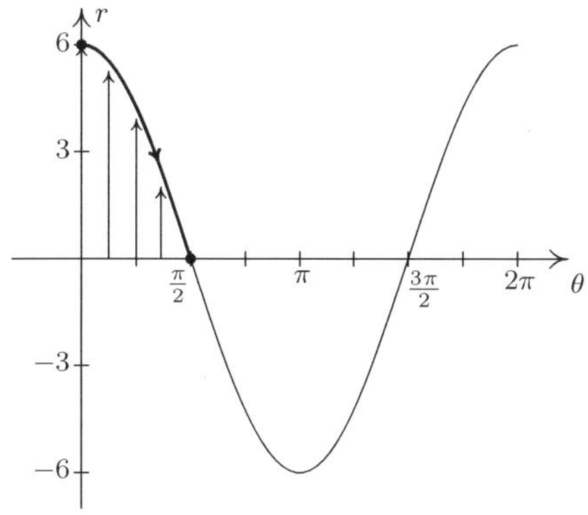

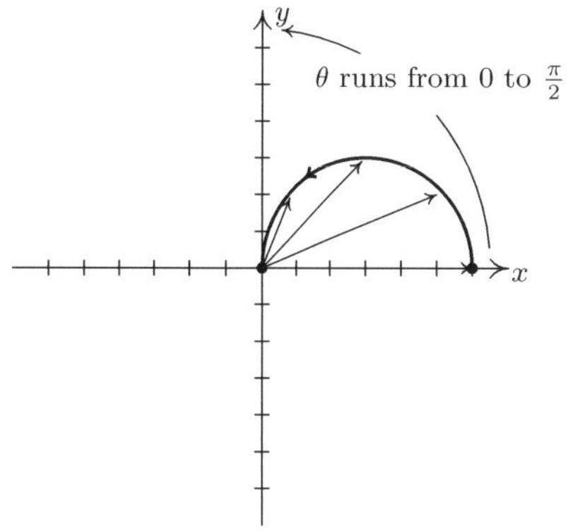

Next, we repeat the process as θ ranges from $\frac{\pi}{2}$ to π. Here, the r values are all negative. This means that in the xy-plane, instead of graphing in Quadrant II, we graph in Quadrant IV, with all of the angle rotations starting from the negative x-axis.

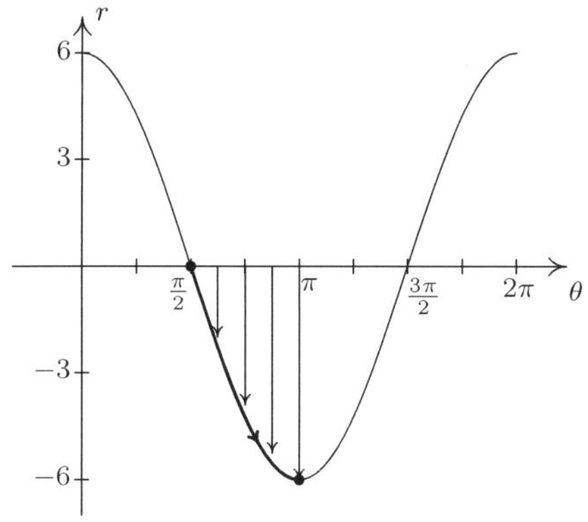

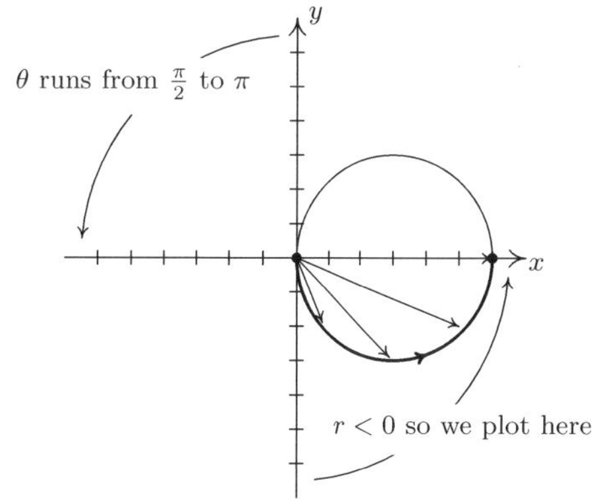

As θ ranges from π to $\frac{3\pi}{2}$, the r values are still negative, which means the graph is traced out in Quadrant I instead of Quadrant III. Since the $|r|$ for these values of θ match the r values for θ in

[3]The graph looks exactly like $y = 6\cos(x)$ in the xy-plane, and for good reason. At this stage, we are just graphing the relationship between r and θ before we interpret them as polar coordinates (r, θ) on the xy-plane.

$\left[0, \frac{\pi}{2}\right]$, we have that the curve begins to retrace itself at this point. Proceeding further, we find that when $\frac{3\pi}{2} \leq \theta \leq 2\pi$, we retrace the portion of the curve in Quadrant IV that we first traced out as $\frac{\pi}{2} \leq \theta \leq \pi$. The reader is invited to verify that plotting any range of θ outside the interval $[0, \pi]$ results in retracting some portion of the curve.[4] We present the final graph below.

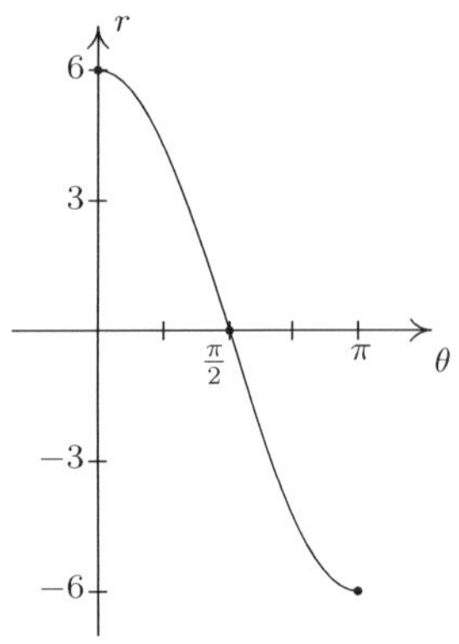

$r = 6\cos(\theta)$ in the θr-plane

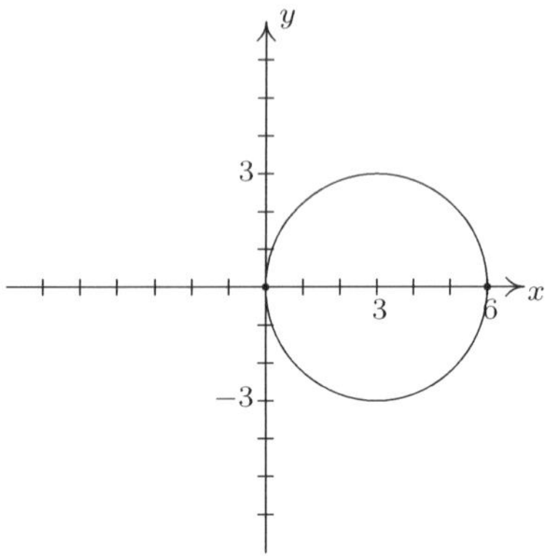

$r = 6\cos(\theta)$ in the xy-plane

Example 11.5.2. Graph the following polar equations.

1. $r = 4 - 2\sin(\theta)$
2. $r = 2 + 4\cos(\theta)$
3. $r = 5\sin(2\theta)$
4. $r^2 = 16\cos(2\theta)$

Solution.

1. We first plot the fundamental cycle of $r = 4 - 2\sin(\theta)$ on the θr-axes. To help us visualize what is going on graphically, we divide up $[0, 2\pi]$ into the usual four subintervals $\left[0, \frac{\pi}{2}\right]$, $\left[\frac{\pi}{2}, \pi\right]$, $\left[\pi, \frac{3\pi}{2}\right]$ and $\left[\frac{3\pi}{2}, 2\pi\right]$, and proceed as we did above. As θ ranges from 0 to $\frac{\pi}{2}$, r decreases from 4 to 2. This means that the curve in the xy-plane starts 4 units from the origin on the positive x-axis and gradually pulls in towards the origin as it moves towards the positive y-axis.

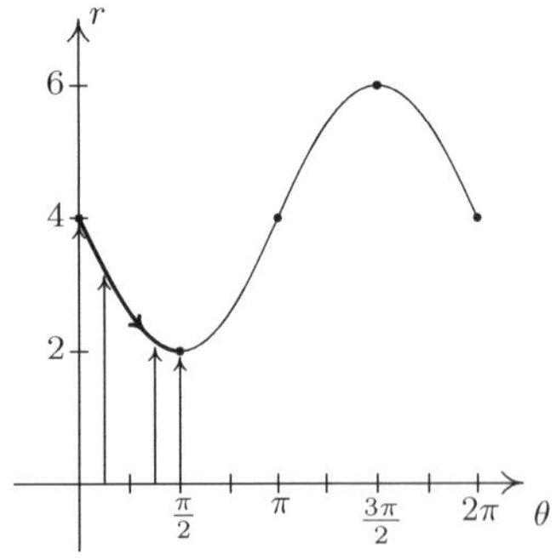

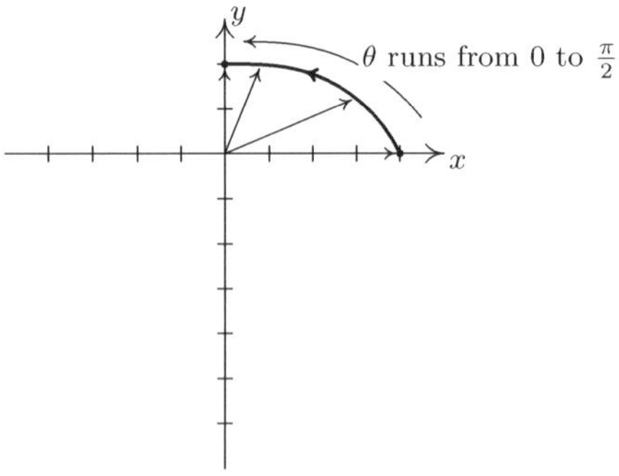

[4]The graph of $r = 6\cos(\theta)$ looks suspiciously like a circle, for good reason. See number 1a in Example 11.4.3.

Next, as θ runs from $\frac{\pi}{2}$ to π, we see that r increases from 2 to 4. Picking up where we left off, we gradually pull the graph away from the origin until we reach the negative x-axis.

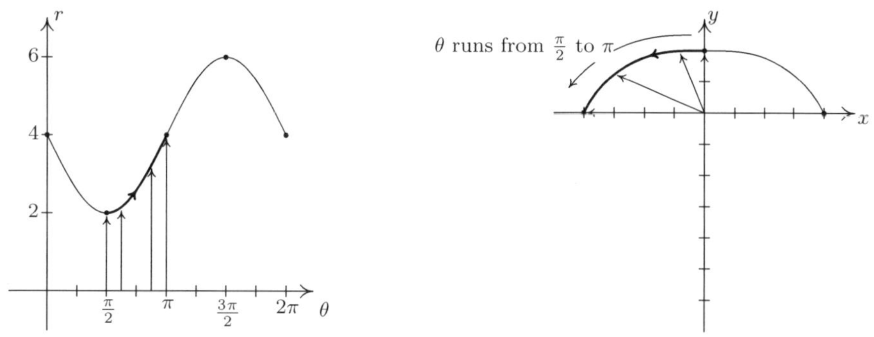

Over the interval $\left[\pi, \frac{3\pi}{2}\right]$, we see that r increases from 4 to 6. On the xy-plane, the curve sweeps out away from the origin as it travels from the negative x-axis to the negative y-axis.

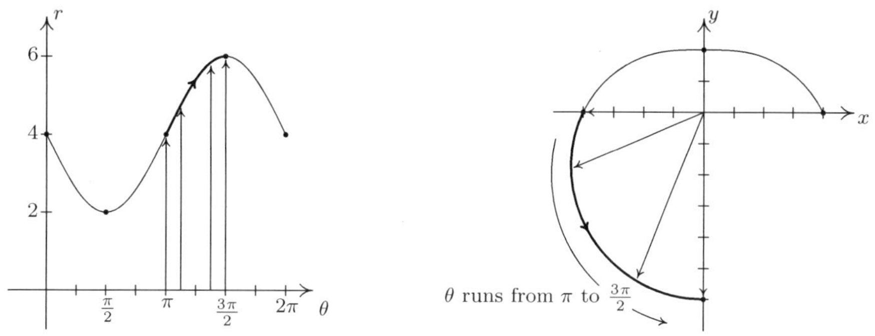

Finally, as θ takes on values from $\frac{3\pi}{2}$ to 2π, r decreases from 6 back to 4. The graph on the xy-plane pulls in from the negative y-axis to finish where we started.

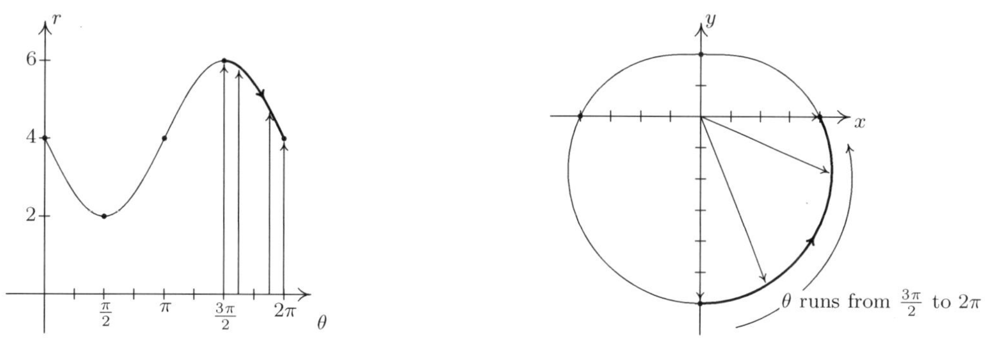

We leave it to the reader to verify that plotting points corresponding to values of θ outside the interval $[0, 2\pi]$ results in retracing portions of the curve, so we are finished.

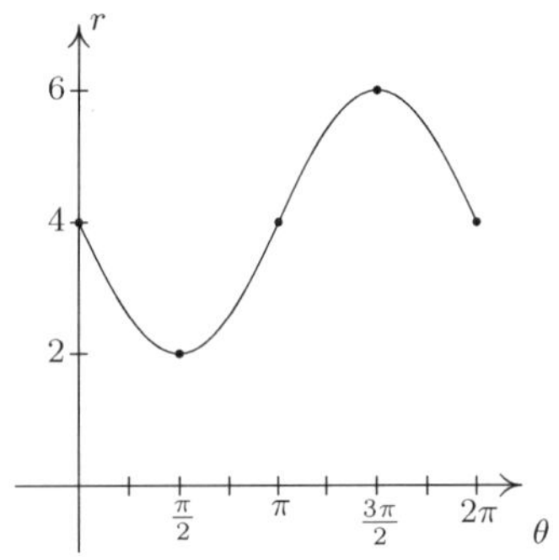

$r = 4 - 2\sin(\theta)$ in the θr-plane

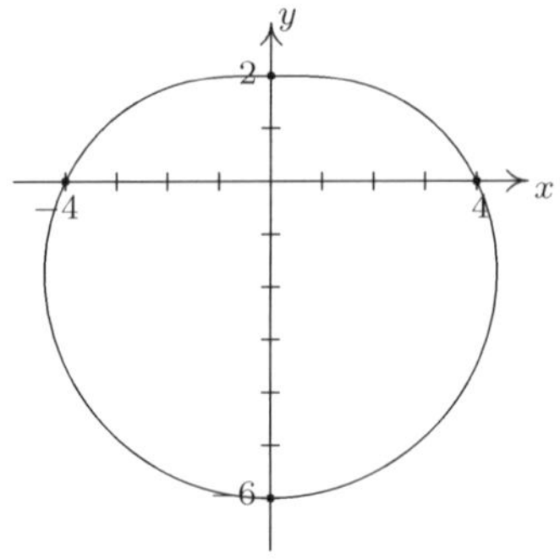

$r = 4 - 2\sin(\theta)$ in the xy-plane.

2. The first thing to note when graphing $r = 2 + 4\cos(\theta)$ on the θr-plane over the interval $[0, 2\pi]$ is that the graph crosses through the θ-axis. This corresponds to the graph of the curve passing through the origin in the xy-plane, and our first task is to determine when this happens. Setting $r = 0$ we get $2 + 4\cos(\theta) = 0$, or $\cos(\theta) = -\frac{1}{2}$. Solving for θ in $[0, 2\pi]$ gives $\theta = \frac{2\pi}{3}$ and $\theta = \frac{4\pi}{3}$. Since these values of θ are important geometrically, we break the interval $[0, 2\pi]$ into six subintervals: $\left[0, \frac{\pi}{2}\right]$, $\left[\frac{\pi}{2}, \frac{2\pi}{3}\right]$, $\left[\frac{2\pi}{3}, \pi\right]$, $\left[\pi, \frac{4\pi}{3}\right]$, $\left[\frac{4\pi}{3}, \frac{3\pi}{2}\right]$ and $\left[\frac{3\pi}{2}, 2\pi\right]$. As θ ranges from 0 to $\frac{\pi}{2}$, r decreases from 6 to 2. Plotting this on the xy-plane, we start 6 units out from the origin on the positive x-axis and slowly pull in towards the positive y-axis.

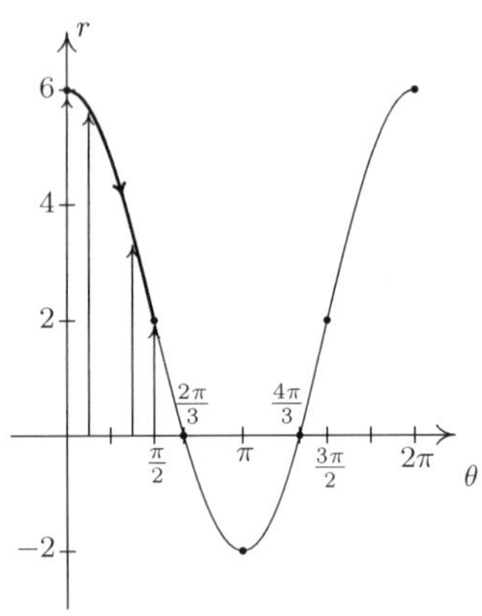

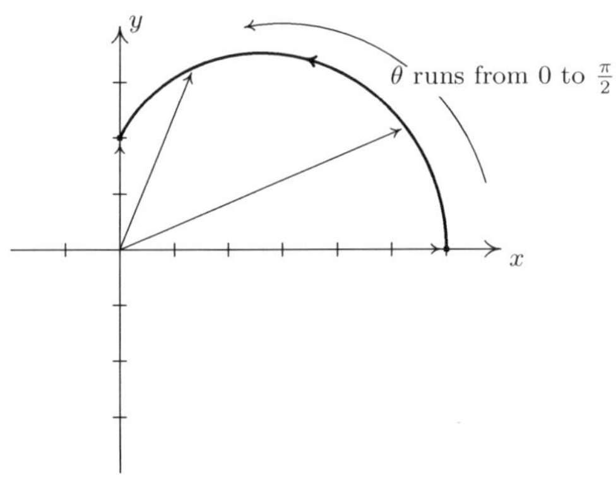

On the interval $\left[\frac{\pi}{2}, \frac{2\pi}{3}\right]$, r decreases from 2 to 0, which means the graph is heading into (and will eventually cross through) the origin. Not only do we reach the origin when $\theta = \frac{2\pi}{3}$, a theorem from Calculus[5] states that the curve hugs the line $\theta = \frac{2\pi}{3}$ as it approaches the origin.

[5]The 'tangents at the pole' theorem from second semester Calculus.

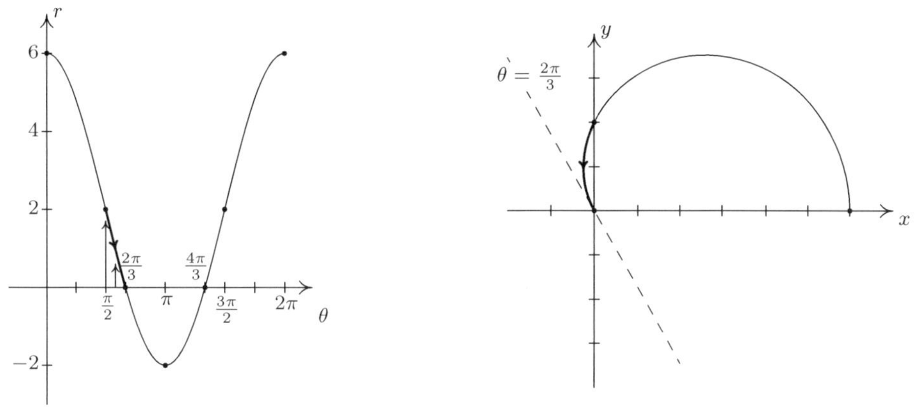

On the interval $\left[\frac{2\pi}{3}, \pi\right]$, r ranges from 0 to -2. Since $r \leq 0$, the curve passes through the origin in the xy-plane, following the line $\theta = \frac{2\pi}{3}$ and continues upwards through Quadrant IV towards the positive x-axis.[6] Since $|r|$ is increasing from 0 to 2, the curve pulls away from the origin to finish at a point on the positive x-axis.

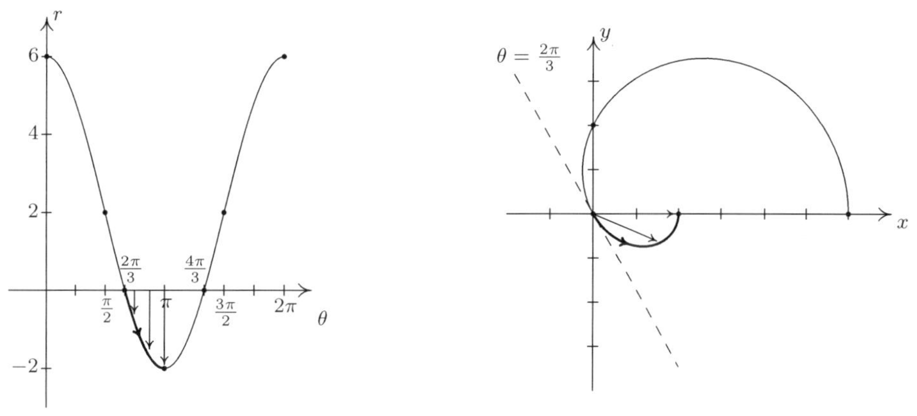

Next, as θ progresses from π to $\frac{4\pi}{3}$, r ranges from -2 to 0. Since $r \leq 0$, we continue our graph in the first quadrant, heading into the origin along the line $\theta = \frac{4\pi}{3}$.

[6]Recall that one way to visualize plotting polar coordinates (r, θ) with $r < 0$ is to start the rotation from the left side of the pole - in this case, the negative x-axis. Rotating between $\frac{2\pi}{3}$ and π radians from the negative x-axis in this case determines the region between the line $\theta = \frac{2\pi}{3}$ and the x-axis in Quadrant IV.

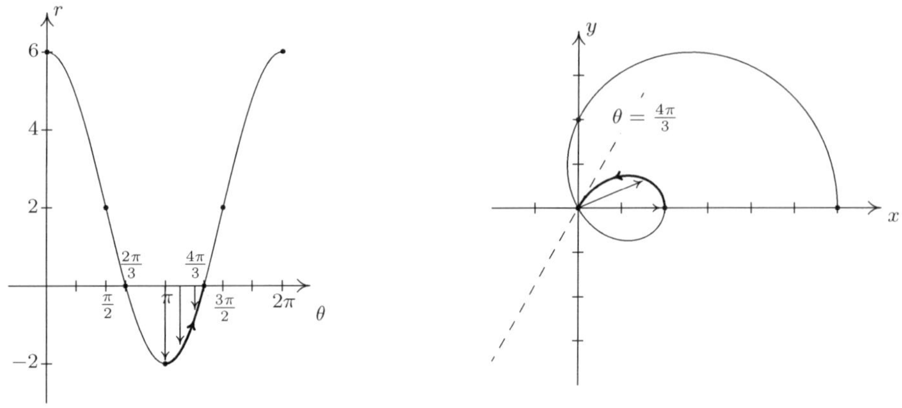

On the interval $\left[\frac{4\pi}{3}, \frac{3\pi}{2}\right]$, r returns to positive values and increases from 0 to 2. We hug the line $\theta = \frac{4\pi}{3}$ as we move through the origin and head towards the negative y-axis.

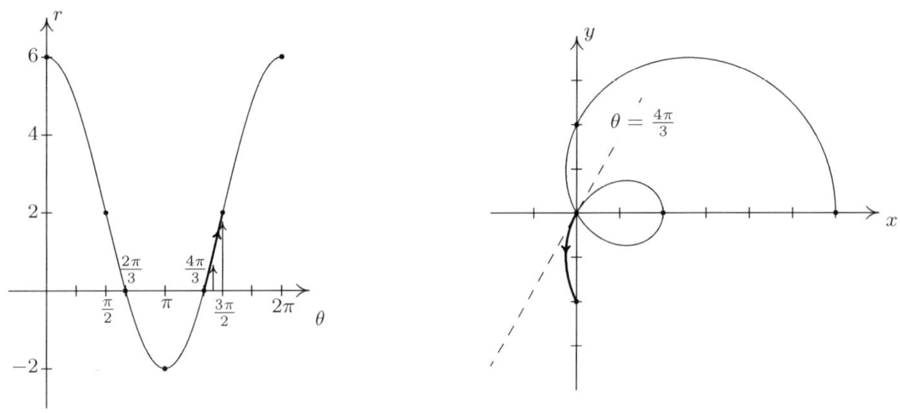

As we round out the interval, we find that as θ runs through $\frac{3\pi}{2}$ to 2π, r increases from 2 out to 6, and we end up back where we started, 6 units from the origin on the positive x-axis.

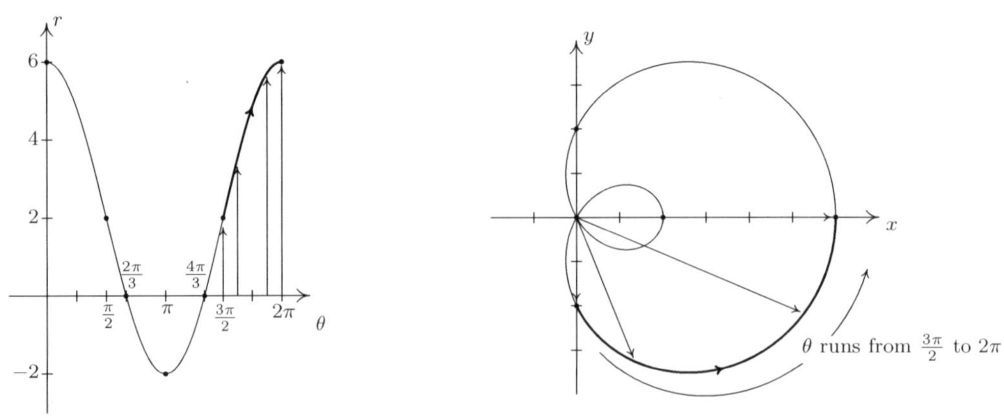

Again, we invite the reader to show that plotting the curve for values of θ outside $[0, 2\pi]$ results in retracing a portion of the curve already traced. Our final graph is below.

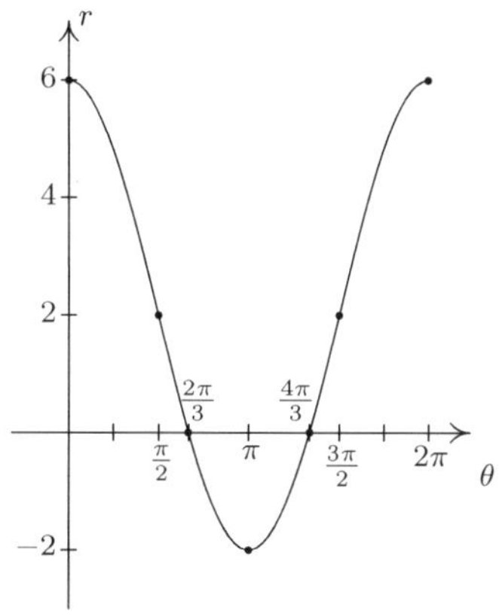

$r = 2 + 4\cos(\theta)$ in the θr-plane

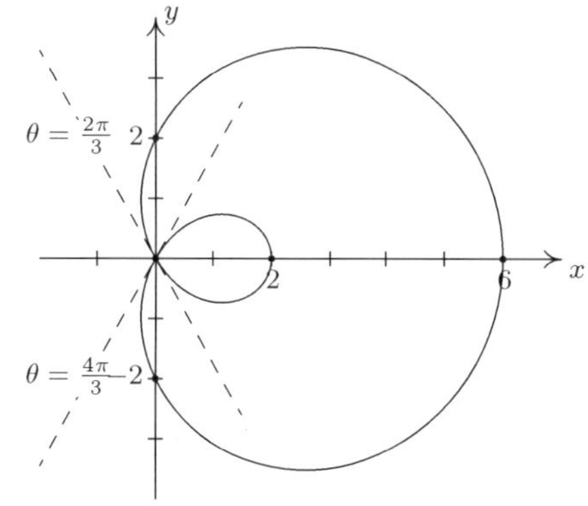

$r = 2 + 4\cos(\theta)$ in the xy-plane

3. As usual, we start by graphing a fundamental cycle of $r = 5\sin(2\theta)$ in the θr-plane, which in this case, occurs as θ ranges from 0 to π. We partition our interval into subintervals to help us with the graphing, namely $\left[0, \frac{\pi}{4}\right]$, $\left[\frac{\pi}{4}, \frac{\pi}{2}\right]$, $\left[\frac{\pi}{2}, \frac{3\pi}{4}\right]$ and $\left[\frac{3\pi}{4}, \pi\right]$. As θ ranges from 0 to $\frac{\pi}{4}$, r increases from 0 to 5. This means that the graph of $r = 5\sin(2\theta)$ in the xy-plane starts at the origin and gradually sweeps out so it is 5 units away from the origin on the line $\theta = \frac{\pi}{4}$.

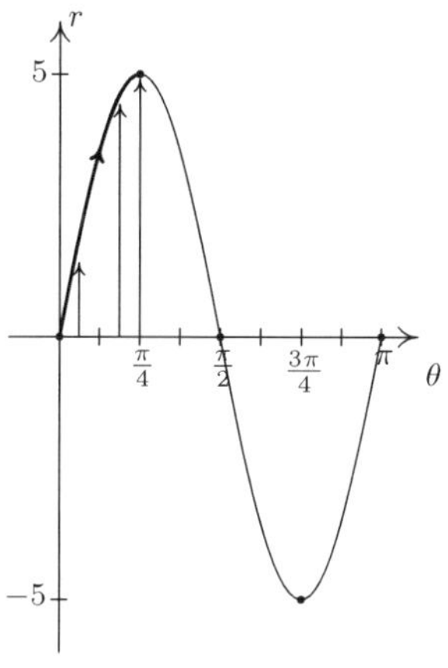

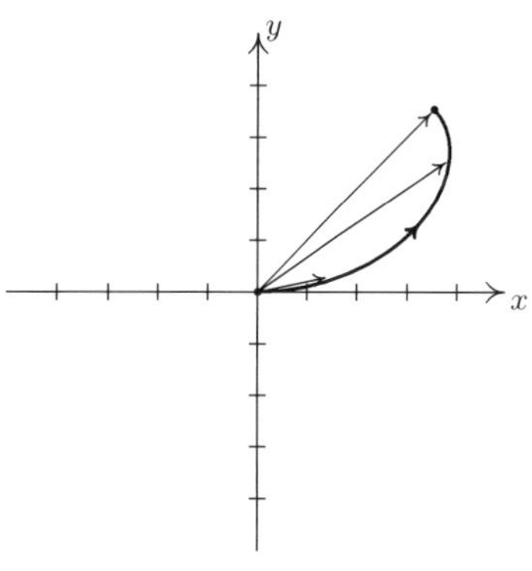

Next, we see that r decreases from 5 to 0 as θ runs through $\left[\frac{\pi}{4}, \frac{\pi}{2}\right]$, and furthermore, r is heading negative as θ crosses $\frac{\pi}{2}$. Hence, we draw the curve hugging the line $\theta = \frac{\pi}{2}$ (the y-axis) as the curve heads to the origin.

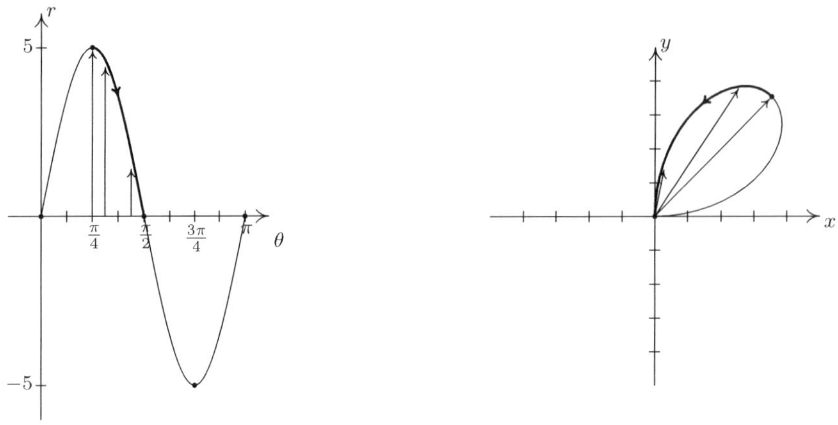

As θ runs from $\frac{\pi}{2}$ to $\frac{3\pi}{4}$, r becomes negative and ranges from 0 to -5. Since $r \leq 0$, the curve pulls away from the negative y-axis into Quadrant IV.

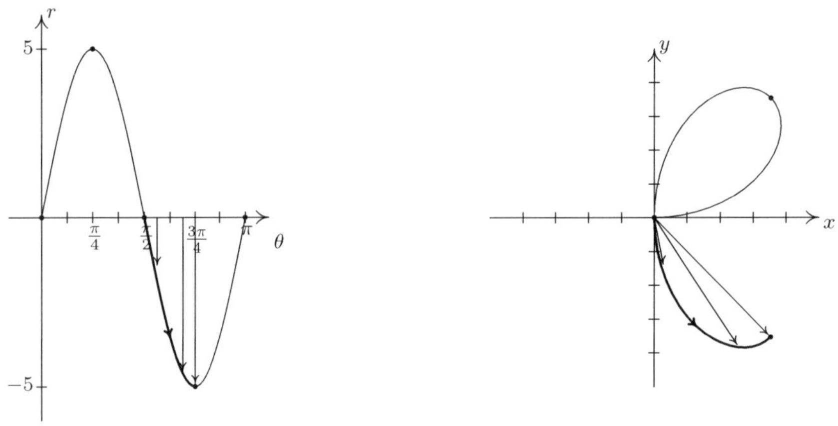

For $\frac{3\pi}{4} \leq \theta \leq \pi$, r increases from -5 to 0, so the curve pulls back to the origin.

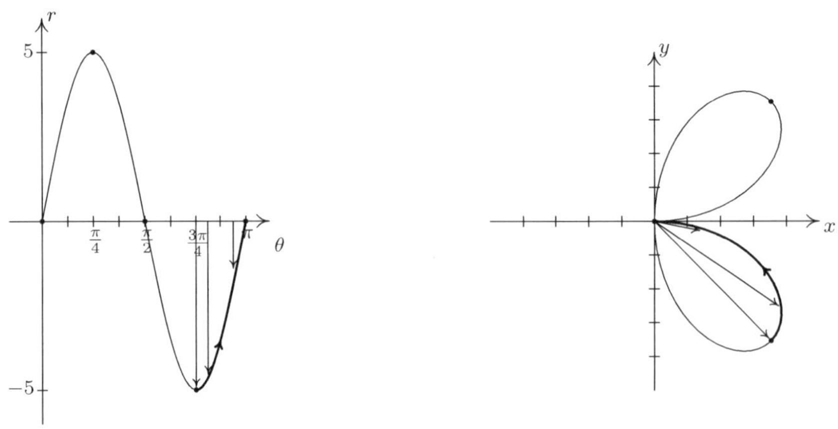

Even though we have finished with one complete cycle of $r = 5\sin(2\theta)$, if we continue plotting beyond $\theta = \pi$, we find that the curve continues into the third quadrant! Below we present a graph of a second cycle of $r = 5\sin(2\theta)$ which continues on from the first. The boxed labels on the θ-axis correspond to the portions with matching labels on the curve in the xy-plane.

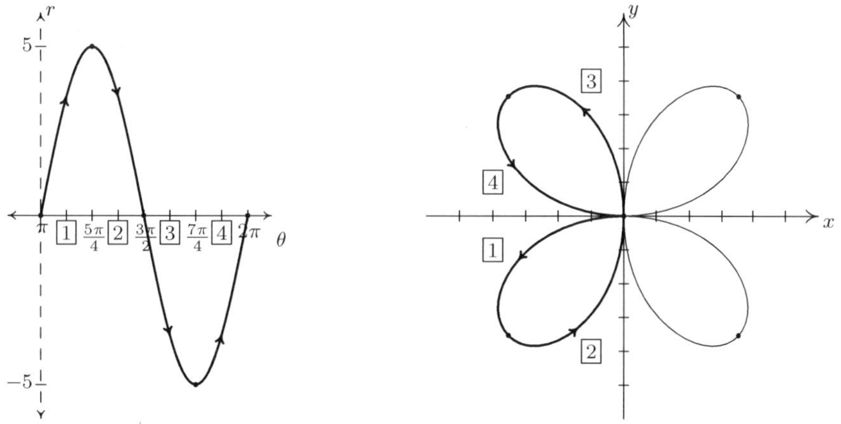

We have the final graph below.

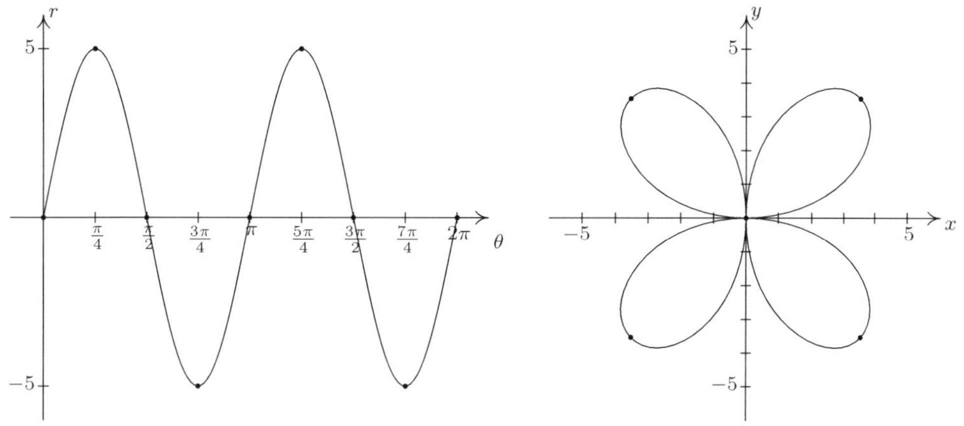

$r = 5\sin(2\theta)$ in the θr-plane $\qquad$ $r = 5\sin(2\theta)$ in the xy-plane

4. Graphing $r^2 = 16\cos(2\theta)$ is complicated by the r^2, so we solve to get $r = \pm\sqrt{16\cos(2\theta)} = \pm 4\sqrt{\cos(2\theta)}$. How do we sketch such a curve? First off, we sketch a fundamental period of $r = \cos(2\theta)$ which we have dotted in the figure below. When $\cos(2\theta) < 0$, $\sqrt{\cos(2\theta)}$ is undefined, so we don't have any values on the interval $\left(\frac{\pi}{4}, \frac{3\pi}{4}\right)$. On the intervals which remain, $\cos(2\theta)$ ranges from 0 to 1, inclusive. Hence, $\sqrt{\cos(2\theta)}$ ranges from 0 to 1 as well.[7] From this, we know $r = \pm 4\sqrt{\cos(2\theta)}$ ranges continuously from 0 to ± 4, respectively. Below we graph both $r = 4\sqrt{\cos(2\theta)}$ and $r = -4\sqrt{\cos(2\theta)}$ on the θr plane and use them to sketch the corresponding pieces of the curve $r^2 = 16\cos(2\theta)$ in the xy-plane. As we have seen in earlier

[7]Owing to the relationship between $y = x$ and $y = \sqrt{x}$ over $[0, 1]$, we also know $\sqrt{\cos(2\theta)} \geq \cos(2\theta)$ wherever the former is defined.

examples, the lines $\theta = \frac{\pi}{4}$ and $\theta = \frac{3\pi}{4}$, which are the zeros of the functions $r = \pm 4\sqrt{\cos(2\theta)}$, serve as guides for us to draw the curve as is passes through the origin.

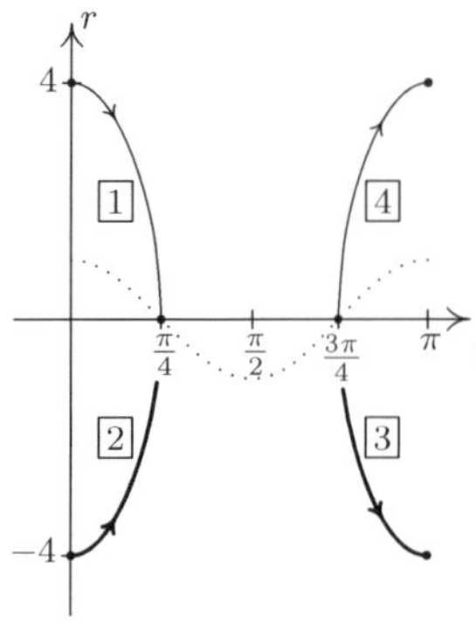

$r = 4\sqrt{\cos(2\theta)}$ and
$\boldsymbol{r = -4\sqrt{\cos(2\theta)}}$

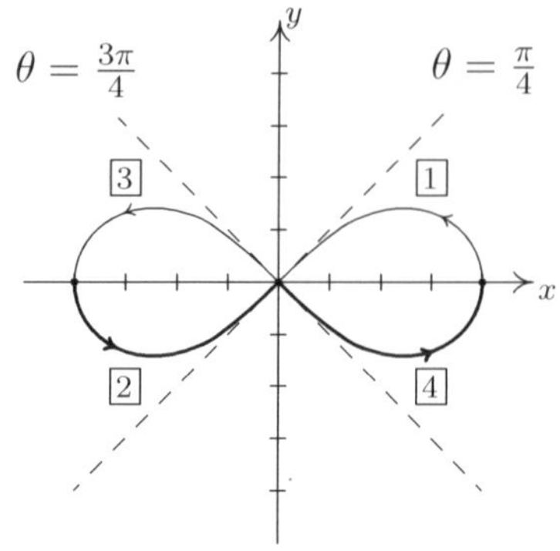

As we plot points corresponding to values of θ outside of the interval $[0, \pi]$, we find ourselves retracing parts of the curve,[8] so our final answer is below.

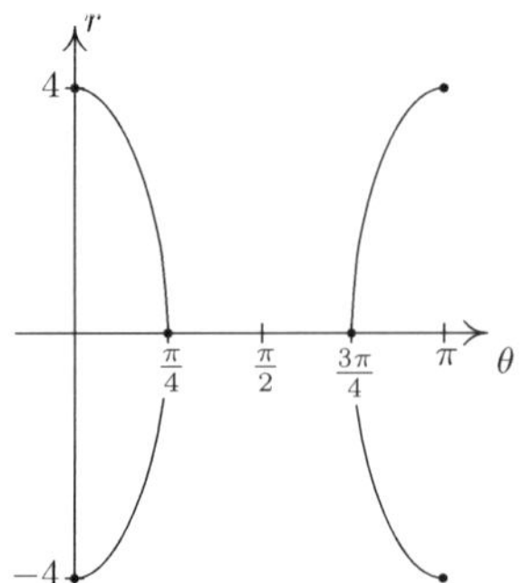

$r = \pm 4\sqrt{\cos(2\theta)}$
in the θr-plane

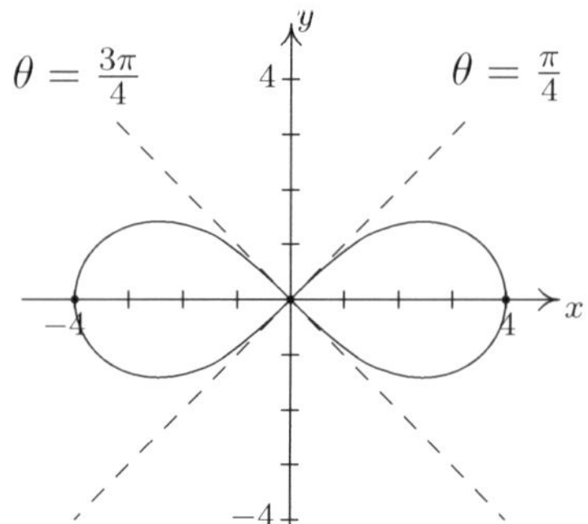

$r^2 = 16\cos(2\theta)$
in the xy-plane

□

A few remarks are in order. First, there is no relation, in general, between the period of the function $f(\theta)$ and the length of the interval required to sketch the complete graph of $r = f(\theta)$ in the xy-plane. As we saw on page 941, despite the fact that the period of $f(\theta) = 6\cos(\theta)$ is 2π, we sketched the complete graph of $r = 6\cos(\theta)$ in the xy-plane just using the values of θ as θ ranged from 0 to π. In Example 11.5.2, number 3, the period of $f(\theta) = 5\sin(2\theta)$ is π, but in order to obtain the complete graph of $r = 5\sin(2\theta)$, we needed to run θ from 0 to 2π. While many of the 'common' polar graphs can be grouped into families,[9] the authors truly feel that taking the time to work through each graph in the manner presented here is the best way to not only understand the polar

[8]In this case, we could have generated the entire graph by using just the plot $r = 4\sqrt{\cos(2\theta)}$, but graphed over the interval $[0, 2\pi]$ in the θr-plane. We leave the details to the reader.

[9]Numbers 1 and 2 in Example 11.5.2 are examples of 'limaçons,' number 3 is an example of a 'polar rose,' and number 4 is the famous 'Lemniscate of Bernoulli.'

coordinate system, but also prepare you for what is needed in Calculus. Second, the symmetry seen in the examples is also a common occurrence when graphing polar equations. In addition to the usual kinds of symmetry discussed up to this point in the text (symmetry about each axis and the origin), it is possible to talk about *rotational* symmetry. We leave the discussion of symmetry to the Exercises. In our next example, we are given the task of finding the intersection points of polar curves. According to the Fundamental Graphing Principle for Polar Equations on page 938, in order for a point P to be on the graph of a polar equation, it must have a *representation* $P(r, \theta)$ which satisfies the equation. What complicates matters in polar coordinates is that any given point has infinitely many representations. As a result, if a point P is on the graph of two different polar equations, it is entirely possible that the representation $P(r, \theta)$ which satisfies one of the equations does not satisfy the other equation. Here, more than ever, we need to rely on the Geometry as much as the Algebra to find our solutions.

Example 11.5.3. Find the points of intersection of the graphs of the following polar equations.

1. $r = 2\sin(\theta)$ and $r = 2 - 2\sin(\theta)$
2. $r = 2$ and $r = 3\cos(\theta)$
3. $r = 3$ and $r = 6\cos(2\theta)$
4. $r = 3\sin\left(\frac{\theta}{2}\right)$ and $r = 3\cos\left(\frac{\theta}{2}\right)$

Solution.

1. Following the procedure in Example 11.5.2, we graph $r = 2\sin(\theta)$ and find it to be a circle centered at the point with rectangular coordinates $(0, 1)$ with a radius of 1. The graph of $r = 2 - 2\sin(\theta)$ is a special kind of limaçon called a 'cardioid.'[10]

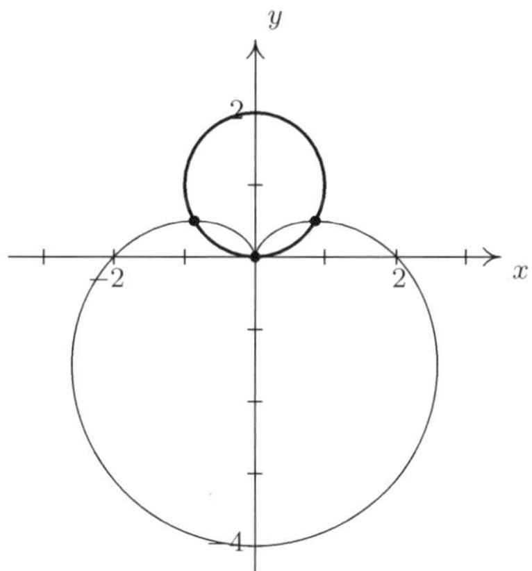

$r = 2 - 2\sin(\theta)$ and $\boldsymbol{r = 2\sin(\theta)}$

It appears as if there are three intersection points: one in the first quadrant, one in the second quadrant, and the origin. Our next task is to find polar representations of these points. In

[10] Presumably, the name is derived from its resemblance to a stylized human heart.

order for a point P to be on the graph of $r = 2\sin(\theta)$, it must have a representation $P(r,\theta)$ which satisfies $r = 2\sin(\theta)$. If P is also on the graph of $r = 2-2\sin(\theta)$, then P has a (possibly different) representation $P(r',\theta')$ which satisfies $r' = 2\sin(\theta')$. We first try to see if we can find any points which have a single representation $P(r,\theta)$ that satisfies both $r = 2\sin(\theta)$ and $r = 2-2\sin(\theta)$. Assuming such a pair (r,θ) exists, then equating[11] the expressions for r gives $2\sin(\theta) = 2-2\sin(\theta)$ or $\sin(\theta) = \frac{1}{2}$. From this, we get $\theta = \frac{\pi}{6} + 2\pi k$ or $\theta = \frac{5\pi}{6} + 2\pi k$ for integers k. Plugging $\theta = \frac{\pi}{6}$ into $r = 2\sin(\theta)$, we get $r = 2\sin\left(\frac{\pi}{6}\right) = 2\left(\frac{1}{2}\right) = 1$, which is also the value we obtain when we substitute it into $r = 2-2\sin(\theta)$. Hence, $\left(1, \frac{\pi}{6}\right)$ is one representation for the point of intersection in the first quadrant. For the point of intersection in the second quadrant, we try $\theta = \frac{5\pi}{6}$. Both equations give us the point $\left(1, \frac{5\pi}{6}\right)$, so this is our answer here. What about the origin? We know from Section 11.4 that the pole may be represented as $(0,\theta)$ for any angle θ. On the graph of $r = 2\sin(\theta)$, we start at the origin when $\theta = 0$ and return to it at $\theta = \pi$, and as the reader can verify, we are at the origin exactly when $\theta = \pi k$ for integers k. On the curve $r = 2-2\sin(\theta)$, however, we reach the origin when $\theta = \frac{\pi}{2}$, and more generally, when $\theta = \frac{\pi}{2} + 2\pi k$ for integers k. There is no integer value of k for which $\pi k = \frac{\pi}{2} + 2\pi k$ which means while the origin is on both graphs, the point is never reached simultaneously. In any case, we have determined the three points of intersection to be $\left(1, \frac{\pi}{6}\right)$, $\left(1, \frac{5\pi}{6}\right)$ and the origin.

2. As before, we make a quick sketch of $r = 2$ and $r = 3\cos(\theta)$ to get feel for the number and location of the intersection points. The graph of $r = 2$ is a circle, centered at the origin, with a radius of 2. The graph of $r = 3\cos(\theta)$ is also a circle - but this one is centered at the point with rectangular coordinates $\left(\frac{3}{2}, 0\right)$ and has a radius of $\frac{3}{2}$.

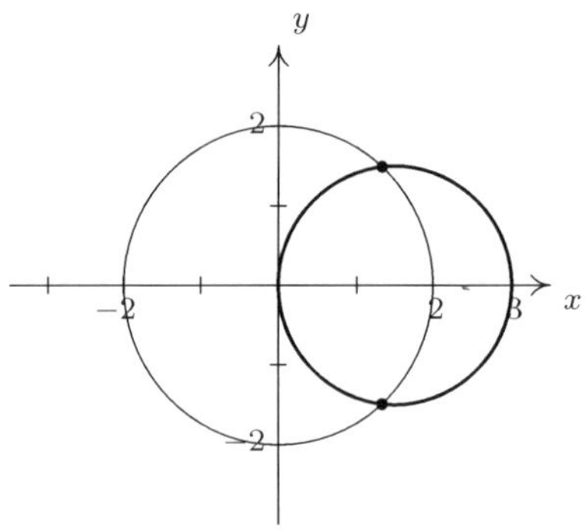

$r = 2$ and $\boldsymbol{r = 3\cos(\theta)}$

We have two intersection points to find, one in Quadrant I and one in Quadrant IV. Proceeding as above, we first determine if any of the intersection points P have a representation (r,θ) which satisfies both $r = 2$ and $r = 3\cos(\theta)$. Equating these two expressions for r, we get $\cos(\theta) = \frac{2}{3}$. To solve this equation, we need the arccosine function. We get

[11]We are really using the technique of substitution to solve the system of equations $\left\{ \begin{array}{rcl} r & = & 2\sin(\theta) \\ r & = & 2-2\sin(\theta) \end{array} \right.$

$\theta = \arccos\left(\frac{2}{3}\right) + 2\pi k$ or $\theta = 2\pi - \arccos\left(\frac{2}{3}\right) + 2\pi k$ for integers k. From these solutions, we get $\left(2, \arccos\left(\frac{2}{3}\right)\right)$ as one representation for our answer in Quadrant I, and $\left(2, 2\pi - \arccos\left(\frac{2}{3}\right)\right)$ as one representation for our answer in Quadrant IV. The reader is encouraged to check these results algebraically and geometrically.

3. Proceeding as above, we first graph $r = 3$ and $r = 6\cos(2\theta)$ to get an idea of how many intersection points to expect and where they lie. The graph of $r = 3$ is a circle centered at the origin with a radius of 3 and the graph of $r = 6\cos(2\theta)$ is another four-leafed rose.[12]

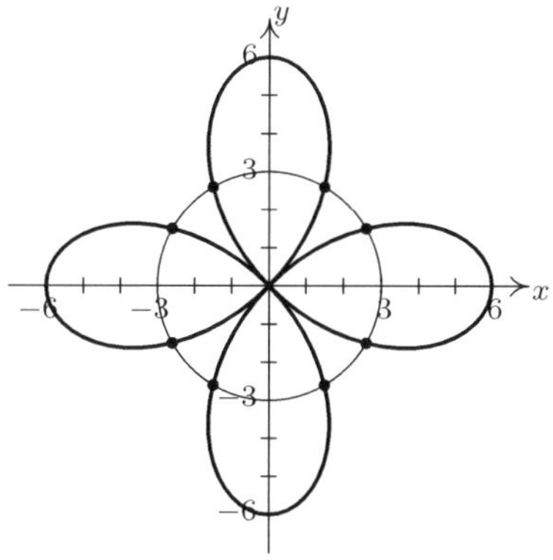

$r = 3$ and $\boldsymbol{r = 6\cos(2\theta)}$

It appears as if there are eight points of intersection - two in each quadrant. We first look to see if there any points $P(r, \theta)$ with a representation that satisfies both $r = 3$ and $r = 6\cos(2\theta)$. For these points, $6\cos(2\theta) = 3$ or $\cos(2\theta) = \frac{1}{2}$. Solving, we get $\theta = \frac{\pi}{6} + \pi k$ or $\theta = \frac{5\pi}{6} + \pi k$ for integers k. Out of all of these solutions, we obtain just four distinct points represented by $\left(3, \frac{\pi}{6}\right)$, $\left(3, \frac{5\pi}{6}\right)$, $\left(3, \frac{7\pi}{6}\right)$ and $\left(3, \frac{11\pi}{6}\right)$. To determine the coordinates of the remaining four points, we have to consider how the representations of the points of intersection can differ. We know from Section 11.4 that if (r, θ) and (r', θ') represent the same point and $r \neq 0$, then either $r = r'$ or $r = -r'$. If $r = r'$, then $\theta' = \theta + 2\pi k$, so one possibility is that an intersection point P has a representation (r, θ) which satisfies $r = 3$ and another representation $(r, \theta + 2\pi k)$ for some integer, k which satisfies $r = 6\cos(2\theta)$. At this point,[13] if we replace every occurrence of θ in the equation $r = 6\cos(2\theta)$ with $(\theta + 2\pi k)$ and then see if, by equating the resulting expressions for r, we get any more solutions for θ. Since $\cos(2(\theta + 2\pi k)) = \cos(2\theta + 4\pi k) = \cos(2\theta)$ for every integer k, however, the equation $r = 6\cos(2(\theta + 2\pi k))$ reduces to the same equation we had before, $r = 6\cos(2\theta)$, which means we get no additional solutions. Moving on to the case where $r = -r'$, we have that $\theta' = \theta + (2k+1)\pi$ for integers k. We look to see if we can find points P which have a representation (r, θ) that satisfies $r = 3$ and another,

[12]See Example 11.5.2 number 3.

[13]The authors have chosen to replace θ with $\theta + 2\pi k$ in the equation $r = 6\cos(2\theta)$ for illustration purposes only. We could have just as easily chosen to do this substitution in the equation $r = 3$. Since there is no θ in $r = 3$, however, this case would reduce to the previous case instantly. The reader is encouraged to follow this latter procedure in the interests of efficiency.

$(-r, \theta + (2k+1)\pi)$, that satisfies $r = 6\cos(2\theta)$. To do this, we substitute[14] $(-r)$ for r and $(\theta + (2k+1)\pi)$ for θ in the equation $r = 6\cos(2\theta)$ and get $-r = 6\cos(2(\theta + (2k+1)\pi))$. Since $\cos(2(\theta + (2k+1)\pi)) = \cos(2\theta + (2k+1)(2\pi)) = \cos(2\theta)$ for all integers k, the equation $-r = 6\cos(2(\theta + (2k+1)\pi))$ reduces to $-r = 6\cos(2\theta)$, or $r = -6\cos(2\theta)$. Coupling this equation with $r = 3$ gives $-6\cos(2\theta) = 3$ or $\cos(2\theta) = -\frac{1}{2}$. We get $\theta = \frac{\pi}{3} + \pi k$ or $\theta = \frac{2\pi}{3} + \pi k$. From these solutions, we obtain[15] the remaining four intersection points with representations $\left(-3, \frac{\pi}{3}\right)$, $\left(-3, \frac{2\pi}{3}\right)$, $\left(-3, \frac{4\pi}{3}\right)$ and $\left(-3, \frac{5\pi}{3}\right)$, which we can readily check graphically.

4. As usual, we begin by graphing $r = 3\sin\left(\frac{\theta}{2}\right)$ and $r = 3\cos\left(\frac{\theta}{2}\right)$. Using the techniques presented in Example 11.5.2, we find that we need to plot both functions as θ ranges from 0 to 4π to obtain the complete graph. To our surprise and/or delight, it appears as if these two equations describe the *same* curve!

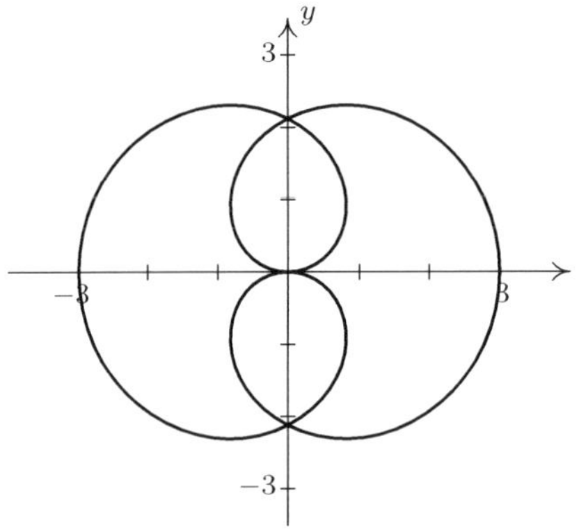

$r = 3\sin\left(\frac{\theta}{2}\right)$ and $\boldsymbol{r = 3\cos\left(\frac{\theta}{2}\right)}$
appear to determine the same curve in the $\boldsymbol{xy}$-plane

To verify this incredible claim,[16] we need to show that, in fact, the graphs of these two equations intersect at all points on the plane. Suppose P has a representation (r, θ) which satisfies both $r = 3\sin\left(\frac{\theta}{2}\right)$ and $r = 3\cos\left(\frac{\theta}{2}\right)$. Equating these two expressions for r gives the equation $3\sin\left(\frac{\theta}{2}\right) = 3\cos\left(\frac{\theta}{2}\right)$. While normally we discourage dividing by a variable expression (in case it could be 0), we note here that if $3\cos\left(\frac{\theta}{2}\right) = 0$, then for our equation to hold, $3\sin\left(\frac{\theta}{2}\right) = 0$ as well. Since no angles have both cosine and sine equal to zero, we are safe to divide both sides of the equation $3\sin\left(\frac{\theta}{2}\right) = 3\cos\left(\frac{\theta}{2}\right)$ by $3\cos\left(\frac{\theta}{2}\right)$ to get $\tan\left(\frac{\theta}{2}\right) = 1$ which gives $\theta = \frac{\pi}{2} + 2\pi k$ for integers k. From these solutions, however, we

[14] Again, we could have easily chosen to substitute these into $r = 3$ which would give $-r = 3$, or $r = -3$.

[15] We obtain these representations by substituting the values for θ into $r = 6\cos(2\theta)$, once again, for illustration purposes. Again, in the interests of efficiency, we could 'plug' these values for θ into $r = 3$ (where there is no θ) and get the list of points: $\left(3, \frac{\pi}{3}\right)$, $\left(3, \frac{2\pi}{3}\right)$, $\left(3, \frac{4\pi}{3}\right)$ and $\left(3, \frac{5\pi}{3}\right)$. While it is not true that $\left(3, \frac{\pi}{3}\right)$ represents the same point as $\left(-3, \frac{\pi}{3}\right)$, we still get the same *set* of solutions.

[16] A quick sketch of $r = 3\sin\left(\frac{\theta}{2}\right)$ and $r = 3\cos\left(\frac{\theta}{2}\right)$ in the θr-plane will convince you that, viewed as functions of r, these are two different animals.

get only *one* intersection point which can be represented by $\left(\frac{3\sqrt{2}}{2}, \frac{\pi}{2}\right)$. We now investigate other representations for the intersection points. Suppose P is an intersection point with a representation (r, θ) which satisfies $r = 3\sin\left(\frac{\theta}{2}\right)$ and the same point P has a different representation $(r, \theta + 2\pi k)$ for some integer k which satisfies $r = 3\cos\left(\frac{\theta}{2}\right)$. Substituting into the latter, we get $r = 3\cos\left(\frac{1}{2}\left[\theta + 2\pi k\right]\right) = 3\cos\left(\frac{\theta}{2} + \pi k\right)$. Using the sum formula for cosine, we expand $3\cos\left(\frac{\theta}{2} + \pi k\right) = 3\cos\left(\frac{\theta}{2}\right)\cos(\pi k) - 3\sin\left(\frac{\theta}{2}\right)\sin(\pi k) = \pm 3\cos\left(\frac{\theta}{2}\right)$, since $\sin(\pi k) = 0$ for all integers k, and $\cos(\pi k) = \pm 1$ for all integers k. If k is an even integer, we get the same equation $r = 3\cos\left(\frac{\theta}{2}\right)$ as before. If k is odd, we get $r = -3\cos\left(\frac{\theta}{2}\right)$. This latter expression for r leads to the equation $3\sin\left(\frac{\theta}{2}\right) = -3\cos\left(\frac{\theta}{2}\right)$, or $\tan\left(\frac{\theta}{2}\right) = -1$. Solving, we get $\theta = -\frac{\pi}{2} + 2\pi k$ for integers k, which gives the intersection point $\left(\frac{3\sqrt{2}}{2}, -\frac{\pi}{2}\right)$. Next, we assume P has a representation (r, θ) which satisfies $r = 3\sin\left(\frac{\theta}{2}\right)$ and a representation $(-r, \theta + (2k+1)\pi)$ which satisfies $r = 3\cos\left(\frac{\theta}{2}\right)$ for some integer k. Substituting $(-r)$ for r and $(\theta + (2k+1)\pi)$ in for θ into $r = 3\cos\left(\frac{\theta}{2}\right)$ gives $-r = 3\cos\left(\frac{1}{2}\left[\theta + (2k+1)\pi\right]\right)$. Once again, we use the sum formula for cosine to get

$$\begin{aligned}
\cos\left(\tfrac{1}{2}\left[\theta + (2k+1)\pi\right]\right) &= \cos\left(\frac{\theta}{2} + \frac{(2k+1)\pi}{2}\right) \\
&= \cos\left(\tfrac{\theta}{2}\right)\cos\left(\frac{(2k+1)\pi}{2}\right) - \sin\left(\tfrac{\theta}{2}\right)\sin\left(\frac{(2k+1)\pi}{2}\right) \\
&= \pm\sin\left(\tfrac{\theta}{2}\right)
\end{aligned}$$

where the last equality is true since $\cos\left(\frac{(2k+1)\pi}{2}\right) = 0$ and $\sin\left(\frac{(2k+1)\pi}{2}\right) = \pm 1$ for integers k. Hence, $-r = 3\cos\left(\frac{1}{2}\left[\theta + (2k+1)\pi\right]\right)$ can be rewritten as $r = \pm 3\sin\left(\frac{\theta}{2}\right)$. If we choose $k = 0$, then $\sin\left(\frac{(2k+1)\pi}{2}\right) = \sin\left(\frac{\pi}{2}\right) = 1$, and the equation $-r = 3\cos\left(\frac{1}{2}\left[\theta + (2k+1)\pi\right]\right)$ in this case reduces to $-r = -3\sin\left(\frac{\theta}{2}\right)$, or $r = 3\sin\left(\frac{\theta}{2}\right)$ which is the other equation under consideration! What this means is that if a polar representation (r, θ) for the point P satisfies $r = 3\sin(\frac{\theta}{2})$, then the representation $(-r, \theta + \pi)$ for P automatically satisfies $r = 3\cos\left(\frac{\theta}{2}\right)$. Hence the equations $r = 3\sin(\frac{\theta}{2})$ and $r = 3\cos(\frac{\theta}{2})$ determine the same set of points in the plane. □

Our work in Example 11.5.3 justifies the following.

Guidelines for Finding Points of Intersection of Graphs of Polar Equations

To find the points of intersection of the graphs of two polar equations E_1 and E_2:

- Sketch the graphs of E_1 and E_2. Check to see if the curves intersect at the origin (pole).
- Solve for pairs (r, θ) which satisfy both E_1 and E_2.
- Substitute $(\theta + 2\pi k)$ for θ in either one of E_1 or E_2 (but not both) and solve for pairs (r, θ) which satisfy both equations. Keep in mind that k is an integer.
- Substitute $(-r)$ for r and $(\theta + (2k+1)\pi)$ for θ in either one of E_1 or E_2 (but not both) and solve for pairs (r, θ) which satisfy both equations. Keep in mind that k is an integer.

Our last example ties together graphing and points of intersection to describe regions in the plane.

Example 11.5.4. Sketch the region in the xy-plane described by the following sets.

1. $\left\{(r,\theta)\,|\,0 \leq r \leq 5\sin(2\theta),\, 0 \leq \theta \leq \frac{\pi}{2}\right\}$
2. $\left\{(r,\theta)\,|\,3 \leq r \leq 6\cos(2\theta),\, 0 \leq \theta \leq \frac{\pi}{6}\right\}$
3. $\left\{(r,\theta)\,|\,2+4\cos(\theta) \leq r \leq 0,\, \frac{2\pi}{3} \leq \theta \leq \frac{4\pi}{3}\right\}$
4. $\left\{(r,\theta)\,|\,0 \leq r \leq 2\sin(\theta),\, 0 \leq \theta \leq \frac{\pi}{6}\right\} \cup \left\{(r,\theta)\,|\,0 \leq r \leq 2-2\sin(\theta),\, \frac{\pi}{6} \leq \theta \leq \frac{\pi}{2}\right\}$

Solution. Our first step in these problems is to sketch the graphs of the polar equations involved to get a sense of the geometric situation. Since all of the equations in this example are found in either Example 11.5.2 or Example 11.5.3, most of the work is done for us.

1. We know from Example 11.5.2 number 3 that the graph of $r = 5\sin(2\theta)$ is a rose. Moreover, we know from our work there that as $0 \leq \theta \leq \frac{\pi}{2}$, we are tracing out the 'leaf' of the rose which lies in the first quadrant. The inequality $0 \leq r \leq 5\sin(2\theta)$ means we want to capture all of the points between the origin ($r = 0$) and the curve $r = 5\sin(2\theta)$ as θ runs through $\left[0, \frac{\pi}{2}\right]$. Hence, the region we seek is the leaf itself.

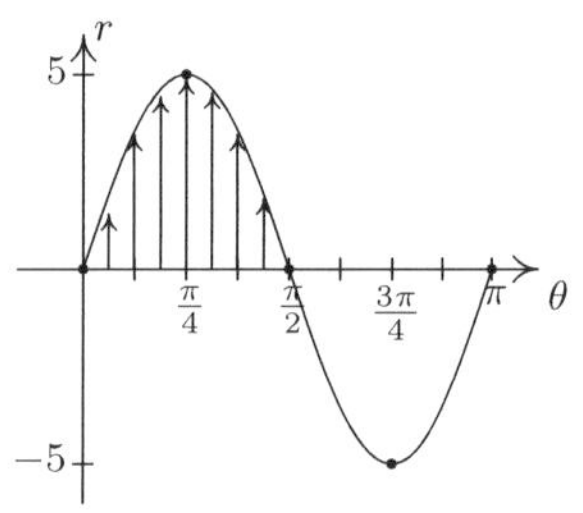

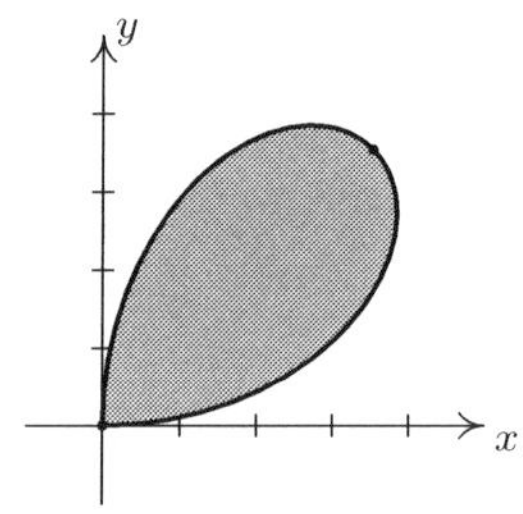

$\left\{(r,\theta)\,|\,0 \leq r \leq 5\sin(2\theta), 0 \leq \theta \leq \frac{\pi}{2}\right\}$

2. We know from Example 11.5.3 number 3 that $r = 3$ and $r = 6\cos(2\theta)$ intersect at $\theta = \frac{\pi}{6}$, so the region that is being described here is the set of points whose directed distance r from the origin is at least 3 but no more than $6\cos(2\theta)$ as θ runs from 0 to $\frac{\pi}{6}$. In other words, we are looking at the points outside or on the circle (since $r \geq 3$) but inside or on the rose (since $r \leq 6\cos(2\theta)$). We shade the region below.

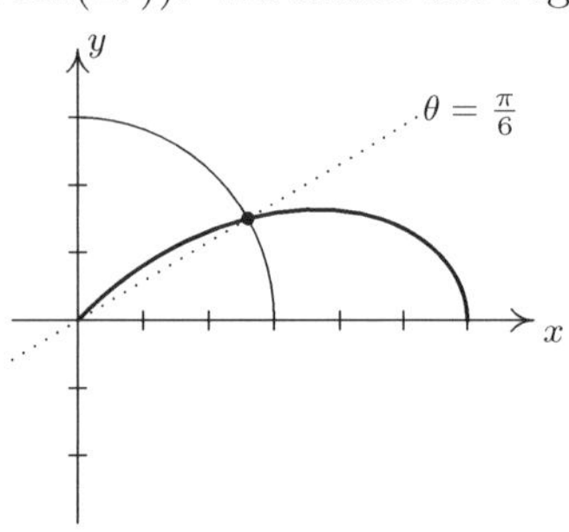

$r = 3$ and $\boldsymbol{r = 6\cos(2\theta)}$

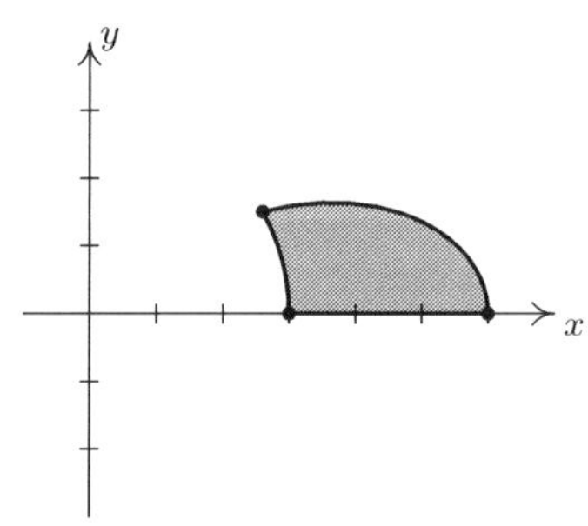

$\left\{(r,\theta)\,|\,3 \leq r \leq 6\cos(2\theta), 0 \leq \theta \leq \frac{\pi}{6}\right\}$

3. From Example 11.5.2 number 2, we know that the graph of $r = 2 + 4\cos(\theta)$ is a limaçon whose 'inner loop' is traced out as θ runs through the given values $\frac{2\pi}{3}$ to $\frac{4\pi}{3}$. Since the values r takes on in this interval are non-positive, the inequality $2 + 4\cos(\theta) \leq r \leq 0$ makes sense, and we are looking for all of the points between the pole $r = 0$ and the limaçon as θ ranges over the interval $\left[\frac{2\pi}{3}, \frac{4\pi}{3}\right]$. In other words, we shade in the inner loop of the limaçon.

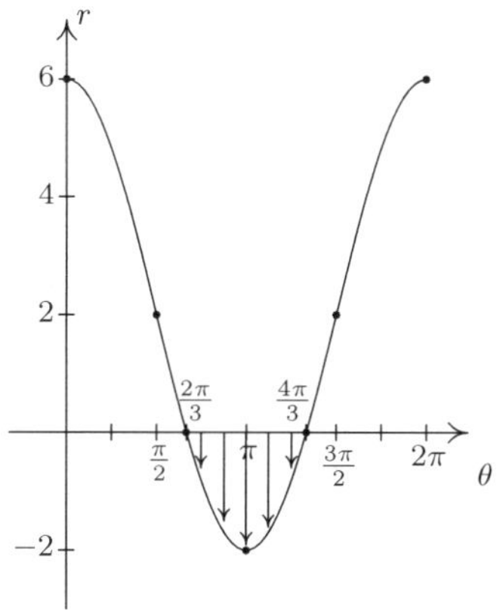

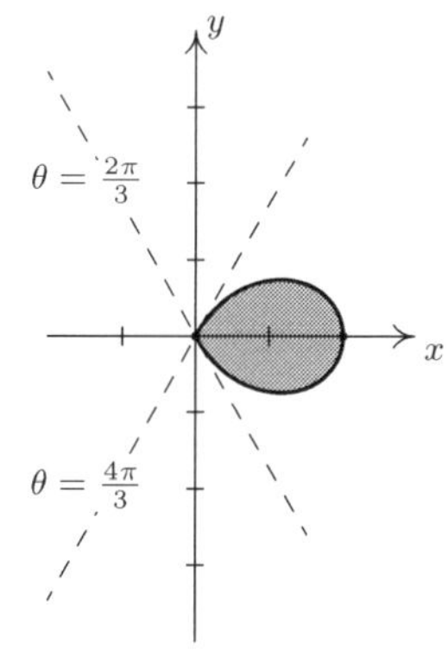

$\left\{(r,\theta) \,|\, 2 + 4\cos(\theta) \leq r \leq 0, \frac{2\pi}{3} \leq \theta \leq \frac{4\pi}{3}\right\}$

4. We have two regions described here connected with the union symbol '$\cup$.' We shade each in turn and find our final answer by combining the two. In Example 11.5.3, number 1, we found that the curves $r = 2\sin(\theta)$ and $r = 2 - 2\sin(\theta)$ intersect when $\theta = \frac{\pi}{6}$. Hence, for the first region, $\left\{(r,\theta) \,|\, 0 \leq r \leq 2\sin(\theta), 0 \leq \theta \leq \frac{\pi}{6}\right\}$, we are shading the region between the origin $(r = 0)$ out to the circle $(r = 2\sin(\theta))$ as θ ranges from 0 to $\frac{\pi}{6}$, which is the angle of intersection of the two curves. For the second region, $\left\{(r,\theta) \,|\, 0 \leq r \leq 2 - 2\sin(\theta), \frac{\pi}{6} \leq \theta \leq \frac{\pi}{2}\right\}$, θ picks up where it left off at $\frac{\pi}{6}$ and continues to $\frac{\pi}{2}$. In this case, however, we are shading from the origin $(r = 0)$ out to the cardioid $r = 2 - 2\sin(\theta)$ which pulls into the origin at $\theta = \frac{\pi}{2}$. Putting these two regions together gives us our final answer.

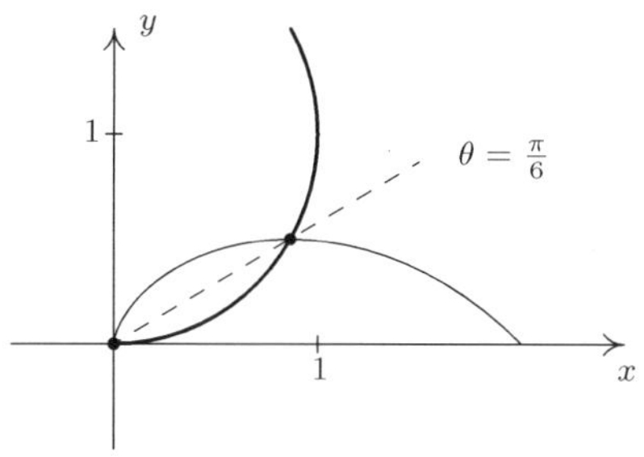

$r = 2 - 2\sin(\theta)$ and $\boldsymbol{r = 2\sin(\theta)}$

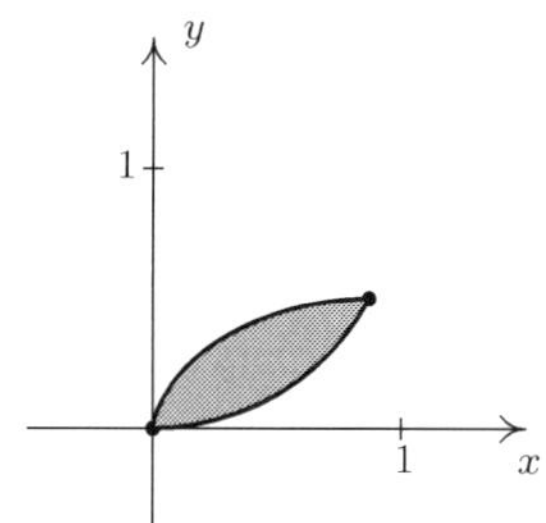

$\left\{(r,\theta) \,|\, 0 \leq r \leq 2\sin(\theta),\ 0 \leq \theta \leq \frac{\pi}{6}\right\} \cup$
$\left\{(r,\theta) \,|\, 0 \leq r \leq 2 - 2\sin(\theta),\ \frac{\pi}{6} \leq \theta \leq \frac{\pi}{2}\right\}$

□

11.5.1 EXERCISES

In Exercises 1 - 20, plot the graph of the polar equation by hand. Carefully label your graphs.

1. Circle: $r = 6\sin(\theta)$
2. Circle: $r = 2\cos(\theta)$
3. Rose: $r = 2\sin(2\theta)$
4. Rose: $r = 4\cos(2\theta)$
5. Rose: $r = 5\sin(3\theta)$
6. Rose: $r = \cos(5\theta)$
7. Rose: $r = \sin(4\theta)$
8. Rose: $r = 3\cos(4\theta)$
9. Cardioid: $r = 3 - 3\cos(\theta)$
10. Cardioid: $r = 5 + 5\sin(\theta)$
11. Cardioid: $r = 2 + 2\cos(\theta)$
12. Cardioid: $r = 1 - \sin(\theta)$
13. Limaçon: $r = 1 - 2\cos(\theta)$
14. Limaçon: $r = 1 - 2\sin(\theta)$
15. Limaçon: $r = 2\sqrt{3} + 4\cos(\theta)$
16. Limaçon: $r = 3 - 5\cos(\theta)$
17. Limaçon: $r = 3 - 5\sin(\theta)$
18. Limaçon: $r = 2 + 7\sin(\theta)$
19. Lemniscate: $r^2 = \sin(2\theta)$
20. Lemniscate: $r^2 = 4\cos(2\theta)$

In Exercises 21 - 30, find the exact polar coordinates of the points of intersection of graphs of the polar equations. Remember to check for intersection at the pole (origin).

21. $r = 3\cos(\theta)$ and $r = 1 + \cos(\theta)$
22. $r = 1 + \sin(\theta)$ and $r = 1 - \cos(\theta)$
23. $r = 1 - 2\sin(\theta)$ and $r = 2$
24. $r = 1 - 2\cos(\theta)$ and $r = 1$
25. $r = 2\cos(\theta)$ and $r = 2\sqrt{3}\sin(\theta)$
26. $r = 3\cos(\theta)$ and $r = \sin(\theta)$
27. $r^2 = 4\cos(2\theta)$ and $r = \sqrt{2}$
28. $r^2 = 2\sin(2\theta)$ and $r = 1$
29. $r = 4\cos(2\theta)$ and $r = 2$
30. $r = 2\sin(2\theta)$ and $r = 1$

In Exercises 31 - 40, sketch the region in the xy-plane described by the given set.

31. $\{(r,\theta) \,|\, 0 \leq r \leq 3,\ 0 \leq \theta \leq 2\pi\}$
32. $\{(r,\theta) \,|\, 0 \leq r \leq 4\sin(\theta),\ 0 \leq \theta \leq \pi\}$
33. $\left\{(r,\theta) \,|\, 0 \leq r \leq 3\cos(\theta),\ -\frac{\pi}{2} \leq \theta \leq \frac{\pi}{2}\right\}$
34. $\left\{(r,\theta) \,|\, 0 \leq r \leq 2\sin(2\theta),\ 0 \leq \theta \leq \frac{\pi}{2}\right\}$

35. $\left\{(r,\theta)\,|\,0 \leq r \leq 4\cos(2\theta),\ -\frac{\pi}{4} \leq \theta \leq \frac{\pi}{4}\right\}$

36. $\left\{(r,\theta)\,|\,1 \leq r \leq 1-2\cos(\theta),\ \frac{\pi}{2} \leq \theta \leq \frac{3\pi}{2}\right\}$

37. $\left\{(r,\theta)\,|\,1+\cos(\theta) \leq r \leq 3\cos(\theta),\ -\frac{\pi}{3} \leq \theta \leq \frac{\pi}{3}\right\}$

38. $\left\{(r,\theta)\,|\,1 \leq r \leq \sqrt{2\sin(2\theta)},\ \frac{13\pi}{12} \leq \theta \leq \frac{17\pi}{12}\right\}$

39. $\left\{(r,\theta)\,|\,0 \leq r \leq 2\sqrt{3}\sin(\theta),\ 0 \leq \theta \leq \frac{\pi}{6}\right\} \cup \left\{(r,\theta)\,|\,0 \leq r \leq 2\cos(\theta),\ \frac{\pi}{6} \leq \theta \leq \frac{\pi}{2}\right\}$

40. $\left\{(r,\theta)\,|\,0 \leq r \leq 2\sin(2\theta),\ 0 \leq \theta \leq \frac{\pi}{12}\right\} \cup \left\{(r,\theta)\,|\,0 \leq r \leq 1,\ \frac{\pi}{12} \leq \theta \leq \frac{\pi}{4}\right\}$

In Exercises 41 - 50, use set-builder notation to describe the polar region. Assume that the region contains its bounding curves.

41. The region inside the circle $r = 5$.

42. The region inside the circle $r = 5$ which lies in Quadrant III.

43. The region inside the left half of the circle $r = 6\sin(\theta)$.

44. The region inside the circle $r = 4\cos(\theta)$ which lies in Quadrant IV.

45. The region inside the top half of the cardioid $r = 3 - 3\cos(\theta)$

46. The region inside the cardioid $r = 2 - 2\sin(\theta)$ which lies in Quadrants I and IV.

47. The inside of the petal of the rose $r = 3\cos(4\theta)$ which lies on the positive x-axis

48. The region inside the circle $r = 5$ but outside the circle $r = 3$.

49. The region which lies inside of the circle $r = 3\cos(\theta)$ but outside of the circle $r = \sin(\theta)$

50. The region in Quadrant I which lies inside both the circle $r = 3$ as well as the rose $r = 6\sin(2\theta)$

While the authors truly believe that graphing polar curves by hand is fundamental to your understanding of the polar coordinate system, we would be derelict in our duties if we totally ignored the graphing calculator. Indeed, there are some important polar curves which are simply too difficult to graph by hand and that makes the calculator an important tool for your further studies in Mathematics, Science and Engineering. We now give a brief demonstration of how to use the graphing calculator to plot polar curves. The first thing you must do is switch the MODE of your calculator to POL, which stands for "polar".

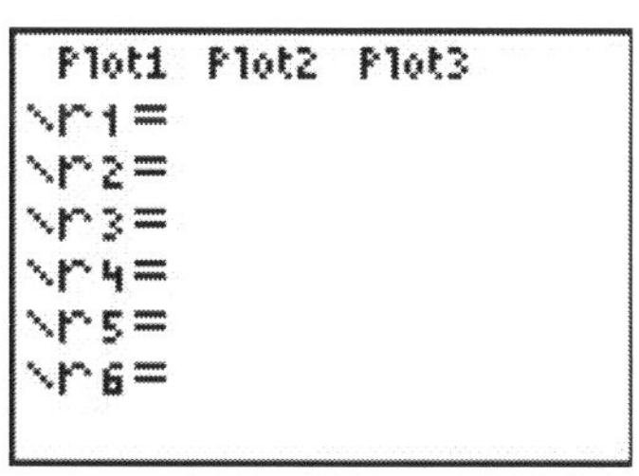

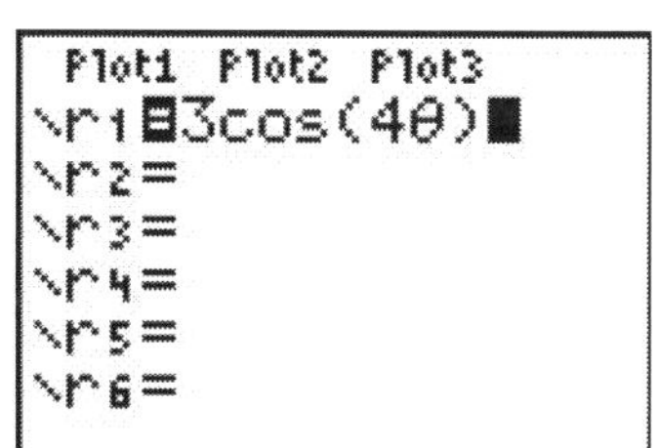

This changes the "Y=" menu as seen above in the middle. Let's plot the polar rose given by $r = 3\cos(4\theta)$ from Exercise 8 above. We type the function into the "r=" menu as seen above on the right. We need to set the viewing window so that the curve displays properly, but when we look at the WINDOW menu, we find three extra lines.

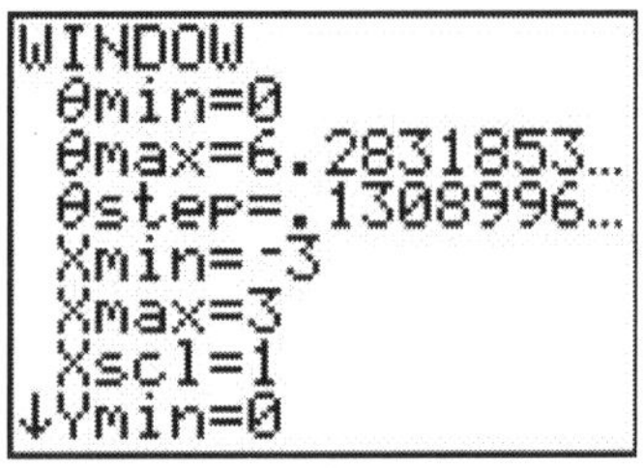

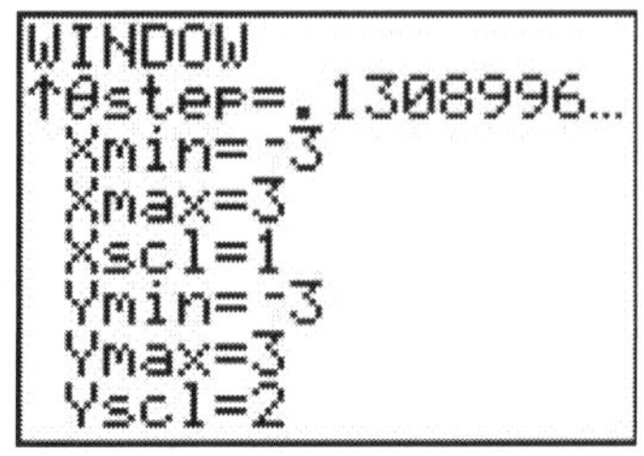

In order for the calculator to be able to plot $r = 3\cos(4\theta)$ in the xy-plane, we need to tell it not only the dimensions which x and y will assume, but we also what values of θ to use. From our previous work, we know that we need $0 \leq \theta \leq 2\pi$, so we enter the data you see above. (I'll say more about the θ-step in just a moment.) Hitting GRAPH yields the curve below on the left which doesn't look quite right. The issue here is that the calculator screen is 96 pixels wide but only 64 pixels tall. To get a true geometric perspective, we need to hit ZOOM SQUARE (seen below in the middle) to produce a more accurate graph which we present below on the right.

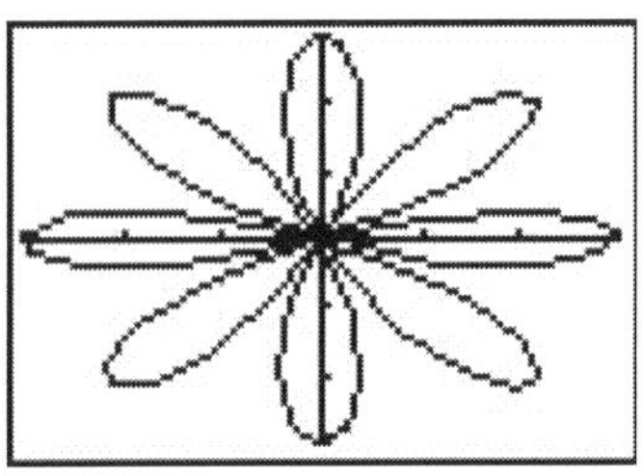

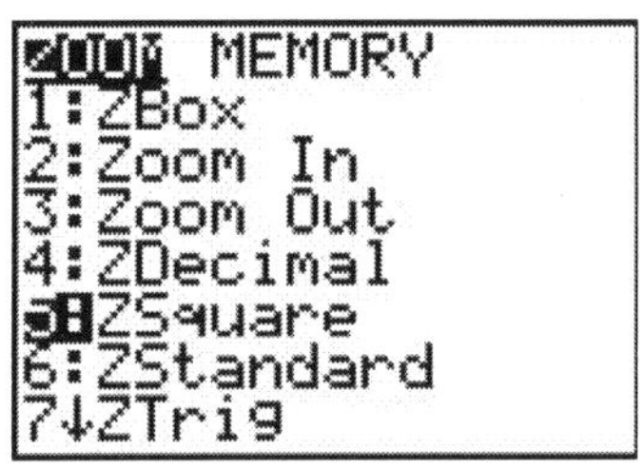

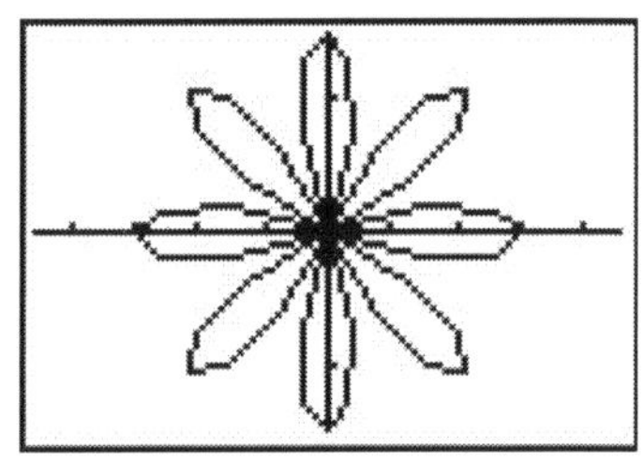

In function mode, the calculator automatically divided the interval [Xmin, Xmax] into 96 equal subintervals. In polar mode, however, we must specify how to split up the interval [θmin, θmax] using the θstep. For most graphs, a θstep of 0.1 is fine. If you make it too small then the calculator takes a long time to graph. It you make it too big, you get chunky garbage like this.

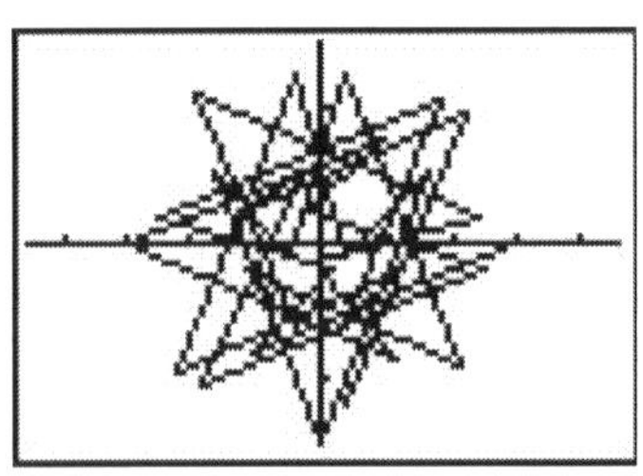

You will need to experiment with the settings in order to get a nice graph. Exercises 51 - 60 give you some curves to graph using your calculator. Notice that some of them have explicit bounds on θ and others do not.

51. $r = \theta,\ 0 \leq \theta \leq 12\pi$

52. $r = \ln(\theta),\ 1 \leq \theta \leq 12\pi$

53. $r = e^{.1\theta},\ 0 \leq \theta \leq 12\pi$

54. $r = \theta^3 - \theta,\ -1.2 \leq \theta \leq 1.2$

55. $r = \sin(5\theta) - 3\cos(\theta)$

56. $r = \sin^3\left(\frac{\theta}{2}\right) + \cos^2\left(\frac{\theta}{3}\right)$

57. $r = \arctan(\theta),\ -\pi \leq \theta \leq \pi$

58. $r = \dfrac{1}{1 - \cos(\theta)}$

59. $r = \dfrac{1}{2 - \cos(\theta)}$

60. $r = \dfrac{1}{2 - 3\cos(\theta)}$

61. How many petals does the polar rose $r = \sin(2\theta)$ have? What about $r = \sin(3\theta)$, $r = \sin(4\theta)$ and $r = \sin(5\theta)$? With the help of your classmates, make a conjecture as to how many petals the polar rose $r = \sin(n\theta)$ has for any natural number n. Replace sine with cosine and repeat the investigation. How many petals does $r = \cos(n\theta)$ have for each natural number n?

Looking back through the graphs in the section, it's clear that many polar curves enjoy various forms of symmetry. However, classifying symmetry for polar curves is not as straight-forward as it was for equations back on page 26. In Exercises 62 - 64, we have you and your classmates explore some of the more basic forms of symmetry seen in common polar curves.

62. Show that if f is even[17] then the graph of $r = f(\theta)$ is symmetric about the x-axis.

    (a) Show that $f(\theta) = 2 + 4\cos(\theta)$ is even and verify that the graph of $r = 2 + 4\cos(\theta)$ is indeed symmetric about the x-axis. (See Example 11.5.2 number 2.)

    (b) Show that $f(\theta) = 3\sin\left(\frac{\theta}{2}\right)$ is **not** even, yet the graph of $r = 3\sin\left(\frac{\theta}{2}\right)$ **is** symmetric about the x-axis. (See Example 11.5.3 number 4.)

63. Show that if f is odd[18] then the graph of $r = f(\theta)$ is symmetric about the origin.

    (a) Show that $f(\theta) = 5\sin(2\theta)$ is odd and verify that the graph of $r = 5\sin(2\theta)$ is indeed symmetric about the origin. (See Example 11.5.2 number 3.)

    (b) Show that $f(\theta) = 3\cos\left(\frac{\theta}{2}\right)$ is **not** odd, yet the graph of $r = 3\cos\left(\frac{\theta}{2}\right)$ **is** symmetric about the origin. (See Example 11.5.3 number 4.)

64. Show that if $f(\pi - \theta) = f(\theta)$ for all θ in the domain of f then the graph of $r = f(\theta)$ is symmetric about the y-axis.

    (a) For $f(\theta) = 4 - 2\sin(\theta)$, show that $f(\pi - \theta) = f(\theta)$ and the graph of $r = 4 - 2\sin(\theta)$ is symmetric about the y-axis, as required. (See Example 11.5.2 number 1.)

[17] Recall that this means $f(-\theta) = f(\theta)$ for θ in the domain of f.

[18] Recall that this means $f(-\theta) = -f(\theta)$ for θ in the domain of f.

(b) For $f(\theta) = 5\sin(2\theta)$, show that $f\left(\pi - \frac{\pi}{4}\right) \neq f\left(\frac{\pi}{4}\right)$, yet the graph of $r = 5\sin(2\theta)$ **is** symmetric about the y-axis. (See Example 11.5.2 number 3.)

In Section 1.7, we discussed transformations of graphs. In Exercise 65 we have you and your classmates explore transformations of polar graphs.

65. For Exercises 65a and 65b below, let $f(\theta) = \cos(\theta)$ and $g(\theta) = 2 - \sin(\theta)$.

(a) Using your graphing calculator, compare the graph of $r = f(\theta)$ to each of the graphs of $r = f\left(\theta + \frac{\pi}{4}\right)$, $r = f\left(\theta + \frac{3\pi}{4}\right)$, $r = f\left(\theta - \frac{\pi}{4}\right)$ and $r = f\left(\theta - \frac{3\pi}{4}\right)$. Repeat this process for $g(\theta)$. In general, how do you think the graph of $r = f(\theta + \alpha)$ compares with the graph of $r = f(\theta)$?

(b) Using your graphing calculator, compare the graph of $r = f(\theta)$ to each of the graphs of $r = 2f(\theta)$, $r = \frac{1}{2}f(\theta)$, $r = -f(\theta)$ and $r = -3f(\theta)$. Repeat this process for $g(\theta)$. In general, how do you think the graph of $r = k \cdot f(\theta)$ compares with the graph of $r = f(\theta)$? (Does it matter if $k > 0$ or $k < 0$?)

66. In light of Exercises 62 - 64, how would the graph of $r = f(-\theta)$ compare with the graph of $r = f(\theta)$ for a generic function f? What about the graphs of $r = -f(\theta)$ and $r = f(\theta)$? What about $r = f(\theta)$ and $r = f(\pi - \theta)$? Test out your conjectures using a variety of polar functions found in this section with the help of a graphing utility.

67. With the help of your classmates, research cardioid microphones.

68. Back in Section 1.2, in the paragraph before Exercise 53, we gave you this link to a fascinating list of curves. Some of these curves have polar representations which we invite you and your classmates to research.

11.5.2 Answers

1. Circle: $r = 6\sin(\theta)$

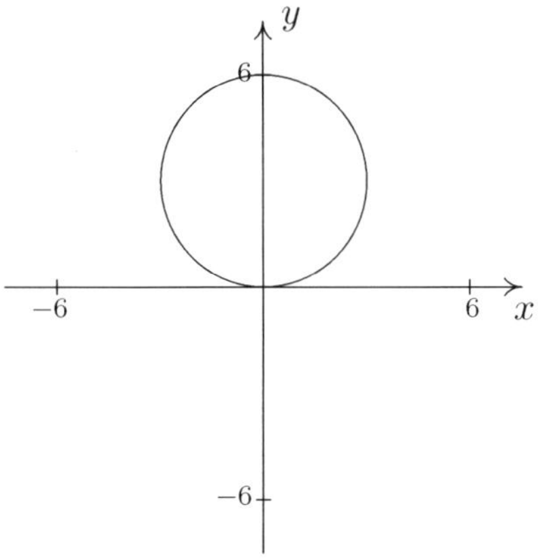

2. Circle: $r = 2\cos(\theta)$

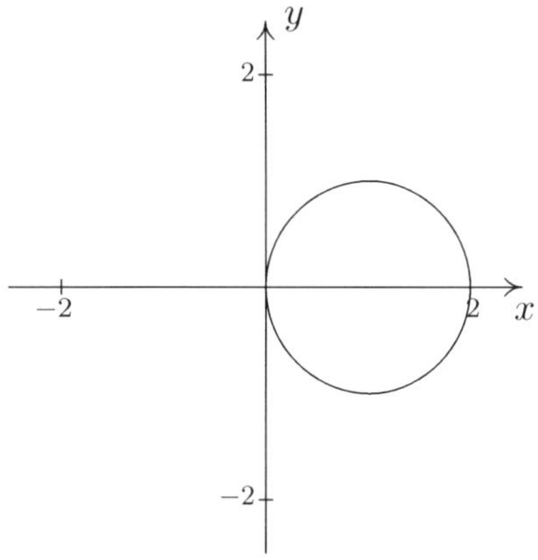

3. Rose: $r = 2\sin(2\theta)$

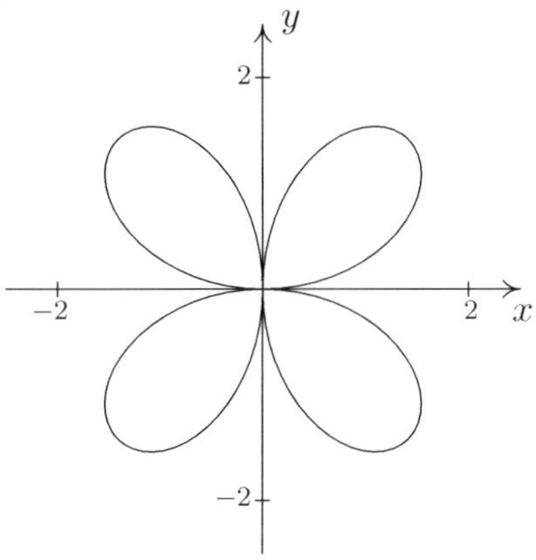

4. Rose: $r = 4\cos(2\theta)$

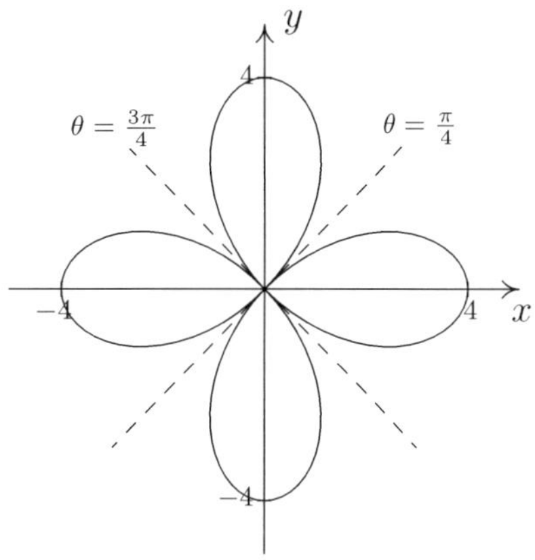

5. Rose: $r = 5\sin(3\theta)$

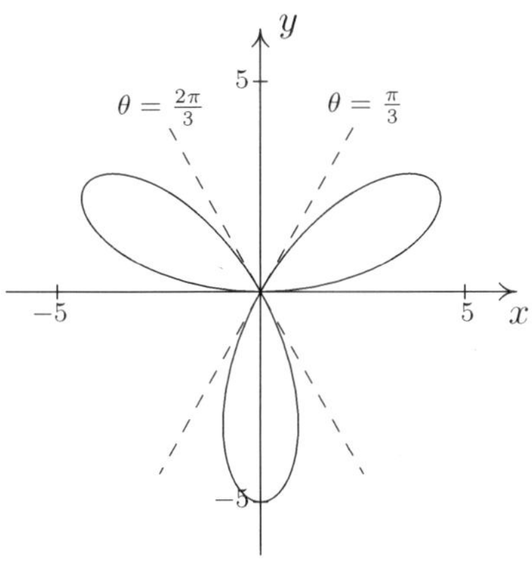

6. Rose: $r = \cos(5\theta)$

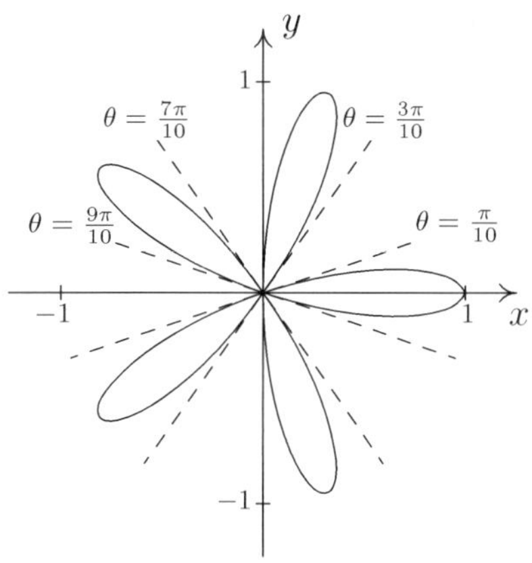

7. Rose: $r = \sin(4\theta)$

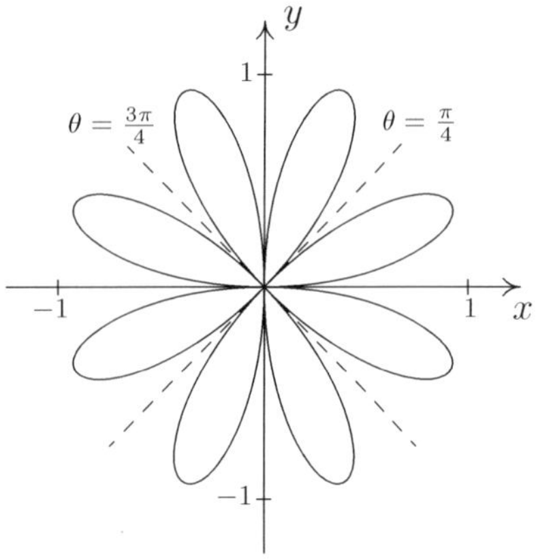

8. Rose: $r = 3\cos(4\theta)$

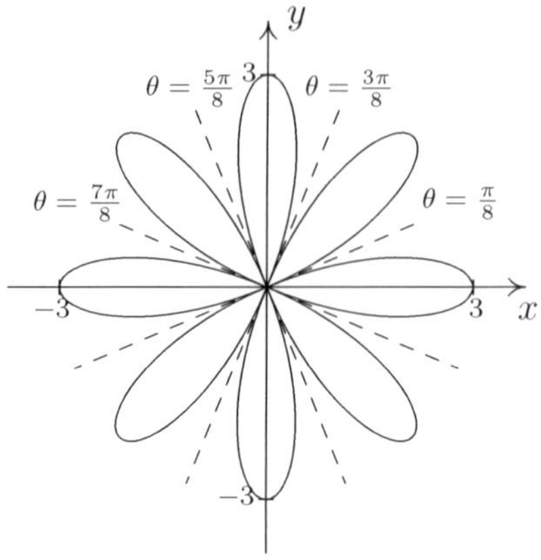

9. Cardioid: $r = 3 - 3\cos(\theta)$

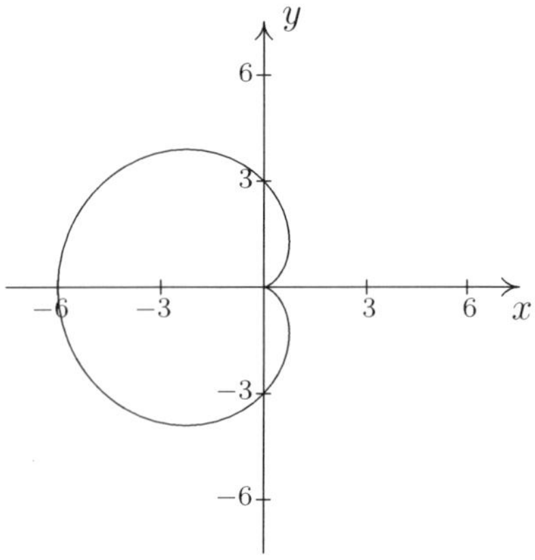

10. Cardioid: $r = 5 + 5\sin(\theta)$

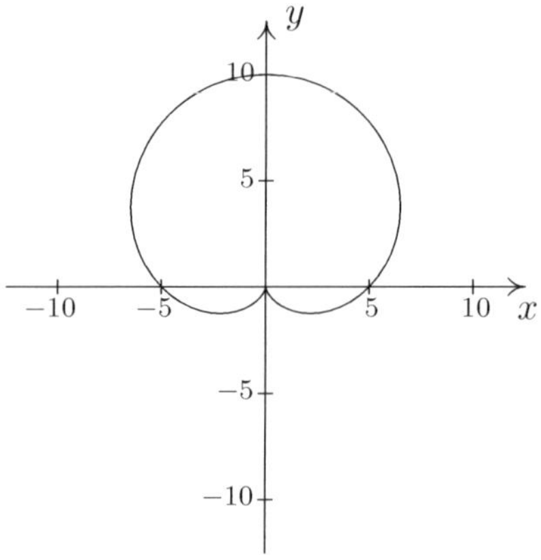

11. Cardioid: $r = 2 + 2\cos(\theta)$

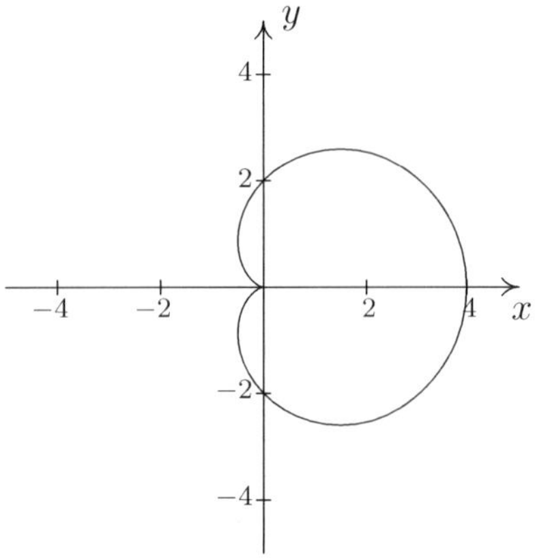

12. Cardioid: $r = 1 - \sin(\theta)$

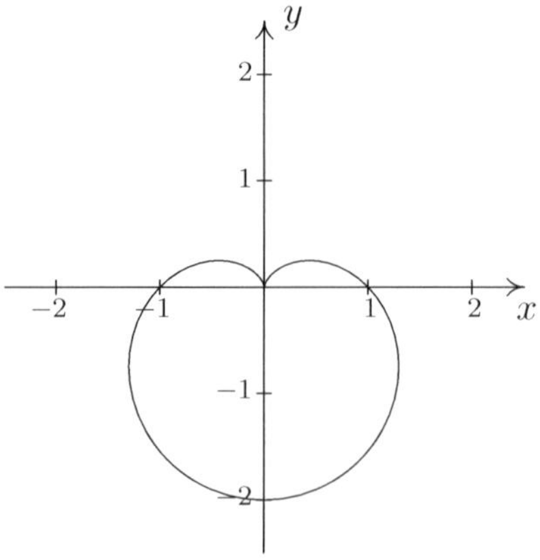

13. Limaçon: $r = 1 - 2\cos(\theta)$

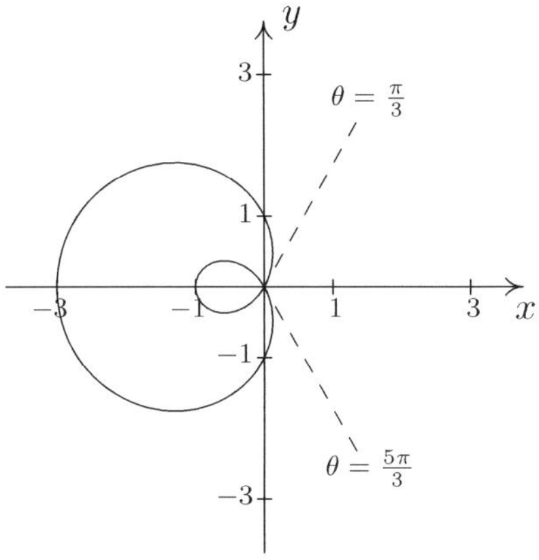

14. Limaçon: $r = 1 - 2\sin(\theta)$

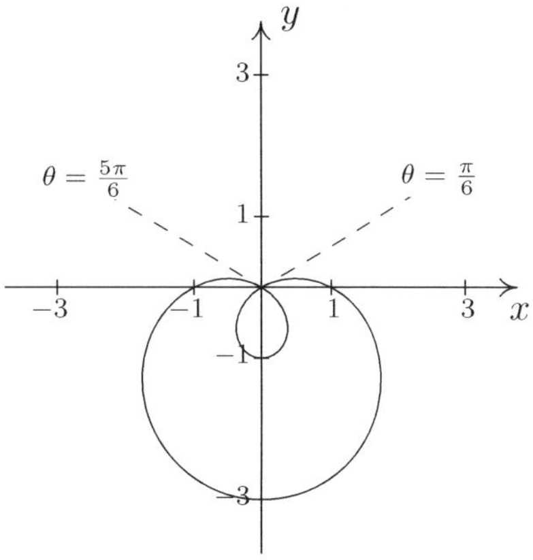

15. Limaçon: $r = 2\sqrt{3} + 4\cos(\theta)$

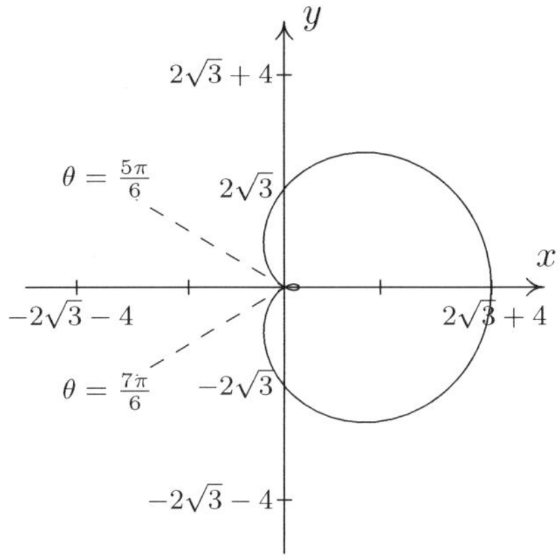

16. Limaçon: $r = 3 - 5\cos(\theta)$

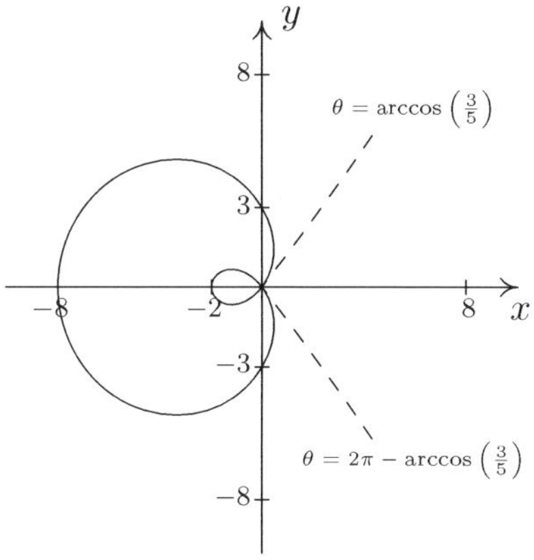

17. Limaçon: $r = 3 - 5\sin(\theta)$

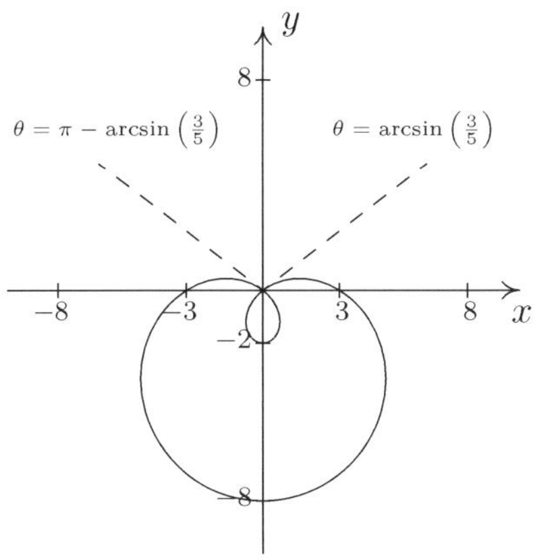

18. Limaçon: $r = 2 + 7\sin(\theta)$

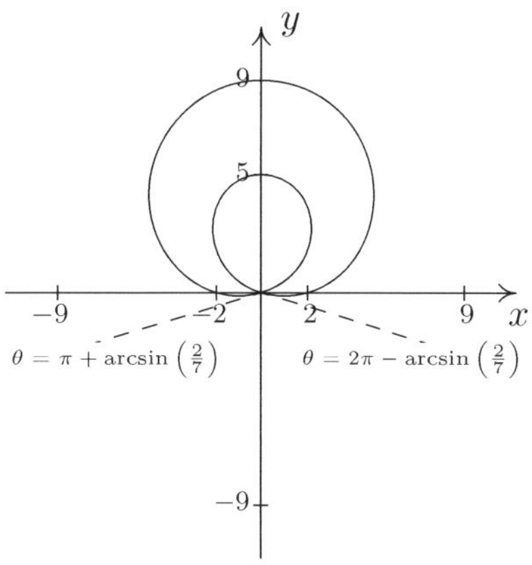

19. Lemniscate: $r^2 = \sin(2\theta)$

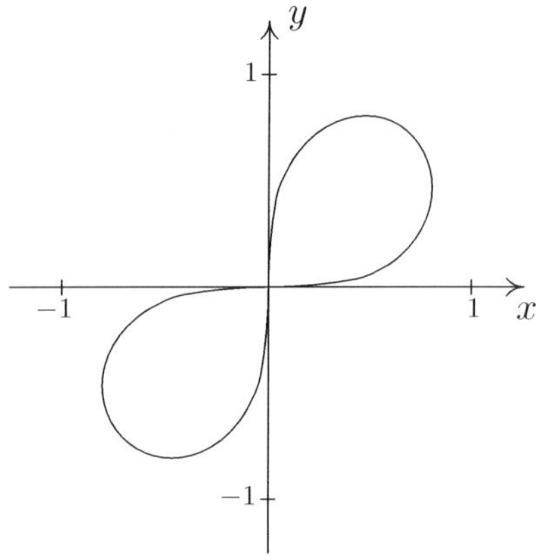

20. Lemniscate: $r^2 = 4\cos(2\theta)$

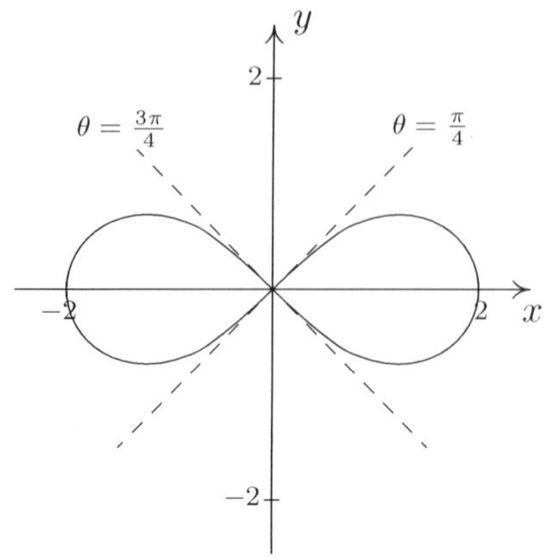

21. $r = 3\cos(\theta)$ and $r = 1 + \cos(\theta)$

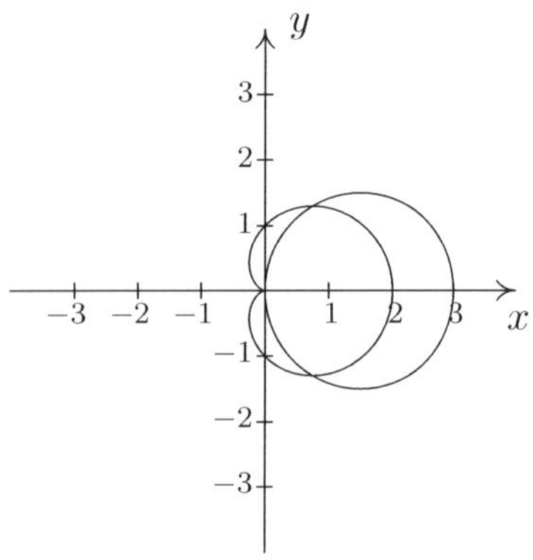

$\left(\frac{3}{2}, \frac{\pi}{3}\right)$, $\left(\frac{3}{2}, \frac{5\pi}{3}\right)$, pole

22. $r = 1 + \sin(\theta)$ and $r = 1 - \cos(\theta)$

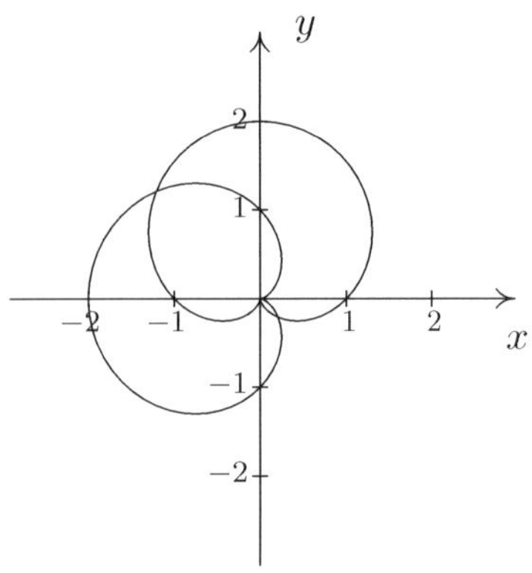

$\left(\frac{2+\sqrt{2}}{2}, \frac{3\pi}{4}\right)$, $\left(\frac{2-\sqrt{2}}{2}, \frac{7\pi}{4}\right)$, pole

23. $r = 1 - 2\sin(\theta)$ and $r = 2$ $\qquad \left(2, \frac{7\pi}{6}\right), \left(2, \frac{11\pi}{6}\right)$

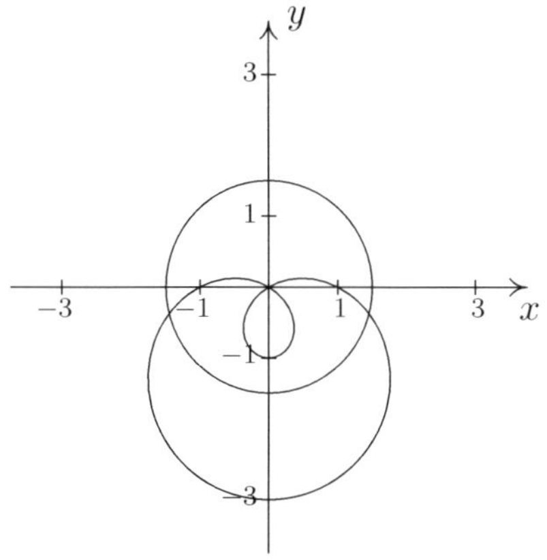

24. $r = 1 - 2\cos(\theta)$ and $r = 1$ $\qquad \left(1, \frac{\pi}{2}\right), \left(1, \frac{3\pi}{2}\right), (-1, 0)$

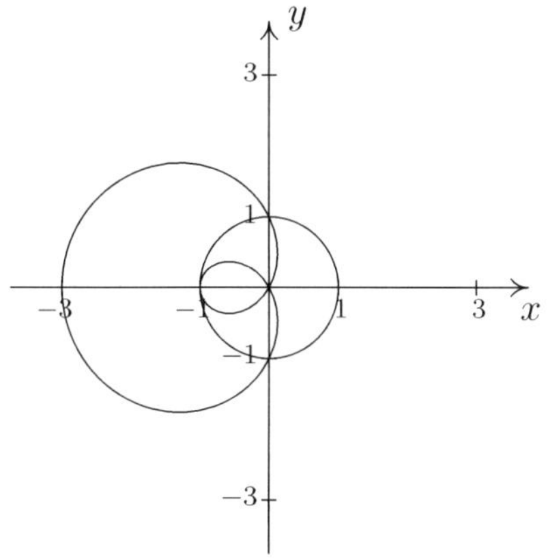

25. $r = 2\cos(\theta)$ and $r = 2\sqrt{3}\sin(\theta)$ $\qquad \left(\sqrt{3}, \frac{\pi}{6}\right)$, pole

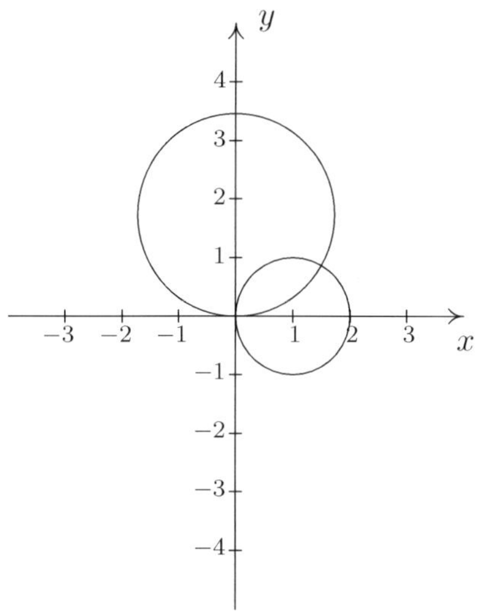

26. $r = 3\cos(\theta)$ and $r = \sin(\theta)$

$\left(\dfrac{3\sqrt{10}}{10}, \arctan(3)\right)$, pole

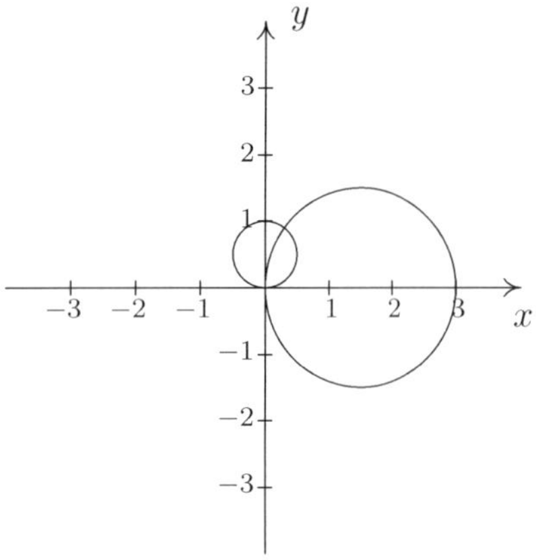

27. $r^2 = 4\cos(2\theta)$ and $r = \sqrt{2}$

$\left(\sqrt{2}, \dfrac{\pi}{6}\right)$, $\left(\sqrt{2}, \dfrac{5\pi}{6}\right)$, $\left(\sqrt{2}, \dfrac{7\pi}{6}\right)$, $\left(\sqrt{2}, \dfrac{11\pi}{6}\right)$

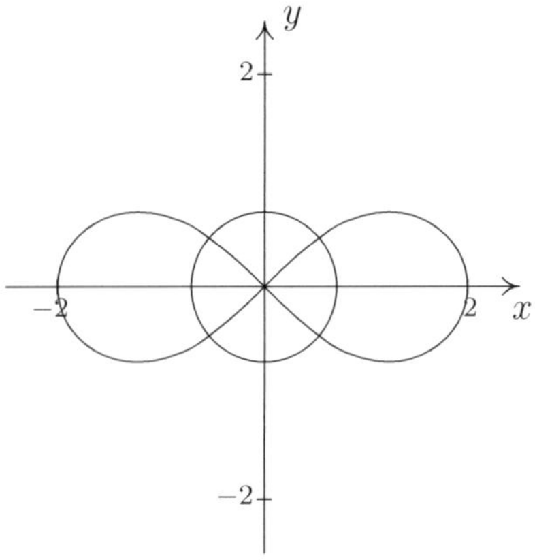

28. $r^2 = 2\sin(2\theta)$ and $r = 1$

$\left(1, \dfrac{\pi}{12}\right)$, $\left(1, \dfrac{5\pi}{12}\right)$, $\left(1, \dfrac{13\pi}{12}\right)$, $\left(1, \dfrac{17\pi}{12}\right)$

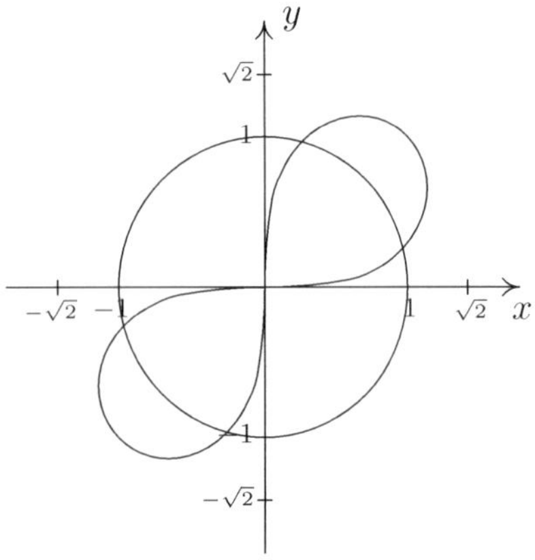

29. $r = 4\cos(2\theta)$ and $r = 2$

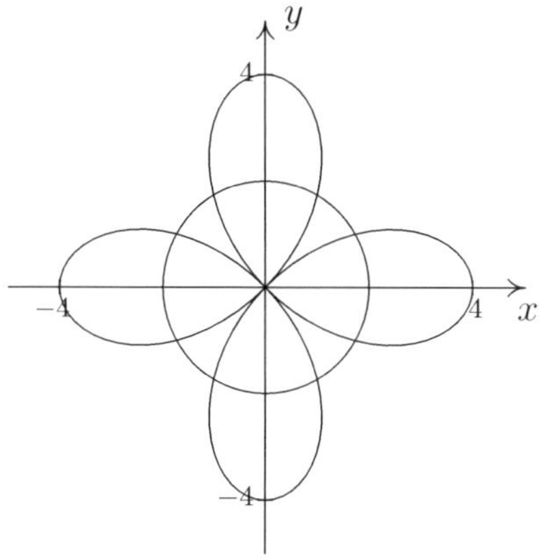

$\left(2, \frac{\pi}{6}\right)$, $\left(2, \frac{5\pi}{6}\right)$, $\left(2, \frac{7\pi}{6}\right)$, $\left(2, \frac{11\pi}{6}\right)$, $\left(-2, \frac{\pi}{3}\right)$, $\left(-2, \frac{2\pi}{3}\right)$, $\left(-2, \frac{4\pi}{3}\right)$, $\left(-2, \frac{5\pi}{3}\right)$

30. $r = 2\sin(2\theta)$ and $r = 1$

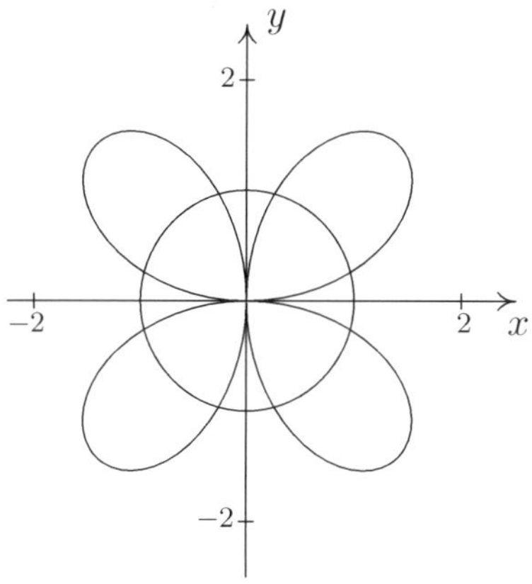

$\left(1, \frac{\pi}{12}\right)$, $\left(1, \frac{5\pi}{12}\right)$, $\left(1, \frac{13\pi}{12}\right)$, $\left(1, \frac{17\pi}{12}\right)$, $\left(-1, \frac{7\pi}{12}\right)$, $\left(-1, \frac{11\pi}{12}\right)$, $\left(-1, \frac{19\pi}{12}\right)$, $\left(-1, \frac{23\pi}{12}\right)$

31. $\{(r, \theta) \mid 0 \leq r \leq 3,\, 0 \leq \theta \leq 2\pi\}$

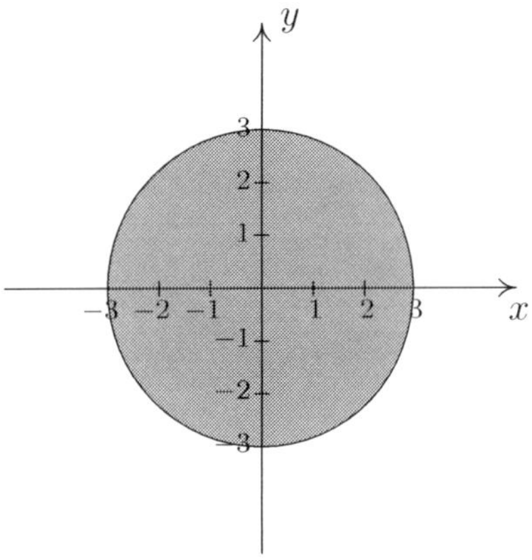

32. $\{(r, \theta) \mid 0 \leq r \leq 4\sin(\theta),\, 0 \leq \theta \leq \pi\}$

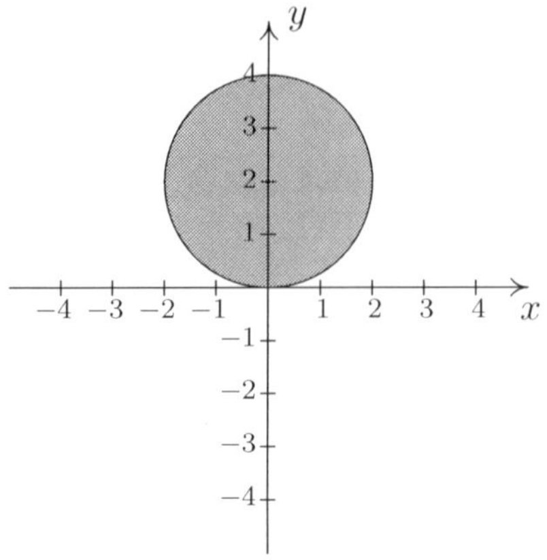

33. $\left\{(r, \theta) \mid 0 \leq r \leq 3\cos(\theta),\, -\frac{\pi}{2} \leq \theta \leq \frac{\pi}{2}\right\}$

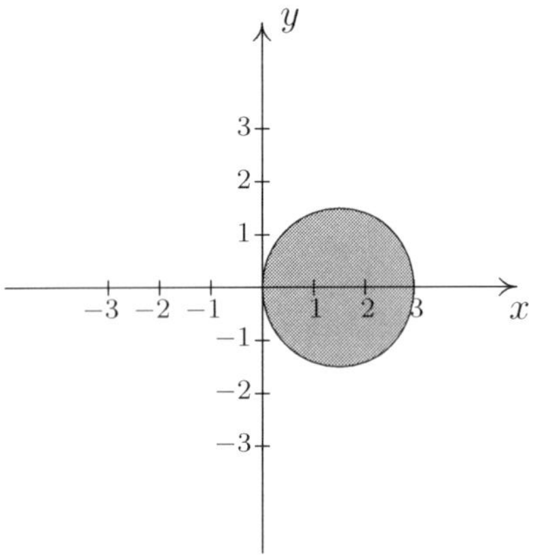

34. $\left\{(r, \theta) \mid 0 \leq r \leq 2\sin(2\theta),\, 0 \leq \theta \leq \frac{\pi}{2}\right\}$

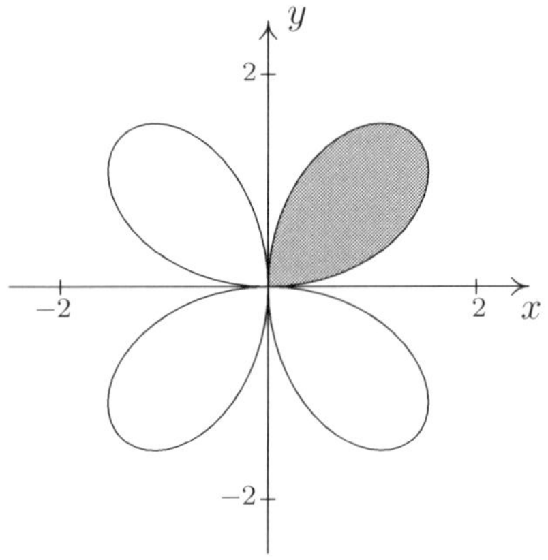

35. $\left\{(r, \theta) \mid 0 \leq r \leq 4\cos(2\theta),\, -\frac{\pi}{4} \leq \theta \leq \frac{\pi}{4}\right\}$

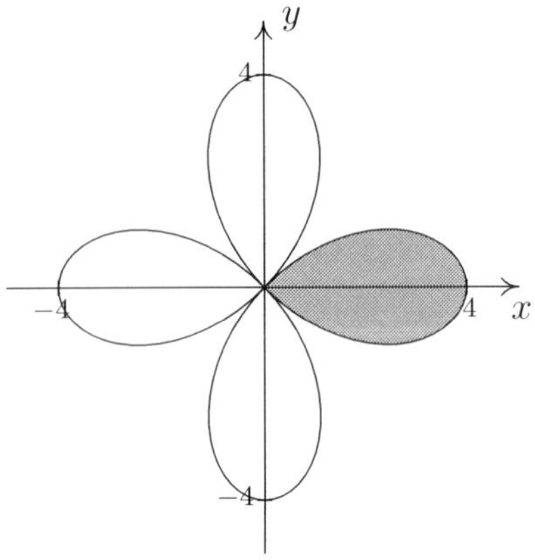

36. $\left\{(r, \theta) \mid 1 \leq r \leq 1 - 2\cos(\theta),\, \frac{\pi}{2} \leq \theta \leq \frac{3\pi}{2}\right\}$

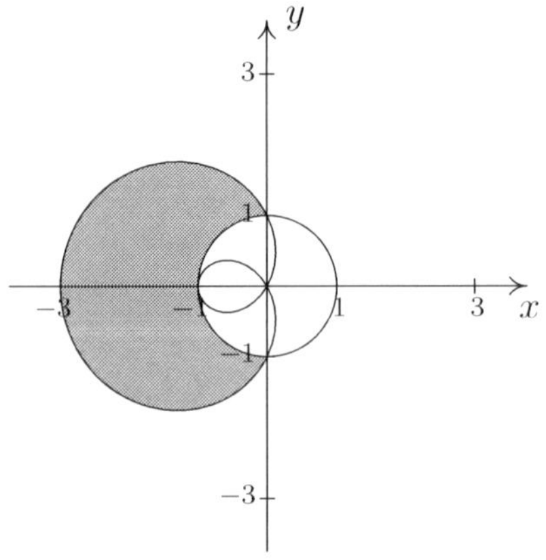

37. $\left\{(r, \theta) \,|\, 1 + \cos(\theta) \leq r \leq 3\cos(\theta),\ -\frac{\pi}{3} \leq \theta \leq \frac{\pi}{3}\right\}$

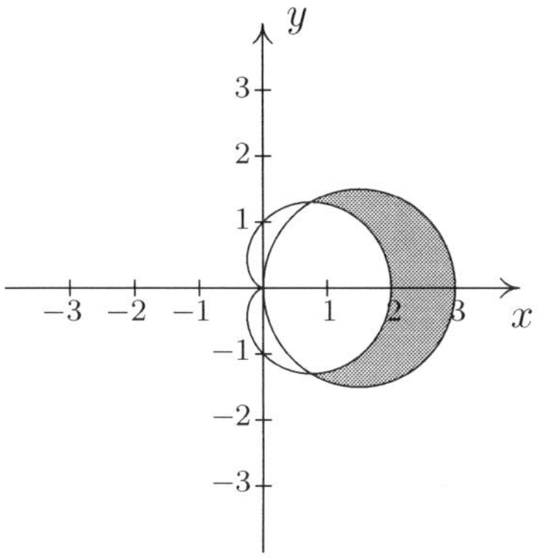

38. $\left\{(r, \theta) \,|\, 1 \leq r \leq \sqrt{2\sin(2\theta)},\ \frac{13\pi}{12} \leq \theta \leq \frac{17\pi}{12}\right\}$

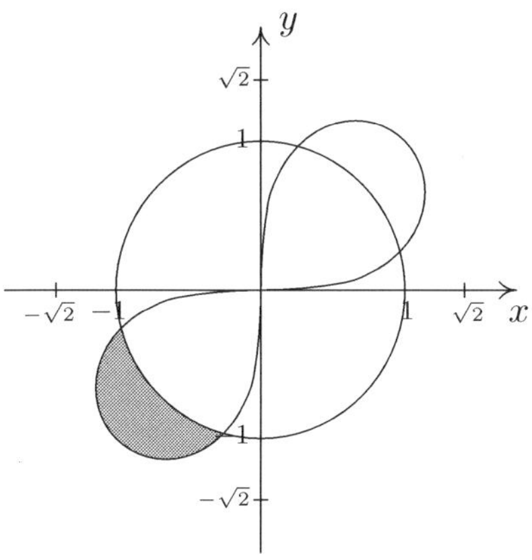

39. $\left\{(r, \theta) \,|\, 0 \leq r \leq 2\sqrt{3}\sin(\theta),\ 0 \leq \theta \leq \frac{\pi}{6}\right\} \cup \left\{(r, \theta) \,|\, 0 \leq r \leq 2\cos(\theta),\ \frac{\pi}{6} \leq \theta \leq \frac{\pi}{2}\right\}$

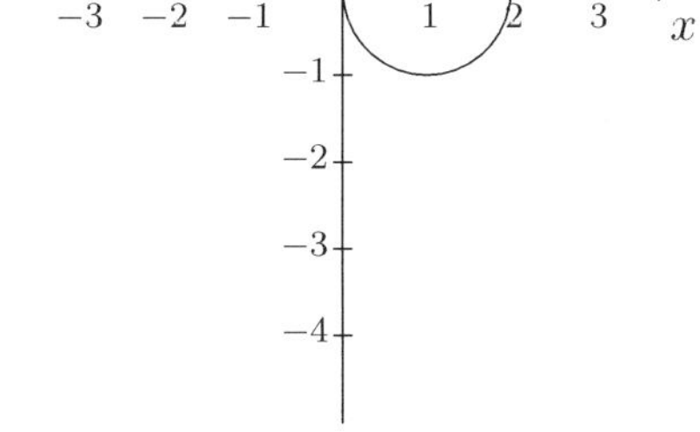

40. $\left\{(r,\theta)\,|\,0 \leq r \leq 2\sin(2\theta),\, 0 \leq \theta \leq \frac{\pi}{12}\right\} \cup \left\{(r,\theta)\,|\,0 \leq r \leq 1,\, \frac{\pi}{12} \leq \theta \leq \frac{\pi}{4}\right\}$

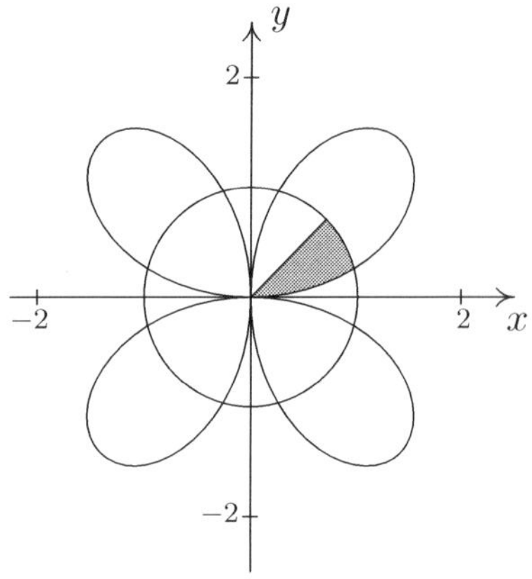

41. $\{(r,\theta)\,|\,0 \leq r \leq 5,\, 0 \leq \theta \leq 2\pi\}$

42. $\left\{(r,\theta)\,|\,0 \leq r \leq 5,\, \pi \leq \theta \leq \frac{3\pi}{2}\right\}$

43. $\left\{(r,\theta)\,|\,0 \leq r \leq 6\sin(\theta),\, \frac{\pi}{2} \leq \theta \leq \pi\right\}$

44. $\left\{(r,\theta)\,|\,4\cos(\theta) \leq r \leq 0,\, \frac{\pi}{2} \leq \theta \leq \pi\right\}$

45. $\{(r,\theta)\,|\,0 \leq r \leq 3 - 3\cos(\theta),\, 0 \leq \theta \leq \pi\}$

46. $\left\{(r,\theta)\,|\,0 \leq r \leq 2 - 2\sin(\theta),\, 0 \leq \theta \leq \frac{\pi}{2}\right\} \cup \left\{(r,\theta)\,|\,0 \leq r \leq 2 - 2\sin(\theta),\, \frac{3\pi}{2} \leq \theta \leq 2\pi\right\}$
or $\left\{(r,\theta)\,|\,0 \leq r \leq 2 - 2\sin(\theta),\, \frac{3\pi}{2} \leq \theta \leq \frac{5\pi}{2}\right\}$

47. $\left\{(r,\theta)\,|\,0 \leq r \leq 3\cos(4\theta),\, 0 \leq \theta \leq \frac{\pi}{8}\right\} \cup \left\{(r,\theta)\,|\,0 \leq r \leq 3\cos(4\theta),\, \frac{15\pi}{8} \leq \theta \leq 2\pi\right\}$
or $\left\{(r,\theta)\,|\,0 \leq r \leq 3\cos(4\theta),\, -\frac{\pi}{8} \leq \theta \leq \frac{\pi}{8}\right\}$

48. $\{(r,\theta)\,|\,3 \leq r \leq 5,\, 0 \leq \theta \leq 2\pi\}$

49. $\left\{(r,\theta)\,|\,0 \leq r \leq 3\cos(\theta),\, -\frac{\pi}{2} \leq \theta \leq 0\right\} \cup \{(r,\theta)\,|\,\sin(\theta) \leq r \leq 3\cos(\theta),\, 0 \leq \theta \leq \arctan(3)\}$

50. $\left\{(r,\theta)\,|\,0 \leq r \leq 6\sin(2\theta),\, 0 \leq \theta \leq \frac{\pi}{12}\right\} \cup \left\{(r,\theta)\,|\,0 \leq r \leq 3,\, \frac{\pi}{12} \leq \theta \leq \frac{5\pi}{12}\right\} \cup$
$\left\{(r,\theta)\,|\,0 \leq r \leq 6\sin(2\theta),\, \frac{5\pi}{12} \leq \theta \leq \frac{\pi}{2}\right\}$

11.6 Hooked on Conics Again

In this section, we revisit our friends the Conic Sections which we began studying in Chapter 7. Our first task is to formalize the notion of rotating axes so this subsection is actually a follow-up to Example 8.3.3 in Section 8.3. In that example, we saw that the graph of $y = \frac{2}{x}$ is actually a hyperbola. More specifically, it is the hyperbola obtained by rotating the graph of $x^2 - y^2 = 4$ counter-clockwise through a 45° angle. Armed with polar coordinates, we can generalize the process of rotating axes as shown below.

11.6.1 Rotation of Axes

Consider the x- and y-axes below along with the dashed x'- and y'-axes obtained by rotating the x- and y-axes counter-clockwise through an angle θ and consider the point $P(x, y)$. The coordinates (x, y) are rectangular coordinates and are based on the x- and y-axes. Suppose we wished to find rectangular coordinates based on the x'- and y'-axes. That is, we wish to determine $P(x', y')$. While this seems like a formidable challenge, it is nearly trivial if we use polar coordinates. Consider the angle ϕ whose initial side is the positive x'-axis and whose terminal side contains the point P.

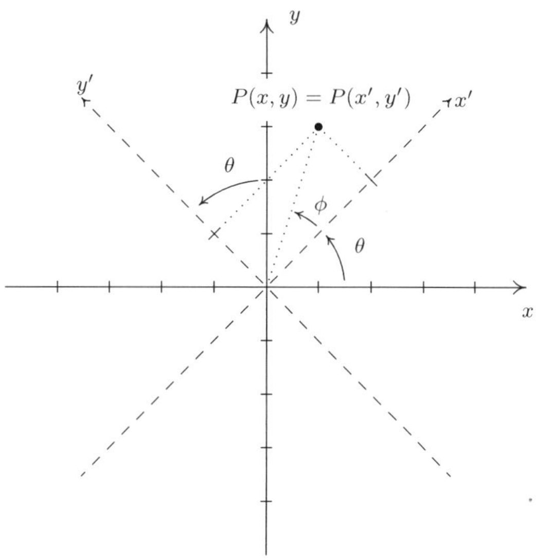

We relate $P(x, y)$ and $P(x', y')$ by converting them to polar coordinates. Converting $P(x, y)$ to polar coordinates with $r > 0$ yields $x = r\cos(\theta + \phi)$ and $y = r\sin(\theta + \phi)$. To convert the point $P(x', y')$ into polar coordinates, we first match the polar axis with the positive x'-axis, choose the same $r > 0$ (since the origin is the same in both systems) and get $x' = r\cos(\phi)$ and $y' = r\sin(\phi)$. Using the sum formulas for sine and cosine, we have

$$\begin{array}{rcll}
x & = & r\cos(\theta + \phi) & \\
 & = & r\cos(\theta)\cos(\phi) - r\sin(\theta)\sin(\phi) & \text{Sum formula for cosine} \\
 & = & (r\cos(\phi))\cos(\theta) - (r\sin(\phi))\sin(\theta) & \\
 & = & x'\cos(\theta) - y'\sin(\theta) & \text{Since } x' = r\cos(\phi) \text{ and } y' = r\sin(\phi)
\end{array}$$

Similarly, using the sum formula for sine we get $y = x'\sin(\theta) + y'\cos(\theta)$. These equations enable us to easily convert points with $x'y'$-coordinates back into xy-coordinates. They also enable us to easily convert equations in the variables x and y into equations in the variables in terms of x' and y'.[1] If we want equations which enable us to convert points with xy-coordinates into $x'y'$-coordinates, we need to solve the system

$$\left\{ \begin{array}{rcl} x'\cos(\theta) - y'\sin(\theta) & = & x \\ x'\sin(\theta) + y'\cos(\theta) & = & y \end{array} \right.$$

for x' and y'. Perhaps the cleanest way[2] to solve this system is to write it as a matrix equation. Using the machinery developed in Section 8.4, we write the above system as the matrix equation $AX' = X$ where

$$A = \left[\begin{array}{rr} \cos(\theta) & -\sin(\theta) \\ \sin(\theta) & \cos(\theta) \end{array} \right], \quad X' = \left[\begin{array}{c} x' \\ y' \end{array} \right], \quad X = \left[\begin{array}{c} x \\ y \end{array} \right]$$

Since $\det(A) = (\cos(\theta))(\cos(\theta)) - (-\sin(\theta))(\sin(\theta)) = \cos^2(\theta) + \sin^2(\theta) = 1$, the determinant of A is not zero so A is invertible and $X' = A^{-1}X$. Using the formula given in Equation 8.2 with $\det(A) = 1$, we find

$$A^{-1} = \left[\begin{array}{rr} \cos(\theta) & \sin(\theta) \\ -\sin(\theta) & \cos(\theta) \end{array} \right]$$

so that

$$\begin{array}{rcl} X' & = & A^{-1}X \\ \left[\begin{array}{c} x' \\ y' \end{array} \right] & = & \left[\begin{array}{rr} \cos(\theta) & \sin(\theta) \\ -\sin(\theta) & \cos(\theta) \end{array} \right] \left[\begin{array}{c} x \\ y \end{array} \right] \\ \left[\begin{array}{c} x' \\ y' \end{array} \right] & = & \left[\begin{array}{c} x\cos(\theta) + y\sin(\theta) \\ -x\sin(\theta) + y\cos(\theta) \end{array} \right] \end{array}$$

From which we get $x' = x\cos(\theta) + y\sin(\theta)$ and $y' = -x\sin(\theta) + y\cos(\theta)$. To summarize,

Theorem 11.9. Rotation of Axes: Suppose the positive x and y axes are rotated counter-clockwise through an angle θ to produce the axes x' and y', respectively. Then the coordinates $P(x, y)$ and $P(x', y')$ are related by the following systems of equations

$$\left\{ \begin{array}{rcl} x & = & x'\cos(\theta) - y'\sin(\theta) \\ y & = & x'\sin(\theta) + y'\cos(\theta) \end{array} \right. \quad \text{and} \quad \left\{ \begin{array}{rcl} x' & = & x\cos(\theta) + y\sin(\theta) \\ y' & = & -x\sin(\theta) + y\cos(\theta) \end{array} \right.$$

We put the formulas in Theorem 11.9 to good use in the following example.

[1]Sound familiar? In Section 11.4, the equations $x = r\cos(\theta)$ and $y = r\sin(\theta)$ make it easy to convert *points* from polar coordinates into rectangular coordinates, and they make it easy to convert *equations* from rectangular coordinates into polar coordinates.

[2]We could, of course, interchange the roles of x and x', y and y' and replace ϕ with $-\phi$ to get x' and y' in terms of x and y, but that seems like cheating. The matrix A introduced here is revisited in the Exercises.

Example 11.6.1. Suppose the x- and y- axes are both rotated counter-clockwise through the angle $\theta = \frac{\pi}{3}$ to produce the x'- and y'- axes, respectively.

1. Let $P(x, y) = (2, -4)$ and find $P(x', y')$. Check your answer algebraically and graphically.
2. Convert the equation $21x^2 + 10xy\sqrt{3} + 31y^2 = 144$ to an equation in x' and y' and graph.

Solution.

1. If $P(x, y) = (2, -4)$ then $x = 2$ and $y = -4$. Using these values for x and y along with $\theta = \frac{\pi}{3}$, Theorem 11.9 gives $x' = x\cos(\theta) + y\sin(\theta) = 2\cos\left(\frac{\pi}{3}\right) + (-4)\sin\left(\frac{\pi}{3}\right)$ which simplifies to $x' = 1 - 2\sqrt{3}$. Similarly, $y' = -x\sin(\theta) + y\cos(\theta) = (-2)\sin\left(\frac{\pi}{3}\right) + (-4)\cos\left(\frac{\pi}{3}\right)$ which gives $y' = -\sqrt{3} - 2 = -2 - \sqrt{3}$. Hence $P(x', y') = \left(1 - 2\sqrt{3}, -2 - \sqrt{3}\right)$. To check our answer algebraically, we use the formulas in Theorem 11.9 to convert $P(x', y') = \left(1 - 2\sqrt{3}, -2 - \sqrt{3}\right)$ back into x and y coordinates. We get

$$\begin{aligned} x &= x'\cos(\theta) - y'\sin(\theta) \\ &= (1 - 2\sqrt{3})\cos\left(\tfrac{\pi}{3}\right) - (-2 - \sqrt{3})\sin\left(\tfrac{\pi}{3}\right) \\ &= \left(\tfrac{1}{2} - \sqrt{3}\right) - \left(-\sqrt{3} - \tfrac{3}{2}\right) \\ &= 2 \end{aligned}$$

Similarly, using $y = x'\sin(\theta) + y'\cos(\theta)$, we obtain $y = -4$ as required. To check our answer graphically, we sketch in the x'-axis and y'-axis to see if the new coordinates $P(x', y') = \left(1 - 2\sqrt{3}, -2 - \sqrt{3}\right) \approx (-2.46, -3.73)$ seem reasonable. Our graph is below.

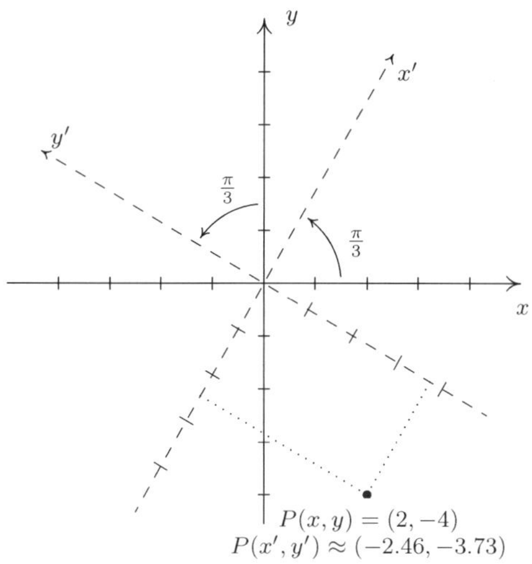

2. To convert the equation $21x^2+10xy\sqrt{3}+31y^2 = 144$ to an equation in the variables x' and y', we substitute $x = x'\cos\left(\frac{\pi}{3}\right) - y'\sin\left(\frac{\pi}{3}\right) = \frac{x'}{2} - \frac{y'\sqrt{3}}{2}$ and $y = x'\sin\left(\frac{\pi}{3}\right) + y'\cos\left(\frac{\pi}{3}\right) = \frac{x'\sqrt{3}}{2} + \frac{y'}{2}$

and simplify. While this is by no means a trivial task, it is nothing more than a hefty dose of Beginning Algebra. We will not go through the entire computation, but rather, the reader should take the time to do it. Start by verifying that

$$x^2 = \frac{(x')^2}{4} - \frac{x'y'\sqrt{3}}{2} + \frac{3(y')^2}{4}, \quad xy = \frac{(x')^2\sqrt{3}}{4} - \frac{x'y'}{2} - \frac{(y')^2\sqrt{3}}{4}, \quad y^2 = \frac{3(x')^2}{4} + \frac{x'y'\sqrt{3}}{2} + \frac{(y')^2}{4}$$

To our surprise and delight, the equation $21x^2 + 10xy\sqrt{3} + 31y^2 = 144$ in xy-coordinates reduces to $36(x')^2 + 16(y')^2 = 144$, or $\frac{(x')^2}{4} + \frac{(y')^2}{9} = 1$ in $x'y'$-coordinates. The latter is an ellipse centered at $(0,0)$ with vertices along the y'-axis with ($x'y'$-coordinates) $(0, \pm 3)$ and whose minor axis has endpoints with ($x'y'$-coordinates) $(\pm 2, 0)$. We graph it below.

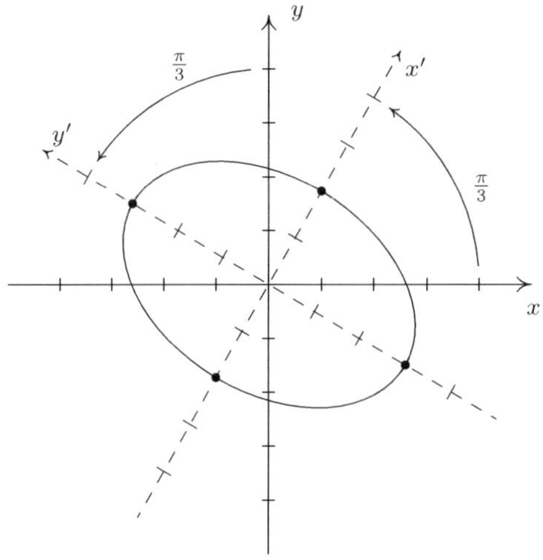

$21x^2 + 10xy\sqrt{3} + 31y^2 = 144$

□

The elimination of the troublesome 'xy' term from the equation $21x^2 + 10xy\sqrt{3} + 31y^2 = 144$ in Example 11.6.1 number 2 allowed us to graph the equation by hand using what we learned in Chapter 7. It is natural to wonder if we can always do this. That is, given an equation of the form $Ax^2+Bxy+Cy^2+Dx+Ey+F = 0$, with $B \neq 0$, is there an angle θ so that if we rotate the x and y-axes counter-clockwise through that angle θ, the equation in the rotated variables x' and y' contains no $x'y'$ term? To explore this conjecture, we make the usual substitutions $x = x'\cos(\theta) - y'\sin(\theta)$ and $y = x'\sin(\theta) + y'\cos(\theta)$ into the equation $Ax^2 + Bxy + Cy^2 + Dx + Ey + F = 0$ and set the coefficient of the $x'y'$ term equal to 0. Terms containing $x'y'$ in this expression will come from the first three terms of the equation: Ax^2, Bxy and Cy^2. We leave it to the reader to verify that

$$\begin{aligned} x^2 &= (x')^2\cos^2(\theta) - 2x'y'\cos(\theta)\sin(\theta) + (y')^2\sin(\theta) \\ xy &= (x')^2\cos(\theta)\sin(\theta) + x'y'\left(\cos^2(\theta) - \sin^2(\theta)\right) - (y')^2\cos(\theta)\sin(\theta) \\ y^2 &= (x')^2\sin^2(\theta) + 2x'y'\cos(\theta)\sin(\theta) + (y')^2\cos^2(\theta) \end{aligned}$$

The contribution to the $x'y'$-term from Ax^2 is $-2A\cos(\theta)\sin(\theta)$, from Bxy it is $B\left(\cos^2(\theta)-\sin^2(\theta)\right)$, and from Cy^2 it is $2C\cos(\theta)\sin(\theta)$. Equating the $x'y'$-term to 0, we get

$$\begin{aligned} -2A\cos(\theta)\sin(\theta)+B\left(\cos^2(\theta)-\sin^2(\theta)\right)+2C\cos(\theta)\sin(\theta) &= 0 \\ -A\sin(2\theta)+B\cos(2\theta)+C\sin(2\theta) &= 0 \quad \text{Double Angle Identities} \end{aligned}$$

From this, we get $B\cos(2\theta)=(A-C)\sin(2\theta)$, and our goal is to solve for θ in terms of the coefficients A, B and C. Since we are assuming $B\neq 0$, we can divide both sides of this equation by B. To solve for θ we would like to divide both sides of the equation by $\sin(2\theta)$, provided of course that we have assurances that $\sin(2\theta)\neq 0$. If $\sin(2\theta)=0$, then we would have $B\cos(2\theta)=0$, and since $B\neq 0$, this would force $\cos(2\theta)=0$. Since no angle θ can have both $\sin(2\theta)=0$ and $\cos(2\theta)=0$, we can safely assume[3] $\sin(2\theta)\neq 0$. We get $\frac{\cos(2\theta)}{\sin(2\theta)}=\frac{A-C}{B}$, or $\cot(2\theta)=\frac{A-C}{B}$. We have just proved the following theorem.

Theorem 11.10. The equation $Ax^2+Bxy+Cy^2+Dx+Ey+F=0$ with $B\neq 0$ can be transformed into an equation in variables x' and y' without any $x'y'$ terms by rotating the x- and y- axes counter-clockwise through an angle θ which satisfies $\cot(2\theta)=\frac{A-C}{B}$.

We put Theorem 11.10 to good use in the following example.

Example 11.6.2. Graph the following equations.

1. $5x^2+26xy+5y^2-16x\sqrt{2}+16y\sqrt{2}-104=0$

2. $16x^2+24xy+9y^2+15x-20y=0$

Solution.

1. Since the equation $5x^2+26xy+5y^2-16x\sqrt{2}+16y\sqrt{2}-104=0$ is already given to us in the form required by Theorem 11.10, we identify $A=5$, $B=26$ and $C=5$ so that $\cot(2\theta)=\frac{A-C}{B}=\frac{5-5}{26}=0$. This means $\cot(2\theta)=0$ which gives $\theta=\frac{\pi}{4}+\frac{\pi}{2}k$ for integers k. We choose $\theta=\frac{\pi}{4}$ so that our rotation equations are $x=\frac{x'\sqrt{2}}{2}-\frac{y'\sqrt{2}}{2}$ and $y=\frac{x'\sqrt{2}}{2}+\frac{y'\sqrt{2}}{2}$. The reader should verify that

$$x^2=\frac{(x')^2}{2}-x'y'+\frac{(y')^2}{2}, \quad xy=\frac{(x')^2}{2}-\frac{(y')^2}{2}, \quad y^2=\frac{(x')^2}{2}+x'y'+\frac{(y')^2}{2}$$

Making the other substitutions, we get that $5x^2+26xy+5y^2-16x\sqrt{2}+16y\sqrt{2}-104=0$ reduces to $18(x')^2-8(y')^2+32y'-104=0$, or $\frac{(x')^2}{4}-\frac{(y'-2)^2}{9}=1$. The latter is the equation of a hyperbola centered at the $x'y'$-coordinates $(0,2)$ opening in the x' direction with vertices $(\pm 2,2)$ (in $x'y'$-coordinates) and asymptotes $y'=\pm\frac{3}{2}x'+2$. We graph it below.

[3] The reader is invited to think about the case $\sin(2\theta)=0$ geometrically. What happens to the axes in this case?

2. From $16x^2 + 24xy + 9y^2 + 15x - 20y = 0$, we get $A = 16$, $B = 24$ and $C = 9$ so that $\cot(2\theta) = \frac{7}{24}$. Since this isn't one of the values of the common angles, we will need to use inverse functions. Ultimately, we need to find $\cos(\theta)$ and $\sin(\theta)$, which means we have two options. If we use the arccotangent function immediately, after the usual calculations we get $\theta = \frac{1}{2}\text{arccot}\left(\frac{7}{24}\right)$. To get $\cos(\theta)$ and $\sin(\theta)$ from this, we would need to use half angle identities. Alternatively, we can start with $\cot(2\theta) = \frac{7}{24}$, use a double angle identity, and then go after $\cos(\theta)$ and $\sin(\theta)$. We adopt the second approach. From $\cot(2\theta) = \frac{7}{24}$, we have $\tan(2\theta) = \frac{24}{7}$. Using the double angle identity for tangent, we have $\frac{2\tan(\theta)}{1-\tan^2(\theta)} = \frac{24}{7}$, which gives $24\tan^2(\theta) + 14\tan(\theta) - 24 = 0$. Factoring, we get $2(3\tan(\theta)+4)(4\tan(\theta)-3) = 0$ which gives $\tan(\theta) = -\frac{4}{3}$ or $\tan(\theta) = \frac{3}{4}$. While either of these values of $\tan(\theta)$ satisfies the equation $\cot(2\theta) = \frac{7}{24}$, we choose $\tan(\theta) = \frac{3}{4}$, since this produces an acute angle,[4] $\theta = \arctan\left(\frac{3}{4}\right)$. To find the rotation equations, we need $\cos(\theta) = \cos\left(\arctan\left(\frac{3}{4}\right)\right)$ and $\sin(\theta) = \sin\left(\arctan\left(\frac{3}{4}\right)\right)$. Using the techniques developed in Section 10.6 we get $\cos(\theta) = \frac{4}{5}$ and $\sin(\theta) = \frac{3}{5}$. Our rotation equations are $x = x'\cos(\theta) - y'\sin(\theta) = \frac{4x'}{5} - \frac{3y'}{5}$ and $y = x'\sin(\theta) + y'\cos(\theta) = \frac{3x'}{5} + \frac{4y'}{5}$. As usual, we now substitute these quantities into $16x^2 + 24xy + 9y^2 + 15x - 20y = 0$ and simplify. As a first step, the reader can verify

$$x^2 = \frac{16(x')^2}{25} - \frac{24x'y'}{25} + \frac{9(y')^2}{25}, \quad xy = \frac{12(x')^2}{25} + \frac{7x'y'}{25} - \frac{12(y')^2}{25}, \quad y^2 = \frac{9(x')^2}{25} + \frac{24x'y'}{25} + \frac{16(y')^2}{25}$$

Once the dust settles, we get $25(x')^2 - 25y' = 0$, or $y' = (x')^2$, whose graph is a parabola opening along the positive y'-axis with vertex $(0,0)$. We graph this equation below.

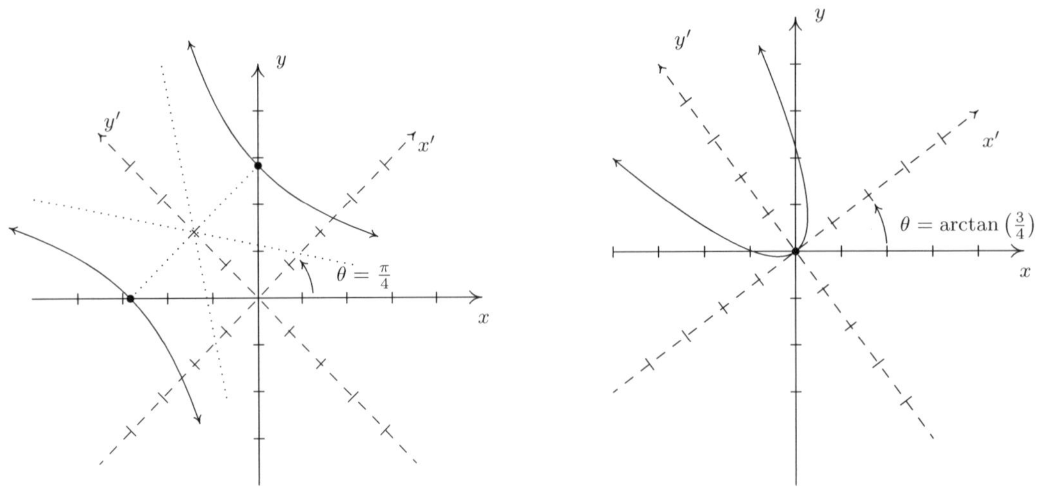

$5x^2 + 26xy + 5y^2 - 16x\sqrt{2} + 16y\sqrt{2} - 104 = 0$ $\qquad$ $16x^2 + 24xy + 9y^2 + 15x - 20y = 0$

□

[4] As usual, there are infinitely many solutions to $\tan(\theta) = \frac{3}{4}$. We choose the acute angle $\theta = \arctan\left(\frac{3}{4}\right)$. The reader is encouraged to think about why there is always at least one acute answer to $\cot(2\theta) = \frac{A-C}{B}$ and what this means geometrically in terms of what we are trying to accomplish by rotating the axes. The reader is also encouraged to keep a sharp lookout for the angles which satisfy $\tan(\theta) = -\frac{4}{3}$ in our final graph. (Hint: $\left(\frac{3}{4}\right)\left(-\frac{4}{3}\right) = -1$.)

We note that even though the coefficients of x^2 and y^2 were both positive numbers in parts 1 and 2 of Example 11.6.2, the graph in part 1 turned out to be a hyperbola and the graph in part 2 worked out to be a parabola. Whereas in Chapter 7, we could easily pick out which conic section we were dealing with based on the presence (or absence) of quadratic terms and their coefficients, Example 11.6.2 demonstrates that all bets are off when it comes to conics with an xy term which require rotation of axes to put them into a more standard form. Nevertheless, it is possible to determine which conic section we have by looking at a special, familiar *combination* of the coefficients of the quadratic terms. We have the following theorem.

Theorem 11.11. Suppose the equation $Ax^2 + Bxy + Cy^2 + Dx + Ey + F = 0$ describes a non-degenerate conic section.[a]

- If $B^2 - 4AC > 0$ then the graph of the equation is a hyperbola.
- If $B^2 - 4AC = 0$ then the graph of the equation is a parabola.
- If $B^2 - 4AC < 0$ then the graph of the equation is an ellipse or circle.

[a]Recall that this means its graph is either a circle, parabola, ellipse or hyperbola. See page 497.

As you may expect, the quantity B^2-4AC mentioned in Theorem 11.11 is called the **discriminant** of the conic section. While we will not attempt to explain the deep Mathematics which produces this 'coincidence', we will at least work through the proof of Theorem 11.11 mechanically to show that it is true.[5] First note that if the coefficient $B = 0$ in the equation $Ax^2+Bxy+Cy^2+Dx+Ey+F = 0$, Theorem 11.11 reduces to the result presented in Exercise 34 in Section 7.5, so we proceed here under the assumption that $B \neq 0$. We rotate the xy-axes counter-clockwise through an angle θ which satisfies $\cot(2\theta) = \frac{A-C}{B}$ to produce an equation with no $x'y'$-term in accordance with Theorem 11.10: $A'(x')^2 + C(y')^2 + Dx' + Ey' + F' = 0$. In this form, we can invoke Exercise 34 in Section 7.5 once more using the product $A'C'$. Our goal is to find the product $A'C'$ in terms of the coefficients A, B and C in the original equation. To that end, we make the usual substitutions $x = x'\cos(\theta) - y'\sin(\theta)$ $y = x'\sin(\theta) + y'\cos(\theta)$ into $Ax^2 + Bxy + Cy^2 + Dx + Ey + F = 0$. We leave it to the reader to show that, after gathering like terms, the coefficient A' on $(x')^2$ and the coefficient C' on $(y')^2$ are

$$\begin{aligned} A' &= A\cos^2(\theta) + B\cos(\theta)\sin(\theta) + C\sin^2(\theta) \\ C' &= A\sin^2(\theta) - B\cos(\theta)\sin(\theta) + C\cos^2(\theta) \end{aligned}$$

In order to make use of the condition $\cot(2\theta) = \frac{A-C}{B}$, we rewrite our formulas for A' and C' using the power reduction formulas. After some regrouping, we get

$$\begin{aligned} 2A' &= [(A+C) + (A-C)\cos(2\theta)] + B\sin(2\theta) \\ 2C' &= [(A+C) - (A-C)\cos(2\theta)] - B\sin(2\theta) \end{aligned}$$

Next, we try to make sense of the product

$$(2A')(2C') = \{[(A+C) + (A-C)\cos(2\theta)] + B\sin(2\theta)\}\{[(A+C) - (A-C)\cos(2\theta)] - B\sin(2\theta)\}$$

[5]We hope that someday you get to see *why* this works the way it does.

We break this product into pieces. First, we use the difference of squares to multiply the 'first' quantities in each factor to get

$$[(A+C)+(A-C)\cos(2\theta)]\,[(A+C)-(A-C)\cos(2\theta)] \quad = \quad (A+C)^2-(A-C)^2\cos^2(2\theta)$$

Next, we add the product of the 'outer' and 'inner' quantities in each factor to get

$$\begin{array}{rcl} -B\sin(2\theta)\,[(A+C)+(A-C)\cos(2\theta)] & & \\ +B\sin(2\theta)\,[(A+C)-(A-C)\cos(2\theta)] & = & -2B(A-C)\cos(2\theta)\sin(2\theta) \end{array}$$

The product of the 'last' quantity in each factor is $(B\sin(2\theta))(-B\sin(2\theta)) = -B^2\sin^2(2\theta)$. Putting all of this together yields

$$4A'C' \quad = \quad (A+C)^2-(A-C)^2\cos^2(2\theta)-2B(A-C)\cos(2\theta)\sin(2\theta)-B^2\sin^2(2\theta)$$

From $\cot(2\theta) = \frac{A-C}{B}$, we get $\frac{\cos(2\theta)}{\sin(2\theta)} = \frac{A-C}{B}$, or $(A-C)\sin(2\theta) = B\cos(2\theta)$. We use this substitution twice along with the Pythagorean Identity $\cos^2(2\theta) = 1-\sin^2(2\theta)$ to get

$$\begin{array}{rcl} 4A'C' & = & (A+C)^2-(A-C)^2\cos^2(2\theta)-2B(A-C)\cos(2\theta)\sin(2\theta)-B^2\sin^2(2\theta) \\ & = & (A+C)^2-(A-C)^2\left[1-\sin^2(2\theta)\right]-2B\cos(2\theta)B\cos(2\theta)-B^2\sin^2(2\theta) \\ & = & (A+C)^2-(A-C)^2+(A-C)^2\sin^2(2\theta)-2B^2\cos^2(2\theta)-B^2\sin^2(2\theta) \\ & = & (A+C)^2-(A-C)^2+[(A-C)\sin(2\theta)]^2-2B^2\cos^2(2\theta)-B^2\sin^2(2\theta) \\ & = & (A+C)^2-(A-C)^2+[B\cos(2\theta)]^2-2B^2\cos^2(2\theta)-B^2\sin^2(2\theta) \\ & = & (A+C)^2-(A-C)^2+B^2\cos^2(2\theta)-2B^2\cos^2(2\theta)-B^2\sin^2(2\theta) \\ & = & (A+C)^2-(A-C)^2-B^2\cos^2(2\theta)-B^2\sin^2(2\theta) \\ & = & (A+C)^2-(A-C)^2-B^2\left[\cos^2(2\theta)+\sin^2(2\theta)\right] \\ & = & (A+C)^2-(A-C)^2-B^2 \\ & = & \left(A^2+2AC+C^2\right)-\left(A^2-2AC+C^2\right)-B^2 \\ & = & 4AC-B^2 \end{array}$$

Hence, $B^2-4AC = -4A'C'$, so the quantity B^2-4AC has the opposite sign of $A'C'$. The result now follows by applying Exercise 34 in Section 7.5.

Example 11.6.3. Use Theorem 11.11 to classify the graphs of the following non-degenerate conics.

1. $21x^2+10xy\sqrt{3}+31y^2 = 144$
2. $5x^2+26xy+5y^2-16x\sqrt{2}+16y\sqrt{2}-104 = 0$
3. $16x^2+24xy+9y^2+15x-20y = 0$

Solution. This is a straightforward application of Theorem 11.11.

1. We have $A = 21$, $B = 10\sqrt{3}$ and $C = 31$ so $B^2 - 4AC = (10\sqrt{3})^2 - 4(21)(31) = -2304 < 0$. Theorem 11.11 predicts the graph is an ellipse, which checks with our work from Example 11.6.1 number 2.

2. Here, $A = 5$, $B = 26$ and $C = 5$, so $B^2 - 4AC = 26^2 - 4(5)(5) = 576 > 0$. Theorem 11.11 classifies the graph as a hyperbola, which matches our answer to Example 11.6.2 number 1.

3. Finally, we have $A = 16$, $B = 24$ and $C = 9$ which gives $24^2 - 4(16)(9) = 0$. Theorem 11.11 tells us that the graph is a parabola, matching our result from Example 11.6.2 number 2. □

11.6.2 The Polar Form of Conics

In this subsection, we start from scratch to reintroduce the conic sections from a more unified perspective. We have our 'new' definition below.

Definition 11.1. Given a fixed line L, a point F not on L, and a positive number e, a conic section is the set of all points P such that

$$\frac{\text{the distance from } P \text{ to } F}{\text{the distance from } P \text{ to } L} = e$$

The line L is called the **directrix** of the conic section, the point F is called a **focus** of the conic section, and the constant e is called the **eccentricity** of the conic section.

We have seen the notions of focus and directrix before in the definition of a parabola, Definition 7.3. There, a parabola is defined as the set of points equidistant from the focus and directrix, giving an eccentricity $e = 1$ according to Definition 11.1. We have also seen the concept of eccentricity before. It was introduced for ellipses in Definition 7.5 in Section 7.4, and later extended to hyperbolas in Exercise 31 in Section 7.5. There, e was also defined as a ratio of distances, though in these cases the distances involved were measurements from the center to a focus and from the center to a vertex. One way to reconcile the 'old' ideas of focus, directrix and eccentricity with the 'new' ones presented in Definition 11.1 is to derive equations for the conic sections using Definition 11.1 and compare these parameters with what we know from Chapter 7. We begin by assuming the conic section has eccentricity e, a focus F at the origin and that the directrix is the vertical line $x = -d$ as in the figure below.

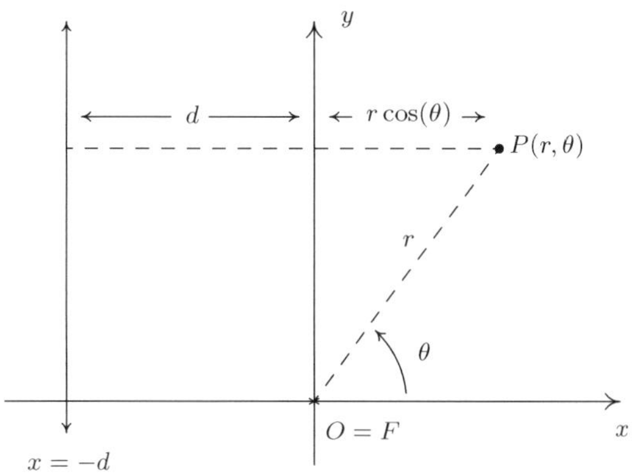

Using a polar coordinate representation $P(r, \theta)$ for a point on the conic with $r > 0$, we get

$$e = \frac{\text{the distance from } P \text{ to } F}{\text{the distance from } P \text{ to } L} = \frac{r}{d + r\cos(\theta)}$$

so that $r = e(d + r\cos(\theta))$. Solving this equation for r, yields

$$r = \frac{ed}{1 - e\cos(\theta)}$$

At this point, we convert the equation $r = e(d + r\cos(\theta))$ back into a rectangular equation in the variables x and y. If $e > 0$, but $e \neq 1$, the usual conversion process outlined in Section 11.4 gives[6]

$$\left(\frac{\left(1-e^2\right)^2}{e^2d^2}\right)\left(x - \frac{e^2d}{1-e^2}\right)^2 + \left(\frac{1-e^2}{e^2d^2}\right)y^2 = 1$$

We leave it to the reader to show if $0 < e < 1$, this is the equation of an ellipse centered at $\left(\frac{e^2d}{1-e^2}, 0\right)$ with major axis along the x-axis. Using the notation from Section 7.4, we have $a^2 = \frac{e^2d^2}{(1-e^2)^2}$ and $b^2 = \frac{e^2d^2}{1-e^2}$, so the major axis has length $\frac{2ed}{1-e^2}$ and the minor axis has length $\frac{2ed}{\sqrt{1-e^2}}$. Moreover, we find that one focus is $(0, 0)$ and working through the formula given in Definition 7.5 gives the eccentricity to be e, as required. If $e > 1$, then the equation generates a hyperbola with center $\left(\frac{e^2d}{1-e^2}, 0\right)$ whose transverse axis lies along the x-axis. Since such hyperbolas have the form $\frac{(x-h)^2}{a^2} - \frac{y^2}{b^2} = 1$, we need to take the *opposite* reciprocal of the coefficient of y^2 to find b^2. We get[7] $a^2 = \frac{e^2d^2}{(1-e^2)^2} = \frac{e^2d^2}{(e^2-1)^2}$ and $b^2 = -\frac{e^2d^2}{1-e^2} = \frac{e^2d^2}{e^2-1}$, so the transverse axis has length $\frac{2ed}{e^2-1}$ and the conjugate axis has length $\frac{2ed}{\sqrt{e^2-1}}$. Additionally, we verify that one focus is at $(0, 0)$, and the formula given in Exercise 31 in Section 7.5 gives the eccentricity is e in this case as well. If $e = 1$, the equation $r = \frac{ed}{1-e\cos(\theta)}$ reduces to $r = \frac{d}{1-\cos(\theta)}$ which gives the rectangular equation $y^2 = 2d\left(x + \frac{d}{2}\right)$. This is a parabola with vertex $\left(-\frac{d}{2}, 0\right)$ opening to the right. In the language of Section 7.3, $4p = 2d$ so $p = \frac{d}{2}$, the focus is $(0, 0)$, the focal diameter is $2d$ and the directrix is $x = -d$, as required. Hence, we have shown that in all cases, our 'new' understanding of 'conic section', 'focus', 'eccentricity' and 'directrix' as presented in Definition 11.1 correspond with the 'old' definitions given in Chapter 7.

Before we summarize our findings, we note that in order to arrive at our general equation of a conic $r = \frac{ed}{1-e\cos(\theta)}$, we assumed that the directrix was the line $x = -d$ for $d > 0$. We could have just as easily chosen the directrix to be $x = d$, $y = -d$ or $y = d$. As the reader can verify, in these cases we obtain the forms $r = \frac{ed}{1+e\cos(\theta)}$, $r = \frac{ed}{1-e\sin(\theta)}$ and $r = \frac{ed}{1+e\sin(\theta)}$, respectively. The key thing to remember is that in any of these cases, the directrix is always perpendicular to the major axis of an ellipse and it is always perpendicular to the transverse axis of the hyperbola. For parabolas, knowing the focus is $(0, 0)$ and the directrix also tells us which way the parabola opens. We have established the following theorem.

[6]Turn $r = e(d + r\cos(\theta))$ into $r = e(d + x)$ and square both sides to get $r^2 = e^2(d + x)^2$. Replace r^2 with $x^2 + y^2$, expand $(d + x)^2$, combine like terms, complete the square on x and clean things up.

[7]Since $e > 1$ in this case, $1 - e^2 < 0$. Hence, we rewrite $\left(1 - e^2\right)^2 = \left(e^2 - 1\right)^2$ to help simplify things later on.

Theorem 11.12. Suppose e and d are positive numbers. Then

- the graph of $r = \frac{ed}{1-e\cos(\theta)}$ is the graph of a conic section with directrix $x = -d$.
- the graph of $r = \frac{ed}{1+e\cos(\theta)}$ is the graph of a conic section with directrix $x = d$.
- the graph of $r = \frac{ed}{1-e\sin(\theta)}$ is the graph of a conic section with directrix $y = -d$.
- the graph of $r = \frac{ed}{1+e\sin(\theta)}$ is the graph of a conic section with directrix $y = d$.

In each case above, $(0,0)$ is a focus of the conic and the number e is the eccentricity of the conic.

- If $0 < e < 1$, the graph is an ellipse whose major axis has length $\frac{2ed}{1-e^2}$ and whose minor axis has length $\frac{2ed}{\sqrt{1-e^2}}$
- If $e = 1$, the graph is a parabola whose focal diameter is $2d$.
- If $e > 1$, the graph is a hyperbola whose transverse axis has length $\frac{2ed}{e^2-1}$ and whose conjugate axis has length $\frac{2ed}{\sqrt{e^2-1}}$.

We test out Theorem 11.12 in the next example.

Example 11.6.4. Sketch the graphs of the following equations.

1. $r = \dfrac{4}{1-\sin(\theta)}$
2. $r = \dfrac{12}{3-\cos(\theta)}$
3. $r = \dfrac{6}{1+2\sin(\theta)}$

Solution.

1. From $r = \frac{4}{1-\sin(\theta)}$, we first note $e = 1$ which means we have a parabola on our hands. Since $ed = 4$, we have $d = 4$ and considering the form of the equation, this puts the directrix at $y = -4$. Since the focus is at $(0,0)$, we know that the vertex is located at the point (in rectangular coordinates) $(0,-2)$ and must open upwards. With $d = 4$, we have a focal diameter of $2d = 8$, so the parabola contains the points $(\pm 4, 0)$. We graph $r = \frac{4}{1-\sin(\theta)}$ below.

2. We first rewrite $r = \frac{12}{3-\cos(\theta)}$ in the form found in Theorem 11.12, namely $r = \frac{4}{1-(1/3)\cos(\theta)}$. Since $e = \frac{1}{3}$ satisfies $0 < e < 1$, we know that the graph of this equation is an ellipse. Since $ed = 4$, we have $d = 12$ and, based on the form of the equation, the directrix is $x = -12$. This means that the ellipse has its major axis along the x-axis. We can find the vertices of the ellipse by finding the points of the ellipse which lie on the x-axis. We find $r(0) = 6$ and $r(\pi) = 3$ which correspond to the rectangular points $(-3,0)$ and $(6,0)$, so these are our vertices. The center of the ellipse is the midpoint of the vertices, which in this case is $\left(\frac{3}{2},0\right)$.[8] We know one focus is $(0,0)$, which is $\frac{3}{2}$ from the center $\left(\frac{3}{2},0\right)$ and this allows us to find the

[8]As a quick check, we have from Theorem 11.12 the major axis should have length $\frac{2ed}{1-e^2} = \frac{(2)(4)}{1-(1/3)^2} = 9$.

other focus $(3, 0)$, even though we are not asked to do so. Finally, we know from Theorem 11.12 that the length of the minor axis is $\frac{2ed}{\sqrt{1-e^2}} = \frac{4}{\sqrt{1-(1/3)^2}} = 6\sqrt{3}$ which means the endpoints of the minor axis are $\left(\frac{3}{2}, \pm 3\sqrt{2}\right)$. We now have everything we need to graph $r = \frac{12}{3-\cos(\theta)}$.

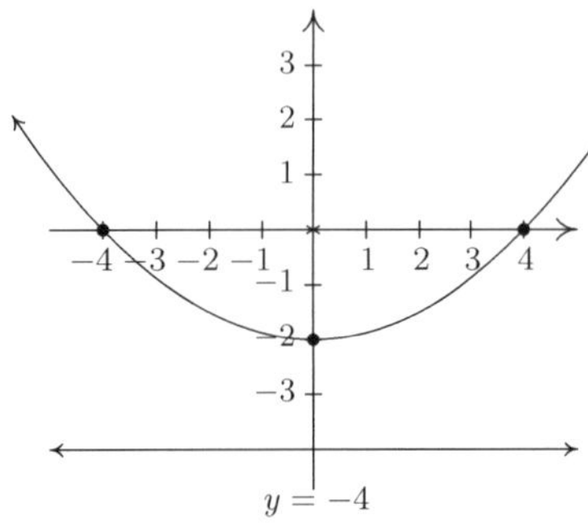

$r = \frac{4}{1-\sin(\theta)}$

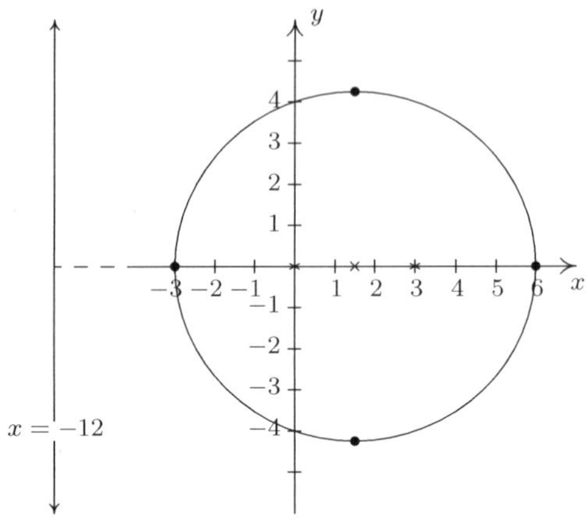

$r = \frac{12}{3-\cos(\theta)}$

3. From $r = \frac{6}{1+2\sin(\theta)}$ we get $e = 2 > 1$ so the graph is a hyperbola. Since $ed = 6$, we get $d = 3$, and from the form of the equation, we know the directrix is $y = 3$. This means the transverse axis of the hyperbola lies along the y-axis, so we can find the vertices by looking where the hyperbola intersects the y-axis. We find $r\left(\frac{\pi}{2}\right) = 2$ and $r\left(\frac{3\pi}{2}\right) = -6$. These two points correspond to the rectangular points $(0, 2)$ and $(0, 6)$ which puts the center of the hyperbola at $(0, 4)$. Since one focus is at $(0, 0)$, which is 4 units away from the center, we know the other focus is at $(0, 8)$. According to Theorem 11.12, the conjugate axis has a length of $\frac{2ed}{\sqrt{e^2-1}} = \frac{(2)(6)}{\sqrt{2^2-1}} = 4\sqrt{3}$. Putting this together with the location of the vertices, we get that the asymptotes of the hyperbola have slopes $\pm\frac{2}{2\sqrt{3}} = \pm\frac{\sqrt{3}}{3}$. Since the center of the hyperbola is $(0, 4)$, the asymptotes are $y = \pm\frac{\sqrt{3}}{3}x + 4$. We graph the hyperbola below.

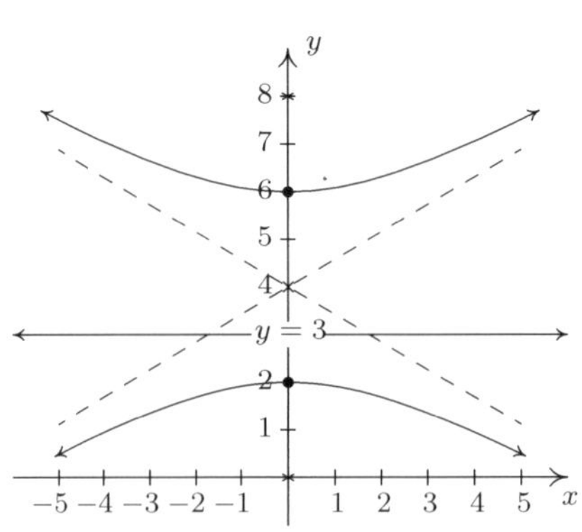

$r = \frac{6}{1+2\sin(\theta)}$

□

In light of Section 11.6.1, the reader may wonder what the rotated form of the conic sections would look like in polar form. We know from Exercise 65 in Section 11.5 that replacing θ with $(\theta - \phi)$ in an expression $r = f(\theta)$ rotates the graph of $r = f(\theta)$ counter-clockwise by an angle ϕ. For instance, to graph $r = \frac{4}{1-\sin\left(\theta - \frac{\pi}{4}\right)}$ all we need to do is rotate the graph of $r = \frac{4}{1-\sin(\theta)}$, which we obtained in Example 11.6.4 number 1, counter-clockwise by $\frac{\pi}{4}$ radians, as shown below.

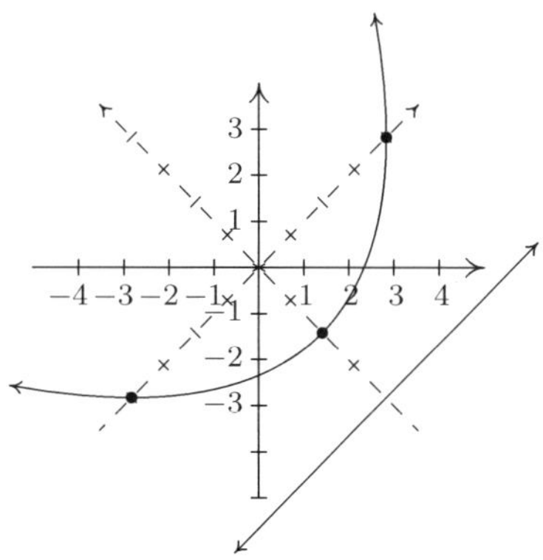

$r = \frac{4}{1-\sin\left(\theta - \frac{\pi}{4}\right)}$

Using rotations, we can greatly simplify the form of the conic sections presented in Theorem 11.12, since any three of the forms given there can be obtained from the fourth by rotating through some multiple of $\frac{\pi}{2}$. Since rotations do not affect lengths, all of the formulas for lengths Theorem 11.12 remain intact. In the theorem below, we also generalize our formula for conic sections to include circles centered at the origin by extending the concept of eccentricity to include $e = 0$. We conclude this section with the statement of the following theorem.

Theorem 11.13. Given constants $\ell > 0$, $e \geq 0$ and ϕ, the graph of the equation

$$r = \frac{\ell}{1 - e\cos(\theta - \phi)}$$

is a conic section with eccentricity e and one focus at $(0, 0)$.

- If $e = 0$, the graph is a circle centered at $(0, 0)$ with radius ℓ.
- If $e \neq 0$, then the conic has a focus at $(0, 0)$ and the directrix contains the point with polar coordinates $(-d, \phi)$ where $d = \frac{\ell}{e}$.
  - If $0 < e < 1$, the graph is an ellipse whose major axis has length $\frac{2ed}{1-e^2}$ and whose minor axis has length $\frac{2ed}{\sqrt{1-e^2}}$
  - If $e = 1$, the graph is a parabola whose focal diameter is $2d$.
  - If $e > 1$, the graph is a hyperbola whose transverse axis has length $\frac{2ed}{e^2-1}$ and whose conjugate axis has length $\frac{2ed}{\sqrt{e^2-1}}$.

11.6.3 EXERCISES

Graph the following equations.

1. $x^2 + 2xy + y^2 - x\sqrt{2} + y\sqrt{2} - 6 = 0$
2. $7x^2 - 4xy\sqrt{3} + 3y^2 - 2x - 2y\sqrt{3} - 5 = 0$
3. $5x^2 + 6xy + 5y^2 - 4\sqrt{2}x + 4\sqrt{2}y = 0$
4. $x^2 + 2\sqrt{3}xy + 3y^2 + 2\sqrt{3}x - 2y - 16 = 0$
5. $13x^2 - 34xy\sqrt{3} + 47y^2 - 64 = 0$
6. $x^2 - 2\sqrt{3}xy - y^2 + 8 = 0$
7. $x^2 - 4xy + 4y^2 - 2x\sqrt{5} - y\sqrt{5} = 0$
8. $8x^2 + 12xy + 17y^2 - 20 = 0$

Graph the following equations.

9. $r = \dfrac{2}{1 - \cos(\theta)}$
10. $r = \dfrac{3}{2 + \sin(\theta)}$
11. $r = \dfrac{3}{2 - \cos(\theta)}$
12. $r = \dfrac{2}{1 + \sin(\theta)}$
13. $r = \dfrac{4}{1 + 3\cos(\theta)}$
14. $r = \dfrac{2}{1 - 2\sin(\theta)}$
15. $r = \dfrac{2}{1 + \sin(\theta - \frac{\pi}{3})}$
16. $r = \dfrac{6}{3 - \cos\left(\theta + \frac{\pi}{4}\right)}$

The matrix $A(\theta) = \begin{bmatrix} \cos(\theta) & -\sin(\theta) \\ \sin(\theta) & \cos(\theta) \end{bmatrix}$ is called a **rotation matrix**. We've seen this matrix most recently in the proof of used in the proof of Theorem 11.9.

17. Show the matrix from Example 8.3.3 in Section 8.3 is none other than $A\left(\frac{\pi}{4}\right)$.
18. Discuss with your classmates how to use $A(\theta)$ to rotate points in the plane.
19. Using the even / odd identities for cosine and sine, show $A(\theta)^{-1} = A(-\theta)$. Interpret this geometrically.

11.6.4 Answers

1. $x^2 + 2xy + y^2 - x\sqrt{2} + y\sqrt{2} - 6 = 0$ becomes $(x')^2 = -(y' - 3)$ after rotating counter-clockwise through $\theta = \frac{\pi}{4}$.

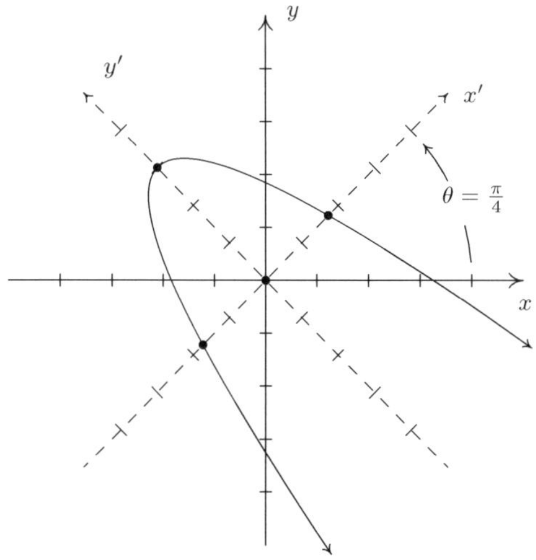

$x^2 + 2xy + y^2 - x\sqrt{2} + y\sqrt{2} - 6 = 0$

2. $7x^2 - 4xy\sqrt{3} + 3y^2 - 2x - 2y\sqrt{3} - 5 = 0$ becomes $\frac{(x'-2)^2}{9} + (y')^2 = 1$ after rotating counter-clockwise through $\theta = \frac{\pi}{3}$

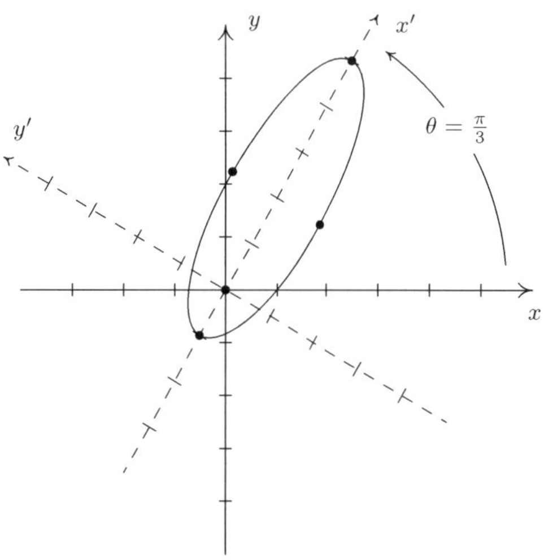

$7x^2 - 4xy\sqrt{3} + 3y^2 - 2x - 2y\sqrt{3} - 5 = 0$

3. $5x^2 + 6xy + 5y^2 - 4\sqrt{2}x + 4\sqrt{2}y = 0$ becomes $(x')^2 + \frac{(y'+2)^2}{4} = 1$ after rotating counter-clockwise through $\theta = \frac{\pi}{4}$.

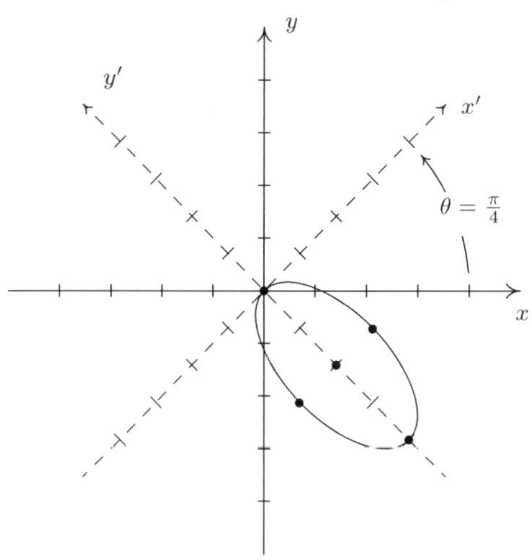

$5x^2 + 6xy + 5y^2 - 4\sqrt{2}x + 4\sqrt{2}y = 0$

4. $x^2 + 2\sqrt{3}xy + 3y^2 + 2\sqrt{3}x - 2y - 16 = 0$ becomes$(x')^2 = y' + 4$ after rotating counter-clockwise through $\theta = \frac{\pi}{3}$

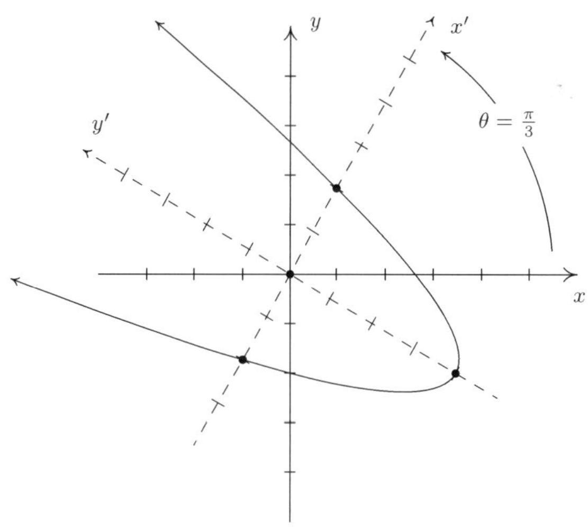

$x^2 + 2\sqrt{3}xy + 3y^2 + 2\sqrt{3}x - 2y - 16 = 0$

5. $13x^2 - 34xy\sqrt{3} + 47y^2 - 64 = 0$
becomes $(y')^2 - \frac{(x')^2}{16} = 1$ after rotating counter-clockwise through $\theta = \frac{\pi}{6}$.

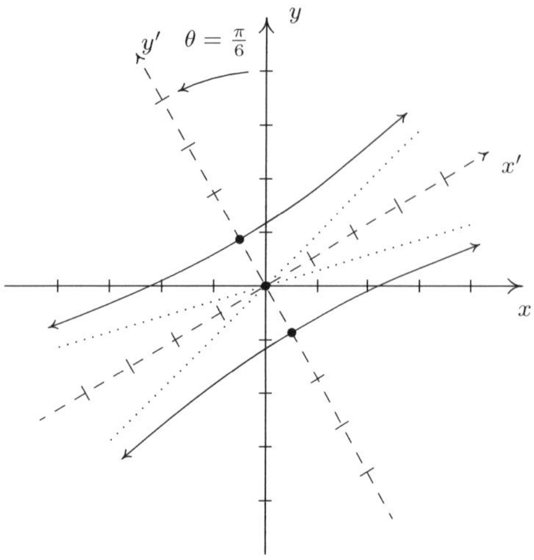

$13x^2 - 34xy\sqrt{3} + 47y^2 - 64 = 0$

6. $x^2 - 2\sqrt{3}xy - y^2 + 8 = 0$
becomes $\frac{(x')^2}{4} - \frac{(y')^2}{4} = 1$ after rotating counter-clockwise through $\theta = \frac{\pi}{3}$

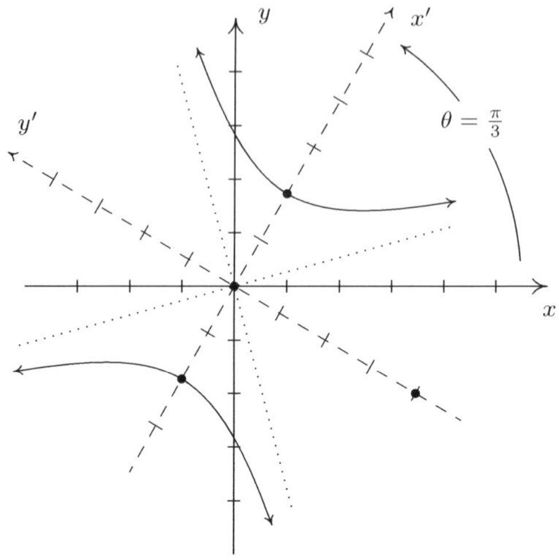

$x^2 - 2\sqrt{3}xy - y^2 + 8 = 0$

7. $x^2 - 4xy + 4y^2 - 2x\sqrt{5} - y\sqrt{5} = 0$
becomes $(y')^2 = x$ after rotating counter-clockwise through $\theta = \arctan\left(\frac{1}{2}\right)$.

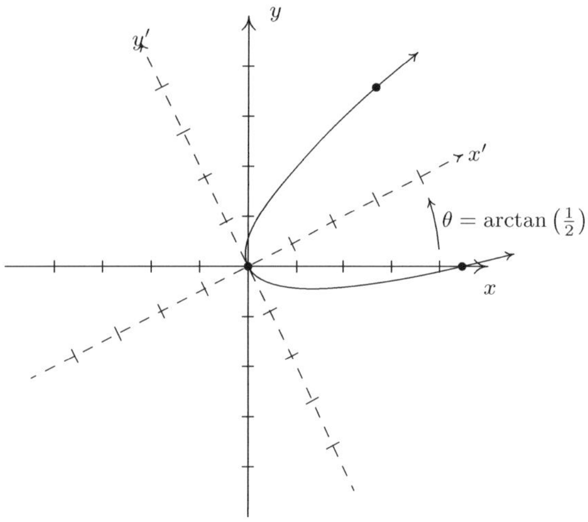

$x^2 - 4xy + 4y^2 - 2x\sqrt{5} - y\sqrt{5} = 0$

8. $8x^2 + 12xy + 17y^2 - 20 = 0$
becomes $(x')^2 + \frac{(y')^2}{4} = 1$ after rotating counter-clockwise through $\theta = \arctan(2)$

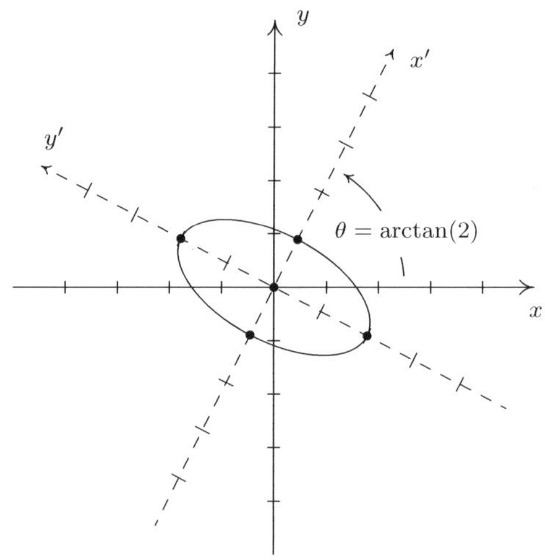

$8x^2 + 12xy + 17y^2 - 20 = 0$

9. $r = \frac{2}{1-\cos(\theta)}$ is a parabola
directrix $x = -2$, vertex $(-1, 0)$
focus $(0, 0)$, focal diameter 4

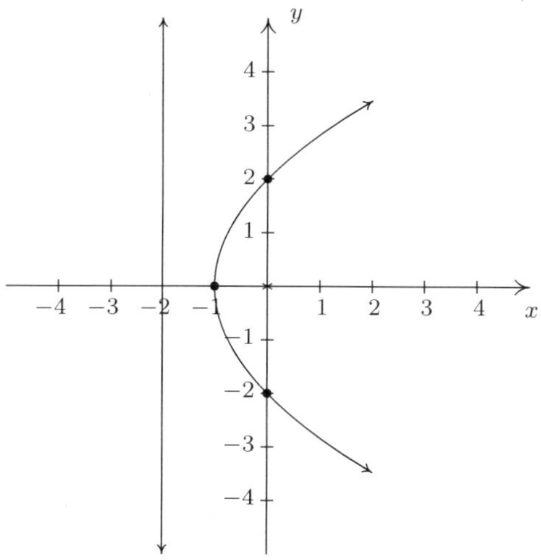

10. $r = \frac{3}{2+\sin(\theta)} = \frac{\frac{3}{2}}{1+\frac{1}{2}\sin(\theta)}$ is an ellipse
directrix $y = 3$, vertices $(0, 1)$, $(0, -3)$
center $(0, -2)$, foci $(0, 0)$, $(0, -2)$
minor axis length $2\sqrt{3}$

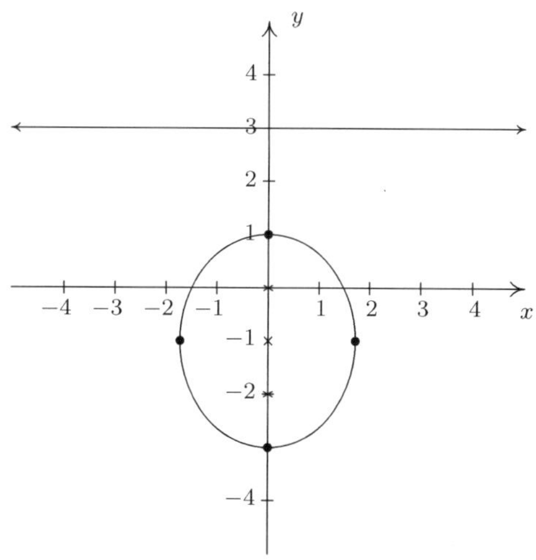

11. $r = \frac{3}{2-\cos(\theta)} = \frac{\frac{3}{2}}{1-\frac{1}{2}\cos(\theta)}$ is an ellipse
directrix $x = -3$, vertices $(-1, 0)$, $(3, 0)$
center $(1, 0)$, foci $(0, 0)$, $(2, 0)$
minor axis length $2\sqrt{3}$

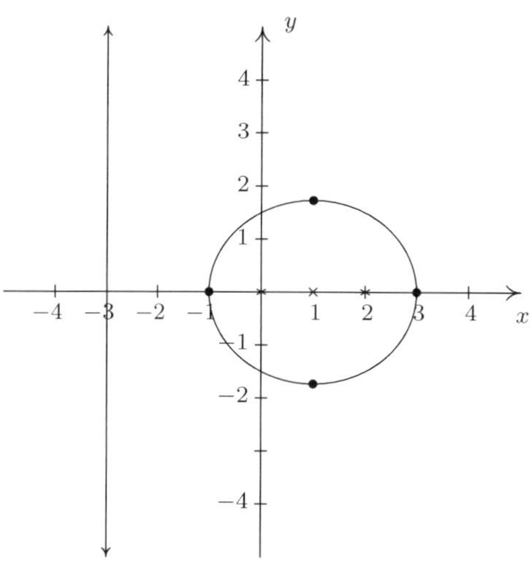

12. $r = \frac{2}{1+\sin(\theta)}$ is a parabola
directrix $y = 2$, vertex $(0, 1)$
focus $(0, 0)$, focal diameter 4

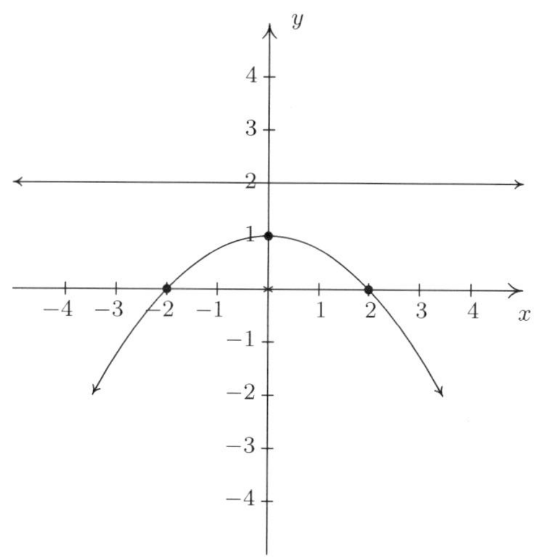

13. $r = \frac{4}{1+3\cos(\theta)}$ is a hyperbola
directrix $x = \frac{4}{3}$, vertices $(1,0)$, $(2,0)$
center $\left(\frac{3}{2},0\right)$, foci $(0,0)$, $(3,0)$
conjugate axis length $2\sqrt{2}$

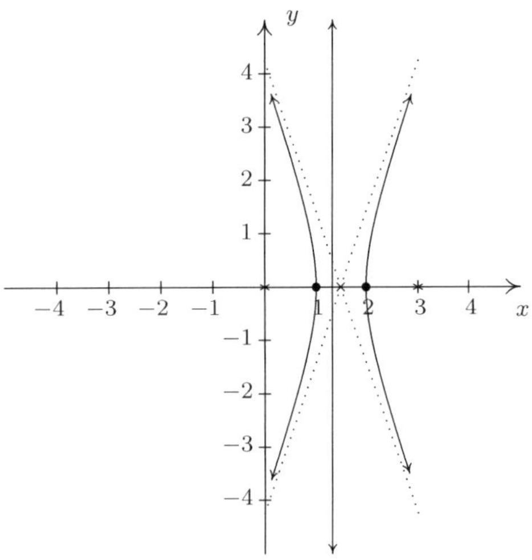

14. $r = \frac{2}{1-2\sin(\theta)}$ is a hyperbola
directrix $y = -1$, vertices $\left(0,-\frac{2}{3}\right)$, $(0,-2)$
center $\left(0,-\frac{4}{3}\right)$, foci $(0,0)$, $\left(0,-\frac{8}{3}\right)$
conjugate axis length $\frac{2\sqrt{3}}{3}$

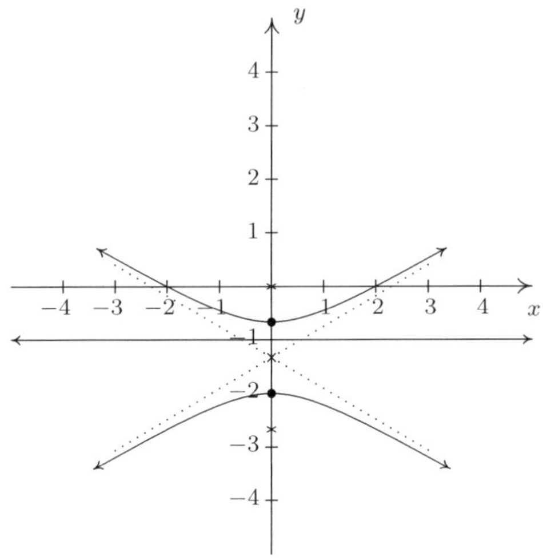

15. $r = \frac{2}{1+\sin\left(\theta-\frac{\pi}{3}\right)}$ is
the parabola $r = \frac{2}{1+\sin(\theta)}$
rotated through $\phi = \frac{\pi}{3}$

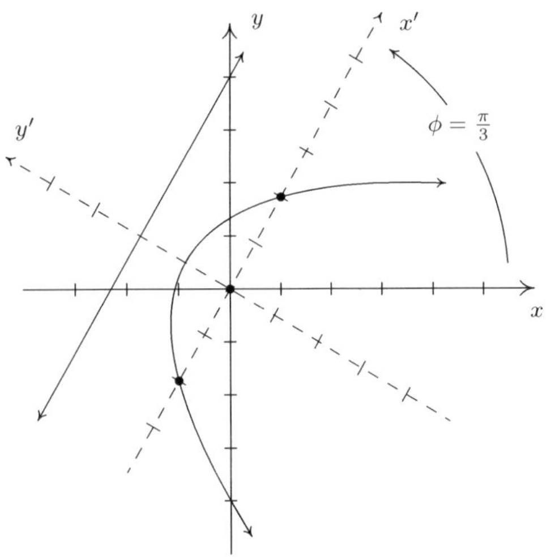

16. $r = \frac{6}{3-\cos\left(\theta+\frac{\pi}{4}\right)}$ is the ellipse
$r = \frac{6}{3-\cos(\theta)} = \frac{2}{1-\frac{1}{3}\cos(\theta)}$
rotated through $\phi = -\frac{\pi}{4}$

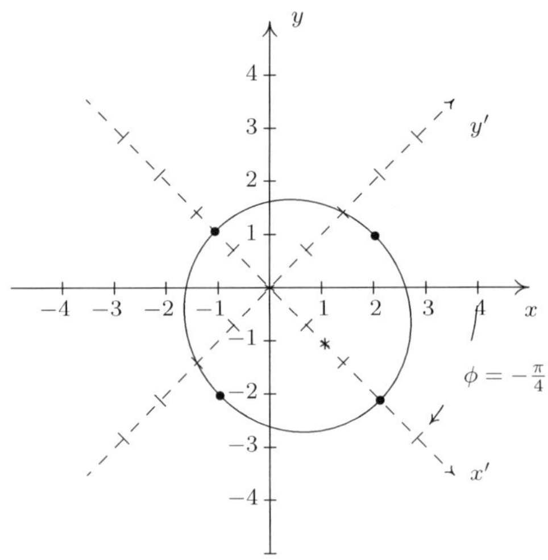

11.7 Polar Form of Complex Numbers

In this section, we return to our study of complex numbers which were first introduced in Section 3.4. Recall that a **complex number** is a number of the form $z = a + bi$ where a and b are real numbers and i is the imaginary unit defined by $i = \sqrt{-1}$. The number a is called the **real part** of z, denoted $\text{Re}(z)$, while the real number b is called the **imaginary part** of z, denoted $\text{Im}(z)$. From Intermediate Algebra, we know that if $z = a + bi = c + di$ where a, b, c and d are real numbers, then $a = c$ and $b = d$, which means $\text{Re}(z)$ and $\text{Im}(z)$ are well-defined.[1] To start off this section, we associate each complex number $z = a + bi$ with the point (a, b) on the coordinate plane. In this case, the x-axis is relabeled as the **real axis**, which corresponds to the real number line as usual, and the y-axis is relabeled as the **imaginary axis**, which is demarcated in increments of the imaginary unit i. The plane determined by these two axes is called the **complex plane**.

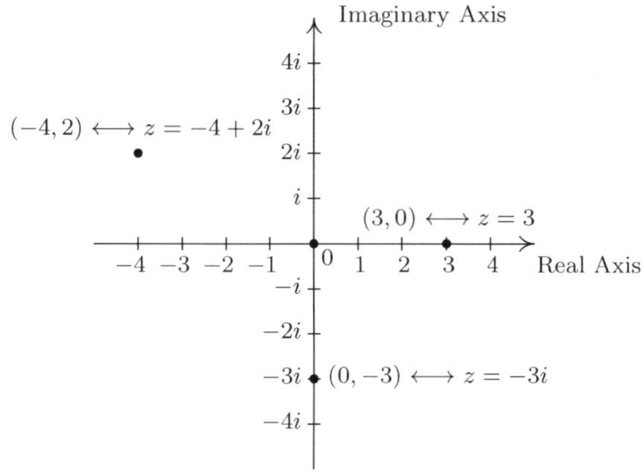

The Complex Plane

Since the ordered pair (a, b) gives the *rectangular* coordinates associated with the complex number $z = a + bi$, the expression $z = a + bi$ is called the **rectangular form** of z. Of course, we could just as easily associate z with a pair of *polar* coordinates (r, θ). Although it is not as straightforward as the definitions of $\text{Re}(z)$ and $\text{Im}(z)$, we can still give r and θ special names in relation to z.

Definition 11.2. The Modulus and Argument of Complex Numbers: Let $z = a + bi$ be a complex number with $a = \text{Re}(z)$ and $b = \text{Im}(z)$. Let (r, θ) be a polar representation of the point with rectangular coordinates (a, b) where $r \geq 0$.

- The **modulus** of z, denoted $|z|$, is defined by $|z| = r$.
- The angle θ is an **argument** of z. The set of all arguments of z is denoted $\arg(z)$.
- If $z \neq 0$ and $-\pi < \theta \leq \pi$, then θ is the **principal argument** of z, written $\theta = \text{Arg}(z)$.

[1] 'Well-defined' means that no matter how we express z, the number $\text{Re}(z)$ is always the same, and the number $\text{Im}(z)$ is always the same. In other words, Re and Im are *functions* of complex numbers.

Some remarks about Definition 11.2 are in order. We know from Section 11.4 that every point in the plane has infinitely many polar coordinate representations (r, θ) which means it's worth our time to make sure the quantities 'modulus', 'argument' and 'principal argument' are well-defined. Concerning the modulus, if $z = 0$ then the point associated with z is the origin. In this case, the *only* r-value which can be used here is $r = 0$. Hence for $z = 0$, $|z| = 0$ is well-defined. If $z \neq 0$, then the point associated with z is not the origin, and there are *two* possibilities for r: one positive and one negative. However, we stipulated $r \geq 0$ in our definition so this pins down the value of $|z|$ to one and only one number. Thus the modulus is well-defined in this case, too.[2] Even with the requirement $r \geq 0$, there are infinitely many angles θ which can be used in a polar representation of a point (r, θ). If $z \neq 0$ then the point in question is not the origin, so all of these angles θ are coterminal. Since coterminal angles are exactly 2π radians apart, we are guaranteed that only one of them lies in the interval $(-\pi, \pi]$, and this angle is what we call the principal argument of z, $\text{Arg}(z)$. In fact, the set $\arg(z)$ of all arguments of z can be described using set-builder notation as $\arg(z) = \{\text{Arg}(z) + 2\pi k \,|\, k \text{ is an integer}\}$. Note that since $\arg(z)$ is a *set*, we will write '$\theta \in \arg(z)$' to mean 'θ is in[3] the set of arguments of z'. If $z = 0$ then the point in question is the origin, which we know can be represented in polar coordinates as $(0, \theta)$ for *any* angle θ. In this case, we have $\arg(0) = (-\infty, \infty)$ and since there is no one value of θ which lies $(-\pi, \pi]$, we leave $\text{Arg}(0)$ undefined.[4] It is time for an example.

Example 11.7.1. For each of the following complex numbers find $\text{Re}(z)$, $\text{Im}(z)$, $|z|$, $\arg(z)$ and $\text{Arg}(z)$. Plot z in the complex plane.

1. $z = \sqrt{3} - i$
2. $z = -2 + 4i$
3. $z = 3i$
4. $z = -117$

Solution.

1. For $z = \sqrt{3} - i = \sqrt{3} + (-1)i$, we have $\text{Re}(z) = \sqrt{3}$ and $\text{Im}(z) = -1$. To find $|z|$, $\arg(z)$ and $\text{Arg}(z)$, we need to find a polar representation (r, θ) with $r \geq 0$ for the point $P(\sqrt{3}, -1)$ associated with z. We know $r^2 = (\sqrt{3})^2 + (-1)^2 = 4$, so $r = \pm 2$. Since we require $r \geq 0$, we choose $r = 2$, so $|z| = 2$. Next, we find a corresponding angle θ. Since $r > 0$ and P lies in Quadrant IV, θ is a Quadrant IV angle. We know $\tan(\theta) = \frac{-1}{\sqrt{3}} = -\frac{\sqrt{3}}{3}$, so $\theta = -\frac{\pi}{6} + 2\pi k$ for integers k. Hence, $\arg(z) = \left\{-\frac{\pi}{6} + 2\pi k \,|\, k \text{ is an integer}\right\}$. Of these values, only $\theta = -\frac{\pi}{6}$ satisfies the requirement that $-\pi < \theta \leq \pi$, hence $\text{Arg}(z) = -\frac{\pi}{6}$.

2. The complex number $z = -2 + 4i$ has $\text{Re}(z) = -2$, $\text{Im}(z) = 4$, and is associated with the point $P(-2, 4)$. Our next task is to find a polar representation (r, θ) for P where $r \geq 0$. Running through the usual calculations gives $r = 2\sqrt{5}$, so $|z| = 2\sqrt{5}$. To find θ, we get $\tan(\theta) = -2$, and since $r > 0$ and P lies in Quadrant II, we know θ is a Quadrant II angle. We find $\theta = \pi + \arctan(-2) + 2\pi k$, or, more succinctly $\theta = \pi - \arctan(2) + 2\pi k$ for integers k. Hence $\arg(z) = \{\pi - \arctan(2) + 2\pi k \,|\, k \text{ is an integer}\}$. Only $\theta = \pi - \arctan(2)$ satisfies the requirement $-\pi < \theta \leq \pi$, so $\text{Arg}(z) = \pi - \arctan(2)$.

[2]In case you're wondering, the use of the absolute value notation $|z|$ for modulus will be explained shortly.

[3]Recall the symbol being used here, '$\in$,' is the mathematical symbol which denotes membership in a set.

[4]If we had Calculus, we would regard Arg(0) as an 'indeterminate form.' But we don't, so we won't.

3. We rewrite $z = 3i$ as $z = 0 + 3i$ to find $\text{Re}(z) = 0$ and $\text{Im}(z) = 3$. The point in the plane which corresponds to z is $(0, 3)$ and while we could go through the usual calculations to find the required polar form of this point, we can almost 'see' the answer. The point $(0, 3)$ lies 3 units away from the origin on the positive y-axis. Hence, $r = |z| = 3$ and $\theta = \frac{\pi}{2} + 2\pi k$ for integers k. We get $\arg(z) = \left\{\frac{\pi}{2} + 2\pi k \,|\, k \text{ is an integer}\right\}$ and $\text{Arg}(z) = \frac{\pi}{2}$.

4. As in the previous problem, we write $z = -117 = -117 + 0i$ so $\text{Re}(z) = -117$ and $\text{Im}(z) = 0$. The number $z = -117$ corresponds to the point $(-117, 0)$, and this is another instance where we can determine the polar form 'by eye'. The point $(-117, 0)$ is 117 units away from the origin along the negative x-axis. Hence, $r = |z| = 117$ and $\theta = \pi + 2\pi = (2k + 1)\pi k$ for integers k. We have $\arg(z) = \{(2k + 1)\pi \,|\, k \text{ is an integer}\}$. Only one of these values, $\theta = \pi$, just barely lies in the interval $(-\pi, \pi]$ which means and $\text{Arg}(z) = \pi$. We plot z along with the other numbers in this example below.

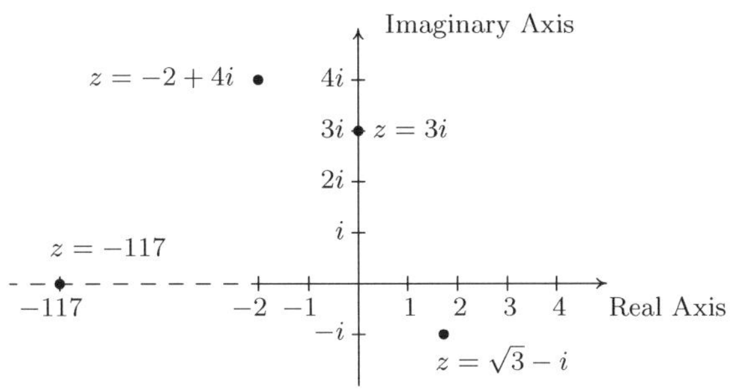

□

Now that we've had some practice computing the modulus and argument of some complex numbers, it is time to explore their properties. We have the following theorem.

Theorem 11.14. Properties of the Modulus: Let z and w be complex numbers.

- $|z|$ is the distance from z to 0 in the complex plane
- $|z| \geq 0$ and $|z| = 0$ if and only if $z = 0$
- $|z| = \sqrt{\text{Re}(z)^2 + \text{Im}(z)^2}$
- **Product Rule:** $|zw| = |z||w|$
- **Power Rule:** $|z^n| = |z|^n$ for all natural numbers, n
- **Quotient Rule:** $\left|\dfrac{z}{w}\right| = \dfrac{|z|}{|w|}$, provided $w \neq 0$

To prove the first three properties in Theorem 11.14, suppose $z = a + bi$ where a and b are real numbers. To determine $|z|$, we find a polar representation (r, θ) with $r \geq 0$ for the point (a, b). From Section 11.4, we know $r^2 = a^2 + b^2$ so that $r = \pm\sqrt{a^2 + b^2}$. Since we require $r \geq 0$, then it must be that $r = \sqrt{a^2 + b^2}$, which means $|z| = \sqrt{a^2 + b^2}$. Using the distance formula, we find the distance from $(0,0)$ to (a,b) is also $\sqrt{a^2+b^2}$, establishing the first property.[5] For the second property, note that since $|z|$ is a distance, $|z| \geq 0$. Furthermore, $|z| = 0$ if and only if the distance from z to 0 is 0, and the latter happens if and only if $z = 0$, which is what we were asked to show.[6] For the third property, we note that since $a = \text{Re}(z)$ and $b = \text{Im}(z)$, $z = \sqrt{a^2+b^2} = \sqrt{\text{Re}(z)^2 + \text{Im}(z)^2}$.

To prove the product rule, suppose $z = a + bi$ and $w = c + di$ for real numbers a, b, c and d. Then $zw = (a+bi)(c+di)$. After the usual arithmetic[7] we get $zw = (ac - bd) + (ad + bc)i$. Therefore,

$$\begin{aligned}
|zw| &= \sqrt{(ac-bd)^2 + (ad+bc)^2} \\
&= \sqrt{a^2c^2 - 2abcd + b^2d^2 + a^2d^2 + 2abcd + b^2c^2} && \text{Expand} \\
&= \sqrt{a^2c^2 + a^2d^2 + b^2c^2 + b^2d^2} && \text{Rearrange terms} \\
&= \sqrt{a^2\left(c^2+d^2\right) + b^2\left(c^2+d^2\right)} && \text{Factor} \\
&= \sqrt{\left(a^2+b^2\right)\left(c^2+d^2\right)} && \text{Factor} \\
&= \sqrt{a^2+b^2}\sqrt{c^2+d^2} && \text{Product Rule for Radicals} \\
&= |z||w| && \text{Definition of } |z| \text{ and } |w|
\end{aligned}$$

Hence $|zw| = |z||w|$ as required.

Now that the Product Rule has been established, we use it and the Principle of Mathematical Induction[8] to prove the power rule. Let $P(n)$ be the statement $|z^n| = |z|^n$. Then $P(1)$ is true since $\left|z^1\right| = |z| = |z|^1$. Next, assume $P(k)$ is true. That is, assume $\left|z^k\right| = |z|^k$ for some $k \geq 1$. Our job is to show that $P(k+1)$ is true, namely $\left|z^{k+1}\right| = |z|^{k+1}$. As is customary with induction proofs, we first try to reduce the problem in such a way as to use the Induction Hypothesis.

$$\begin{aligned}
\left|z^{k+1}\right| &= \left|z^k z\right| && \text{Properties of Exponents} \\
&= \left|z^k\right||z| && \text{Product Rule} \\
&= |z|^k|z| && \text{Induction Hypothesis} \\
&= |z|^{k+1} && \text{Properties of Exponents}
\end{aligned}$$

Hence, $P(k+1)$ is true, which means $|z^n| = |z|^n$ is true for all natural numbers n.

Like the Power Rule, the Quotient Rule can also be established with the help of the Product Rule. We assume $w \neq 0$ (so $|w| \neq 0$) and we get

$$\begin{aligned}
\left|\frac{z}{w}\right| &= \left|(z)\left(\frac{1}{w}\right)\right| \\
&= |z|\left|\frac{1}{w}\right| && \text{Product Rule.}
\end{aligned}$$

[5]Since the absolute value $|x|$ of a real number x can be viewed as the distance from x to 0 on the number line, this first property justifies the notation $|z|$ for modulus. We leave it to the reader to show that if z is real, then the definition of modulus coincides with absolute value so the notation $|z|$ is unambiguous.

[6]This may be considered by some to be a bit of a cheat, so we work through the underlying Algebra to see this is true. We know $|z| = 0$ if and only if $\sqrt{a^2+b^2} = 0$ if and only if $a^2 + b^2 = 0$, which is true if and only if $a = b = 0$. The latter happens if and only if $z = a + bi = 0$. There.

[7]See Example 3.4.1 in Section 3.4 for a review of complex number arithmetic.

[8]See Section 9.3 for a review of this technique.

Hence, the proof really boils down to showing $\left|\frac{1}{w}\right| = \frac{1}{|w|}$. This is left as an exercise.

Next, we characterize the argument of a complex number in terms of its real and imaginary parts.

Theorem 11.15. Properties of the Argument: Let z be a complex number.

- If $\text{Re}(z) \neq 0$ and $\theta \in \arg(z)$, then $\tan(\theta) = \frac{\text{Im}(z)}{\text{Re}(z)}$.
- If $\text{Re}(z) = 0$ and $\text{Im}(z) > 0$, then $\arg(z) = \left\{\frac{\pi}{2} + 2\pi k \,|\, k \text{ is an integer}\right\}$.
- If $\text{Re}(z) = 0$ and $\text{Im}(z) < 0$, then $\arg(z) = \left\{-\frac{\pi}{2} + 2\pi k \,|\, k \text{ is an integer}\right\}$.
- If $\text{Re}(z) = \text{Im}(z) = 0$, then $z = 0$ and $\arg(z) = (-\infty, \infty)$.

To prove Theorem 11.15, suppose $z = a + bi$ for real numbers a and b. By definition, $a = \text{Re}(z)$ and $b = \text{Im}(z)$, so the point associated with z is $(a, b) = (\text{Re}(z), \text{Im}(z))$. From Section 11.4, we know that if (r, θ) is a polar representation for $(\text{Re}(z), \text{Im}(z))$, then $\tan(\theta) = \frac{\text{Im}(z)}{\text{Re}(z)}$, provided $\text{Re}(z) \neq 0$. If $\text{Re}(z) = 0$ and $\text{Im}(z) > 0$, then z lies on the positive imaginary axis. Since we take $r > 0$, we have that θ is coterminal with $\frac{\pi}{2}$, and the result follows. If $\text{Re}(z) = 0$ and $\text{Im}(z) < 0$, then z lies on the negative imaginary axis, and a similar argument shows θ is coterminal with $-\frac{\pi}{2}$. The last property in the theorem was already discussed in the remarks following Definition 11.2.

Our next goal is to completely marry the Geometry and the Algebra of the complex numbers. To that end, consider the figure below.

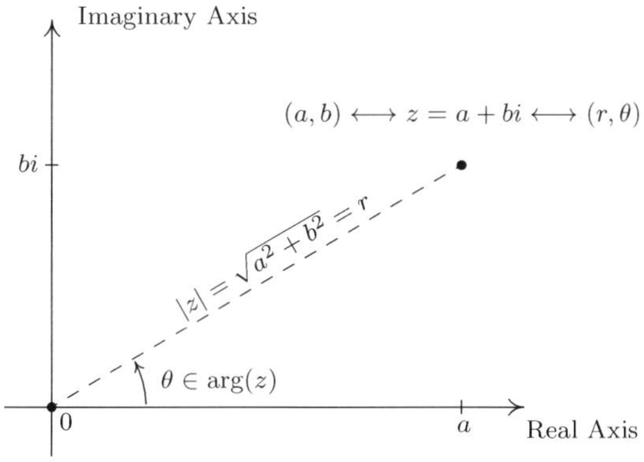

Polar coordinates, (r, θ) associated with $z = a + bi$ with $r \geq 0$.

We know from Theorem 11.7 that $a = r\cos(\theta)$ and $b = r\sin(\theta)$. Making these substitutions for a and b gives $z = a + bi = r\cos(\theta) + r\sin(\theta)i = r\left[\cos(\theta) + i\sin(\theta)\right]$. The expression '$\cos(\theta) + i\sin(\theta)$' is abbreviated $\text{cis}(\theta)$ so we can write $z = r\text{cis}(\theta)$. Since $r = |z|$ and $\theta \in \arg(z)$, we get

Definition 11.3. A Polar Form of a Complex Number: Suppose z is a complex number and $\theta \in \arg(z)$. The expression:

$$|z|\text{cis}(\theta) = |z|\left[\cos(\theta) + i\sin(\theta)\right]$$

is called a polar form for z.

Since there are infinitely many choices for $\theta \in \arg(z)$, there infinitely many polar forms for z, so we used the indefinite article 'a' in Definition 11.3. It is time for an example.

Example 11.7.2.

1. Find the rectangular form of the following complex numbers. Find $\text{Re}(z)$ and $\text{Im}(z)$.

   (a) $z = 4\text{cis}\left(\frac{2\pi}{3}\right)$ (b) $z = 2\text{cis}\left(-\frac{3\pi}{4}\right)$ (c) $z = 3\text{cis}(0)$ (d) $z = \text{cis}\left(\frac{\pi}{2}\right)$

2. Use the results from Example 11.7.1 to find a polar form of the following complex numbers.

   (a) $z = \sqrt{3} - i$ (b) $z = -2 + 4i$ (c) $z = 3i$ (d) $z = -117$

Solution.

1. The key to this problem is to write out $\text{cis}(\theta)$ as $\cos(\theta) + i\sin(\theta)$.

   (a) By definition, $z = 4\text{cis}\left(\frac{2\pi}{3}\right) = 4\left[\cos\left(\frac{2\pi}{3}\right) + i\sin\left(\frac{2\pi}{3}\right)\right]$. After some simplifying, we get $z = -2 + 2i\sqrt{3}$, so that $\text{Re}(z) = -2$ and $\text{Im}(z) = 2\sqrt{3}$.

   (b) Expanding, we get $z = 2\text{cis}\left(-\frac{3\pi}{4}\right) = 2\left[\cos\left(-\frac{3\pi}{4}\right) + i\sin\left(-\frac{3\pi}{4}\right)\right]$. From this, we find $z = -\sqrt{2} - i\sqrt{2}$, so $\text{Re}(z) = -\sqrt{2} = \text{Im}(z)$.

   (c) We get $z = 3\text{cis}(0) = 3\left[\cos(0) + i\sin(0)\right] = 3$. Writing $3 = 3 + 0i$, we get $\text{Re}(z) = 3$ and $\text{Im}(z) = 0$, which makes sense seeing as 3 is a real number.

   (d) Lastly, we have $z = \text{cis}\left(\frac{\pi}{2}\right) = \cos\left(\frac{\pi}{2}\right) + i\sin\left(\frac{\pi}{2}\right) = i$. Since $i = 0 + 1i$, we get $\text{Re}(z) = 0$ and $\text{Im}(z) = 1$. Since i is called the 'imaginary unit,' these answers make perfect sense.

2. To write a polar form of a complex number z, we need two pieces of information: the modulus $|z|$ and an argument (not necessarily the principal argument) of z. We shamelessly mine our solution to Example 11.7.1 to find what we need.

   (a) For $z = \sqrt{3} - i$, $|z| = 2$ and $\theta = -\frac{\pi}{6}$, so $z = 2\text{cis}\left(-\frac{\pi}{6}\right)$. We can check our answer by converting it back to rectangular form to see that it simplifies to $z = \sqrt{3} - i$.

   (b) For $z = -2 + 4i$, $|z| = 2\sqrt{5}$ and $\theta = \pi - \arctan(2)$. Hence, $z = 2\sqrt{5}\text{cis}(\pi - \arctan(2))$. It is a good exercise to actually show that this polar form reduces to $z = -2 + 4i$.

   (c) For $z = 3i$, $|z| = 3$ and $\theta = \frac{\pi}{2}$. In this case, $z = 3\text{cis}\left(\frac{\pi}{2}\right)$. This can be checked geometrically. Head out 3 units from 0 along the positive real axis. Rotating $\frac{\pi}{2}$ radians counter-clockwise lands you exactly 3 units above 0 on the imaginary axis at $z = 3i$.

   (d) Last but not least, for $z = -117$, $|z| = 117$ and $\theta = \pi$. We get $z = 117\text{cis}(\pi)$. As with the previous problem, our answer is easily checked geometrically. □

The following theorem summarizes the advantages of working with complex numbers in polar form.

Theorem 11.16. Products, Powers and Quotients Complex Numbers in Polar Form: Suppose z and w are complex numbers with polar forms $z = |z|\text{cis}(\alpha)$ and $w = |w|\text{cis}(\beta)$. Then

- **Product Rule:** $zw = |z||w|\text{cis}(\alpha + \beta)$
- **Power Rule (DeMoivre's Theorem) :** $z^n = |z|^n\text{cis}(n\theta)$ for every natural number n
- **Quotient Rule:** $\dfrac{z}{w} = \dfrac{|z|}{|w|}\text{cis}(\alpha - \beta)$, provided $|w| \neq 0$

The proof of Theorem 11.16 requires a healthy mix of definition, arithmetic and identities. We first start with the product rule.

$$\begin{aligned} zw &= [|z|\text{cis}(\alpha)]\,[|w|\text{cis}(\beta)] \\ &= |z||w|\,[\cos(\alpha) + i\sin(\alpha)]\,[\cos(\beta) + i\sin(\beta)] \end{aligned}$$

We now focus on the quantity in brackets on the right hand side of the equation.

$$\begin{aligned} [\cos(\alpha) + i\sin(\alpha)]\,[\cos(\beta) + i\sin(\beta)] &= \cos(\alpha)\cos(\beta) + i\cos(\alpha)\sin(\beta) \\ &\quad + i\sin(\alpha)\cos(\beta) + i^2\sin(\alpha)\sin(\beta) \\ &= \cos(\alpha)\cos(\beta) + i^2\sin(\alpha)\sin(\beta) && \text{Rearranging terms} \\ &\quad + i\sin(\alpha)\cos(\beta) + i\cos(\alpha)\sin(\beta) \\ &= (\cos(\alpha)\cos(\beta) - \sin(\alpha)\sin(\beta)) && \text{Since } i^2 = -1 \\ &\quad + i\,(\sin(\alpha)\cos(\beta) + \cos(\alpha)\sin(\beta)) && \text{Factor out } i \\ &= \cos(\alpha + \beta) + i\sin(\alpha + \beta) && \text{Sum identities} \\ &= \text{cis}(\alpha + \beta) && \text{Definition of `cis'} \end{aligned}$$

Putting this together with our earlier work, we get $zw = |z||w|\text{cis}(\alpha + \beta)$, as required.

Moving right along, we next take aim at the Power Rule, better known as DeMoivre's Theorem.[9] We proceed by induction on n. Let $P(n)$ be the sentence $z^n = |z|^n\text{cis}(n\theta)$. Then $P(1)$ is true, since $z^1 = z = |z|\text{cis}(\theta) = |z|^1\text{cis}(1 \cdot \theta)$. We now assume $P(k)$ is true, that is, we assume $z^k = |z|^k\text{cis}(k\theta)$ for some $k \geq 1$. Our goal is to show that $P(k+1)$ is true, or that $z^{k+1} = |z|^{k+1}\text{cis}((k+1)\theta)$. We have

$$\begin{aligned} z^{k+1} &= z^k z && \text{Properties of Exponents} \\ &= \left(|z|^k\text{cis}(k\theta)\right)\left(|z|\text{cis}(\theta)\right) && \text{Induction Hypothesis} \\ &= \left(|z|^k|z|\right)\text{cis}(k\theta + \theta) && \text{Product Rule} \\ &= |z|^{k+1}\text{cis}((k+1)\theta) \end{aligned}$$

[9]Compare this proof with the proof of the Power Rule in Theorem 11.14.

Hence, assuming $P(k)$ is true, we have that $P(k+1)$ is true, so by the Principle of Mathematical Induction, $z^n = |z|^n \text{cis}(n\theta)$ for all natural numbers n.

The last property in Theorem 11.16 to prove is the quotient rule. Assuming $|w| \neq 0$ we have

$$\begin{aligned} \frac{z}{w} &= \frac{|z|\text{cis}(\alpha)}{|w|\text{cis}(\beta)} \\ &= \left(\frac{|z|}{|w|}\right) \frac{\cos(\alpha) + i\sin(\alpha)}{\cos(\beta) + i\sin(\beta)} \end{aligned}$$

Next, we multiply both the numerator and denominator of the right hand side by $(\cos(\beta) - i\sin(\beta))$ which is the complex conjugate of $(\cos(\beta) + i\sin(\beta))$ to get

$$\frac{z}{w} = \left(\frac{|z|}{|w|}\right) \frac{\cos(\alpha) + i\sin(\alpha)}{\cos(\beta) + i\sin(\beta)} \cdot \frac{\cos(\beta) - i\sin(\beta)}{\cos(\beta) - i\sin(\beta)}$$

If we let the numerator be $N = [\cos(\alpha) + i\sin(\alpha)]\,[\cos(\beta) - i\sin(\beta)]$ and simplify we get

$$\begin{aligned} N &= [\cos(\alpha) + i\sin(\alpha)]\,[\cos(\beta) - i\sin(\beta)] & \\ &= \cos(\alpha)\cos(\beta) - i\cos(\alpha)\sin(\beta) + i\sin(\alpha)\cos(\beta) - i^2\sin(\alpha)\sin(\beta) & \text{Expand} \\ &= [\cos(\alpha)\cos(\beta) + \sin(\alpha)\sin(\beta)] + i\,[\sin(\alpha)\cos(\beta) - \cos(\alpha)\sin(\beta)] & \text{Rearrange and Factor} \\ &= \cos(\alpha - \beta) + i\sin(\alpha - \beta) & \text{Difference Identities} \\ &= \text{cis}(\alpha - \beta) & \text{Definition of 'cis'} \end{aligned}$$

If we call the denominator D then we get

$$\begin{aligned} D &= [\cos(\beta) + i\sin(\beta)]\,[\cos(\beta) - i\sin(\beta)] & \\ &= \cos^2(\beta) - i\cos(\beta)\sin(\beta) + i\cos(\beta)\sin(\beta) - i^2\sin^2(\beta) & \text{Expand} \\ &= \cos^2(\beta) - i^2\sin^2(\beta) & \text{Simplify} \\ &= \cos^2(\beta) + \sin^2(\beta) & \text{Again, } i^2 = -1 \\ &= 1 & \text{Pythagorean Identity} \end{aligned}$$

Putting it all together, we get

$$\begin{aligned} \frac{z}{w} &= \left(\frac{|z|}{|w|}\right) \frac{\cos(\alpha) + i\sin(\alpha)}{\cos(\beta) + i\sin(\beta)} \cdot \frac{\cos(\beta) - i\sin(\beta)}{\cos(\beta) - i\sin(\beta)} \\ &= \left(\frac{|z|}{|w|}\right) \frac{\text{cis}(\alpha - \beta)}{1} \\ &= \frac{|z|}{|w|}\text{cis}(\alpha - \beta) \end{aligned}$$

and we are done. The next example makes good use of Theorem 11.16.

Example 11.7.3. Let $z = 2\sqrt{3} + 2i$ and $w = -1 + i\sqrt{3}$. Use Theorem 11.16 to find the following.

1. zw
2. w^5
3. $\dfrac{z}{w}$

Write your final answers in rectangular form.

Solution. In order to use Theorem 11.16, we need to write z and w in polar form. For $z = 2\sqrt{3}+2i$, we find $|z| = \sqrt{(2\sqrt{3})^2 + (2)^2} = \sqrt{16} = 4$. If $\theta \in \arg(z)$, we know $\tan(\theta) = \frac{\text{Im}(z)}{\text{Re}(z)} = \frac{2}{2\sqrt{3}} = \frac{\sqrt{3}}{3}$. Since z lies in Quadrant I, we have $\theta = \frac{\pi}{6} + 2\pi k$ for integers k. Hence, $z = 4\text{cis}\left(\frac{\pi}{6}\right)$. For $w = -1 + i\sqrt{3}$, we have $|w| = \sqrt{(-1)^2 + (\sqrt{3})^2} = 2$. For an argument θ of w, we have $\tan(\theta) = \frac{\sqrt{3}}{-1} = -\sqrt{3}$. Since w lies in Quadrant II, $\theta = \frac{2\pi}{3} + 2\pi k$ for integers k and $w = 2\text{cis}\left(\frac{2\pi}{3}\right)$. We can now proceed.

1. We get $zw = \left(4\text{cis}\left(\frac{\pi}{6}\right)\right)\left(2\text{cis}\left(\frac{2\pi}{3}\right)\right) = 8\text{cis}\left(\frac{\pi}{6} + \frac{2\pi}{3}\right) = 8\text{cis}\left(\frac{5\pi}{6}\right) = 8\left[\cos\left(\frac{5\pi}{6}\right) + i\sin\left(\frac{5\pi}{6}\right)\right]$. After simplifying, we get $zw = -4\sqrt{3} + 4i$.

2. We use DeMoivre's Theorem which yields $w^5 = \left[2\text{cis}\left(\frac{2\pi}{3}\right)\right]^5 = 2^5\text{cis}\left(5 \cdot \frac{2\pi}{3}\right) = 32\text{cis}\left(\frac{10\pi}{3}\right)$. Since $\frac{10\pi}{3}$ is coterminal with $\frac{4\pi}{3}$, we get $w^5 = 32\left[\cos\left(\frac{4\pi}{3}\right) + i\sin\left(\frac{4\pi}{3}\right)\right] = -16 - 16i\sqrt{3}$.

3. Last, but not least, we have $\dfrac{z}{w} = \dfrac{4\text{cis}\left(\frac{\pi}{6}\right)}{2\text{cis}\left(\frac{2\pi}{3}\right)} = \frac{4}{2}\text{cis}\left(\frac{\pi}{6} - \frac{2\pi}{3}\right) = 2\text{cis}\left(-\frac{\pi}{2}\right)$. Since $-\frac{\pi}{2}$ is a quadrantal angle, we can 'see' the rectangular form by moving out 2 units along the positive real axis, then rotating $\frac{\pi}{2}$ radians *clockwise* to arrive at the point 2 units below 0 on the imaginary axis. The long and short of it is that $\frac{z}{w} = -2i$. □

Some remarks are in order. First, the reader may not be sold on using the polar form of complex numbers to multiply complex numbers – especially if they aren't given in polar form to begin with. Indeed, a lot of work was needed to convert the numbers z and w in Example 11.7.3 into polar form, compute their product, and convert back to rectangular form – certainly more work than is required to multiply out $zw = (2\sqrt{3} + 2i)(-1 + i\sqrt{3})$ the old-fashioned way. However, Theorem 11.16 pays huge dividends when computing powers of complex numbers. Consider how we computed w^5 above and compare that to using the Binomial Theorem, Theorem 9.4, to accomplish the same feat by expanding $(-1 + i\sqrt{3})^5$. Division is tricky in the best of times, and we saved ourselves a lot of time and effort using Theorem 11.16 to find and simplify $\frac{z}{w}$ using their polar forms as opposed to starting with $\frac{2\sqrt{3}+2i}{-1+i\sqrt{3}}$, rationalizing the denominator, and so forth.

There is geometric reason for studying these polar forms and we would be derelict in our duties if we did not mention the Geometry hidden in Theorem 11.16. Take the product rule, for instance. If $z = |z|\text{cis}(\alpha)$ and $w = |w|\text{cis}(\beta)$, the formula $zw = |z||w|\text{cis}(\alpha + \beta)$ can be viewed geometrically as a two step process. The multiplication of $|z|$ by $|w|$ can be interpreted as magnifying[10] the distance $|z|$ from z to 0, by the factor $|w|$. Adding the argument of w to the argument of z can be interpreted geometrically as a rotation of β radians counter-clockwise.[11] Focusing on z and w from Example

[10] Assuming $|w| > 1$.
[11] Assuming $\beta > 0$.

11.7.3, we can arrive at the product zw by plotting z, doubling its distance from 0 (since $|w| = 2$), and rotating $\frac{2\pi}{3}$ radians counter-clockwise. The sequence of diagrams below attempts to describe this process geometrically.

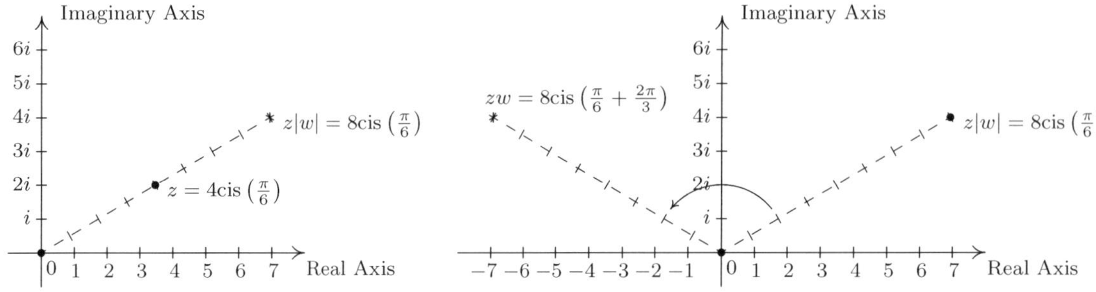

Multiplying z by $|w| = 2$.

Rotating counter-clockwise by $\text{Arg}(w) = \frac{2\pi}{3}$ radians.

Visualizing zw for $z = 4\text{cis}\left(\frac{\pi}{6}\right)$ and $w = 2\text{cis}\left(\frac{2\pi}{3}\right)$.

We may also visualize division similarly. Here, the formula $\frac{z}{w} = \frac{|z|}{|w|}\text{cis}(\alpha - \beta)$ may be interpreted as shrinking[12] the distance from 0 to z by the factor $|w|$, followed up by a *clockwise*[13] rotation of β radians. In the case of z and w from Example 11.7.3, we arrive at $\frac{z}{w}$ by first halving the distance from 0 to z, then rotating clockwise $\frac{2\pi}{3}$ radians.

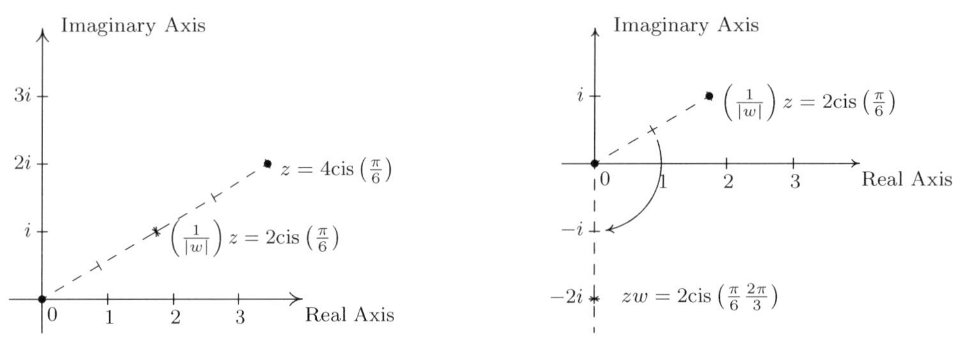

Dividing z by $|w| = 2$.

Rotating clockwise by $\text{Arg}(w) = \frac{2\pi}{3}$ radians.

Visualizing $\frac{z}{w}$ for $z = 4\text{cis}\left(\frac{\pi}{6}\right)$ and $w = 2\text{cis}\left(\frac{2\pi}{3}\right)$.

Our last goal of the section is to reverse DeMoivre's Theorem to extract roots of complex numbers.

Definition 11.4. Let z and w be complex numbers. If there is a natural number n such that $w^n = z$, then w is an $\boldsymbol{n}^{\textbf{th}}$ **root** of z.

Unlike Definition 5.4 in Section 5.3, we do not specify one particular *prinicpal* n^{th} root, hence the use of the indefinite article 'an' as in 'an nth root of z'. Using this definition, both 4 and -4 are

[12]Again, assuming $|w| > 1$.
[13]Again, assuming $\beta > 0$.

square roots of 16, while $\sqrt{16}$ means the principal square root of 16 as in $\sqrt{16} = 4$. Suppose we wish to find all complex third (cube) roots of 8. Algebraically, we are trying to solve $w^3 = 8$. We know that there is only one *real* solution to this equation, namely $w = \sqrt[3]{8} = 2$, but if we take the time to rewrite this equation as $w^3 - 8 = 0$ and factor, we get $(w-2)\left(w^2+2w+4\right) = 0$. The quadratic factor gives two more cube roots $w = -1 \pm i\sqrt{3}$, for a total of three cube roots of 8. In accordance with Theorem 3.14, since the degree of $p(w) = w^3 - 8$ is three, there are three complex zeros, counting multiplicity. Since we have found three distinct zeros, we know these are all of the zeros, so there are exactly three distinct cube roots of 8. Let us now solve this same problem using the machinery developed in this section. To do so, we express $z = 8$ in polar form. Since $z = 8$ lies 8 units away on the positive real axis, we get $z = 8\text{cis}(0)$. If we let $w = |w|\text{cis}(\alpha)$ be a polar form of w, the equation $w^3 = 8$ becomes

$$\begin{array}{rcll} w^3 & = & 8 & \\ \left(|w|\text{cis}(\alpha)\right)^3 & = & 8\text{cis}(0) & \\ |w|^3\text{cis}(3\alpha) & = & 8\text{cis}(0) & \text{DeMoivre's Theorem} \end{array}$$

The complex number on the left hand side of the equation corresponds to the point with polar coordinates $\left(|w|^3, 3\alpha\right)$, while the complex number on the right hand side corresponds to the point with polar coordinates $(8, 0)$. Since $|w| \geq 0$, so is $|w|^3$, which means $\left(|w|^3, 3\alpha\right)$ and $(8,0)$ are two polar representations corresponding to the same complex number, both with positive r values. From Section 11.4, we know $|w|^3 = 8$ and $3\alpha = 0 + 2\pi k$ for integers k. Since $|w|$ is a real number, we solve $|w|^3 = 8$ by extracting the principal cube root to get $|w| = \sqrt[3]{8} = 2$. As for α, we get $\alpha = \frac{2\pi k}{3}$ for integers k. This produces three distinct points with polar coordinates corresponding to $k = 0$, 1 and 2: specifically $(2, 0)$, $\left(2, \frac{2\pi}{3}\right)$ and $\left(2, \frac{4\pi}{3}\right)$. These correspond to the complex numbers $w_0 = 2\text{cis}(0)$, $w_1 = 2\text{cis}\left(\frac{2\pi}{3}\right)$ and $w_2 = 2\text{cis}\left(\frac{4\pi}{3}\right)$, respectively. Writing these out in rectangular form yields $w_0 = 2$, $w_1 = -1 + i\sqrt{3}$ and $w_2 = -1 - i\sqrt{3}$. While this process seems a tad more involved than our previous factoring approach, this procedure can be generalized to find, for example, all of the fifth roots of 32. (Try using Chapter 3 techniques on that!) If we start with a generic complex number in polar form $z = |z|\text{cis}(\theta)$ and solve $w^n = z$ in the same manner as above, we arrive at the following theorem.

Theorem 11.17. The nth roots of a Complex Number: Let $z \neq 0$ be a complex number with polar form $z = r\text{cis}(\theta)$. For each natural number n, z has n distinct n^{th} roots, which we denote by $w_0, w_1, \ldots, w_{n-1}$, and they are given by the formula

$$w_k = \sqrt[n]{r}\text{cis}\left(\frac{\theta}{n} + \frac{2\pi}{n}k\right)$$

The proof of Theorem 11.17 breaks into to two parts: first, showing that each w_k is an nth root, and second, showing that the set $\{w_k \,|\, k = 0, 1, \ldots, (n-1)\}$ consists of n different complex numbers. To show w_k is an nth root of z, we use DeMoivre's Theorem to show $(w_k)^n = z$.

$$\begin{aligned}(w_k)^n &= \left(\sqrt[n]{r}\operatorname{cis}\left(\tfrac{\theta}{n}+\tfrac{2\pi}{n}k\right)\right)^n \\ &= \left(\sqrt[n]{r}\right)^n\operatorname{cis}\left(n\cdot\left[\tfrac{\theta}{n}+\tfrac{2\pi}{n}k\right]\right) \quad \text{DeMoivre's Theorem} \\ &= r\operatorname{cis}\left(\theta+2\pi k\right)\end{aligned}$$

Since k is a whole number, $\cos(\theta + 2\pi k) = \cos(\theta)$ and $\sin(\theta + 2\pi k) = \sin(\theta)$. Hence, it follows that $\operatorname{cis}(\theta + 2\pi k) = \operatorname{cis}(\theta)$, so $(w_k)^n = r\operatorname{cis}(\theta) = z$, as required. To show that the formula in Theorem 11.17 generates n distinct numbers, we assume $n \geq 2$ (or else there is nothing to prove) and note that the modulus of each of the w_k is the same, namely $\sqrt[n]{r}$. Therefore, the only way any two of these polar forms correspond to the same number is if their arguments are coterminal – that is, if the arguments differ by an integer multiple of 2π. Suppose k and j are whole numbers between 0 and $(n-1)$, inclusive, with $k \neq j$. Since k and j are different, let's assume for the sake of argument that $k > j$. Then $\left(\frac{\theta}{n}+\frac{2\pi}{n}k\right) - \left(\frac{\theta}{n}+\frac{2\pi}{n}j\right) = 2\pi\left(\frac{k-j}{n}\right)$. For this to be an integer multiple of 2π, $(k-j)$ must be a multiple of n. But because of the restrictions on k and j, $0 < k - j \leq n - 1$. (Think this through.) Hence, $(k - j)$ is a positive number less than n, so it cannot be a multiple of n. As a result, w_k and w_j are different complex numbers, and we are done. By Theorem 3.14, we know there at most n distinct solutions to $w^n = z$, and we have just found all of them. We illustrate Theorem 11.17 in the next example.

Example 11.7.4. Use Theorem 11.17 to find the following:

1. both square roots of $z = -2 + 2i\sqrt{3}$
2. the four fourth roots of $z = -16$
3. the three cube roots of $z = \sqrt{2} + i\sqrt{2}$
4. the five fifth roots of $z = 1$.

Solution.

1. We start by writing $z = -2 + 2i\sqrt{3} = 4\operatorname{cis}\left(\frac{2\pi}{3}\right)$. To use Theorem 11.17, we identify $r = 4$, $\theta = \frac{2\pi}{3}$ and $n = 2$. We know that z has two square roots, and in keeping with the notation in Theorem 11.17, we'll call them w_0 and w_1. We get $w_0 = \sqrt{4}\operatorname{cis}\left(\frac{(2\pi/3)}{2}+\frac{2\pi}{2}(0)\right) = 2\operatorname{cis}\left(\frac{\pi}{3}\right)$ and $w_1 = \sqrt{4}\operatorname{cis}\left(\frac{(2\pi/3)}{2}+\frac{2\pi}{2}(1)\right) = 2\operatorname{cis}\left(\frac{4\pi}{3}\right)$. In rectangular form, the two square roots of z are $w_0 = 1 + i\sqrt{3}$ and $w_1 = -1 - i\sqrt{3}$. We can check our answers by squaring them and showing that we get $z = -2 + 2i\sqrt{3}$.

2. Proceeding as above, we get $z = -16 = 16\operatorname{cis}(\pi)$. With $r = 16$, $\theta = \pi$ and $n = 4$, we get the four fourth roots of z to be $w_0 = \sqrt[4]{16}\operatorname{cis}\left(\frac{\pi}{4}+\frac{2\pi}{4}(0)\right) = 2\operatorname{cis}\left(\frac{\pi}{4}\right)$, $w_1 = \sqrt[4]{16}\operatorname{cis}\left(\frac{\pi}{4}+\frac{2\pi}{4}(1)\right) = 2\operatorname{cis}\left(\frac{3\pi}{4}\right)$, $w_2 = \sqrt[4]{16}\operatorname{cis}\left(\frac{\pi}{4}+\frac{2\pi}{4}(2)\right) = 2\operatorname{cis}\left(\frac{5\pi}{4}\right)$ and $w_3 = \sqrt[4]{16}\operatorname{cis}\left(\frac{\pi}{4}+\frac{2\pi}{4}(3)\right) = 2\operatorname{cis}\left(\frac{7\pi}{4}\right)$. Converting these to rectangular form gives $w_0 = \sqrt{2}+i\sqrt{2}$, $w_1 = -\sqrt{2}+i\sqrt{2}$, $w_2 = -\sqrt{2}-i\sqrt{2}$ and $w_3 = \sqrt{2} - i\sqrt{2}$.

3. For $z=\sqrt{2}+i\sqrt{2}$, we have $z=2\text{cis}\left(\frac{\pi}{4}\right)$. With $r=2$, $\theta=\frac{\pi}{4}$ and $n=3$ the usual computations yield $w_0=\sqrt[3]{2}\text{cis}\left(\frac{\pi}{12}\right)$, $w_1=\sqrt[3]{2}\text{cis}\left(\frac{9\pi}{12}\right)=\sqrt[3]{2}\text{cis}\left(\frac{3\pi}{4}\right)$ and $w_2=\sqrt[3]{2}\text{cis}\left(\frac{17\pi}{12}\right)$. If we were to convert these to rectangular form, we would need to use either the Sum and Difference Identities in Theorem 10.16 or the Half-Angle Identities in Theorem 10.19 to evaluate w_0 and w_2. Since we are not explicitly told to do so, we leave this as a good, but messy, exercise.

4. To find the five fifth roots of 1, we write $1=1\text{cis}(0)$. We have $r=1$, $\theta=0$ and $n=5$. Since $\sqrt[5]{1}=1$, the roots are $w_0=\text{cis}(0)=1$, $w_1=\text{cis}\left(\frac{2\pi}{5}\right)$, $w_2=\text{cis}\left(\frac{4\pi}{5}\right)$, $w_3=\text{cis}\left(\frac{6\pi}{5}\right)$ and $w_4=\text{cis}\left(\frac{8\pi}{5}\right)$. The situation here is even graver than in the previous example, since we have not developed any identities to help us determine the cosine or sine of $\frac{2\pi}{5}$. At this stage, we could approximate our answers using a calculator, and we leave this as an exercise. □

Now that we have done some computations using Theorem 11.17, we take a step back to look at things geometrically. Essentially, Theorem 11.17 says that to find the nth roots of a complex number, we first take the nth root of the modulus and divide the argument by n. This gives the first root w_0. Each succeessive root is found by adding $\frac{2\pi}{n}$ to the argument, which amounts to rotating w_0 by $\frac{2\pi}{n}$ radians. This results in n roots, spaced equally around the complex plane. As an example of this, we plot our answers to number 2 in Example 11.7.4 below.

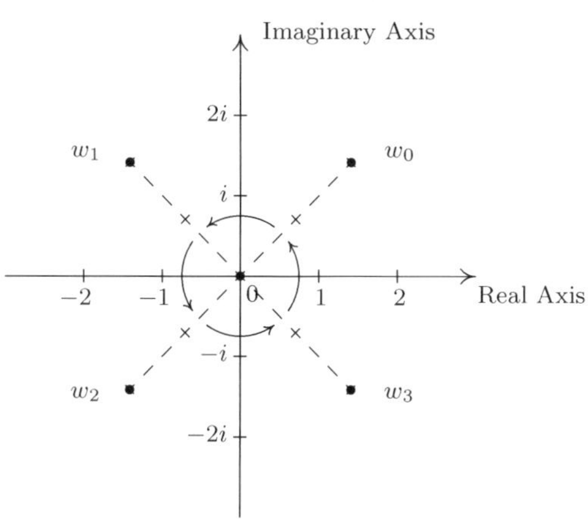

The four fourth roots of $z=-16$ equally spaced $\frac{2\pi}{4}=\frac{\pi}{2}$ around the plane.

We have only glimpsed at the beauty of the complex numbers in this section. The complex plane is without a doubt one of the most important mathematical constructs ever devised. Coupled with Calculus, it is the venue for incredibly important Science and Engineering applications.[14] For now, the following exercises will have to suffice.

[14]For more on this, see the beautifully written epilogue to Section 3.4 found on page 294.

11.7.1 Exercises

In Exercises 1 - 20, find a polar representation for the complex number z and then identify $\text{Re}(z)$, $\text{Im}(z)$, $|z|$, $\arg(z)$ and $\text{Arg}(z)$.

1. $z = 9 + 9i$
2. $z = 5 + 5i\sqrt{3}$
3. $z = 6i$
4. $z = -3\sqrt{2} + 3i\sqrt{2}$
5. $z = -6\sqrt{3} + 6i$
6. $z = -2$
7. $z = -\dfrac{\sqrt{3}}{2} - \dfrac{1}{2}i$
8. $z = -3 - 3i$
9. $z = -5i$
10. $z = 2\sqrt{2} - 2i\sqrt{2}$
11. $z = 6$
12. $z = i\sqrt[3]{7}$
13. $z = 3 + 4i$
14. $z = \sqrt{2} + i$
15. $z = -7 + 24i$
16. $z = -2 + 6i$
17. $z = -12 - 5i$
18. $z = -5 - 2i$
19. $z = 4 - 2i$
20. $z = 1 - 3i$

In Exercises 21 - 40, find the rectangular form of the given complex number. Use whatever identities are necessary to find the exact values.

21. $z = 6\text{cis}(0)$
22. $z = 2\text{cis}\left(\dfrac{\pi}{6}\right)$
23. $z = 7\sqrt{2}\text{cis}\left(\dfrac{\pi}{4}\right)$
24. $z = 3\text{cis}\left(\dfrac{\pi}{2}\right)$
25. $z = 4\text{cis}\left(\dfrac{2\pi}{3}\right)$
26. $z = \sqrt{6}\text{cis}\left(\dfrac{3\pi}{4}\right)$
27. $z = 9\text{cis}\left(\pi\right)$
28. $z = 3\text{cis}\left(\dfrac{4\pi}{3}\right)$
29. $z = 7\text{cis}\left(-\dfrac{3\pi}{4}\right)$
30. $z = \sqrt{13}\text{cis}\left(\dfrac{3\pi}{2}\right)$
31. $z = \dfrac{1}{2}\text{cis}\left(\dfrac{7\pi}{4}\right)$
32. $z = 12\text{cis}\left(-\dfrac{\pi}{3}\right)$
33. $z = 8\text{cis}\left(\dfrac{\pi}{12}\right)$
34. $z = 2\text{cis}\left(\dfrac{7\pi}{8}\right)$
35. $z = 5\text{cis}\left(\arctan\left(\dfrac{4}{3}\right)\right)$
36. $z = \sqrt{10}\text{cis}\left(\arctan\left(\dfrac{1}{3}\right)\right)$
37. $z = 15\text{cis}\left(\arctan\left(-2\right)\right)$
38. $z = \sqrt{3}\left(\arctan\left(-\sqrt{2}\right)\right)$
39. $z = 50\text{cis}\left(\pi - \arctan\left(\dfrac{7}{24}\right)\right)$
40. $z = \dfrac{1}{2}\text{cis}\left(\pi + \arctan\left(\dfrac{5}{12}\right)\right)$

For Exercises 41 - 52, use $z = -\dfrac{3\sqrt{3}}{2} + \dfrac{3}{2}i$ and $w = 3\sqrt{2} - 3i\sqrt{2}$ to compute the quantity. Express your answers in polar form using the principal argument.

41. zw
42. $\dfrac{z}{w}$
43. $\dfrac{w}{z}$
44. z^4
45. w^3
46. z^5w^2
47. z^3w^2
48. $\dfrac{z^2}{w}$

49. $\dfrac{w}{z^2}$

50. $\dfrac{z^3}{w^2}$

51. $\dfrac{w^2}{z^3}$

52. $\left(\dfrac{w}{z}\right)^6$

In Exercises 53 - 64, use DeMoivre's Theorem to find the indicated power of the given complex number. Express your final answers in rectangular form.

53. $\left(-2+2i\sqrt{3}\right)^3$

54. $(-\sqrt{3}-i)^3$

55. $(-3+3i)^4$

56. $(\sqrt{3}+i)^4$

57. $\left(\dfrac{5}{2}+\dfrac{5}{2}i\right)^3$

58. $\left(-\dfrac{1}{2}-\dfrac{\sqrt{3}}{2}i\right)^6$

59. $\left(\dfrac{3}{2}-\dfrac{3}{2}i\right)^3$

60. $\left(\dfrac{\sqrt{3}}{3}-\dfrac{1}{3}i\right)^4$

61. $\left(\dfrac{\sqrt{2}}{2}+\dfrac{\sqrt{2}}{2}i\right)^4$

62. $(2+2i)^5$

63. $(\sqrt{3}-i)^5$

64. $(1-i)^8$

In Exercises 65 - 76, find the indicated complex roots. Express your answers in polar form and then convert them into rectangular form.

65. the two square roots of $z=4i$

66. the two square roots of $z=-25i$

67. the two square roots of $z=1+i\sqrt{3}$

68. the two square roots of $\frac{5}{2}-\frac{5\sqrt{3}}{2}i$

69. the three cube roots of $z=64$

70. the three cube roots of $z=-125$

71. the three cube roots of $z=i$

72. the three cube roots of $z=-8i$

73. the four fourth roots of $z=16$

74. the four fourth roots of $z=-81$

75. the six sixth roots of $z=64$

76. the six sixth roots of $z=-729$

77. Use the Sum and Difference Identities in Theorem 10.16 or the Half Angle Identities in Theorem 10.19 to express the three cube roots of $z=\sqrt{2}+i\sqrt{2}$ in rectangular form. (See Example 11.7.4, number 3.)

78. Use a calculator to approximate the five fifth roots of 1. (See Example 11.7.4, number 4.)

79. According to Theorem 3.16 in Section 3.4, the polynomial $p(x)=x^4+4$ can be factored into the product linear and irreducible quadratic factors. In Exercise 28 in Section 8.7, we showed you how to factor this polynomial into the product of two irreducible quadratic factors using a system of non-linear equations. Now that we can compute the complex fourth roots of -4 directly, we can simply apply the Complex Factorization Theorem, Theorem 3.14, to obtain the linear factorization $p(x)=(x-(1+i))(x-(1-i))(x-(-1+i))(x-(-1-i))$. By multiplying the first two factors together and then the second two factors together, thus pairing up the complex conjugate pairs of zeros Theorem 3.15 told us we'd get, we have that $p(x)=(x^2-2x+2)(x^2+2x+2)$. Use the 12 complex 12th roots of 4096 to factor $p(x)=x^{12}-4096$ into a product of linear and irreducible quadratic factors.

80. Complete the proof of Theorem 11.14 by showing that if $w \neq 0$ than $\left|\frac{1}{w}\right| = \frac{1}{|w|}$.

81. Recall from Section 3.4 that given a complex number $z = a+bi$ its complex conjugate, denoted $\overline{z}$, is given by $\overline{z} = a - bi$.

    (a) Prove that $|\overline{z}| = |z|$.

    (b) Prove that $|z| = \sqrt{z\overline{z}}$

    (c) Show that $\text{Re}(z) = \dfrac{z + \overline{z}}{2}$ and $\text{Im}(z) = \dfrac{z - \overline{z}}{2i}$

    (d) Show that if $\theta \in \arg(z)$ then $-\theta \in \arg(\overline{z})$. Interpret this result geometrically.

    (e) Is it always true that $\text{Arg}(\overline{z}) = -\text{Arg}(z)$?

82. Given any natural number $n \geq 2$, the n complex n^{th} roots of the number $z = 1$ are called the **n^{th} Roots of Unity**. In the following exercises, assume that n is a fixed, but arbitrary, natural number such that $n \geq 2$.

    (a) Show that $w = 1$ is an nth root of unity.

    (b) Show that if both w_j and w_k are n^{th} roots of unity then so is their product $w_j w_k$.

    (c) Show that if w_j is an nth root of unity then there exists another n^{th} root of unity $w_{j'}$ such that $w_j w_{j'} = 1$. Hint: If $w_j = \text{cis}(\theta)$ let $w_{j'} = \text{cis}(2\pi - \theta)$. You'll need to verify that $w_{j'} = \text{cis}(2\pi - \theta)$ is indeed an nth root of unity.

83. Another way to express the polar form of a complex number is to use the exponential function. For real numbers t, Euler's Formula defines $e^{it} = \cos(t) + i\sin(t)$.

    (a) Use Theorem 11.16 to show that $e^{ix}e^{iy} = e^{i(x+y)}$ for all real numbers x and y.

    (b) Use Theorem 11.16 to show that $\left(e^{ix}\right)^n = e^{i(nx)}$ for any real number x and any natural number n.

    (c) Use Theorem 11.16 to show that $\dfrac{e^{ix}}{e^{iy}} = e^{i(x-y)}$ for all real numbers x and y.

    (d) If $z = r\text{cis}(\theta)$ is the polar form of z, show that $z = re^{it}$ where $\theta = t$ radians.

    (e) Show that $e^{i\pi} + 1 = 0$. (This famous equation relates the five most important constants in all of Mathematics with the three most fundamental operations in Mathematics.)

    (f) Show that $\cos(t) = \dfrac{e^{it} + e^{-it}}{2}$ and that $\sin(t) = \dfrac{e^{it} - e^{-it}}{2i}$ for all real numbers t.

11.7.2 ANSWERS

1. $z = 9 + 9i = 9\sqrt{2}\text{cis}\left(\frac{\pi}{4}\right), \quad \text{Re}(z) = 9, \quad \text{Im}(z) = 9, \quad |z| = 9\sqrt{2}$
   $\arg(z) = \left\{\frac{\pi}{4} + 2\pi k \,|\, k \text{ is an integer}\right\}$ and $\text{Arg}(z) = \frac{\pi}{4}$.

2. $z = 5 + 5i\sqrt{3} = 10\text{cis}\left(\frac{\pi}{3}\right), \quad \text{Re}(z) = 5, \quad \text{Im}(z) = 5\sqrt{3}, \quad |z| = 10$
   $\arg(z) = \left\{\frac{\pi}{3} + 2\pi k \,|\, k \text{ is an integer}\right\}$ and $\text{Arg}(z) = \frac{\pi}{3}$.

3. $z = 6i = 6\text{cis}\left(\frac{\pi}{2}\right), \quad \text{Re}(z) = 0, \quad \text{Im}(z) = 6, \quad |z| = 6$
   $\arg(z) = \left\{\frac{\pi}{2} + 2\pi k \,|\, k \text{ is an integer}\right\}$ and $\text{Arg}(z) = \frac{\pi}{2}$.

4. $z = -3\sqrt{2} + 3i\sqrt{2} = 6\text{cis}\left(\frac{3\pi}{4}\right), \quad \text{Re}(z) = -3\sqrt{2}, \quad \text{Im}(z) = 3\sqrt{2}, \quad |z| = 6$
   $\arg(z) = \left\{\frac{3\pi}{4} + 2\pi k \,|\, k \text{ is an integer}\right\}$ and $\text{Arg}(z) = \frac{3\pi}{4}$.

5. $z = -6\sqrt{3} + 6i = 12\text{cis}\left(\frac{5\pi}{6}\right), \quad \text{Re}(z) = -6\sqrt{3}, \quad \text{Im}(z) = 6, \quad |z| = 12$
   $\arg(z) = \left\{\frac{5\pi}{6} + 2\pi k \,|\, k \text{ is an integer}\right\}$ and $\text{Arg}(z) = \frac{5\pi}{6}$.

6. $z = -2 = 2\text{cis}(\pi), \quad \text{Re}(z) = -2, \quad \text{Im}(z) = 0, \quad |z| = 2$
   $\arg(z) = \{(2k+1)\pi \,|\, k \text{ is an integer}\}$ and $\text{Arg}(z) = \pi$.

7. $z = -\frac{\sqrt{3}}{2} - \frac{1}{2}i = \text{cis}\left(\frac{7\pi}{6}\right), \quad \text{Re}(z) = -\frac{\sqrt{3}}{2}, \quad \text{Im}(z) = -\frac{1}{2}, \quad |z| = 1$
   $\arg(z) = \left\{\frac{7\pi}{6} + 2\pi k \,|\, k \text{ is an integer}\right\}$ and $\text{Arg}(z) = -\frac{5\pi}{6}$.

8. $z = -3 - 3i = 3\sqrt{2}\text{cis}\left(\frac{5\pi}{4}\right), \quad \text{Re}(z) = -3, \quad \text{Im}(z) = -3, \quad |z| = 3\sqrt{2}$
   $\arg(z) = \left\{\frac{5\pi}{4} + 2\pi k \,|\, k \text{ is an integer}\right\}$ and $\text{Arg}(z) = -\frac{3\pi}{4}$.

9. $z = -5i = 5\text{cis}\left(\frac{3\pi}{2}\right), \quad \text{Re}(z) = 0, \quad \text{Im}(z) = -5, \quad |z| = 5$
   $\arg(z) = \left\{\frac{3\pi}{2} + 2\pi k \,|\, k \text{ is an integer}\right\}$ and $\text{Arg}(z) = -\frac{\pi}{2}$.

10. $z = 2\sqrt{2} - 2i\sqrt{2} = 4\text{cis}\left(\frac{7\pi}{4}\right), \quad \text{Re}(z) = 2\sqrt{2}, \quad \text{Im}(z) = -2\sqrt{2}, \quad |z| = 4$
    $\arg(z) = \left\{\frac{7\pi}{4} + 2\pi k \,|\, k \text{ is an integer}\right\}$ and $\text{Arg}(z) = -\frac{\pi}{4}$.

11. $z = 6 = 6\text{cis}(0), \quad \text{Re}(z) = 6, \quad \text{Im}(z) = 0, \quad |z| = 6$
    $\arg(z) = \{2\pi k \,|\, k \text{ is an integer}\}$ and $\text{Arg}(z) = 0$.

12. $z = i\sqrt[3]{7} = \sqrt[3]{7}\text{cis}\left(\frac{\pi}{2}\right), \quad \text{Re}(z) = 0, \quad \text{Im}(z) = \sqrt[3]{7}, \quad |z| = \sqrt[3]{7}$
    $\arg(z) = \left\{\frac{\pi}{2} + 2\pi k \,|\, k \text{ is an integer}\right\}$ and $\text{Arg}(z) = \frac{\pi}{2}$.

13. $z = 3 + 4i = 5\text{cis}\left(\arctan\left(\frac{4}{3}\right)\right), \quad \text{Re}(z) = 3, \quad \text{Im}(z) = 4, \quad |z| = 5$
    $\arg(z) = \left\{\arctan\left(\frac{4}{3}\right) + 2\pi k \,|\, k \text{ is an integer}\right\}$ and $\text{Arg}(z) = \arctan\left(\frac{4}{3}\right)$.

14. $z = \sqrt{2} + i = \sqrt{3}\text{cis}\left(\arctan\left(\frac{\sqrt{2}}{2}\right)\right)$, $\text{Re}(z) = \sqrt{2}$, $\text{Im}(z) = 1$, $|z| = \sqrt{3}$
$\arg(z) = \left\{\arctan\left(\frac{\sqrt{2}}{2}\right) + 2\pi k \,|\, k \text{ is an integer}\right\}$ and $\text{Arg}(z) = \arctan\left(\frac{\sqrt{2}}{2}\right)$.

15. $z = -7 + 24i = 25\text{cis}\left(\pi - \arctan\left(\frac{24}{7}\right)\right)$, $\text{Re}(z) = -7$, $\text{Im}(z) = 24$, $|z| = 25$
$\arg(z) = \left\{\pi - \arctan\left(\frac{24}{7}\right) + 2\pi k \,|\, k \text{ is an integer}\right\}$ and $\text{Arg}(z) = \pi - \arctan\left(\frac{24}{7}\right)$.

16. $z = -2 + 6i = 2\sqrt{10}\text{cis}\left(\pi - \arctan(3)\right)$, $\text{Re}(z) = -2$, $\text{Im}(z) = 6$, $|z| = 2\sqrt{10}$
$\arg(z) = \{\pi - \arctan(3) + 2\pi k \,|\, k \text{ is an integer}\}$ and $\text{Arg}(z) = \pi - \arctan(3)$.

17. $z = -12 - 5i = 13\text{cis}\left(\pi + \arctan\left(\frac{5}{12}\right)\right)$, $\text{Re}(z) = -12$, $\text{Im}(z) = -5$, $|z| = 13$
$\arg(z) = \left\{\pi + \arctan\left(\frac{5}{12}\right) + 2\pi k \,|\, k \text{ is an integer}\right\}$ and $\text{Arg}(z) = \arctan\left(\frac{5}{12}\right) - \pi$.

18. $z = -5 - 2i = \sqrt{29}\text{cis}\left(\pi + \arctan\left(\frac{2}{5}\right)\right)$, $\text{Re}(z) = -5$, $\text{Im}(z) = -2$, $|z| = \sqrt{29}$
$\arg(z) = \left\{\pi + \arctan\left(\frac{2}{5}\right) + 2\pi k \,|\, k \text{ is an integer}\right\}$ and $\text{Arg}(z) = \arctan\left(\frac{2}{5}\right) - \pi$.

19. $z = 4 - 2i = 2\sqrt{5}\text{cis}\left(\arctan\left(-\frac{1}{2}\right)\right)$, $\text{Re}(z) = 4$, $\text{Im}(z) = 2$, $|z| = 2\sqrt{5}$
$\arg(z) = \left\{\arctan\left(-\frac{1}{2}\right) + 2\pi k \,|\, k \text{ is an integer}\right\}$ and $\text{Arg}(z) = \arctan\left(-\frac{1}{2}\right) = -\arctan\left(\frac{1}{2}\right)$.

20. $z = 1 - 3i = \sqrt{10}\text{cis}\left(\arctan(-3)\right)$, $\text{Re}(z) = 1$, $\text{Im}(z) = -3$, $|z| = \sqrt{10}$
$\arg(z) = \{\arctan(-3) + 2\pi k \,|\, k \text{ is an integer}\}$ and $\text{Arg}(z) = \arctan(-3) = -\arctan(3)$.

21. $z = 6\text{cis}(0) = 6$

22. $z = 2\text{cis}\left(\frac{\pi}{6}\right) = \sqrt{3} + i$

23. $z = 7\sqrt{2}\text{cis}\left(\frac{\pi}{4}\right) = 7 + 7i$

24. $z = 3\text{cis}\left(\frac{\pi}{2}\right) = 3i$

25. $z = 4\text{cis}\left(\frac{2\pi}{3}\right) = -2 + 2i\sqrt{3}$

26. $z = \sqrt{6}\text{cis}\left(\frac{3\pi}{4}\right) = -\sqrt{3} + i\sqrt{3}$

27. $z = 9\text{cis}(\pi) = -9$

28. $z = 3\text{cis}\left(\frac{4\pi}{3}\right) = -\frac{3}{2} - \frac{3i\sqrt{3}}{2}$

29. $z = 7\text{cis}\left(-\frac{3\pi}{4}\right) = -\frac{7\sqrt{2}}{2} - \frac{7\sqrt{2}}{2}i$

30. $z = \sqrt{13}\text{cis}\left(\frac{3\pi}{2}\right) = -i\sqrt{13}$

31. $z = \frac{1}{2}\text{cis}\left(\frac{7\pi}{4}\right) = \frac{\sqrt{2}}{4} - i\frac{\sqrt{2}}{4}$

32. $z = 12\text{cis}\left(-\frac{\pi}{3}\right) = 6 - 6i\sqrt{3}$

33. $z = 8\text{cis}\left(\frac{\pi}{12}\right) = 4\sqrt{2 + \sqrt{3}} + 4i\sqrt{2 - \sqrt{3}}$

34. $z = 2\text{cis}\left(\frac{7\pi}{8}\right) = -\sqrt{2 + \sqrt{2}} + i\sqrt{2 - \sqrt{2}}$

35. $z = 5\text{cis}\left(\arctan\left(\frac{4}{3}\right)\right) = 3 + 4i$

36. $z = \sqrt{10}\text{cis}\left(\arctan\left(\frac{1}{3}\right)\right) = 3 + i$

37. $z = 15\text{cis}\left(\arctan(-2)\right) = 3\sqrt{5} - 6i\sqrt{5}$

38. $z = \sqrt{3}\text{cis}\left(\arctan\left(-\sqrt{2}\right)\right) = 1 - i\sqrt{2}$

39. $z = 50\text{cis}\left(\pi - \arctan\left(\frac{7}{24}\right)\right) = -48 + 14i$

40. $z = \frac{1}{2}\text{cis}\left(\pi + \arctan\left(\frac{5}{12}\right)\right) = -\frac{6}{13} - \frac{5i}{26}$

In Exercises 41 - 52, we have that $z = -\frac{3\sqrt{3}}{2} + \frac{3}{2}i = 3\text{cis}\left(\frac{5\pi}{6}\right)$ and $w = 3\sqrt{2} - 3i\sqrt{2} = 6\text{cis}\left(-\frac{\pi}{4}\right)$ so we get the following.

41. $zw = 18\text{cis}\left(\frac{7\pi}{12}\right)$
42. $\frac{z}{w} = \frac{1}{2}\text{cis}\left(-\frac{11\pi}{12}\right)$
43. $\frac{w}{z} = 2\text{cis}\left(\frac{11\pi}{12}\right)$
44. $z^4 = 81\text{cis}\left(-\frac{2\pi}{3}\right)$
45. $w^3 = 216\text{cis}\left(-\frac{3\pi}{4}\right)$
46. $z^5w^2 = 8748\text{cis}\left(-\frac{\pi}{3}\right)$
47. $z^3w^2 = 972\text{cis}(0)$
48. $\frac{z^2}{w} = \frac{3}{2}\text{cis}\left(-\frac{\pi}{12}\right)$
49. $\frac{w}{z^2} = \frac{2}{3}\text{cis}\left(\frac{\pi}{12}\right)$
50. $\frac{z^3}{w^2} = \frac{3}{4}\text{cis}(\pi)$
51. $\frac{w^2}{z^3} = \frac{4}{3}\text{cis}(\pi)$
52. $\left(\frac{w}{z}\right)^6 = 64\text{cis}\left(-\frac{\pi}{2}\right)$
53. $\left(-2 + 2i\sqrt{3}\right)^3 = 64$
54. $(-\sqrt{3} - i)^3 = -8i$
55. $(-3 + 3i)^4 = -324$
56. $(\sqrt{3} + i)^4 = -8 + 8i\sqrt{3}$
57. $\left(\frac{5}{2} + \frac{5}{2}i\right)^3 = -\frac{125}{4} + \frac{125}{4}i$
58. $\left(-\frac{1}{2} - \frac{i\sqrt{3}}{2}\right)^6 = 1$
59. $\left(\frac{3}{2} - \frac{3}{2}i\right)^3 = -\frac{27}{4} - \frac{27}{4}i$
60. $\left(\frac{\sqrt{3}}{3} - \frac{1}{3}i\right)^4 = -\frac{8}{81} - \frac{8i\sqrt{3}}{81}$
61. $\left(\frac{\sqrt{2}}{2} + \frac{\sqrt{2}}{2}i\right)^4 = -1$
62. $(2 + 2i)^5 = -128 - 128i$
63. $(\sqrt{3} - i)^5 = -16\sqrt{3} - 16i$
64. $(1 - i)^8 = 16$
65. Since $z = 4i = 4\text{cis}\left(\frac{\pi}{2}\right)$ we have

   $w_0 = 2\text{cis}\left(\frac{\pi}{4}\right) = \sqrt{2} + i\sqrt{2}$ $\quad$ $w_1 = 2\text{cis}\left(\frac{5\pi}{4}\right) = -\sqrt{2} - i\sqrt{2}$

66. Since $z = -25i = 25\text{cis}\left(\frac{3\pi}{2}\right)$ we have

   $w_0 = 5\text{cis}\left(\frac{3\pi}{4}\right) = -\frac{5\sqrt{2}}{2} + \frac{5\sqrt{2}}{2}i$ $\quad$ $w_1 = 5\text{cis}\left(\frac{7\pi}{4}\right) = \frac{5\sqrt{2}}{2} - \frac{5\sqrt{2}}{2}i$

67. Since $z = 1 + i\sqrt{3} = 2\text{cis}\left(\frac{\pi}{3}\right)$ we have

   $w_0 = \sqrt{2}\text{cis}\left(\frac{\pi}{6}\right) = \frac{\sqrt{6}}{2} + \frac{\sqrt{2}}{2}i$ $\quad$ $w_1 = \sqrt{2}\text{cis}\left(\frac{7\pi}{6}\right) = -\frac{\sqrt{6}}{2} - \frac{\sqrt{2}}{2}i$

68. Since $z = \frac{5}{2} - \frac{5\sqrt{3}}{2}i = 5\text{cis}\left(\frac{5\pi}{3}\right)$ we have

   $w_0 = \sqrt{5}\text{cis}\left(\frac{5\pi}{6}\right) = -\frac{\sqrt{15}}{2} + \frac{\sqrt{5}}{2}i$ $\quad$ $w_1 = \sqrt{5}\text{cis}\left(\frac{11\pi}{6}\right) = \frac{\sqrt{15}}{2} - \frac{\sqrt{5}}{2}i$

69. Since $z = 64 = 64\text{cis}(0)$ we have

   $w_0 = 4\text{cis}(0) = 4$ $\quad$ $w_1 = 4\text{cis}\left(\frac{2\pi}{3}\right) = -2 + 2i\sqrt{3}$ $\quad$ $w_2 = 4\text{cis}\left(\frac{4\pi}{3}\right) = -2 - 2i\sqrt{3}$

70. Since $z = -125 = 125\text{cis}(\pi)$ we have

$w_0 = 5\text{cis}\left(\frac{\pi}{3}\right) = \frac{5}{2} + \frac{5\sqrt{3}}{2}i$ $\quad w_1 = 5\text{cis}(\pi) = -5$ $\quad w_2 = 5\text{cis}\left(\frac{5\pi}{3}\right) = \frac{5}{2} - \frac{5\sqrt{3}}{2}i$

71. Since $z = i = \text{cis}\left(\frac{\pi}{2}\right)$ we have

$w_0 = \text{cis}\left(\frac{\pi}{6}\right) = \frac{\sqrt{3}}{2} + \frac{1}{2}i$ $\quad w_1 = \text{cis}\left(\frac{5\pi}{6}\right) = -\frac{\sqrt{3}}{2} + \frac{1}{2}i$ $\quad w_2 = \text{cis}\left(\frac{3\pi}{2}\right) = -i$

72. Since $z = -8i = 8\text{cis}\left(\frac{3\pi}{2}\right)$ we have

$w_0 = 2\text{cis}\left(\frac{\pi}{2}\right) = 2i$ $\quad w_1 = 2\text{cis}\left(\frac{7\pi}{6}\right) = -\sqrt{3} - i$ $\quad w_2 = \text{cis}\left(\frac{11\pi}{6}\right) = \sqrt{3} - i$

73. Since $z = 16 = 16\text{cis}(0)$ we have

$w_0 = 2\text{cis}(0) = 2$ $\quad w_1 = 2\text{cis}\left(\frac{\pi}{2}\right) = 2i$

$w_2 = 2\text{cis}(\pi) = -2$ $\quad w_3 = 2\text{cis}\left(\frac{3\pi}{2}\right) = -2i$

74. Since $z = -81 = 81\text{cis}(\pi)$ we have

$w_0 = 3\text{cis}\left(\frac{\pi}{4}\right) = \frac{3\sqrt{2}}{2} + \frac{3\sqrt{2}}{2}i$ $\quad w_1 = 3\text{cis}\left(\frac{3\pi}{4}\right) = -\frac{3\sqrt{2}}{2} + \frac{3\sqrt{2}}{2}i$

$w_2 = 3\text{cis}\left(\frac{5\pi}{4}\right) = -\frac{3\sqrt{2}}{2} - \frac{3\sqrt{2}}{2}i$ $\quad w_3 = 3\text{cis}\left(\frac{7\pi}{4}\right) = \frac{3\sqrt{2}}{2} - \frac{3\sqrt{2}}{2}i$

75. Since $z = 64 = 64\text{cis}(0)$ we have

$w_0 = 2\text{cis}(0) = 2$ $\quad w_1 = 2\text{cis}\left(\frac{\pi}{3}\right) = 1 + \sqrt{3}i$ $\quad w_2 = 2\text{cis}\left(\frac{2\pi}{3}\right) = -1 + \sqrt{3}i$

$w_3 = 2\text{cis}(\pi) = -2$ $\quad w_4 = 2\text{cis}\left(-\frac{2\pi}{3}\right) = -1 - \sqrt{3}i$ $\quad w_5 = 2\text{cis}\left(-\frac{\pi}{3}\right) = 1 - \sqrt{3}i$

76. Since $z = -729 = 729\text{cis}(\pi)$ we have

$w_0 = 3\text{cis}\left(\frac{\pi}{6}\right) = \frac{3\sqrt{3}}{2} + \frac{3}{2}i$ $\quad w_1 = 3\text{cis}\left(\frac{\pi}{2}\right) = 3i$ $\quad w_2 = 3\text{cis}\left(\frac{5\pi}{6}\right) = -\frac{3\sqrt{3}}{2} + \frac{3}{2}i$

$w_3 = 3\text{cis}\left(\frac{7\pi}{6}\right) = -\frac{3\sqrt{3}}{2} - \frac{3}{2}i$ $\quad w_4 = 3\text{cis}\left(-\frac{3\pi}{2}\right) = -3i$ $\quad w_5 = 3\text{cis}\left(-\frac{11\pi}{6}\right) = \frac{3\sqrt{3}}{2} - \frac{3}{2}i$

77. Note: In the answers for w_0 and w_2 the first rectangular form comes from applying the appropriate Sum or Difference Identity ($\frac{\pi}{12} = \frac{\pi}{3} - \frac{\pi}{4}$ and $\frac{17\pi}{12} = \frac{2\pi}{3} + \frac{3\pi}{4}$, respectively) and the second comes from using the Half-Angle Identities.

$$w_0 = \sqrt[3]{2}\text{cis}\left(\frac{\pi}{12}\right) = \sqrt[3]{2}\left(\frac{\sqrt{6}+\sqrt{2}}{4} + i\left(\frac{\sqrt{6}-\sqrt{2}}{4}\right)\right) = \sqrt[3]{2}\left(\frac{\sqrt{2+\sqrt{3}}}{2} + i\frac{\sqrt{2-\sqrt{3}}}{2}\right)$$

$$w_1 = \sqrt[3]{2}\text{cis}\left(\frac{3\pi}{4}\right) = \sqrt[3]{2}\left(-\frac{\sqrt{2}}{2} + \frac{\sqrt{2}}{2}i\right)$$

$$w_2 = \sqrt[3]{2}\text{cis}\left(\frac{17\pi}{12}\right) = \sqrt[3]{2}\left(\frac{\sqrt{2}-\sqrt{6}}{4} + i\left(\frac{-\sqrt{2}-\sqrt{6}}{4}\right)\right) = \sqrt[3]{2}\left(\frac{\sqrt{2-\sqrt{3}}}{2} + i\frac{\sqrt{2+\sqrt{3}}}{2}\right)$$

78. $w_0 = \text{cis}(0) = 1$
$w_1 = \text{cis}\left(\frac{2\pi}{5}\right) \approx 0.309 + 0.951i$
$w_2 = \text{cis}\left(\frac{4\pi}{5}\right) \approx -0.809 + 0.588i$
$w_3 = \text{cis}\left(\frac{6\pi}{5}\right) \approx -0.809 - 0.588i$
$w_4 = \text{cis}\left(\frac{8\pi}{5}\right) \approx 0.309 - 0.951i$

79. $p(x) = x^{12} - 4096 = (x-2)(x+2)(x^2+4)(x^2-2x+4)(x^2+2x+4)(x^2-2\sqrt{3}x+4)(x^2+2\sqrt{3}+4)$

11.8 Vectors

As we have seen numerous times in this book, Mathematics can be used to model and solve real-world problems. For many applications, real numbers suffice; that is, real numbers with the appropriate units attached can be used to answer questions like "How close is the nearest Sasquatch nest?" There are other times though, when these kinds of quantities do not suffice. Perhaps it is important to know, for instance, how close the nearest Sasquatch nest is as well as the direction in which it lies. (Foreshadowing the use of bearings in the exercises, perhaps?) To answer questions like these which involve both a quantitative answer, or *magnitude*, along with a *direction*, we use the mathematical objects called **vectors**.[1] A vector is represented geometrically as a directed line segment where the magnitude of the vector is taken to be the length of the line segment and the direction is made clear with the use of an arrow at one endpoint of the segment. When referring to vectors in this text, we shall adopt[2] the 'arrow' notation, so the symbol $\vec{v}$ is read as 'the vector v'. Below is a typical vector $\vec{v}$ with endpoints $P(1,2)$ and $Q(4,6)$. The point P is called the *initial point* or *tail* of $\vec{v}$ and the point Q is called the *terminal point* or *head* of $\vec{v}$. Since we can reconstruct $\vec{v}$ completely from P and Q, we write $\vec{v} = \overrightarrow{PQ}$, where the order of points P (initial point) and Q (terminal point) is important. (Think about this before moving on.)

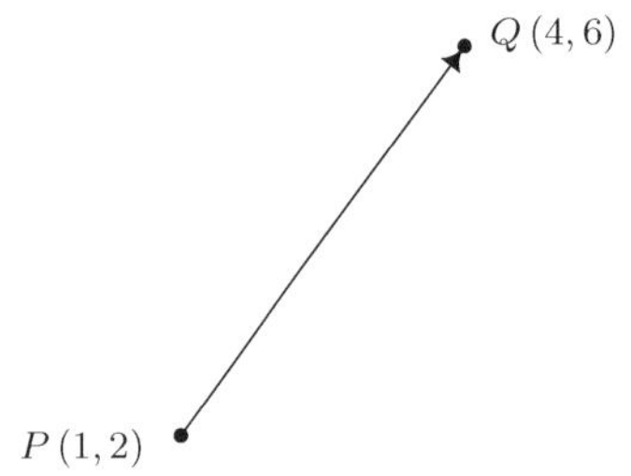

$\vec{v} = \overrightarrow{PQ}$

While it is true that P and Q completely determine $\vec{v}$, it is important to note that since vectors are defined in terms of their two characteristics, magnitude and direction, any directed line segment with the same length and direction as $\vec{v}$ is considered to be the same vector as $\vec{v}$, regardless of its initial point. In the case of our vector $\vec{v}$ above, any vector which moves three units to the right and four up[3] from its initial point to arrive at its terminal point is considered the same vector as $\vec{v}$. The notation we use to capture this idea is the *component form* of the vector, $\vec{v} = \langle 3, 4 \rangle$, where the first number, 3, is called the *x-component* of $\vec{v}$ and the second number, 4, is called the *y-component* of $\vec{v}$. If we wanted to reconstruct $\vec{v} = \langle 3, 4 \rangle$ with initial point $P'(-2,3)$, then we would find the terminal point of $\vec{v}$ by adding 3 to the x-coordinate and adding 4 to the y-coordinate to obtain the terminal point $Q'(1,7)$, as seen below.

---

[1] The word 'vector' comes from the Latin *vehere* meaning 'to convey' or 'to carry.'

[2] Other textbook authors use bold vectors such as $\boldsymbol{v}$. We find that writing in bold font on the chalkboard is inconvenient at best, so we have chosen the 'arrow' notation.

[3] If this idea of 'over' and 'up' seems familiar, it should. The slope of the line segment containing $\vec{v}$ is $\frac{4}{3}$.

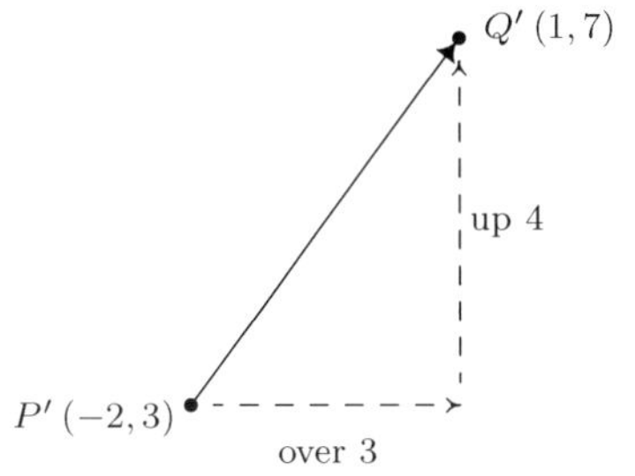

$\vec{v} = \langle 3, 4\rangle$ with initial point $P'(-2,3)$.

The component form of a vector is what ties these very geometric objects back to Algebra and ultimately Trigonometry. We generalize our example in our definition below.

Definition 11.5. Suppose $\vec{v}$ is represented by a directed line segment with initial point $P(x_0, y_0)$ and terminal point $Q(x_1, y_1)$. The **component form** of $\vec{v}$ is given by

$$\vec{v} = \overrightarrow{PQ} = \langle x_1 - x_0, y_1 - y_0\rangle$$

Using the language of components, we have that two vectors are equal if and only if their corresponding components are equal. That is, $\langle v_1, v_2\rangle = \langle v_1', v_2'\rangle$ if and only if $v_1 = v_1'$ and $v_2 = v_2'$. (Again, think about this before reading on.) We now set about defining operations on vectors. Suppose we are given two vectors $\vec{v}$ and $\vec{w}$. The sum, or *resultant* vector $\vec{v} + \vec{w}$ is obtained as follows. First, plot $\vec{v}$. Next, plot $\vec{w}$ so that its initial point is the terminal point of $\vec{v}$. To plot the vector $\vec{v} + \vec{w}$ we begin at the initial point of $\vec{v}$ and end at the terminal point of $\vec{w}$. It is helpful to think of the vector $\vec{v} + \vec{w}$ as the 'net result' of moving along $\vec{v}$ then moving along $\vec{w}$.

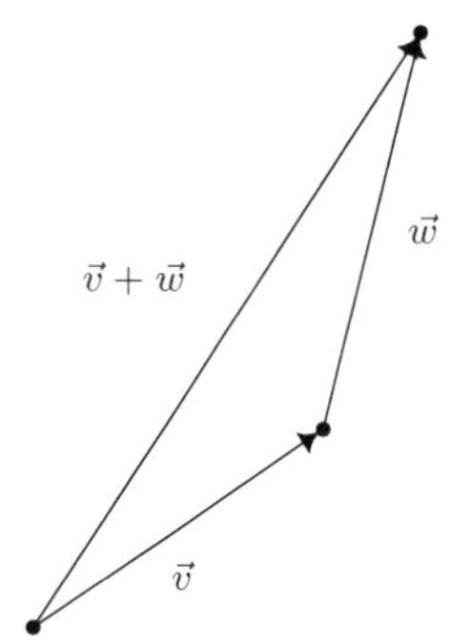

$\vec{v}$, $\vec{w}$, and $\vec{v} + \vec{w}$

Our next example makes good use of resultant vectors and reviews bearings and the Law of Cosines.[4]

Example 11.8.1. A plane leaves an airport with an airspeed[5] of 175 miles per hour at a bearing of N40°E. A 35 mile per hour wind is blowing at a bearing of S60°E. Find the true speed of the plane, rounded to the nearest mile per hour, and the true bearing of the plane, rounded to the nearest degree.

[4]If necessary, review page 905 and Section 11.3.

[5]That is, the speed of the plane relative to the air around it. If there were no wind, plane's airspeed would be the same as its speed as observed from the ground. How does wind affect this? Keep reading!

Solution: For both the plane and the wind, we are given their speeds and their directions. Coupling speed (as a magnitude) with direction is the concept of *velocity* which we've seen a few times before in this textbook.[6] We let $\vec{v}$ denote the plane's velocity and $\vec{w}$ denote the wind's velocity in the diagram below. The 'true' speed and bearing is found by analyzing the resultant vector, $\vec{v}+\vec{w}$. From the vector diagram, we get a triangle, the lengths of whose sides are the magnitude of $\vec{v}$, which is 175, the magnitude of $\vec{w}$, which is 35, and the magnitude of $\vec{v}+\vec{w}$, which we'll call c. From the given bearing information, we go through the usual geometry to determine that the angle between the sides of length 35 and 175 measures 100°.

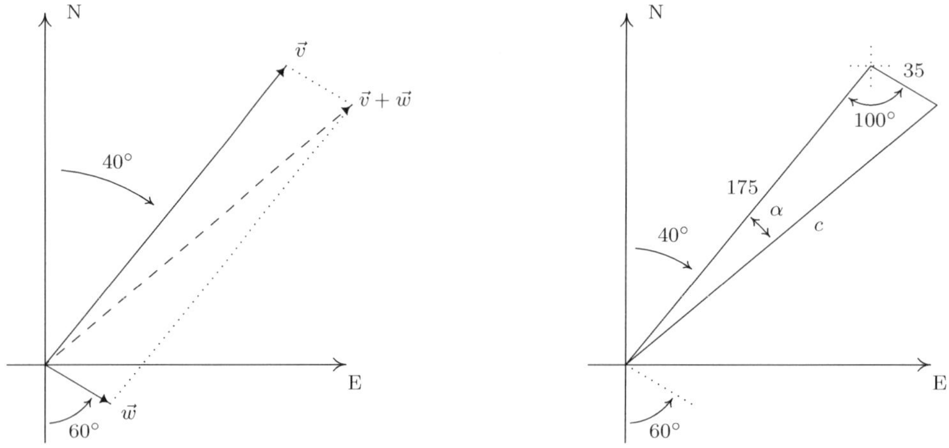

From the Law of Cosines, we determine $c = \sqrt{31850 - 12250\cos(100^\circ)} \approx 184$, which means the true speed of the plane is (approximately) 184 miles per hour. To determine the true bearing of the plane, we need to determine the angle α. Using the Law of Cosines once more,[7] we find $\cos(\alpha) = \frac{c^2+29400}{350c}$ so that $\alpha \approx 11^\circ$. Given the geometry of the situation, we add α to the given 40° and find the true bearing of the plane to be (approximately) N51°E. □

Our next step is to define addition of vectors component-wise to match the geometric action.[8]

Definition 11.6. Suppose $\vec{v} = \langle v_1, v_2\rangle$ and $\vec{w} = \langle w_1, w_2\rangle$. The vector $\vec{v}+\vec{w}$ is defined by

$$\vec{v}+\vec{w} = \langle v_1 + w_1, v_2 + w_2\rangle$$

Example 11.8.2. Let $\vec{v} = \langle 3, 4\rangle$ and suppose $\vec{w} = \overrightarrow{PQ}$ where $P(-3, 7)$ and $Q(-2, 5)$. Find $\vec{v}+\vec{w}$ and interpret this sum geometrically.

Solution. Before can add the vectors using Definition 11.6, we need to write $\vec{w}$ in component form. Using Definition 11.5, we get $\vec{w} = \langle -2-(-3), 5-7\rangle = \langle 1, -2\rangle$. Thus

[6] See Section 10.1.1, for instance.

[7] Or, since our given angle, 100°, is obtuse, we could use the Law of Sines without any ambiguity here.

[8] Adding vectors 'component-wise' should seem hauntingly familiar. Compare this with how matrix addition was defined in section 8.3. In fact, in more advanced courses such as Linear Algebra, vectors are defined as $1 \times n$ or $n \times 1$ matrices, depending on the situation.

$$\begin{aligned}\vec{v}+\vec{w} &= \langle 3,4\rangle + \langle 1,-2\rangle \\ &= \langle 3+1, 4+(-2)\rangle \\ &= \langle 4,2\rangle\end{aligned}$$

To visualize this sum, we draw $\vec{v}$ with its initial point at $(0,0)$ (for convenience) so that its terminal point is $(3,4)$. Next, we graph $\vec{w}$ with its initial point at $(3,4)$. Moving one to the right and two down, we find the terminal point of $\vec{w}$ to be $(4,2)$. We see that the vector $\vec{v}+\vec{w}$ has initial point $(0,0)$ and terminal point $(4,2)$ so its component form is $\langle 4,2\rangle$, as required.

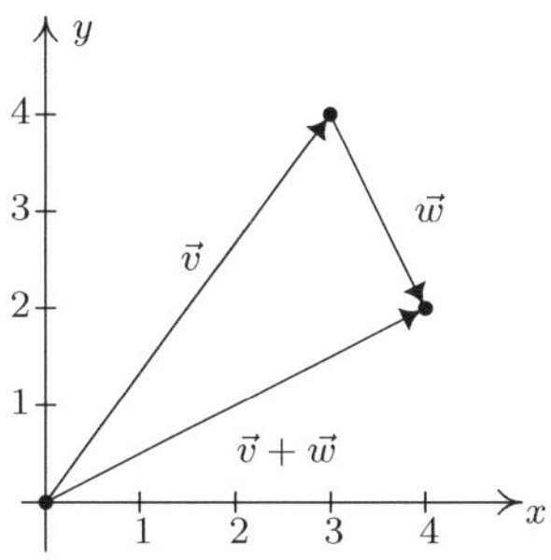

□

In order for vector addition to enjoy the same kinds of properties as real number addition, it is necessary to extend our definition of vectors to include a 'zero vector', $\vec{0} = \langle 0,0\rangle$. Geometrically, $\vec{0}$ represents a point, which we can think of as a directed line segment with the same initial and terminal points. The reader may well object to the inclusion of $\vec{0}$, since after all, vectors are supposed to have both a magnitude (length) and a direction. While it seems clear that the magnitude of $\vec{0}$ should be 0, it is not clear what its direction is. As we shall see, the direction of $\vec{0}$ is in fact undefined, but this minor hiccup in the natural flow of things is worth the benefits we reap by including $\vec{0}$ in our discussions. We have the following theorem.

Theorem 11.18. Properties of Vector Addition

- **Commutative Property:** For all vectors $\vec{v}$ and $\vec{w}$, $\vec{v}+\vec{w} = \vec{w}+\vec{v}$.
- **Associative Property:** For all vectors $\vec{u}, \vec{v}$ and $\vec{w}$, $(\vec{u}+\vec{v})+\vec{w} = \vec{u}+(\vec{v}+\vec{w})$.
- **Identity Property:** The vector $\vec{0}$ acts as the additive identity for vector addition. That is, for all vectors $\vec{v}$,
$$\vec{v}+\vec{0} = \vec{0}+\vec{v} = \vec{v}.$$
- **Inverse Property:** Every vector $\vec{v}$ has a unique additive inverse, denoted $-\vec{v}$. That is, for every vector $\vec{v}$, there is a vector $-\vec{v}$ so that
$$\vec{v}+(-\vec{v}) = (-\vec{v})+\vec{v} = \vec{0}.$$

The properties in Theorem 11.18 are easily verified using the definition of vector addition.[9] For the commutative property, we note that if $\vec{v} = \langle v_1, v_2 \rangle$ and $\vec{w} = \langle w_1, w_2 \rangle$ then

$$\begin{aligned} \vec{v} + \vec{w} &= \langle v_1, v_2 \rangle + \langle w_1, w_2 \rangle \\ &= \langle v_1 + w_1, v_2 + w_2 \rangle \\ &= \langle w_1 + v_1, w_2 + v_2 \rangle \\ &= \vec{w} + \vec{v} \end{aligned}$$

Geometrically, we can 'see' the commutative property by realizing that the sums $\vec{v} + \vec{w}$ and $\vec{w} + \vec{v}$ are the same directed diagonal determined by the parallelogram below.

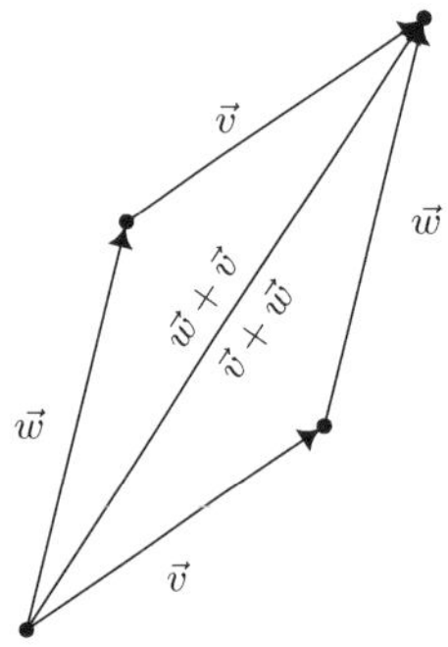

Demonstrating the commutative property of vector addition.

The proofs of the associative and identity properties proceed similarly, and the reader is encouraged to verify them and provide accompanying diagrams. The existence and uniqueness of the additive inverse is yet another property inherited from the real numbers. Given a vector $\vec{v} = \langle v_1, v_2 \rangle$, suppose we wish to find a vector $\vec{w} = \langle w_1, w_2 \rangle$ so that $\vec{v} + \vec{w} = \vec{0}$. By the definition of vector addition, we have $\langle v_1 + w_1, v_2 + w_2 \rangle = \langle 0, 0 \rangle$, and hence, $v_1 + w_1 = 0$ and $v_2 + w_2 = 0$. We get $w_1 = -v_1$ and $w_2 = -v_2$ so that $\vec{w} = \langle -v_1, -v_2 \rangle$. Hence, $\vec{v}$ has an additive inverse, and moreover, it is unique and can be obtained by the formula $-\vec{v} = \langle -v_1, -v_2 \rangle$. Geometrically, the vectors $\vec{v} = \langle v_1, v_2 \rangle$ and $-\vec{v} = \langle -v_1, -v_2 \rangle$ have the same length, but opposite directions. As a result, when adding the vectors geometrically, the sum $\vec{v} + (-\vec{v})$ results in starting at the initial point of $\vec{v}$ and ending back at the initial point of $\vec{v}$, or in other words, the net result of moving $\vec{v}$ then $-\vec{v}$ is not moving at all.

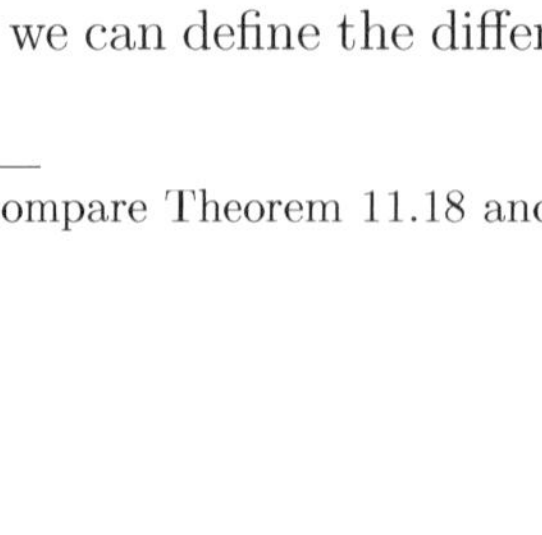

Using the additive inverse of a vector, we can define the difference of two vectors, $\vec{v} - \vec{w} = \vec{v} + (-\vec{w})$. If $\vec{v} = \langle v_1, v_2 \rangle$ and $\vec{w} = \langle w_1, w_2 \rangle$ then

[9] The interested reader is encouraged to compare Theorem 11.18 and the ensuing discussion with Theorem 8.3 in Section 8.3 and the discussion there.

$$\begin{aligned}
\vec{v} - \vec{w} &= \vec{v} + (-\vec{w}) \\
&= \langle v_1, v_2 \rangle + \langle -w_1, -w_2 \rangle \\
&= \langle v_1 + (-w_1), v_2 + (-w_2) \rangle \\
&= \langle v_1 - w_1, v_2 - w_2 \rangle
\end{aligned}$$

In other words, like vector addition, vector subtraction works component-wise. To interpret the vector $\vec{v} - \vec{w}$ geometrically, we note

$$\begin{aligned}
\vec{w} + (\vec{v} - \vec{w}) &= \vec{w} + (\vec{v} + (-\vec{w})) && \text{Definition of Vector Subtraction} \\
&= \vec{w} + ((-\vec{w}) + \vec{v}) && \text{Commutativity of Vector Addition} \\
&= (\vec{w} + (-\vec{w})) + \vec{v} && \text{Associativity of Vector Addition} \\
&= \vec{0} + \vec{v} && \text{Definition of Additive Inverse} \\
&= \vec{v} && \text{Definition of Additive Identity}
\end{aligned}$$

This means that the 'net result' of moving along $\vec{w}$ then moving along $\vec{v} - \vec{w}$ is just $\vec{v}$ itself. From the diagram below, we see that $\vec{v} - \vec{w}$ may be interpreted as the vector whose initial point is the terminal point of $\vec{w}$ and whose terminal point is the terminal point of $\vec{v}$ as depicted below. It is also worth mentioning that in the parallelogram determined by the vectors $\vec{v}$ and $\vec{w}$, the vector $\vec{v} - \vec{w}$ is one of the diagonals – the other being $\vec{v} + \vec{w}$.

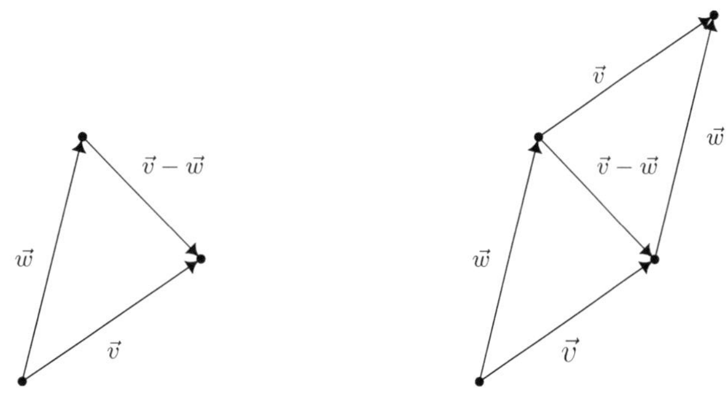

Next, we discuss *scalar* multiplication – that is, taking a real number times a vector. We define scalar multiplication for vectors in the same way we defined it for matrices in Section 8.3.

Definition 11.7. If k is a real number and $\vec{v} = \langle v_1, v_2 \rangle$, we define $k\vec{v}$ by

$$k\vec{v} = k \langle v_1, v_2 \rangle = \langle kv_1, kv_2 \rangle$$

Scalar multiplication by k in vectors can be understood geometrically as scaling the vector (if $k > 0$) or scaling the vector and reversing its direction (if $k < 0$) as demonstrated below.

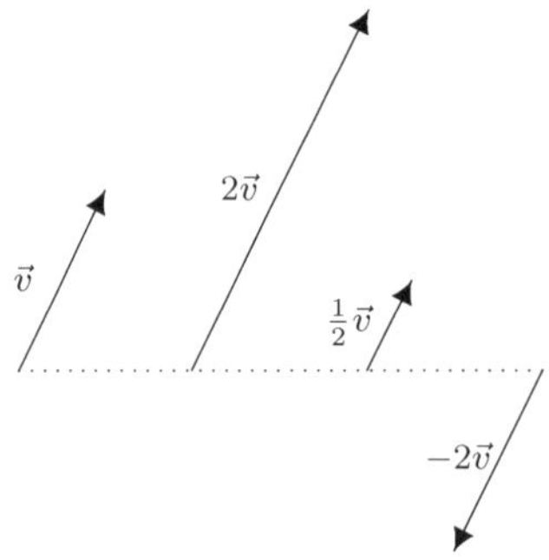

Note that, by definition 11.7, $(-1)\vec{v} = (-1)\langle v_1, v_2\rangle = \langle (-1)v_1, (-1)v_2\rangle = \langle -v_1, -v_2\rangle = -\vec{v}$. This, and other properties of scalar multiplication are summarized below.

Theorem 11.19. Properties of Scalar Multiplication

- **Associative Property:** For every vector $\vec{v}$ and scalars k and r, $(kr)\vec{v} = k(r\vec{v})$.
- **Identity Property:** For all vectors $\vec{v}$, $1\vec{v} = \vec{v}$.
- **Additive Inverse Property:** For all vectors $\vec{v}$, $-\vec{v} = (-1)\vec{v}$.
- **Distributive Property of Scalar Multiplication over Scalar Addition:** For every vector $\vec{v}$ and scalars k and r,
$$(k+r)\vec{v} = k\vec{v} + r\vec{v}$$
- **Distributive Property of Scalar Multiplication over Vector Addition:** For all vectors $\vec{v}$ and $\vec{w}$ and scalars k,
$$k(\vec{v} + \vec{w}) = k\vec{v} + k\vec{w}$$
- **Zero Product Property:** If $\vec{v}$ is vector and k is a scalar, then
$$k\vec{v} = \vec{0} \quad \text{if and only if} \quad k = 0 \quad \text{or} \quad \vec{v} = \vec{0}$$

The proof of Theorem 11.19, like the proof of Theorem 11.18, ultimately boils down to the definition of scalar multiplication and properties of real numbers. For example, to prove the associative property, we let $\vec{v} = \langle v_1, v_2\rangle$. If k and r are scalars then

$$\begin{array}{rcll} (kr)\vec{v} & = & (kr)\langle v_1, v_2\rangle & \\ & = & \langle (kr)v_1, (kr)v_2\rangle & \text{Definition of Scalar Multiplication} \\ & = & \langle k(rv_1), k(rv_2)\rangle & \text{Associative Property of Real Number Multiplication} \\ & = & k\langle rv_1, rv_2\rangle & \text{Definition of Scalar Multiplication} \\ & = & k\left(r\langle v_1, v_2\rangle\right) & \text{Definition of Scalar Multiplication} \\ & = & k(r\vec{v}) & \end{array}$$

The remaining properties are proved similarly and are left as exercises.

Our next example demonstrates how Theorem 11.19 allows us to do the same kind of algebraic manipulations with vectors as we do with variables – multiplication and division of vectors notwithstanding. If the pedantry seems familiar, it should. This is the same treatment we gave Example 8.3.1 in Section 8.3. As in that example, we spell out the solution in excruciating detail to encourage the reader to think carefully about why each step is justified.

Example 11.8.3. Solve $5\vec{v} - 2\left(\vec{v} + \langle 1, -2\rangle\right) = \vec{0}$ for $\vec{v}$.

Solution.

$$\begin{aligned}
5\vec{v} - 2\left(\vec{v} + \langle 1, -2\rangle\right) &= \vec{0} \\
5\vec{v} + (-1)\left[2\left(\vec{v} + \langle 1, -2\rangle\right)\right] &= \vec{0} \\
5\vec{v} + [(-1)(2)]\left(\vec{v} + \langle 1, -2\rangle\right) &= \vec{0} \\
5\vec{v} + (-2)\left(\vec{v} + \langle 1, -2\rangle\right) &= \vec{0} \\
5\vec{v} + \left[(-2)\vec{v} + (-2)\langle 1, -2\rangle\right] &= \vec{0} \\
5\vec{v} + \left[(-2)\vec{v} + \langle (-2)(1), (-2)(-2)\rangle\right] &= \vec{0} \\
\left[5\vec{v} + (-2)\vec{v}\right] + \langle -2, 4\rangle &= \vec{0} \\
(5 + (-2))\vec{v} + \langle -2, 4\rangle &= \vec{0} \\
3\vec{v} + \langle -2, 4\rangle &= \vec{0} \\
\left(3\vec{v} + \langle -2, 4\rangle\right) + \left(-\langle -2, 4\rangle\right) &= \vec{0} + \left(-\langle -2, 4\rangle\right) \\
3\vec{v} + \left[\langle -2, 4\rangle + \left(-\langle -2, 4\rangle\right)\right] &= \vec{0} + (-1)\langle -2, 4\rangle \\
3\vec{v} + \vec{0} &= \vec{0} + \langle (-1)(-2), (-1)(4)\rangle \\
3\vec{v} &= \langle 2, -4\rangle \\
\tfrac{1}{3}\left(3\vec{v}\right) &= \tfrac{1}{3}\left(\langle 2, -4\rangle\right) \\
\left[\left(\tfrac{1}{3}\right)(3)\right]\vec{v} &= \left\langle \left(\tfrac{1}{3}\right)(2), \left(\tfrac{1}{3}\right)(-4)\right\rangle \\
1\vec{v} &= \left\langle \tfrac{2}{3}, -\tfrac{4}{3}\right\rangle \\
\vec{v} &= \left\langle \tfrac{2}{3}, -\tfrac{4}{3}\right\rangle
\end{aligned}$$

□

A vector whose initial point is $(0, 0)$ is said to be in **standard position**. If $\vec{v} = \langle v_1, v_2\rangle$ is plotted in standard position, then its terminal point is necessarily (v_1, v_2). (Once more, think about this before reading on.)

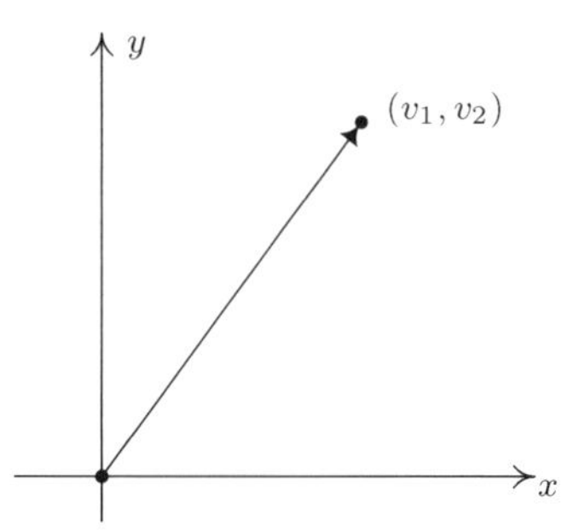

$\vec{v} = \langle v_1, v_2\rangle$ in standard position.

Plotting a vector in standard position enables us to more easily quantify the concepts of magnitude and direction of the vector. We can convert the point (v_1, v_2) in rectangular coordinates to a pair (r, θ) in polar coordinates where $r \geq 0$. The magnitude of $\vec{v}$, which we said earlier was length of the directed line segment, is $r = \sqrt{v_1^2 + v_2^2}$ and is denoted by $\|\vec{v}\|$. From Section 11.4, we know $v_1 = r\cos(\theta) = \|\vec{v}\|\cos(\theta)$ and $v_2 = r\sin(\theta) = \|\vec{v}\|\sin(\theta)$. From the definition of scalar multiplication and vector equality, we get

$$\begin{aligned} \vec{v} &= \langle v_1, v_2 \rangle \\ &= \langle \|\vec{v}\|\cos(\theta), \|\vec{v}\|\sin(\theta) \rangle \\ &= \|\vec{v}\| \langle \cos(\theta), \sin(\theta) \rangle \end{aligned}$$

This motivates the following definition.

Definition 11.8. Suppose $\vec{v}$ is a vector with component form $\vec{v} = \langle v_1, v_2 \rangle$. Let (r, θ) be a polar representation of the point with rectangular coordinates (v_1, v_2) with $r \geq 0$.

- The **magnitude** of $\vec{v}$, denoted $\|\vec{v}\|$, is given by $\|\vec{v}\| = r = \sqrt{v_1^2 + v_2^2}$
- If $\vec{v} \neq \vec{0}$, the **(vector) direction** of $\vec{v}$, denoted $\hat{v}$ is given by $\hat{v} = \langle \cos(\theta), \sin(\theta) \rangle$

Taken together, we get $\vec{v} = \langle \|\vec{v}\|\cos(\theta), \|\vec{v}\|\sin(\theta) \rangle$.

A few remarks are in order. First, we note that if $\vec{v} \neq 0$ then even though there are infinitely many angles θ which satisfy Definition 11.8, the stipulation $r > 0$ means that all of the angles are coterminal. Hence, if θ and θ' both satisfy the conditions of Definition 11.8, then $\cos(\theta) = \cos(\theta')$ and $\sin(\theta) = \sin(\theta')$, and as such, $\langle \cos(\theta), \sin(\theta) \rangle = \langle \cos(\theta'), \sin(\theta') \rangle$ making $\hat{v}$ is well-defined.[10] If $\vec{v} = \vec{0}$, then $\vec{v} = \langle 0, 0 \rangle$, and we know from Section 11.4 that $(0, \theta)$ is a polar representation for the origin for any angle θ. For this reason, $\hat{0}$ is undefined. The following theorem summarizes the important facts about the magnitude and direction of a vector.

Theorem 11.20. Properties of Magnitude and Direction: Suppose $\vec{v}$ is a vector.

- $\|\vec{v}\| \geq 0$ and $\|\vec{v}\| = 0$ if and only if $\vec{v} = \vec{0}$
- For all scalars k, $\|k\,\vec{v}\| = |k|\|\vec{v}\|$.
- If $\vec{v} \neq \vec{0}$ then $\vec{v} = \|\vec{v}\|\hat{v}$, so that $\hat{v} = \left(\frac{1}{\|\vec{v}\|}\right)\vec{v}$.

The proof of the first property in Theorem 11.20 is a direct consequence of the definition of $\|\vec{v}\|$. If $\vec{v} = \langle v_1, v_2 \rangle$, then $\|\vec{v}\| = \sqrt{v_1^2 + v_2^2}$ which is by definition greater than or equal to 0. Moreover, $\sqrt{v_1^2 + v_2^2} = 0$ if and only of $v_1^2 + v_2^2 = 0$ if and only if $v_1 = v_2 = 0$. Hence, $\|\vec{v}\| = 0$ if and only if $\vec{v} = \langle 0, 0 \rangle = \vec{0}$, as required.

The second property is a result of the definition of magnitude and scalar multiplication along with a propery of radicals. If $\vec{v} = \langle v_1, v_2 \rangle$ and k is a scalar then

[10] If this all looks familiar, it should. The interested reader is invited to compare Definition 11.8 to Definition 11.2 in Section 11.7.

$$\begin{aligned}
\|k\,\vec{v}\| &= \|k\,\langle v_1, v_2\rangle\,\| \\
&= \|\,\langle kv_1, kv_2\rangle\,\| && \text{Definition of scalar multiplication} \\
&= \sqrt{(kv_1)^2 + (kv_2)^2} && \text{Definition of magnitude} \\
&= \sqrt{k^2v_1^2 + k^2v_2^2} \\
&= \sqrt{k^2(v_1^2 + v_2^2)} \\
&= \sqrt{k^2}\sqrt{v_1^2 + v_2^2} && \text{Product Rule for Radicals} \\
&= |k|\sqrt{v_1^2 + v_2^2} && \text{Since } \sqrt{k^2} = |k| \\
&= |k|\|\vec{v}\|
\end{aligned}$$

The equation $\vec{v} = \|\vec{v}\|\hat{v}$ in Theorem 11.20 is a consequence of the definitions of $\|\vec{v}\|$ and $\hat{v}$ and was worked out in the discussion just prior to Definition 11.8 on page 1020. In words, the equation $\vec{v} = \|\vec{v}\|\hat{v}$ says that any given vector is the product of its magnitude and its direction – an important concept to keep in mind when studying and using vectors. The equation $\hat{v} = \left(\frac{1}{\|\vec{v}\|}\right)\vec{v}$ is a result of solving $\vec{v} = \|\vec{v}\|\hat{v}$ for $\hat{v}$ by multiplying[11] both sides of the equation by $\frac{1}{\|\vec{v}\|}$ and using the properties of Theorem 11.19. We are overdue for an example.

Example 11.8.4.

1. Find the component form of the vector $\vec{v}$ with $\|\vec{v}\| = 5$ so that when $\vec{v}$ is plotted in standard position, it lies in Quadrant II and makes a 60° angle[12] with the negative x-axis.
2. For $\vec{v} = \langle 3, -3\sqrt{3}\rangle$, find $\|\vec{v}\|$ and θ, $0 \le \theta < 2\pi$ so that $\vec{v} = \|\vec{v}\|\,\langle\cos(\theta), \sin(\theta)\rangle$.
3. For the vectors $\vec{v} = \langle 3, 4\rangle$ and $\vec{w} = \langle 1, -2\rangle$, find the following.

   (a) $\hat{v}$ (b) $\|\vec{v}\| - 2\|\vec{w}\|$ (c) $\|\vec{v} - 2\vec{w}\|$ (d) $\|\hat{w}\|$

Solution.

1. We are told that $\|\vec{v}\| = 5$ and are given information about its direction, so we can use the formula $\vec{v} = \|\vec{v}\|\hat{v}$ to get the component form of $\vec{v}$. To determine $\hat{v}$, we appeal to Definition 11.8. We are told that $\vec{v}$ lies in Quadrant II and makes a 60° angle with the negative x-axis, so the polar form of the terminal point of $\vec{v}$, when plotted in standard position is $(5, 120^\circ)$. (See the diagram below.) Thus $\hat{v} = \langle\cos(120^\circ), \sin(120^\circ)\rangle = \left\langle -\frac{1}{2}, \frac{\sqrt{3}}{2}\right\rangle$, so $\vec{v} = \|\vec{v}\|\hat{v} = 5\left\langle -\frac{1}{2}, \frac{\sqrt{3}}{2}\right\rangle = \left\langle -\frac{5}{2}, \frac{5\sqrt{3}}{2}\right\rangle$.

[11]Of course, to go from $\vec{v} = \|\vec{v}\|\hat{v}$ to $\hat{v} = \left(\frac{1}{\|\vec{v}\|}\right)\vec{v}$, we are essentially 'dividing both sides' of the equation by the scalar $\|\vec{v}\|$. The authors encourage the reader, however, to work out the details carefully to gain an appreciation of the properties in play.

[12]Due to the utility of vectors in 'real-world' applications, we will usually use degree measure for the angle when giving the vector's direction. However, since Carl doesn't want you to forget about radians, he's made sure there are examples and exercises which use them.

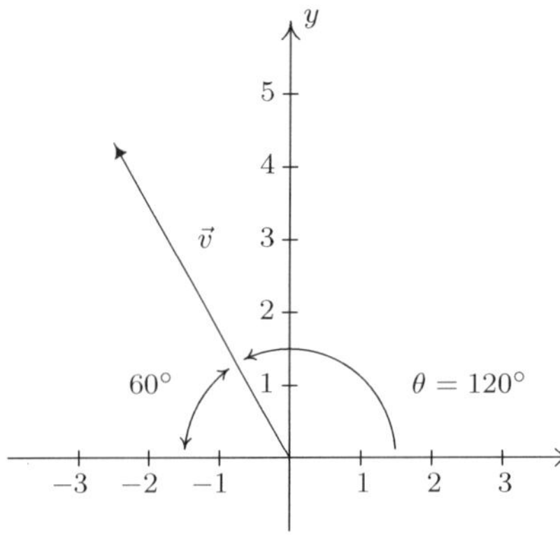

2. For $\vec{v} = \left\langle 3, -3\sqrt{3} \right\rangle$, we get $\|\vec{v}\| = \sqrt{(3)^2 + (-3\sqrt{3})^2} = 6$. In light of Definition 11.8, we can find the θ we're after by converting the point with rectangular coordinates $(3, -3\sqrt{3})$ to polar form (r, θ) where $r = \|\vec{v}\| > 0$. From Section 11.4, we have $\tan(\theta) = \frac{-3\sqrt{3}}{3} = -\sqrt{3}$. Since $(3, -3\sqrt{3})$ is a point in Quadrant IV, θ is a Quadrant IV angle. Hence, we pick $\theta = \frac{5\pi}{3}$. We may check our answer by verifying $\vec{v} = \left\langle 3, -3\sqrt{3} \right\rangle = 6 \left\langle \cos\left(\frac{5\pi}{3}\right), \sin\left(\frac{5\pi}{3}\right) \right\rangle$.

3. (a) Since we are given the component form of $\vec{v}$, we'll use the formula $\hat{v} = \left(\frac{1}{\|\vec{v}\|}\right)\vec{v}$. For $\vec{v} = \langle 3, 4 \rangle$, we have $\|\vec{v}\| = \sqrt{3^2 + 4^2} = \sqrt{25} = 5$. Hence, $\hat{v} = \frac{1}{5}\langle 3, 4 \rangle = \left\langle \frac{3}{5}, \frac{4}{5} \right\rangle$.

   (b) We know from our work above that $\|\vec{v}\| = 5$, so to find $\|\vec{v}\| - 2\|\vec{w}\|$, we need only find $\|\vec{w}\|$. Since $\vec{w} = \langle 1, -2 \rangle$, we get $\|\vec{w}\| = \sqrt{1^2 + (-2)^2} = \sqrt{5}$. Hence, $\|\vec{v}\| - 2\|\vec{w}\| = 5 - 2\sqrt{5}$.

   (c) In the expression $\|\vec{v} - 2\vec{w}\|$, notice that the arithmetic on the vectors comes first, then the magnitude. Hence, our first step is to find the component form of the vector $\vec{v} - 2\vec{w}$. We get $\vec{v} - 2\vec{w} = \langle 3, 4 \rangle - 2\langle 1, -2 \rangle = \langle 1, 8 \rangle$. Hence, $\|\vec{v} - 2\vec{w}\| = \|\langle 1, 8 \rangle\| = \sqrt{1^2 + 8^2} = \sqrt{65}$.

   (d) To find $\|\hat{w}\|$, we first need $\hat{w}$. Using the formula $\hat{w} = \left(\frac{1}{\|\vec{w}\|}\right)\vec{w}$ along with $\|\vec{w}\| = \sqrt{5}$, which we found the in the previous problem, we get $\hat{w} = \frac{1}{\sqrt{5}}\langle 1, -2 \rangle = \left\langle \frac{1}{\sqrt{5}}, -\frac{2}{\sqrt{5}} \right\rangle = \left\langle \frac{\sqrt{5}}{5}, -\frac{2\sqrt{5}}{5} \right\rangle$. Hence, $\|\hat{w}\| = \sqrt{\left(\frac{\sqrt{5}}{5}\right)^2 + \left(-\frac{2\sqrt{5}}{5}\right)^2} = \sqrt{\frac{5}{25} + \frac{20}{25}} = \sqrt{1} = 1$. $\square$

The process exemplified by number 1 in Example 11.8.4 above by which we take information about the magnitude and direction of a vector and find the component form of a vector is called **resolving** a vector into its components. As an application of this process, we revisit Example 11.8.1 below.

Example 11.8.5. A plane leaves an airport with an airspeed of 175 miles per hour with bearing N40°E. A 35 mile per hour wind is blowing at a bearing of S60°E. Find the true speed of the plane, rounded to the nearest mile per hour, and the true bearing of the plane, rounded to the nearest degree.

Solution: We proceed as we did in Example 11.8.1 and let $\vec{v}$ denote the plane's velocity and $\vec{w}$ denote the wind's velocity, and set about determining $\vec{v} + \vec{w}$. If we regard the airport as being at the origin, the positive y-axis acting as due north and the positive x-axis acting as due east, we see that the vectors $\vec{v}$ and $\vec{w}$ are in standard position and their directions correspond to the angles 50° and -30°, respectively. Hence, the component form of $\vec{v} = 175\left\langle \cos(50^\circ), \sin(50^\circ)\right\rangle = \left\langle 175\cos(50^\circ), 175\sin(50^\circ)\right\rangle$ and the component form of $\vec{w} = \left\langle 35\cos(-30^\circ), 35\sin(-30^\circ)\right\rangle$. Since we have no convenient way to express the exact values of cosine and sine of 50°, we leave both vectors in terms of cosines and sines.[13] Adding corresponding components, we find the resultant vector $\vec{v}+\vec{w} = \left\langle 175\cos(50^\circ)+35\cos(-30^\circ), 175\sin(50^\circ)+35\sin(-30^\circ)\right\rangle$. To find the 'true' speed of the plane, we compute the magnitude of this resultant vector

$$\|\vec{v}+\vec{w}\| = \sqrt{(175\cos(50^\circ)+35\cos(-30^\circ))^2 + (175\sin(50^\circ)+35\sin(-30^\circ))^2} \approx 184$$

Hence, the 'true' speed of the plane is approximately 184 miles per hour. To find the true bearing, we need to find the angle θ which corresponds to the polar form (r, θ), $r > 0$, of the point $(x, y) = (175\cos(50^\circ)+35\cos(-30^\circ), 175\sin(50^\circ)+35\sin(-30^\circ))$. Since both of these coordinates are positive,[14] we know θ is a Quadrant I angle, as depicted below. Furthermore,

$$\tan(\theta) = \frac{y}{x} = \frac{175\sin(50^\circ)+35\sin(-30^\circ)}{175\cos(50^\circ)+35\cos(-30^\circ)},$$

so using the arctangent function, we get $\theta \approx 39^\circ$. Since, for the purposes of bearing, we need the angle between $\vec{v}+\vec{w}$ and the positive y-axis, we take the complement of θ and find the 'true' bearing of the plane to be approximately N51°E.

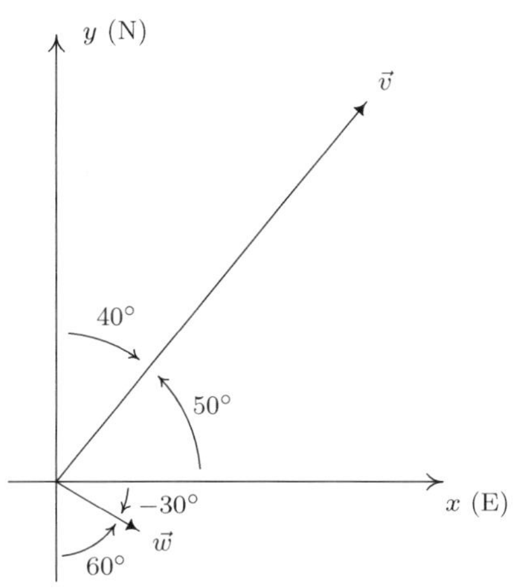

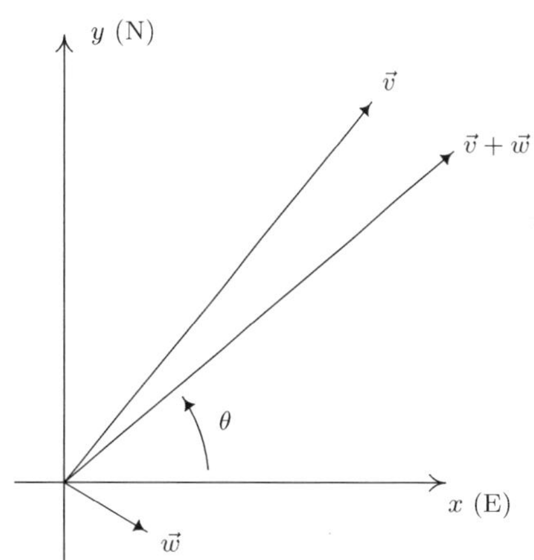

□

In part 3d of Example 11.8.4, we saw that $\|\hat{w}\| = 1$. Vectors with length 1 have a special name and are important in our further study of vectors.

Definition 11.9. Unit Vectors: Let $\vec{v}$ be a vector. If $\|\vec{v}\| = 1$, we say that $\vec{v}$ is a **unit vector**.

[13] Keeping things 'calculator' friendly, for once!

[14] Yes, a calculator approximation is the quickest way to see this, but you can also use good old-fashioned inequalities and the fact that $45^\circ \leq 50^\circ \leq 60^\circ$.

If $\vec{v}$ is a unit vector, then necessarily, $\vec{v} = \|\vec{v}\|\hat{v} = 1 \cdot \hat{v} = \hat{v}$. Conversely, we leave it as an exercise[15] to show that $\hat{v} = \left(\frac{1}{\|\vec{v}\|}\right)\vec{v}$ is a unit vector for any nonzero vector $\vec{v}$. In practice, if $\vec{v}$ is a unit vector we write it as $\hat{v}$ as opposed to $\vec{v}$ because we have reserved the '$\hat{\ }$' notation for unit vectors. The process of multiplying a nonzero vector by the factor $\frac{1}{\|\vec{v}\|}$ to produce a unit vector is called '**normalizing** the vector,' and the resulting vector $\hat{v}$ is called the 'unit vector in the **direction** of $\vec{v}$'. The terminal points of unit vectors, when plotted in standard position, lie on the Unit Circle. (You should take the time to show this.) As a result, we visualize normalizing a nonzero vector $\vec{v}$ as shrinking[16] its terminal point, when plotted in standard position, back to the Unit Circle.

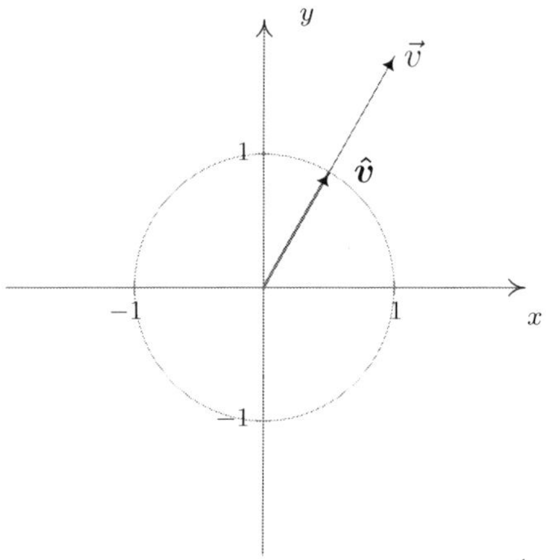

Visualizing vector normalization $\hat{v} = \left(\frac{1}{\|\vec{v}\|}\right)\vec{v}$

Of all of the unit vectors, two deserve special mention.

Definition 11.10. The Principal Unit Vectors:

- The vector $\hat{\imath}$ is defined by $\hat{\imath} = \langle 1, 0\rangle$
- The vector $\hat{\jmath}$ is defined by $\hat{\imath} = \langle 0, 1\rangle$

We can think of the vector $\hat{\imath}$ as representing the positive x-direction, while $\hat{\jmath}$ represents the positive y-direction. We have the following 'decomposition' theorem.[17]

Theorem 11.21. Principal Vector Decomposition Theorem: Let $\vec{v}$ be a vector with component form $\vec{v} = \langle v_1, v_2\rangle$. Then $\vec{v} = v_1\hat{\imath} + v_2\hat{\jmath}$.

The proof of Theorem 11.21 is straightforward. Since $\hat{\imath} = \langle 1, 0\rangle$ and $\hat{\jmath} = \langle 0, 1\rangle$, we have from the definition of scalar multiplication and vector addition that

$$v_1\hat{\imath} + v_2\hat{\jmath} = v_1\langle 1, 0\rangle + v_2\langle 0, 1\rangle = \langle v_1, 0\rangle + \langle 0, v_2\rangle = \langle v_1, v_2\rangle = \vec{v}$$

[15] One proof uses the properties of scalar multiplication and magnitude. If $\vec{v} \neq \vec{0}$, consider $\|\hat{v}\| = \left\|\left(\frac{1}{\|\vec{v}\|}\right)\vec{v}\right\|$. Use the fact that $\|\vec{v}\| \geq 0$ is a scalar and consider factoring.

[16] . . . if $\|\vec{v}\| > 1$ . . .

[17] We will see a generalization of Theorem 11.21 in Section 11.9. Stay tuned!

Geometrically, the situation looks like this:

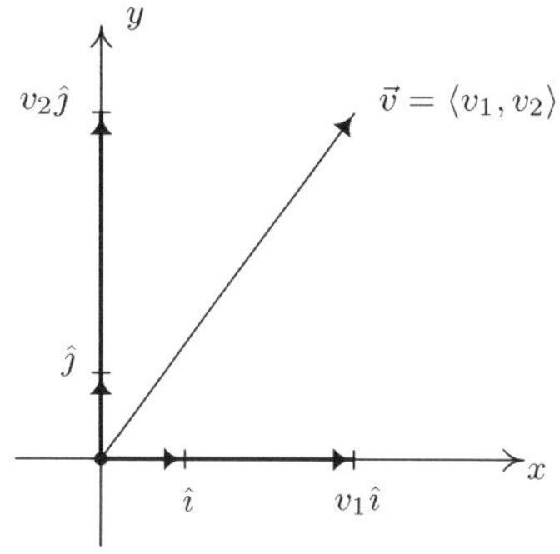

$$\vec{v} = \langle v_1, v_2 \rangle = v_1\hat{\imath} + v_2\hat{\jmath}.$$

We conclude this section with a classic example which demonstrates how vectors are used to model forces. A 'force' is defined as a 'push' or a 'pull.' The intensity of the push or pull is the magnitude of the force, and is measured in Netwons (N) in the SI system or pounds (lbs.) in the English system.[18] The following example uses all of the concepts in this section, and should be studied in great detail.

Example 11.8.6. A 50 pound speaker is suspended from the ceiling by two support braces. If one of them makes a 60° angle with the ceiling and the other makes a 30° angle with the ceiling, what are the tensions on each of the supports?

Solution. We represent the problem schematically below and then provide the corresponding vector diagram.

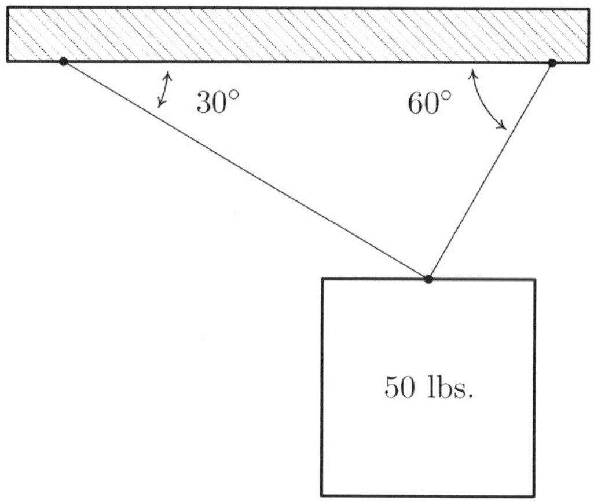

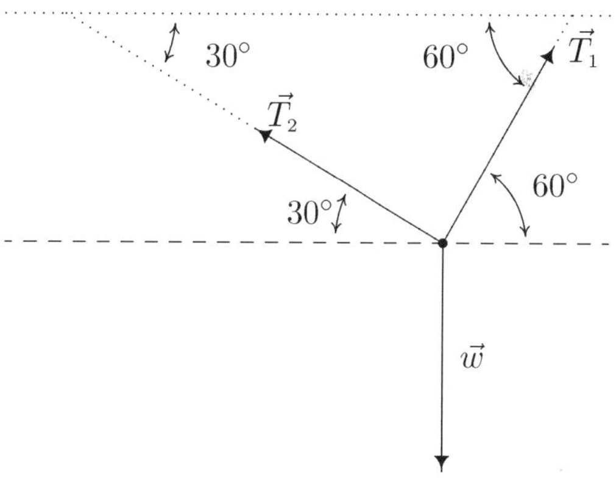

We have three forces acting on the speaker: the weight of the speaker, which we'll call $\vec{w}$, pulling the speaker directly downward, and the forces on the support rods, which we'll call $\vec{T}_1$ and $\vec{T}_2$ (for 'tensions') acting upward at angles 60° and 30°, respectively. We are looking for the tensions on the support, which are the magnitudes $\|\vec{T}_1\|$ and $\|\vec{T}_2\|$. In order for the speaker to remain stationary,[19] we require $\vec{w} + \vec{T}_1 + \vec{T}_2 = \vec{0}$. Viewing the common initial point of these vectors as the

[18] See also Section 11.1.1.

[19] This is the criteria for 'static equilbrium'.

origin and the dashed line as the x-axis, we use Theorem 11.20 to get component representations for the three vectors involved. We can model the weight of the speaker as a vector pointing directly downwards with a magnitude of 50 pounds. That is, $\|\vec{w}\| = 50$ and $\hat{w} = -\hat{\jmath} = \langle 0, -1 \rangle$. Hence, $\vec{w} = 50 \langle 0, -1 \rangle = \langle 0, -50 \rangle$. For the force in the first support, we get

$$\begin{aligned} \vec{T}_1 &= \|\vec{T}_1\| \left\langle \cos\left(60^\circ\right), \sin\left(60^\circ\right) \right\rangle \\ &= \left\langle \frac{\|\vec{T}_1\|}{2}, \frac{\|\vec{T}_1\|\sqrt{3}}{2} \right\rangle \end{aligned}$$

For the second support, we note that the angle 30° is measured from the negative x-axis, so the angle needed to write $\vec{T}_2$ in component form is 150°. Hence

$$\begin{aligned} \vec{T}_2 &= \|\vec{T}_2\| \left\langle \cos\left(150^\circ\right), \sin\left(150^\circ\right) \right\rangle \\ &= \left\langle -\frac{\|\vec{T}_2\|\sqrt{3}}{2}, \frac{\|\vec{T}_2\|}{2} \right\rangle \end{aligned}$$

The requirement $\vec{w} + \vec{T}_1 + \vec{T}_2 = \vec{0}$ gives us this vector equation.

$$\begin{aligned} \vec{w} + \vec{T}_1 + \vec{T}_2 &= \vec{0} \\ \langle 0, -50 \rangle + \left\langle \frac{\|\vec{T}_1\|}{2}, \frac{\|\vec{T}_1\|\sqrt{3}}{2} \right\rangle + \left\langle -\frac{\|\vec{T}_2\|\sqrt{3}}{2}, \frac{\|\vec{T}_2\|}{2} \right\rangle &= \langle 0, 0 \rangle \\ \left\langle \frac{\|\vec{T}_1\|}{2} - \frac{\|\vec{T}_2\|\sqrt{3}}{2}, \frac{\|\vec{T}_1\|\sqrt{3}}{2} + \frac{\|\vec{T}_2\|}{2} - 50 \right\rangle &= \langle 0, 0 \rangle \end{aligned}$$

Equating the corresponding components of the vectors on each side, we get a system of linear equations in the variables $\|\vec{T}_1\|$ and $\|\vec{T}_2\|$.

$$\left\{ \begin{array}{lrcl} (E1) & \dfrac{\|\vec{T}_1\|}{2} - \dfrac{\|\vec{T}_2\|\sqrt{3}}{2} & = & 0 \\ (E2) & \dfrac{\|\vec{T}_1\|\sqrt{3}}{2} + \dfrac{\|\vec{T}_2\|}{2} - 50 & = & 0 \end{array} \right.$$

From $(E1)$, we get $\|\vec{T}_1\| = \|\vec{T}_2\|\sqrt{3}$. Substituting that into $(E2)$ gives $\frac{(\|\vec{T}_2\|\sqrt{3})\sqrt{3}}{2} + \frac{\|\vec{T}_2\|}{2} - 50 = 0$ which yields $2\|\vec{T}_2\| - 50 = 0$. Hence, $\|\vec{T}_2\| = 25$ pounds and $\|\vec{T}_1\| = \|\vec{T}_2\|\sqrt{3} = 25\sqrt{3}$ pounds. □

11.8.1 Exercises

In Exercises 1 - 10, use the given pair of vectors $\vec{v}$ and $\vec{w}$ to find the following quantities. State whether the result is a vector or a scalar.

- $\vec{v} + \vec{w}$
- $\vec{w} - 2\vec{v}$
- $\|\vec{v} + \vec{w}\|$
- $\|\vec{v}\| + \|\vec{w}\|$
- $\|\vec{v}\|\vec{w} - \|\vec{w}\|\vec{v}$
- $\|\vec{w}\|\hat{v}$

Finally, verify that the vectors satisfy the **Parallelogram Law**

$$\|\vec{v}\|^2 + \|\vec{w}\|^2 = \frac{1}{2}\left[\|\vec{v}+\vec{w}\|^2 + \|\vec{v}-\vec{w}\|^2\right]$$

1. $\vec{v} = \langle 12, -5 \rangle$, $\vec{w} = \langle 3, 4 \rangle$
2. $\vec{v} = \langle -7, 24 \rangle$, $\vec{w} = \langle -5, -12 \rangle$
3. $\vec{v} = \langle 2, -1 \rangle$, $\vec{w} = \langle -2, 4 \rangle$
4. $\vec{v} = \langle 10, 4 \rangle$, $\vec{w} = \langle -2, 5 \rangle$
5. $\vec{v} = \langle -\sqrt{3}, 1 \rangle$, $\vec{w} = \langle 2\sqrt{3}, 2 \rangle$
6. $\vec{v} = \left\langle \frac{3}{5}, \frac{4}{5} \right\rangle$, $\vec{w} = \left\langle -\frac{4}{5}, \frac{3}{5} \right\rangle$
7. $\vec{v} = \left\langle \frac{\sqrt{2}}{2}, -\frac{\sqrt{2}}{2} \right\rangle$, $\vec{w} = \left\langle -\frac{\sqrt{2}}{2}, \frac{\sqrt{2}}{2} \right\rangle$
8. $\vec{v} = \left\langle \frac{1}{2}, \frac{\sqrt{3}}{2} \right\rangle$, $\vec{w} = \langle -1, -\sqrt{3} \rangle$
9. $\vec{v} = 3\hat{\imath} + 4\hat{\jmath}$, $\vec{w} = -2\hat{\jmath}$
10. $\vec{v} = \frac{1}{2}(\hat{\imath} + \hat{\jmath})$, $\vec{w} = \frac{1}{2}(\hat{\imath} - \hat{\jmath})$

In Exercises 11 - 25, find the component form of the vector $\vec{v}$ using the information given about its magnitude and direction. Give exact values.

11. $\|\vec{v}\| = 6$; when drawn in standard position $\vec{v}$ lies in Quadrant I and makes a 60° angle with the positive x-axis
12. $\|\vec{v}\| = 3$; when drawn in standard position $\vec{v}$ lies in Quadrant I and makes a 45° angle with the positive x-axis
13. $\|\vec{v}\| = \frac{2}{3}$; when drawn in standard position $\vec{v}$ lies in Quadrant I and makes a 60° angle with the positive y-axis
14. $\|\vec{v}\| = 12$; when drawn in standard position $\vec{v}$ lies along the positive y-axis
15. $\|\vec{v}\| = 4$; when drawn in standard position $\vec{v}$ lies in Quadrant II and makes a 30° angle with the negative x-axis
16. $\|\vec{v}\| = 2\sqrt{3}$; when drawn in standard position $\vec{v}$ lies in Quadrant II and makes a 30° angle with the positive y-axis
17. $\|\vec{v}\| = \frac{7}{2}$; when drawn in standard position $\vec{v}$ lies along the negative x-axis
18. $\|\vec{v}\| = 5\sqrt{6}$; when drawn in standard position $\vec{v}$ lies in Quadrant III and makes a 45° angle with the negative x-axis
19. $\|\vec{v}\| = 6.25$; when drawn in standard position $\vec{v}$ lies along the negative y-axis

20. $\|\vec{v}\| = 4\sqrt{3}$; when drawn in standard position $\vec{v}$ lies in Quadrant IV and makes a 30° angle with the positive x-axis

21. $\|\vec{v}\| = 5\sqrt{2}$; when drawn in standard position $\vec{v}$ lies in Quadrant IV and makes a 45° angle with the negative y-axis

22. $\|\vec{v}\| = 2\sqrt{5}$; when drawn in standard position $\vec{v}$ lies in Quadrant I and makes an angle measuring $\arctan(2)$ with the positive x-axis

23. $\|\vec{v}\| = \sqrt{10}$; when drawn in standard position $\vec{v}$ lies in Quadrant II and makes an angle measuring $\arctan(3)$ with the negative x-axis

24. $\|\vec{v}\| = 5$; when drawn in standard position $\vec{v}$ lies in Quadrant III and makes an angle measuring $\arctan\left(\frac{4}{3}\right)$ with the negative x-axis

25. $\|\vec{v}\| = 26$; when drawn in standard position $\vec{v}$ lies in Quadrant IV and makes an angle measuring $\arctan\left(\frac{5}{12}\right)$ with the positive x-axis

In Exercises 26 - 31, approximate the component form of the vector $\vec{v}$ using the information given about its magnitude and direction. Round your approximations to two decimal places.

26. $\|\vec{v}\| = 392$; when drawn in standard position $\vec{v}$ makes a 117° angle with the positive x-axis

27. $\|\vec{v}\| = 63.92$; when drawn in standard position $\vec{v}$ makes a 78.3° angle with the positive x-axis

28. $\|\vec{v}\| = 5280$; when drawn in standard position $\vec{v}$ makes a 12° angle with the positive x-axis

29. $\|\vec{v}\| = 450$; when drawn in standard position $\vec{v}$ makes a 210.75° angle with the positive x-axis

30. $\|\vec{v}\| = 168.7$; when drawn in standard position $\vec{v}$ makes a 252° angle with the positive x-axis

31. $\|\vec{v}\| = 26$; when drawn in standard position $\vec{v}$ makes a 304.5° angle with the positive x-axis

In Exercises 32 - 52, for the given vector $\vec{v}$, find the magnitude $\|\vec{v}\|$ and an angle θ with $0 \leq \theta < 360^\circ$ so that $\vec{v} = \|\vec{v}\| \langle \cos(\theta), \sin(\theta) \rangle$ (See Definition 11.8.) Round approximations to two decimal places.

32. $\vec{v} = \left\langle 1, \sqrt{3} \right\rangle$

33. $\vec{v} = \langle 5, 5 \rangle$

34. $\vec{v} = \left\langle -2\sqrt{3}, 2 \right\rangle$

35. $\vec{v} = \left\langle -\sqrt{2}, \sqrt{2} \right\rangle$

36. $\vec{v} = \left\langle -\frac{\sqrt{2}}{2}, -\frac{\sqrt{2}}{2} \right\rangle$

37. $\vec{v} = \left\langle -\frac{1}{2}, -\frac{\sqrt{3}}{2} \right\rangle$

38. $\vec{v} = \langle 6, 0 \rangle$

39. $\vec{v} = \langle -2.5, 0 \rangle$

40. $\vec{v} = \left\langle 0, \sqrt{7} \right\rangle$

41. $\vec{v} = -10\hat{\jmath}$

42. $\vec{v} = \langle 3, 4 \rangle$

43. $\vec{v} = \langle 12, 5 \rangle$

44. $\vec{v} = \langle -4, 3 \rangle$

45. $\vec{v} = \langle -7, 24 \rangle$

46. $\vec{v} = \langle -2, -1 \rangle$

47. $\vec{v} = \langle -2, -6 \rangle$

48. $\vec{v} = \hat{\imath} + \hat{\jmath}$

49. $\vec{v} = \hat{\imath} - 4\hat{\jmath}$

50. $\vec{v} = \langle 123.4, -77.05 \rangle$

51. $\vec{v} = \langle 965.15, 831.6 \rangle$

52. $\vec{v} = \langle -114.1, 42.3 \rangle$

53. A small boat leaves the dock at Camp DuNuthin and heads across the Nessie River at 17 miles per hour (that is, with respect to the water) at a bearing of S68°W. The river is flowing due east at 8 miles per hour. What is the boat's true speed and heading? Round the speed to the nearest mile per hour and express the heading as a bearing, rounded to the nearest tenth of a degree.

54. The HMS Sasquatch leaves port with bearing S20°E maintaining a speed of 42 miles per hour (that is, with respect to the water). If the ocean current is 5 miles per hour with a bearing of N60°E, find the HMS Sasquatch's true speed and bearing. Round the speed to the nearest mile per hour and express the heading as a bearing, rounded to the nearest tenth of a degree.

55. If the captain of the HMS Sasquatch in Exercise 54 wishes to reach Chupacabra Cove, an island 100 miles away at a bearing of S20°E from port, in three hours, what speed and heading should she set to take into account the ocean current? Round the speed to the nearest mile per hour and express the heading as a bearing, rounded to the nearest tenth of a degree.

    HINT: If $\vec{v}$ denotes the velocity of the HMS Sasquatch and $\vec{w}$ denotes the velocity of the current, what does $\vec{v} + \vec{w}$ need to be to reach Chupacabra Cove in three hours?

56. In calm air, a plane flying from the Pedimaxus International Airport can reach Cliffs of Insanity Point in two hours by following a bearing of N8.2°E at 96 miles an hour. (The distance between the airport and the cliffs is 192 miles.) If the wind is blowing from the southeast at 25 miles per hour, what speed and bearing should the pilot take so that she makes the trip in two hours along the original heading? Round the speed to the nearest hundredth of a mile per hour and your angle to the nearest tenth of a degree.

57. The SS Bigfoot leaves Yeti Bay on a course of N37°W at a speed of 50 miles per hour. After traveling half an hour, the captain determines he is 30 miles from the bay and his bearing back to the bay is S40°E. What is the speed and bearing of the ocean current? Round the speed to the nearest mile per hour and express the heading as a bearing, rounded to the nearest tenth of a degree.

58. A 600 pound Sasquatch statue is suspended by two cables from a gymnasium ceiling. If each cable makes a 60° angle with the ceiling, find the tension on each cable. Round your answer to the nearest pound.

59. Two cables are to support an object hanging from a ceiling. If the cables are each to make a 42° angle with the ceiling, and each cable is rated to withstand a maximum tension of 100 pounds, what is the heaviest object that can be supported? Round your answer down to the nearest pound.

60. A 300 pound metal star is hanging on two cables which are attached to the ceiling. The left hand cable makes a 72° angle with the ceiling while the right hand cable makes a 18° angle with the ceiling. What is the tension on each of the cables? Round your answers to three decimal places.

61. Two drunken college students have filled an empty beer keg with rocks and tied ropes to it in order to drag it down the street in the middle of the night. The stronger of the two students pulls with a force of 100 pounds at a heading of N77°E and the other pulls at a heading of S68°E. What force should the weaker student apply to his rope so that the keg of rocks heads due east? What resultant force is applied to the keg? Round your answer to the nearest pound.

62. Emboldened by the success of their late night keg pull in Exercise 61 above, our intrepid young scholars have decided to pay homage to the chariot race scene from the movie 'Ben-Hur' by tying three ropes to a couch, loading the couch with all but one of their friends and pulling it due west down the street. The first rope points N80°W, the second points due west and the third points S80°W. The force applied to the first rope is 100 pounds, the force applied to the second rope is 40 pounds and the force applied (by the non-riding friend) to the third rope is 160 pounds. They need the resultant force to be at least 300 pounds otherwise the couch won't move. Does it move? If so, is it heading due west?

63. Let $\vec{v} = \langle v_1, v_2 \rangle$ be any non-zero vector. Show that $\dfrac{1}{\|\vec{v}\|}\vec{v}$ has length 1.

64. We say that two non-zero vectors $\vec{v}$ and $\vec{w}$ are **parallel** if they have same or opposite directions. That is, $\vec{v} \neq \vec{0}$ and $\vec{w} \neq \vec{0}$ are parallel if either $\hat{v} = \hat{w}$ or $\hat{v} = -\hat{w}$. Show that this means $\vec{v} = k\vec{w}$ for some non-zero scalar k and that $k > 0$ if the vectors have the same direction and $k < 0$ if they point in opposite directions.

65. The goal of this exercise is to use vectors to describe non-vertical lines in the plane. To that end, consider the line $y = 2x - 4$. Let $\vec{v}_0 = \langle 0, -4 \rangle$ and let $\vec{s} = \langle 1, 2 \rangle$. Let t be any real number. Show that the vector defined by $\vec{v} = \vec{v}_0 + t\vec{s}$, when drawn in standard position, has its terminal point on the line $y = 2x - 4$. (Hint: Show that $\vec{v}_0 + t\vec{s} = \langle t, 2t - 4 \rangle$ for any real number t.) Now consider the non-vertical line $y = mx + b$. Repeat the previous analysis with $\vec{v}_0 = \langle 0, b \rangle$ and let $\vec{s} = \langle 1, m \rangle$. Thus any non-vertical line can be thought of as a collection of terminal points of the vector sum of $\langle 0, b \rangle$ (the position vector of the y-intercept) and a scalar multiple of the slope vector $\vec{s} = \langle 1, m \rangle$.

66. Prove the associative and identity properties of vector addition in Theorem 11.18.

67. Prove the properties of scalar multiplication in Theorem 11.19.

11.8.2 Answers

1.
   - $\vec{v}+\vec{w}=\langle 15,-1\rangle$, vector
   - $\|\vec{v}+\vec{w}\|=\sqrt{226}$, scalar
   - $\|\vec{v}\|\vec{w}-\|\vec{w}\|\vec{v}=\langle -21,77\rangle$, vector
   - $\vec{w}-2\vec{v}=\langle -21,14\rangle$, vector
   - $\|\vec{v}\|+\|\vec{w}\|=18$, scalar
   - $\|w\|\hat{v}=\left\langle \frac{60}{13},-\frac{25}{13}\right\rangle$, vector
2.
   - $\vec{v}+\vec{w}=\langle -12,12\rangle$, vector
   - $\|\vec{v}+\vec{w}\|=12\sqrt{2}$, scalar
   - $\|\vec{v}\|\vec{w}-\|\vec{w}\|\vec{v}=\langle -34,-612\rangle$, vector
   - $\vec{w}-2\vec{v}=\langle 9,-60\rangle$, vector
   - $\|\vec{v}\|+\|\vec{w}\|=38$, scalar
   - $\|w\|\hat{v}=\left\langle -\frac{91}{25},\frac{312}{25}\right\rangle$, vector
3.
   - $\vec{v}+\vec{w}=\langle 0,3\rangle$, vector
   - $\|\vec{v}+\vec{w}\|=3$, scalar
   - $\|\vec{v}\|\vec{w}-\|\vec{w}\|\vec{v}=\left\langle -6\sqrt{5},6\sqrt{5}\right\rangle$, vector
   - $\vec{w}-2\vec{v}=\langle -6,6\rangle$, vector
   - $\|\vec{v}\|+\|\vec{w}\|=3\sqrt{5}$, scalar
   - $\|w\|\hat{v}=\langle 4,-2\rangle$, vector
4.
   - $\vec{v}+\vec{w}=\langle 8,9\rangle$, vector
   - $\|\vec{v}+\vec{w}\|=\sqrt{145}$, scalar
   - $\|\vec{v}\|\vec{w}-\|\vec{w}\|\vec{v}=\left\langle -14\sqrt{29},6\sqrt{29}\right\rangle$, vector
   - $\vec{w}-2\vec{v}=\langle -22,-3\rangle$, vector
   - $\|\vec{v}\|+\|\vec{w}\|=3\sqrt{29}$, scalar
   - $\|w\|\hat{v}=\langle 5,2\rangle$, vector
5.
   - $\vec{v}+\vec{w}=\left\langle \sqrt{3},3\right\rangle$, vector
   - $\|\vec{v}+\vec{w}\|=2\sqrt{3}$, scalar
   - $\|\vec{v}\|\vec{w}-\|\vec{w}\|\vec{v}=\left\langle 8\sqrt{3},0\right\rangle$, vector
   - $\vec{w}-2\vec{v}=\left\langle 4\sqrt{3},0\right\rangle$, vector
   - $\|\vec{v}\|+\|\vec{w}\|=6$, scalar
   - $\|w\|\hat{v}=\left\langle -2\sqrt{3},2\right\rangle$, vector
6.
   - $\vec{v}+\vec{w}=\left\langle -\frac{1}{5},\frac{7}{5}\right\rangle$, vector
   - $\|\vec{v}+\vec{w}\|=\sqrt{2}$, scalar
   - $\|\vec{v}\|\vec{w}-\|\vec{w}\|\vec{v}=\left\langle -\frac{7}{5},-\frac{1}{5}\right\rangle$, vector
   - $\vec{w}-2\vec{v}=\langle -2,-1\rangle$, vector
   - $\|\vec{v}\|+\|\vec{w}\|=2$, scalar
   - $\|w\|\hat{v}=\left\langle \frac{3}{5},\frac{4}{5}\right\rangle$, vector
7.
   - $\vec{v}+\vec{w}=\langle 0,0\rangle$, vector
   - $\|\vec{v}+\vec{w}\|=0$, scalar
   - $\|\vec{v}\|\vec{w}-\|\vec{w}\|\vec{v}=\left\langle -\sqrt{2},\sqrt{2}\right\rangle$, vector
   - $\vec{w}-2\vec{v}=\left\langle -\frac{3\sqrt{2}}{2},\frac{3\sqrt{2}}{2}\right\rangle$, vector
   - $\|\vec{v}\|+\|\vec{w}\|=2$, scalar
   - $\|w\|\hat{v}=\left\langle \frac{\sqrt{2}}{2},-\frac{\sqrt{2}}{2}\right\rangle$, vector

8.
- $\vec{v} + \vec{w} = \left\langle -\frac{1}{2}, -\frac{\sqrt{3}}{2} \right\rangle$, vector
- $\vec{w} - 2\vec{v} = \langle -2, -2\sqrt{3} \rangle$, vector
- $\|\vec{v} + \vec{w}\| = 1$, scalar
- $\|\vec{v}\| + \|\vec{w}\| = 3$, scalar
- $\|\vec{v}\|\vec{w} - \|\vec{w}\|\vec{v} = \langle -2, -2\sqrt{3} \rangle$, vector
- $\|w\|\hat{v} = \langle 1, \sqrt{3} \rangle$, vector

9.
- $\vec{v} + \vec{w} = \langle 3, 2 \rangle$, vector
- $\vec{w} - 2\vec{v} = \langle -6, -10 \rangle$, vector
- $\|\vec{v} + \vec{w}\| = \sqrt{13}$, scalar
- $\|\vec{v}\| + \|\vec{w}\| = 7$, scalar
- $\|\vec{v}\|\vec{w} - \|\vec{w}\|\vec{v} = \langle -6, -18 \rangle$, vector
- $\|w\|\hat{v} = \langle \frac{6}{5}, \frac{8}{5} \rangle$, vector

10.
- $\vec{v} + \vec{w} = \langle 1, 0 \rangle$, vector
- $\vec{w} - 2\vec{v} = \langle -\frac{1}{2}, -\frac{3}{2} \rangle$, vector
- $\|\vec{v} + \vec{w}\| = 1$, scalar
- $\|\vec{v}\| + \|\vec{w}\| = \sqrt{2}$, scalar
- $\|\vec{v}\|\vec{w} - \|\vec{w}\|\vec{v} = \left\langle 0, -\frac{\sqrt{2}}{2} \right\rangle$, vector
- $\|w\|\hat{v} = \langle \frac{1}{2}, \frac{1}{2} \rangle$, vector

11. $\vec{v} = \langle 3, 3\sqrt{3} \rangle$
12. $\vec{v} = \left\langle \frac{3\sqrt{2}}{2}, \frac{3\sqrt{2}}{2} \right\rangle$
13. $\vec{v} = \left\langle \frac{\sqrt{3}}{3}, \frac{1}{3} \right\rangle$
14. $\vec{v} = \langle 0, 12 \rangle$
15. $\vec{v} = \langle -2\sqrt{3}, 2 \rangle$
16. $\vec{v} = \langle -\sqrt{3}, 3 \rangle$
17. $\vec{v} = \langle -\frac{7}{2}, 0 \rangle$
18. $\vec{v} = \langle -5\sqrt{3}, -5\sqrt{3} \rangle$
19. $\vec{v} = \langle 0, -6.25 \rangle$
20. $\vec{v} = \langle 6, -2\sqrt{3} \rangle$
21. $\vec{v} = \langle 5, -5 \rangle$
22. $\vec{v} = \langle 2, 4 \rangle$
23. $\vec{v} = \langle -1, 3 \rangle$
24. $\vec{v} = \langle -3, -4 \rangle$
25. $\vec{v} = \langle 24, -10 \rangle$
26. $\vec{v} \approx \langle -177.96, 349.27 \rangle$
27. $\vec{v} \approx \langle 12.96, 62.59 \rangle$
28. $\vec{v} \approx \langle 5164.62, 1097.77 \rangle$
29. $\vec{v} \approx \langle -386.73, -230.08 \rangle$
30. $\vec{v} \approx \langle -52.13, -160.44 \rangle$
31. $\vec{v} \approx \langle 14.73, -21.43 \rangle$
32. $\|\vec{v}\| = 2$, $\theta = 60^\circ$
33. $\|\vec{v}\| = 5\sqrt{2}$, $\theta = 45^\circ$
34. $\|\vec{v}\| = 4$, $\theta = 150^\circ$
35. $\|\vec{v}\| = 2$, $\theta = 135^\circ$
36. $\|\vec{v}\| = 1$, $\theta = 225^\circ$
37. $\|\vec{v}\| = 1$, $\theta = 240^\circ$
38. $\|\vec{v}\| = 6$, $\theta = 0^\circ$
39. $\|\vec{v}\| = 2.5$, $\theta = 180^\circ$
40. $\|\vec{v}\| = \sqrt{7}$, $\theta = 90^\circ$
41. $\|\vec{v}\| = 10$, $\theta = 270^\circ$
42. $\|\vec{v}\| = 5$, $\theta \approx 53.13^\circ$
43. $\|\vec{v}\| = 13$, $\theta \approx 22.62^\circ$
44. $\|\vec{v}\| = 5$, $\theta \approx 143.13^\circ$
45. $\|\vec{v}\| = 25$, $\theta \approx 106.26^\circ$
46. $\|\vec{v}\| = \sqrt{5}$, $\theta \approx 206.57^\circ$

47. $\|\vec{v}\| = 2\sqrt{10}$, $\theta \approx 251.57^\circ$
48. $\|\vec{v}\| = \sqrt{2}$, $\theta \approx 45^\circ$
49. $\|\vec{v}\| = \sqrt{17}$, $\theta \approx 284.04^\circ$
50. $\|\vec{v}\| \approx 145.48$, $\theta \approx 328.02^\circ$
51. $\|\vec{v}\| \approx 1274.00$, $\theta \approx 40.75^\circ$
52. $\|\vec{v}\| \approx 121.69$, $\theta \approx 159.66^\circ$
53. The boat's true speed is about 10 miles per hour at a heading of S50.6°W.
54. The HMS Sasquatch's true speed is about 41 miles per hour at a heading of S26.8°E.
55. She should maintain a speed of about 35 miles per hour at a heading of S11.8°E.
56. She should fly at 83.46 miles per hour with a heading of N22.1°E
57. The current is moving at about 10 miles per hour bearing N54.6°W.
58. The tension on each of the cables is about 346 pounds.
59. The maximum weight that can be held by the cables in that configuration is about 133 pounds.
60. The tension on the left hand cable is 285.317 lbs. and on the right hand cable is 92.705 lbs.
61. The weaker student should pull about 60 pounds. The net force on the keg is about 153 pounds.
62. The resultant force is only about 296 pounds so the couch doesn't budge. Even if it did move, the stronger force on the third rope would have made the couch drift slightly to the south as it traveled down the street.

11.9 The Dot Product and Projection

In Section 11.8, we learned how add and subtract vectors and how to multiply vectors by scalars. In this section, we define a product of vectors. We begin with the following definition.

Definition 11.11. Suppose $\vec{v}$ and $\vec{w}$ are vectors whose component forms are $\vec{v} = \langle v_1, v_2 \rangle$ and $\vec{w} = \langle w_1, w_2 \rangle$. The **dot product** of $\vec{v}$ and $\vec{w}$ is given by

$$\vec{v} \cdot \vec{w} = \langle v_1, v_2 \rangle \cdot \langle w_1, w_2 \rangle = v_1 w_1 + v_2 w_2$$

For example, let $\vec{v} = \langle 3, 4 \rangle$ and $\vec{w} = \langle 1, -2 \rangle$. Then $\vec{v} \cdot \vec{w} = \langle 3, 4 \rangle \cdot \langle 1, -2 \rangle = (3)(1) + (4)(-2) = -5$. Note that the dot product takes two *vectors* and produces a *scalar*. For that reason, the quantity $\vec{v} \cdot \vec{w}$ is often called the **scalar product** of $\vec{v}$ and $\vec{w}$. The dot product enjoys the following properties.

Theorem 11.22. Properties of the Dot Product

- **Commutative Property:** For all vectors $\vec{v}$ and $\vec{w}$, $\vec{v} \cdot \vec{w} = \vec{w} \cdot \vec{v}$.
- **Distributive Property:** For all vectors $\vec{u}$, $\vec{v}$ and $\vec{w}$, $\vec{u} \cdot (\vec{v} + \vec{w}) = \vec{u} \cdot \vec{v} + \vec{u} \cdot \vec{w}$.
- **Scalar Property:** For all vectors $\vec{v}$ and $\vec{w}$ and scalars k, $(k\vec{v}) \cdot \vec{w} = k(\vec{v} \cdot \vec{w}) = \vec{v} \cdot (k\vec{w})$.
- **Relation to Magnitude:** For all vectors $\vec{v}$, $\vec{v} \cdot \vec{v} = \|\vec{v}\|^2$.

Like most of the theorems involving vectors, the proof of Theorem 11.22 amounts to using the definition of the dot product and properties of real number arithmetic. To show the commutative property for instance, let $\vec{v} = \langle v_1, v_2 \rangle$ and $\vec{w} = \langle w_1, w_2 \rangle$. Then

$$\begin{array}{rcll}
\vec{v} \cdot \vec{w} & = & \langle v_1, v_2 \rangle \cdot \langle w_1, w_2 \rangle & \\
& = & v_1 w_1 + v_2 w_2 & \text{Definition of Dot Product} \\
& = & w_1 v_1 + w_2 v_2 & \text{Commutativity of Real Number Multiplication} \\
& = & \langle w_1, w_2 \rangle \cdot \langle v_1, v_2 \rangle & \text{Definition of Dot Product} \\
& = & \vec{w} \cdot \vec{v} &
\end{array}$$

The distributive property is proved similarly and is left as an exercise.

For the scalar property, assume that $\vec{v} = \langle v_1, v_2 \rangle$ and $\vec{w} = \langle w_1, w_2 \rangle$ and k is a scalar. Then

$$\begin{array}{rcll}
(k\vec{v}) \cdot \vec{w} & = & (k \langle v_1, v_2 \rangle) \cdot \langle w_1, w_2 \rangle & \\
& = & \langle kv_1, kv_2 \rangle \cdot \langle w_1, w_2 \rangle & \text{Definition of Scalar Multiplication} \\
& = & (kv_1)(w_1) + (kv_2)(w_2) & \text{Definition of Dot Product} \\
& = & k(v_1 w_1) + k(v_2 w_2) & \text{Associativity of Real Number Multiplication} \\
& = & k(v_1 w_1 + v_2 w_2) & \text{Distributive Law of Real Numbers} \\
& = & k \langle v_1, v_2 \rangle \cdot \langle w_1, w_2 \rangle & \text{Definition of Dot Product} \\
& = & k(\vec{v} \cdot \vec{w}) &
\end{array}$$

We leave the proof of $k(\vec{v} \cdot \vec{w}) = \vec{v} \cdot (k\vec{w})$ as an exercise.

For the last property, we note that if $\vec{v} = \langle v_1, v_2 \rangle$, then $\vec{v} \cdot \vec{v} = \langle v_1, v_2 \rangle \cdot \langle v_1, v_2 \rangle = v_1^2 + v_2^2 = \|\vec{v}\|^2$, where the last equality comes courtesy of Definition 11.8.

The following example puts Theorem 11.22 to good use. As in Example 11.8.3, we work out the problem in great detail and encourage the reader to supply the justification for each step.

Example 11.9.1. Prove the identity: $\|\vec{v} - \vec{w}\|^2 = \|\vec{v}\|^2 - 2(\vec{v} \cdot \vec{w}) + \|\vec{w}\|^2$.

Solution. We begin by rewriting $\|\vec{v} - \vec{w}\|^2$ in terms of the dot product using Theorem 11.22.

$$\begin{aligned}
\|\vec{v} - \vec{w}\|^2 &= (\vec{v} - \vec{w}) \cdot (\vec{v} - \vec{w}) \\
&= (\vec{v} + [-\vec{w}]) \cdot (\vec{v} + [-\vec{w}]) \\
&= (\vec{v} + [-\vec{w}]) \cdot \vec{v} + (\vec{v} + [-\vec{w}]) \cdot [-\vec{w}] \\
&= \vec{v} \cdot (\vec{v} + [-\vec{w}]) + [-\vec{w}] \cdot (\vec{v} + [-\vec{w}]) \\
&= \vec{v} \cdot \vec{v} + \vec{v} \cdot [-\vec{w}] + [-\vec{w}] \cdot \vec{v} + [-\vec{w}] \cdot [-\vec{w}] \\
&= \vec{v} \cdot \vec{v} + \vec{v} \cdot [(-1)\vec{w}] + [(-1)\vec{w}] \cdot \vec{v} + [(-1)\vec{w}] \cdot [(-1)\vec{w}] \\
&= \vec{v} \cdot \vec{v} + (-1)(\vec{v} \cdot \vec{w}) + (-1)(\vec{w} \cdot \vec{v}) + [(-1)(-1)](\vec{w} \cdot \vec{w}) \\
&= \vec{v} \cdot \vec{v} + (-1)(\vec{v} \cdot \vec{w}) + (-1)(\vec{v} \cdot \vec{w}) + \vec{w} \cdot \vec{w} \\
&= \vec{v} \cdot \vec{v} - 2(\vec{v} \cdot \vec{w}) + \vec{w} \cdot \vec{w} \\
&= \|\vec{v}\|^2 - 2(\vec{v} \cdot \vec{w}) + \|\vec{w}\|^2
\end{aligned}$$

Hence, $\|\vec{v} - \vec{w}\|^2 = \|\vec{v}\|^2 - 2(\vec{v} \cdot \vec{w}) + \|\vec{w}\|^2$ as required. □

If we take a step back from the pedantry in Example 11.9.1, we see that the bulk of the work is needed to show that $(\vec{v}-\vec{w})\cdot(\vec{v}-\vec{w}) = \vec{v}\cdot\vec{v}-2(\vec{v}\cdot\vec{w})+\vec{w}\cdot\vec{w}$. If this looks familiar, it should. Since the dot product enjoys many of the same properties enjoyed by real numbers, the machinations required to expand $(\vec{v} - \vec{w}) \cdot (\vec{v} - \vec{w})$ for vectors $\vec{v}$ and $\vec{w}$ match those required to expand $(v - w)(v - w)$ for real numbers v and w, and hence we get similar looking results. The identity verified in Example 11.9.1 plays a large role in the development of the geometric properties of the dot product, which we now explore.

Suppose $\vec{v}$ and $\vec{w}$ are two nonzero vectors. If we draw $\vec{v}$ and $\vec{w}$ with the same initial point, we define the **angle between** $\vec{v}$ and $\vec{w}$ to be the angle θ determined by the rays containing the vectors $\vec{v}$ and $\vec{w}$, as illustrated below. We require $0 \leq \theta \leq \pi$. (Think about why this is needed in the definition.)

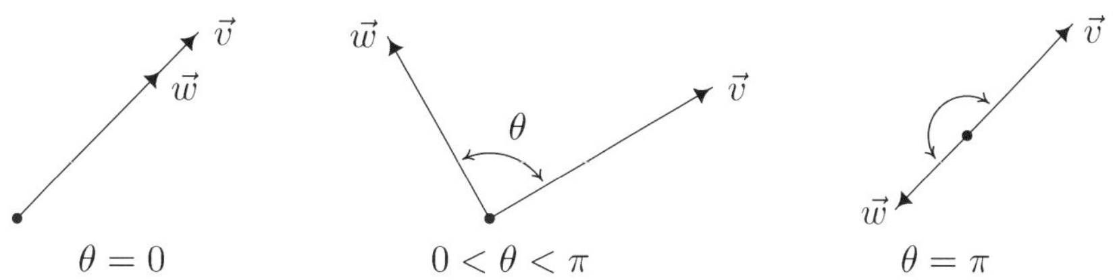

The following theorem gives us some insight into the geometric role the dot product plays.

Theorem 11.23. Geometric Interpretation of Dot Product: If $\vec{v}$ and $\vec{w}$ are nonzero vectors then $\vec{v} \cdot \vec{w} = \|\vec{v}\|\|\vec{w}\|\cos(\theta)$, where θ is the angle between $\vec{v}$ and $\vec{w}$.

We prove Theorem 11.23 in cases. If $\theta = 0$, then $\vec{v}$ and $\vec{w}$ have the same direction. It follows[1] that there is a real number $k > 0$ so that $\vec{w} = k\vec{v}$. Hence, $\vec{v} \cdot \vec{w} = \vec{v} \cdot (k\vec{v}) = k(\vec{v} \cdot \vec{v}) = k\|\vec{v}\|^2 = k\|\vec{v}\|\|\vec{v}\|$. Since $k > 0$, $k = |k|$, so $k\|\vec{v}\| = |k|\|\vec{v}\| = \|k\vec{v}\|$ by Theorem 11.20. Hence, $k\|\vec{v}\|\|\vec{v}\| = \|\vec{v}\|(k\|\vec{v}\|) = \|\vec{v}\|\|k\vec{v}\| = \|\vec{v}\|\|\vec{w}\|$. Since $\cos(0) = 1$, we get $\vec{v} \cdot \vec{w} = k\|\vec{v}\|\|\vec{v}\| = \|\vec{v}\|\|\vec{w}\| = \|\vec{v}\|\|\vec{w}\|\cos(0)$, proving that the formula holds for $\theta = 0$. If $\theta = \pi$, we repeat the argument with the difference being $\vec{w} = k\vec{v}$ where $k < 0$. In this case, $|k| = -k$, so $k\|\vec{v}\| = -|k|\|\vec{v}\| = -\|k\vec{v}\| = -\|\vec{w}\|$. Since $\cos(\pi) = -1$, we get $\vec{v} \cdot \vec{w} = -\|\vec{v}\|\|\vec{w}\| = \|\vec{v}\|\|\vec{w}\|\cos(\pi)$, as required. Next, if $0 < \theta < \pi$, the vectors $\vec{v}$, $\vec{w}$ and $\vec{v} - \vec{w}$ determine a triangle with side lengths $\|\vec{v}\|$, $\|\vec{w}\|$ and $\|\vec{v} - \vec{w}\|$, respectively, as seen below.

The Law of Cosines yields $\|\vec{v} - \vec{w}\|^2 = \|\vec{v}\|^2 + \|\vec{w}\|^2 - 2\|\vec{v}\|\|\vec{w}\|\cos(\theta)$. From Example 11.9.1, we know $\|\vec{v} - \vec{w}\|^2 = \|\vec{v}\|^2 - 2(\vec{v} \cdot \vec{w}) + \|\vec{w}\|^2$. Equating these two expressions for $\|\vec{v} - \vec{w}\|^2$ gives $\|\vec{v}\|^2 + \|\vec{w}\|^2 - 2\|\vec{v}\|\|\vec{w}\|\cos(\theta) = \|\vec{v}\|^2 - 2(\vec{v} \cdot \vec{w}) + \|\vec{w}\|^2$ which reduces to $-2\|\vec{v}\|\|\vec{w}\|\cos(\theta) = -2(\vec{v} \cdot \vec{w})$, or $\vec{v} \cdot \vec{w} = \|\vec{v}\|\|\vec{w}\|\cos(\theta)$, as required. An immediate consequence of Theorem 11.23 is the following.

Theorem 11.24. Let $\vec{v}$ and $\vec{w}$ be nonzero vectors and let θ the angle between $\vec{v}$ and $\vec{w}$. Then

$$\theta = \arccos\left(\frac{\vec{v} \cdot \vec{w}}{\|\vec{v}\|\|\vec{w}\|}\right) = \arccos(\hat{v} \cdot \hat{w})$$

We obtain the formula in Theorem 11.24 by solving the equation given in Theorem 11.23 for θ. Since $\vec{v}$ and $\vec{w}$ are nonzero, so are $\|\vec{v}\|$ and $\|\vec{w}\|$. Hence, we may divide both sides of $\vec{v} \cdot \vec{w} = \|\vec{v}\|\|\vec{w}\|\cos(\theta)$ by $\|\vec{v}\|\|\vec{w}\|$ to get $\cos(\theta) = \frac{\vec{v} \cdot \vec{w}}{\|\vec{v}\|\|\vec{w}\|}$. Since $0 \leq \theta \leq \pi$ by definition, the values of θ exactly match the range of the arccosine function. Hence, $\theta = \arccos\left(\frac{\vec{v} \cdot \vec{w}}{\|\vec{v}\|\|\vec{w}\|}\right)$. Using Theorem 11.22, we can rewrite $\frac{\vec{v} \cdot \vec{w}}{\|\vec{v}\|\|\vec{w}\|} = \left(\frac{1}{\|\vec{v}\|}\vec{v}\right) \cdot \left(\frac{1}{\|\vec{w}\|}\vec{w}\right) = \hat{v} \cdot \hat{w}$, giving us the alternative formula $\theta = \arccos(\hat{v} \cdot \hat{w})$.

We are overdue for an example.

Example 11.9.2. Find the angle between the following pairs of vectors.

1. $\vec{v} = \left\langle 3, -3\sqrt{3} \right\rangle$, and $\vec{w} = \left\langle -\sqrt{3}, 1 \right\rangle$
2. $\vec{v} = \langle 2, 2 \rangle$, and $\vec{w} = \langle 5, -5 \rangle$
3. $\vec{v} = \langle 3, -4 \rangle$, and $\vec{w} = \langle 2, 1 \rangle$

Solution. We use the formula $\theta = \arccos\left(\frac{\vec{v} \cdot \vec{w}}{\|\vec{v}\|\|\vec{w}\|}\right)$ from Theorem 11.24 in each case below.

[1] Since $\vec{v} = \|\vec{v}\|\hat{v}$ and $\vec{w} = \|\vec{w}\|\hat{w}$, if $\hat{v} = \hat{w}$ then $\vec{w} = \|\vec{w}\|\hat{v} = \frac{\|\vec{w}\|}{\|\vec{v}\|}(\|\vec{v}\|\hat{v}) = \frac{\|\vec{w}\|}{\|\vec{v}\|}\vec{v}$. In this case, $k = \frac{\|\vec{w}\|}{\|\vec{v}\|} > 0$.

1. We have $\vec{v}\cdot\vec{w} = \left\langle 3, -3\sqrt{3}\right\rangle \cdot \left\langle -\sqrt{3}, 1\right\rangle = -3\sqrt{3} - 3\sqrt{3} = -6\sqrt{3}$. Since $\|\vec{v}\| = \sqrt{3^2 + (-3\sqrt{3})^2} = \sqrt{36} = 6$ and $\|\vec{w}\| = \sqrt{(-\sqrt{3})^2 + 1^2} = \sqrt{4} = 2$, $\theta = \arccos\left(\frac{-6\sqrt{3}}{12}\right) = \arccos\left(-\frac{\sqrt{3}}{2}\right) = \frac{5\pi}{6}$.

2. For $\vec{v} = \langle 2, 2\rangle$ and $\vec{w} = \langle 5, -5\rangle$, we find $\vec{v}\cdot\vec{w} = \langle 2, 2\rangle \cdot \langle 5, -5\rangle = 10 - 10 = 0$. Hence, it doesn't matter what $\|\vec{v}\|$ and $\|\vec{w}\|$ are,[2] $\theta = \arccos\left(\frac{\vec{v}\cdot\vec{w}}{\|\vec{v}\|\|\vec{w}\|}\right) = \arccos(0) = \frac{\pi}{2}$.

3. We find $\vec{v}\cdot\vec{w} = \langle 3, -4\rangle \cdot \langle 2, 1\rangle = 6 - 4 = 2$. Also $\|\vec{v}\| = \sqrt{3^2 + (-4)^2} = \sqrt{25} = 5$ and $\vec{w} = \sqrt{2^2 + 1^2} = \sqrt{5}$, so $\theta = \arccos\left(\frac{2}{5\sqrt{5}}\right) = \arccos\left(\frac{2\sqrt{5}}{25}\right)$. Since $\frac{2\sqrt{5}}{25}$ isn't the cosine of one of the common angles, we leave our answer as $\theta = \arccos\left(\frac{2\sqrt{5}}{25}\right)$. □

The vectors $\vec{v} = \langle 2, 2\rangle$, and $\vec{w} = \langle 5, -5\rangle$ in Example 11.9.2 are called **orthogonal** and we write $\vec{v} \perp \vec{w}$, because the angle between them is $\frac{\pi}{2}$ radians $= 90^\circ$. Geometrically, when orthogonal vectors are sketched with the same initial point, the lines containing the vectors are perpendicular.

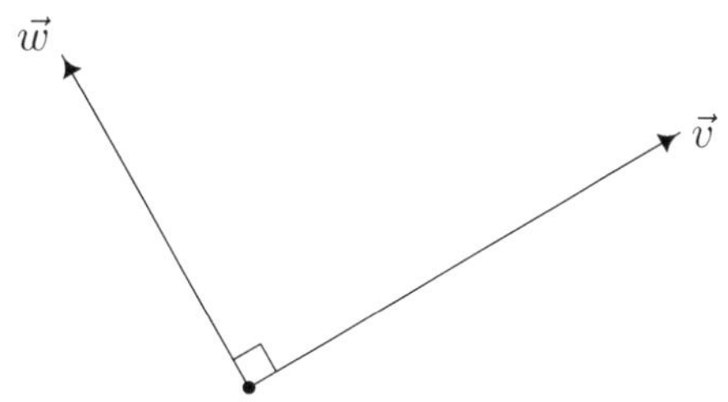

$\vec{v}$ and $\vec{w}$ are orthogonal, $\vec{v} \perp \vec{w}$

We state the relationship between orthogonal vectors and their dot product in the following theorem.

Theorem 11.25. The Dot Product Detects Orthogonality: Let $\vec{v}$ and $\vec{w}$ be nonzero vectors. Then $\vec{v} \perp \vec{w}$ if and only if $\vec{v}\cdot\vec{w} = 0$.

To prove Theorem 11.25, we first assume $\vec{v}$ and $\vec{w}$ are nonzero vectors with $\vec{v} \perp \vec{w}$. By definition, the angle between $\vec{v}$ and $\vec{w}$ is $\frac{\pi}{2}$. By Theorem 11.23, $\vec{v}\cdot\vec{w} = \|\vec{v}\|\|\vec{w}\|\cos\left(\frac{\pi}{2}\right) = 0$. Conversely, if $\vec{v}$ and $\vec{w}$ are nonzero vectors and $\vec{v}\cdot\vec{w} = 0$, then Theorem 11.24 gives $\theta = \arccos\left(\frac{\vec{v}\cdot\vec{w}}{\|\vec{v}\|\|\vec{w}\|}\right) = \arccos\left(\frac{0}{\|\vec{v}\|\|\vec{w}\|}\right) = \arccos(0) = \frac{\pi}{2}$, so $\vec{v} \perp \vec{w}$. We can use Theorem 11.25 in the following example to provide a different proof about the relationship between the slopes of perpendicular lines.[3]

Example 11.9.3. Let L_1 be the line $y = m_1x + b_1$ and let L_2 be the line $y = m_2x + b_2$. Prove that L_1 is perpendicular to L_2 if and only if $m_1 \cdot m_2 = -1$.

Solution. Our strategy is to find two vectors: $\vec{v_1}$, which has the same direction as L_1, and $\vec{v_2}$, which has the same direction as L_2 and show $\vec{v_1} \perp \vec{v_2}$ if and only if $m_1m_2 = -1$. To that end, we substitute $x = 0$ and $x = 1$ into $y = m_1x + b_1$ to find two points which lie on L_1, namely $P(0, b_1)$

[2] Note that there is no 'zero product property' for the dot product since neither $\vec{v}$ nor $\vec{w}$ is $\vec{0}$, yet $\vec{v}\cdot\vec{w} = 0$.
[3] See Exercise 2.1.1 in Section 2.1.

and $Q(1, m_1 + b_1)$. We let $\vec{v}_1 = \overrightarrow{PQ} = \langle 1 - 0, (m_1 + b_1) - b_1 \rangle = \langle 1, m_1 \rangle$, and note that since $\vec{v}_1$ is determined by two points on L_1, it may be viewed as lying on L_1. Hence it has the same direction as L_1. Similarly, we get the vector $\vec{v}_2 = \langle 1, m_2 \rangle$ which has the same direction as the line L_2. Hence, L_1 and L_2 are perpendicular if and only if $\vec{v}_1 \perp \vec{v}_2$. According to Theorem 11.25, $\vec{v}_1 \perp \vec{v}_2$ if and only if $\vec{v}_1 \cdot \vec{v}_2 = 0$. Notice that $\vec{v}_1 \cdot \vec{v}_2 = \langle 1, m_1 \rangle \cdot \langle 1, m_2 \rangle = 1 + m_1 m_2$. Hence, $\vec{v}_1 \cdot \vec{v}_2 = 0$ if and only if $1 + m_1 m_2 = 0$, which is true if and only if $m_1 m_2 = -1$, as required. □

While Theorem 11.25 certainly gives us some insight into what the dot product means geometrically, there is more to the story of the dot product. Consider the two nonzero vectors $\vec{v}$ and $\vec{w}$ drawn with a common initial point O below. For the moment, assume that the angle between $\vec{v}$ and $\vec{w}$, which we'll denote θ, is acute. We wish to develop a formula for the vector $\vec{p}$, indicated below, which is called the **orthogonal projection of $\vec{v}$ onto $\vec{w}$**. The vector $\vec{p}$ is obtained geometrically as follows: drop a perpendicular from the terminal point T of $\vec{v}$ to the vector $\vec{w}$ and call the point of intersection R. The vector $\vec{p}$ is then defined as $\vec{p} = \overrightarrow{OR}$. Like any vector, $\vec{p}$ is determined by its magnitude $\|\vec{p}\|$ and its direction $\hat{p}$ according to the formula $\vec{p} = \|\vec{p}\|\hat{p}$. Since we want $\hat{p}$ to have the same direction as $\vec{w}$, we have $\hat{p} = \hat{w}$. To determine $\|\vec{p}\|$, we make use of Theorem 10.4 as applied to the right triangle $\triangle ORT$. We find $\cos(\theta) = \frac{\|\vec{p}\|}{\|\vec{v}\|}$, or $\|\vec{p}\| = \|\vec{v}\| \cos(\theta)$. To get things in terms of just $\vec{v}$ and $\vec{w}$, we use Theorem 11.23 to get $\|\vec{p}\| = \|\vec{v}\| \cos(\theta) = \frac{\|\vec{v}\|\|\vec{w}\| \cos(\theta)}{\|\vec{w}\|} = \frac{\vec{v} \cdot \vec{w}}{\|\vec{w}\|}$. Using Theorem 11.22, we rewrite $\frac{\vec{v} \cdot \vec{w}}{\|\vec{w}\|} = \vec{v} \cdot \left(\frac{1}{\|\vec{w}\|} \vec{w} \right) = \vec{v} \cdot \hat{w}$. Hence, $\|\vec{p}\| = \vec{v} \cdot \hat{w}$, and since $\hat{p} = \hat{w}$, we now have a formula for $\vec{p}$ completely in terms of $\vec{v}$ and $\vec{w}$, namely $\vec{p} = \|\vec{p}\|\hat{p} = (\vec{v} \cdot \hat{w})\hat{w}$.

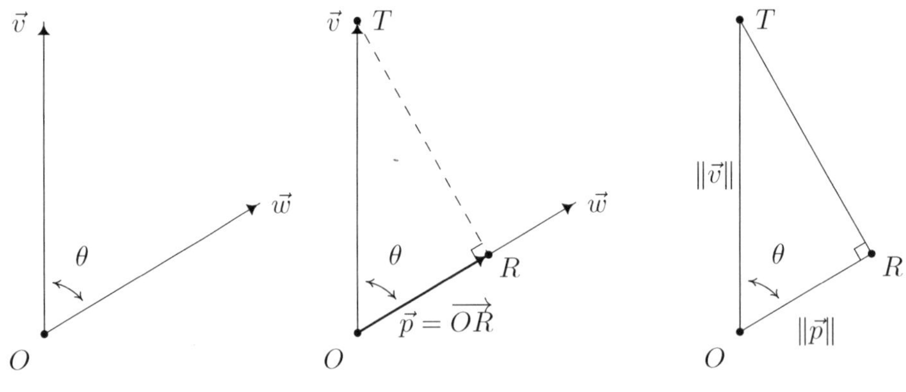

Now suppose that the angle θ between $\vec{v}$ and $\vec{w}$ is obtuse, and consider the diagram below. In this case, we see that $\hat{p} = -\hat{w}$ and using the triangle $\triangle ORT$, we find $\|\vec{p}\| = \|\vec{v}\| \cos(\theta')$. Since $\theta + \theta' = \pi$, it follows that $\cos(\theta') = -\cos(\theta)$, which means $\|\vec{p}\| = \|\vec{v}\| \cos(\theta') = -\|\vec{v}\| \cos(\theta)$. Rewriting this last equation in terms of $\vec{v}$ and $\vec{w}$ as before, we get $\|\vec{p}\| = -(\vec{v} \cdot \hat{w})$. Putting this together with $\hat{p} = -\hat{w}$, we get $\vec{p} = \|\vec{p}\|\hat{p} = -(\vec{v} \cdot \hat{w})(-\hat{w}) = (\vec{v} \cdot \hat{w})\hat{w}$ in this case as well.

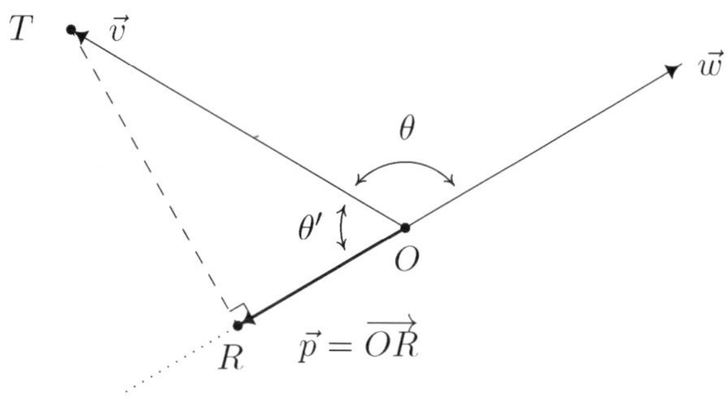

If the angle between $\vec{v}$ and $\vec{w}$ is $\frac{\pi}{2}$ then it is easy to show[4] that $\vec{p} = \vec{0}$. Since $\vec{v} \perp \vec{w}$ in this case, $\vec{v} \cdot \vec{w} = 0$. It follows that $\vec{v} \cdot \hat{w} = 0$ and $\vec{p} = \vec{0} = 0\hat{w} = (\vec{v} \cdot \hat{w})\hat{w}$ in this case, too. This gives us

Definition 11.12. Let $\vec{v}$ and $\vec{w}$ be nonzero vectors. The **orthogonal projection of $\vec{v}$ onto $\vec{w}$**, denoted $\text{proj}_{\vec{w}}(\vec{v})$ is given by $\text{proj}_{\vec{w}}(\vec{v}) = (\vec{v} \cdot \hat{w})\hat{w}$.

Definition 11.12 gives us a good idea what the dot product does. The scalar $\vec{v} \cdot \hat{w}$ is a measure of how much of the vector $\vec{v}$ is in the direction of the vector $\vec{w}$ and is thus called the scalar projection of $\vec{v}$ onto $\vec{w}$. While the formula given in Definition 11.12 is theoretically appealing, because of the presence of the normalized unit vector $\hat{w}$, computing the projection using the formula $\text{proj}_{\vec{w}}(\vec{v}) = (\vec{v} \cdot \hat{w})\hat{w}$ can be messy. We present two other formulas that are often used in practice.

Theorem 11.26. Alternate Formulas for Vector Projections: If $\vec{v}$ and $\vec{w}$ are nonzero vectors then

$$\text{proj}_{\vec{w}}(\vec{v}) = (\vec{v} \cdot \hat{w})\hat{w} = \left(\frac{\vec{v} \cdot \vec{w}}{\|\vec{w}\|^2}\right)\vec{w} = \left(\frac{\vec{v} \cdot \vec{w}}{\vec{w} \cdot \vec{w}}\right)\vec{w}$$

The proof of Theorem 11.26, which we leave to the reader as an exercise, amounts to using the formula $\hat{w} = \left(\frac{1}{\|\vec{w}\|}\right)\vec{w}$ and properties of the dot product. It is time for an example.

Example 11.9.4. Let $\vec{v} = \langle 1, 8 \rangle$ and $\vec{w} = \langle -1, 2 \rangle$. Find $\vec{p} = \text{proj}_{\vec{w}}(\vec{v})$, and plot $\vec{v}$, $\vec{w}$ and $\vec{p}$ in standard position.

Solution. We find $\vec{v} \cdot \vec{w} = \langle 1, 8 \rangle \cdot \langle -1, 2 \rangle = (-1) + 16 = 15$ and $\vec{w} \cdot \vec{w} = \langle -1, 2 \rangle \cdot \langle -1, 2 \rangle = 1 + 4 = 5$. Hence, $\vec{p} = \frac{\vec{v} \cdot \vec{w}}{\vec{w} \cdot \vec{w}}\vec{w} = \frac{15}{5}\langle -1, 2 \rangle = \langle -3, 6 \rangle$. We plot $\vec{v}$, $\vec{w}$ and $\vec{p}$ below.

[4]In this case, the point R coincides with the point O, so $\vec{p} = \overrightarrow{OR} = \overrightarrow{OO} = \vec{0}$.

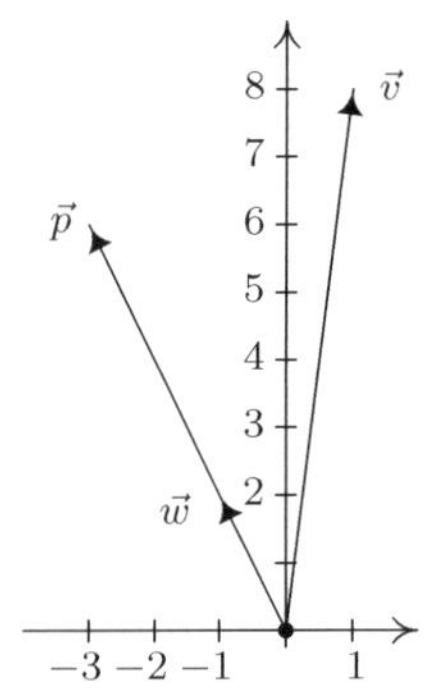

□

Suppose we wanted to verify that our answer $\vec{p}$ in Example 11.9.4 is indeed the orthogonal projection of $\vec{v}$ onto $\vec{w}$. We first note that since $\vec{p}$ is a scalar multiple of $\vec{w}$, it has the correct direction, so what remains to check is the orthogonality condition. Consider the vector $\vec{q}$ whose initial point is the terminal point of $\vec{p}$ and whose terminal point is the terminal point of $\vec{v}$.

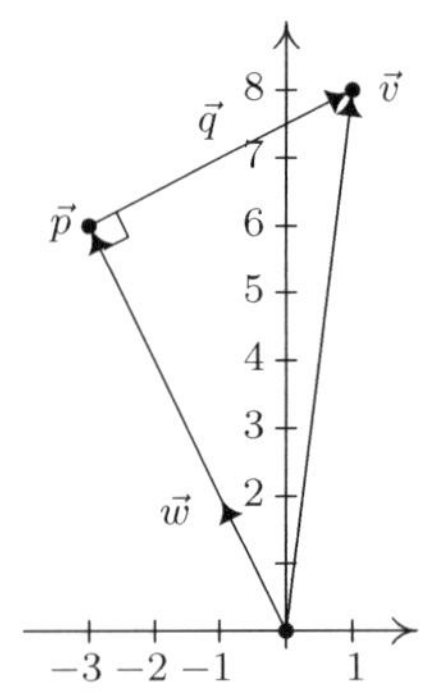

From the definition of vector arithmetic, $\vec{p}+\vec{q}=\vec{v}$, so that $\vec{q}=\vec{v}-\vec{p}$. In the case of Example 11.9.4, $\vec{v}=\langle 1,8\rangle$ and $\vec{p}=\langle -3,6\rangle$, so $\vec{q}=\langle 1,8\rangle-\langle -3,6\rangle=\langle 4,2\rangle$. Then $\vec{q}\cdot\vec{w}=\langle 4,2\rangle\cdot\langle -1,2\rangle=(-4)+4=0$, which shows $\vec{q}\perp\vec{w}$, as required. This result is generalized in the following theorem.

Theorem 11.27. Generalized Decomposition Theorem: Let $\vec{v}$ and $\vec{w}$ be nonzero vectors. There are unique vectors $\vec{p}$ and $\vec{q}$ such that $\vec{v}=\vec{p}+\vec{q}$ where $\vec{p}=k\vec{w}$ for some scalar k, and $\vec{q}\cdot\vec{w}=0$.

Note that if the vectors $\vec{p}$ and $\vec{q}$ in Theorem 11.27 are nonzero, then we can say $\vec{p}$ is *parallel*[5] to $\vec{w}$ and $\vec{q}$ is *orthogonal* to $\vec{w}$. In this case, the vector $\vec{p}$ is sometimes called the 'vector component of $\vec{v}$ parallel to $\vec{w}$' and $\vec{q}$ is called the 'vector component of $\vec{v}$ orthogonal to $\vec{w}$.' To prove Theorem 11.27, we take $\vec{p}=\text{proj}_{\vec{w}}(\vec{v})$ and $\vec{q}=\vec{v}-\vec{p}$. Then $\vec{p}$ is, by definition, a scalar multiple of $\vec{w}$. Next, we compute $\vec{q}\cdot\vec{w}$.

[5] See Exercise 64 in Section 11.8.

$$\begin{aligned}
\vec{q}\cdot\vec{w} &= (\vec{v}-\vec{p})\cdot\vec{w} && \text{Definition of } \vec{q}.\\
&= \vec{v}\cdot\vec{w}-\vec{p}\cdot\vec{w} && \text{Properties of Dot Product}\\
&= \vec{v}\cdot\vec{w}-\left(\frac{\vec{v}\cdot\vec{w}}{\vec{w}\cdot\vec{w}}\vec{w}\right)\cdot\vec{w} && \text{Since } \vec{p}=\text{proj}_{\vec{w}}(\vec{v}).\\
&= \vec{v}\cdot\vec{w}-\left(\frac{\vec{v}\cdot\vec{w}}{\vec{w}\cdot\vec{w}}\right)(\vec{w}\cdot\vec{w}) && \text{Properties of Dot Product.}\\
&= \vec{v}\cdot\vec{w}-\vec{v}\cdot\vec{w}\\
&= 0
\end{aligned}$$

Hence, $\vec{q}\cdot\vec{w}=0$, as required. At this point, we have shown that the vectors $\vec{p}$ and $\vec{q}$ guaranteed by Theorem 11.27 *exist*. Now we need to show that they are *unique*. Suppose $\vec{v}=\vec{p}+\vec{q}=\vec{p}\,'+\vec{q}\,'$ where the vectors $\vec{p}\,'$ and $\vec{q}\,'$ satisfy the same properties described in Theorem 11.27 as $\vec{p}$ and $\vec{q}$. Then $\vec{p}-\vec{p}\,'=\vec{q}\,'-\vec{q}$, so $\vec{w}\cdot(\vec{p}-\vec{p}\,')=\vec{w}\cdot(\vec{q}\,'-\vec{q})=\vec{w}\cdot\vec{q}\,'-\vec{w}\cdot\vec{q}=0-0=0$. Hence, $\vec{w}\cdot(\vec{p}-\vec{p}\,')=0$. Now there are scalars k and k' so that $\vec{p}=k\vec{w}$ and $\vec{p}\,'=k'\vec{w}$. This means $\vec{w}\cdot(\vec{p}-\vec{p}\,')=\vec{w}\cdot(k\vec{w}-k'\vec{w})=\vec{w}\cdot([k-k']\vec{w})=(k-k')(\vec{w}\cdot\vec{w})=(k-k')\|\vec{w}\|^2$. Since $\vec{w}\neq\vec{0}$, $\|\vec{w}\|^2\neq 0$, which means the only way $\vec{w}\cdot(\vec{p}-\vec{p}\,')=(k-k')\|\vec{w}\|^2=0$ is for $k-k'=0$, or $k=k'$. This means $\vec{p}=k\vec{w}=k'\vec{w}=\vec{p}\,'$. With $\vec{q}\,'-\vec{q}=\vec{p}-\vec{p}\,'=\vec{p}-\vec{p}=\vec{0}$, it must be that $\vec{q}\,'=\vec{q}$ as well. Hence, we have shown there is only one way to write $\vec{v}$ as a sum of vectors as described in Theorem 11.27.

We close this section with an application of the dot product. In Physics, if a constant force F is exerted over a distance d, the **work** W done by the force is given by $W=Fd$. Here, we assume the force is being applied in the direction of the motion. If the force applied is not in the direction of the motion, we can use the dot product to find the work done. Consider the scenario below where the constant force $\vec{F}$ is applied to move an object from the point P to the point Q.

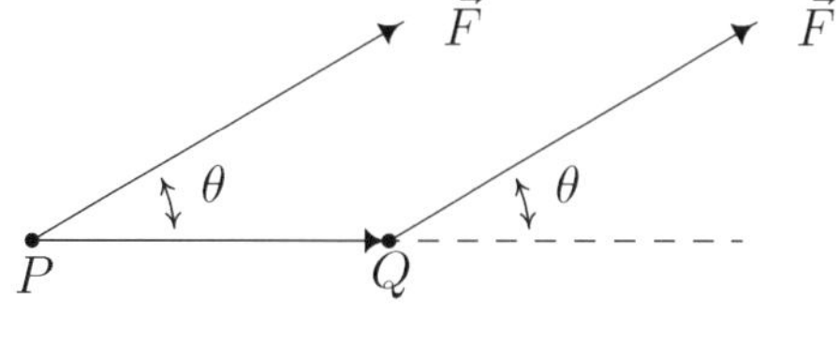

To find the work W done in this scenario, we need to find how much of the force $\vec{F}$ is in the direction of the motion $\overrightarrow{PQ}$. This is precisely what the dot product $\vec{F}\cdot\widehat{PQ}$ represents. Since the distance the object travels is $\|\overrightarrow{PQ}\|$, we get $W=(\vec{F}\cdot\widehat{PQ})\|\overrightarrow{PQ}\|$. Since $\overrightarrow{PQ}=\|\overrightarrow{PQ}\|\widehat{PQ}$, $W=(\vec{F}\cdot\widehat{PQ})\|\overrightarrow{PQ}\|=\vec{F}\cdot(\|\overrightarrow{PQ}\|\widehat{PQ})=\vec{F}\cdot\overrightarrow{PQ}=\|\vec{F}\|\|\overrightarrow{PQ}\|\cos(\theta)$, where θ is the angle between the applied force $\vec{F}$ and the trajectory of the motion $\overrightarrow{PQ}$. We have proved the following.

Theorem 11.28. Work as a Dot Product: Suppose a constant force $\vec{F}$ is applied along the vector $\overrightarrow{PQ}$. The work W done by $\vec{F}$ is given by

$$W = \vec{F} \cdot \overrightarrow{PQ} = \|\vec{F}\| \|\overrightarrow{PQ}\| \cos(\theta),$$

where θ is the angle between $\vec{F}$ and $\overrightarrow{PQ}$.

Example 11.9.5. Taylor exerts a force of 10 pounds to pull her wagon a distance of 50 feet over level ground. If the handle of the wagon makes a 30° angle with the horizontal, how much work did Taylor do pulling the wagon? Assume Taylor exerts the force of 10 pounds at a 30° angle for the duration of the 50 feet.

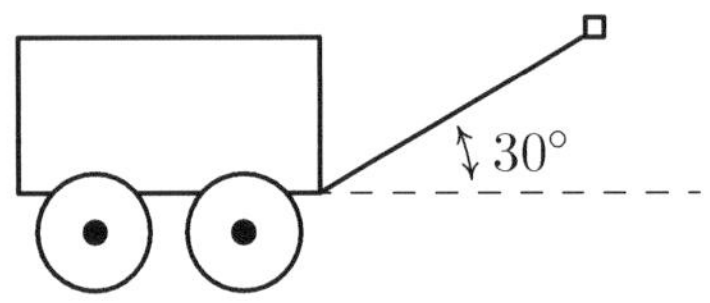

Solution. There are two ways to attack this problem. One way is to find the vectors $\vec{F}$ and $\overrightarrow{PQ}$ mentioned in Theorem 11.28 and compute $W = \vec{F} \cdot \overrightarrow{PQ}$. To do this, we assume the origin is at the point where the handle of the wagon meets the wagon and the positive x-axis lies along the dashed line in the figure above. Since the force applied is a constant 10 pounds, we have $\|\vec{F}\| = 10$. Since it is being applied at a constant angle of $\theta = 30^\circ$ with respect to the positive x-axis, Definition 11.8 gives us $\vec{F} = 10 \left\langle \cos(30^\circ), \sin(30^\circ) \right\rangle = \left\langle 5\sqrt{3}, 5 \right\rangle$. Since the wagon is being pulled along 50 feet in the positive direction, the displacement vector is $\overrightarrow{PQ} = 50\hat{\imath} = 50 \left\langle 1, 0 \right\rangle = \left\langle 50, 0 \right\rangle$. We get $W = \vec{F} \cdot \overrightarrow{PQ} = \left\langle 5\sqrt{3}, 5 \right\rangle \cdot \left\langle 50, 0 \right\rangle = 250\sqrt{3}$. Since force is measured in pounds and distance is measured in feet, we get $W = 250\sqrt{3}$ foot-pounds. Alternatively, we can use the formulation $W = \|\vec{F}\| \|\overrightarrow{PQ}\| \cos(\theta)$ to get $W = (10\,\text{pounds})(50\,\text{feet}) \cos(30^\circ) = 250\sqrt{3}$ foot-pounds of work. □

11.9.1 Exercises

In Exercises 1 - 20, use the pair of vectors $\vec{v}$ and $\vec{w}$ to find the following quantities.

- $\vec{v} \cdot \vec{w}$
- The angle θ (in degrees) between $\vec{v}$ and $\vec{w}$
- $\text{proj}_{\vec{w}}(\vec{v})$
- $\vec{q} = \vec{v} - \text{proj}_{\vec{w}}(\vec{v})$ (Show that $\vec{q} \cdot \vec{w} = 0$.)

1. $\vec{v} = \langle -2, -7 \rangle$ and $\vec{w} = \langle 5, -9 \rangle$
2. $\vec{v} = \langle -6, -5 \rangle$ and $\vec{w} = \langle 10, -12 \rangle$
3. $\vec{v} = \langle 1, \sqrt{3} \rangle$ and $\vec{w} = \langle 1, -\sqrt{3} \rangle$
4. $\vec{v} = \langle 3, 4 \rangle$ and $\vec{w} = \langle -6, -8 \rangle$
5. $\vec{v} = \langle -2, 1 \rangle$ and $\vec{w} = \langle 3, 6 \rangle$
6. $\vec{v} = \langle -3\sqrt{3}, 3 \rangle$ and $\vec{w} = \langle -\sqrt{3}, -1 \rangle$
7. $\vec{v} = \langle 1, 17 \rangle$ and $\vec{w} = \langle -1, 0 \rangle$
8. $\vec{v} = \langle 3, 4 \rangle$ and $\vec{w} = \langle 5, 12 \rangle$
9. $\vec{v} = \langle -4, -2 \rangle$ and $\vec{w} = \langle 1, -5 \rangle$
10. $\vec{v} = \langle -5, 6 \rangle$ and $\vec{w} = \langle 4, -7 \rangle$
11. $\vec{v} = \langle -8, 3 \rangle$ and $\vec{w} = \langle 2, 6 \rangle$
12. $\vec{v} = \langle 34, -91 \rangle$ and $\vec{w} = \langle 0, 1 \rangle$
13. $\vec{v} = 3\hat{\imath} - \hat{\jmath}$ and $\vec{w} = 4\hat{\jmath}$
14. $\vec{v} = -24\hat{\imath} + 7\hat{\jmath}$ and $\vec{w} = 2\hat{\imath}$
15. $\vec{v} = \frac{3}{2}\hat{\imath} + \frac{3}{2}\hat{\jmath}$ and $\vec{w} = \hat{\imath} - \hat{\jmath}$
16. $\vec{v} = 5\hat{\imath} + 12\hat{\jmath}$ and $\vec{w} = -3\hat{\imath} + 4\hat{\jmath}$
17. $\vec{v} = \left\langle \frac{1}{2}, \frac{\sqrt{3}}{2} \right\rangle$ and $\vec{w} = \left\langle -\frac{\sqrt{2}}{2}, \frac{\sqrt{2}}{2} \right\rangle$
18. $\vec{v} = \left\langle \frac{\sqrt{2}}{2}, \frac{\sqrt{2}}{2} \right\rangle$ and $\vec{w} = \left\langle \frac{1}{2}, -\frac{\sqrt{3}}{2} \right\rangle$
19. $\vec{v} = \left\langle \frac{\sqrt{3}}{2}, \frac{1}{2} \right\rangle$ and $\vec{w} = \left\langle -\frac{\sqrt{2}}{2}, -\frac{\sqrt{2}}{2} \right\rangle$
20. $\vec{v} = \left\langle \frac{1}{2}, -\frac{\sqrt{3}}{2} \right\rangle$ and $\vec{w} = \left\langle \frac{\sqrt{2}}{2}, -\frac{\sqrt{2}}{2} \right\rangle$
21. A force of 1500 pounds is required to tow a trailer. Find the work done towing the trailer along a flat stretch of road 300 feet. Assume the force is applied in the direction of the motion.
22. Find the work done lifting a 10 pound book 3 feet straight up into the air. Assume the force of gravity is acting straight downwards.
23. Suppose Taylor fills her wagon with rocks and must exert a force of 13 pounds to pull her wagon across the yard. If she maintains a 15° angle between the handle of the wagon and the horizontal, compute how much work Taylor does pulling her wagon 25 feet. Round your answer to two decimal places.
24. In Exercise 61 in Section 11.8, two drunken college students have filled an empty beer keg with rocks which they drag down the street by pulling on two attached ropes. The stronger of the two students pulls with a force of 100 pounds on a rope which makes a 13° angle with the direction of motion. (In this case, the keg was being pulled due east and the student's heading was N77°E.) Find the work done by this student if the keg is dragged 42 feet.

25. Find the work done pushing a 200 pound barrel 10 feet up a 12.5° incline. Ignore all forces acting on the barrel except gravity, which acts downwards. Round your answer to two decimal places.

    HINT: Since you are working to overcome gravity only, the force being applied acts directly upwards. This means that the angle between the applied force in this case and the motion of the object is *not* the 12.5° of the incline!

26. Prove the distributive property of the dot product in Theorem 11.22.

27. Finish the proof of the scalar property of the dot product in Theorem 11.22.

28. Use the identity in Example 11.9.1 to prove the **Parallelogram Law**

$$\|\vec{v}\|^2 + \|\vec{w}\|^2 = \frac{1}{2}\left[\|\vec{v}+\vec{w}\|^2 + \|\vec{v}-\vec{w}\|^2\right]$$

29. We know that $|x+y| \leq |x| + |y|$ for all real numbers x and y by the Triangle Inequality established in Exercise 36 in Section 2.2. We can now establish a Triangle Inequality for vectors. In this exercise, we prove that $\|\vec{u}+\vec{v}\| \leq \|\vec{u}\| + \|\vec{v}\|$ for all pairs of vectors $\vec{u}$ and $\vec{v}$.

    (a) (Step 1) Show that $\|\vec{u}+\vec{v}\|^2 = \|\vec{u}\|^2 + 2\vec{u}\cdot\vec{v} + \|\vec{v}\|^2$.

    (b) (Step 2) Show that $|\vec{u}\cdot\vec{v}| \leq \|\vec{u}\|\|\vec{v}\|$. This is the celebrated Cauchy-Schwarz Inequality.[6] (Hint: To show this inequality, start with the fact that $|\vec{u}\cdot\vec{v}| = |\ \|\vec{u}\|\|\vec{v}\|\cos(\theta)\ |$ and use the fact that $|\cos(\theta)| \leq 1$ for all θ.)

    (c) (Step 3) Show that $\|\vec{u}+\vec{v}\|^2 = \|\vec{u}\|^2 + 2\vec{u}\cdot\vec{v} + \|\vec{v}\|^2 \leq \|\vec{u}\|^2 + 2|\vec{u}\cdot\vec{v}| + \|\vec{v}\|^2 \leq \|\vec{u}\|^2 + 2\|\vec{u}\|\|\vec{v}\| + \|\vec{v}\|^2 = (\|\vec{u}\| + \|\vec{v}\|)^2$.

    (d) (Step 4) Use Step 3 to show that $\|\vec{u}+\vec{v}\| \leq \|\vec{u}\| + \|\vec{v}\|$ for all pairs of vectors $\vec{u}$ and $\vec{v}$.

    (e) As an added bonus, we can now show that the Triangle Inequality $|z+w| \leq |z| + |w|$ holds for all complex numbers z and w as well. Identify the complex number $z = a + bi$ with the vector $u = \langle a, b\rangle$ and identify the complex number $w = c + di$ with the vector $v = \langle c, d\rangle$ and just follow your nose!

[6] It is also known by other names. Check out this site for details.

11.9.2 Answers

1. $\vec{v} = \langle -2, -7 \rangle$ and $\vec{w} = \langle 5, -9 \rangle$
   $\vec{v} \cdot \vec{w} = 53$
   $\theta = 45^\circ$
   $\text{proj}_{\vec{w}}(\vec{v}) = \left\langle \frac{5}{2}, -\frac{9}{2} \right\rangle$
   $\vec{q} = \left\langle -\frac{9}{2}, -\frac{5}{2} \right\rangle$

2. $\vec{v} = \langle -6, -5 \rangle$ and $\vec{w} = \langle 10, -12 \rangle$
   $\vec{v} \cdot \vec{w} = 0$
   $\theta = 90^\circ$
   $\text{proj}_{\vec{w}}(\vec{v}) = \langle 0, 0 \rangle$
   $\vec{q} = \langle -6, -5 \rangle$

3. $\vec{v} = \langle 1, \sqrt{3} \rangle$ and $\vec{w} = \langle 1, -\sqrt{3} \rangle$
   $\vec{v} \cdot \vec{w} = -2$
   $\theta = 120^\circ$
   $\text{proj}_{\vec{w}}(\vec{v}) = \left\langle -\frac{1}{2}, \frac{\sqrt{3}}{2} \right\rangle$
   $\vec{q} = \left\langle \frac{3}{2}, \frac{\sqrt{3}}{2} \right\rangle$

4. $\vec{v} = \langle 3, 4 \rangle$ and $\vec{w} = \langle -6, -8 \rangle$
   $\vec{v} \cdot \vec{w} = -50$
   $\theta = 180^\circ$
   $\text{proj}_{\vec{w}}(\vec{v}) = \langle 3, 4 \rangle$
   $\vec{q} = \langle 0, 0 \rangle$

5. $\vec{v} = \langle -2, 1 \rangle$ and $\vec{w} = \langle 3, 6 \rangle$
   $\vec{v} \cdot \vec{w} = 0$
   $\theta = 90^\circ$
   $\text{proj}_{\vec{w}}(\vec{v}) = \langle 0, 0 \rangle$
   $\vec{q} = \langle -2, 1 \rangle$

6. $\vec{v} = \langle -3\sqrt{3}, 3 \rangle$ and $\vec{w} = \langle -\sqrt{3}, -1 \rangle$
   $\vec{v} \cdot \vec{w} = 6$
   $\theta = 60^\circ$
   $\text{proj}_{\vec{w}}(\vec{v}) = \left\langle -\frac{3\sqrt{3}}{2}, -\frac{3}{2} \right\rangle$
   $\vec{q} = \left\langle -\frac{3\sqrt{3}}{2}, \frac{9}{2} \right\rangle$

7. $\vec{v} = \langle 1, 17 \rangle$ and $\vec{w} = \langle -1, 0 \rangle$
   $\vec{v} \cdot \vec{w} = -1$
   $\theta \approx 93.37^\circ$
   $\text{proj}_{\vec{w}}(\vec{v}) = \langle 1, 0 \rangle$
   $\vec{q} = \langle 0, 17 \rangle$

8. $\vec{v} = \langle 3, 4 \rangle$ and $\vec{w} = \langle 5, 12 \rangle$
   $\vec{v} \cdot \vec{w} = 63$
   $\theta \approx 14.25^\circ$
   $\text{proj}_{\vec{w}}(\vec{v}) = \left\langle \frac{315}{169}, \frac{756}{169} \right\rangle$
   $\vec{q} = \left\langle \frac{192}{169}, -\frac{80}{169} \right\rangle$

9. $\vec{v} = \langle -4, -2 \rangle$ and $\vec{w} = \langle 1, -5 \rangle$
   $\vec{v} \cdot \vec{w} = 6$
   $\theta \approx 74.74^\circ$
   $\text{proj}_{\vec{w}}(\vec{v}) = \left\langle \frac{3}{13}, -\frac{15}{13} \right\rangle$
   $\vec{q} = \left\langle -\frac{55}{13}, -\frac{11}{13} \right\rangle$

10. $\vec{v} = \langle -5, 6 \rangle$ and $\vec{w} = \langle 4, -7 \rangle$
    $\vec{v} \cdot \vec{w} = -62$
    $\theta \approx 169.94^\circ$
    $\text{proj}_{\vec{w}}(\vec{v}) = \left\langle -\frac{248}{65}, \frac{434}{65} \right\rangle$
    $\vec{q} = \left\langle -\frac{77}{65}, -\frac{44}{65} \right\rangle$

11. $\vec{v} = \langle -8, 3 \rangle$ and $\vec{w} = \langle 2, 6 \rangle$
$\vec{v} \cdot \vec{w} = 2$
$\theta \approx 87.88^\circ$
$\text{proj}_{\vec{w}}(\vec{v}) = \left\langle \frac{1}{10}, \frac{3}{10} \right\rangle$
$\vec{q} = \left\langle -\frac{81}{10}, \frac{27}{10} \right\rangle$

12. $\vec{v} = \langle 34, -91 \rangle$ and $\vec{w} = \langle 0, 1 \rangle$
$\vec{v} \cdot \vec{w} = -91$
$\theta \approx 159.51^\circ$
$\text{proj}_{\vec{w}}(\vec{v}) = \langle 0, -91 \rangle$
$\vec{q} = \langle 34, 0 \rangle$

13. $\vec{v} = 3\hat{\imath} - \hat{\jmath}$ and $\vec{w} = 4\hat{\jmath}$
$\vec{v} \cdot \vec{w} = -4$
$\theta \approx 108.43^\circ$
$\text{proj}_{\vec{w}}(\vec{v}) = \langle 0, -1 \rangle$
$\vec{q} = \langle 3, 0 \rangle$

14. $\vec{v} = -24\hat{\imath} + 7\hat{\jmath}$ and $\vec{w} = 2\hat{\imath}$
$\vec{v} \cdot \vec{w} = -48$
$\theta \approx 163.74^\circ$
$\text{proj}_{\vec{w}}(\vec{v}) = \langle -24, 0 \rangle$
$\vec{q} = \langle 0, 7 \rangle$

15. $\vec{v} = \frac{3}{2}\hat{\imath} + \frac{3}{2}\hat{\jmath}$ and $\vec{w} = \hat{\imath} - \hat{\jmath}$
$\vec{v} \cdot \vec{w} = 0$
$\theta = 90^\circ$
$\text{proj}_{\vec{w}}(\vec{v}) = \langle 0, 0 \rangle$
$\vec{q} = \left\langle \frac{3}{2}, \frac{3}{2} \right\rangle$

16. $\vec{v} = 5\hat{\imath} + 12\hat{\jmath}$ and $\vec{w} = -3\hat{\imath} + 4\hat{\jmath}$
$\vec{v} \cdot \vec{w} = 33$
$\theta \approx 59.49^\circ$
$\text{proj}_{\vec{w}}(\vec{v}) = \left\langle -\frac{99}{25}, \frac{132}{25} \right\rangle$
$\vec{q} = \left\langle \frac{224}{25}, \frac{168}{25} \right\rangle$

17. $\vec{v} = \left\langle \frac{1}{2}, \frac{\sqrt{3}}{2} \right\rangle$ and $\vec{w} = \left\langle -\frac{\sqrt{2}}{2}, \frac{\sqrt{2}}{2} \right\rangle$
$\vec{v} \cdot \vec{w} = \frac{\sqrt{6}-\sqrt{2}}{4}$
$\theta = 75^\circ$
$\text{proj}_{\vec{w}}(\vec{v}) = \left\langle \frac{1-\sqrt{3}}{4}, \frac{\sqrt{3}-1}{4} \right\rangle$
$\vec{q} = \left\langle \frac{1+\sqrt{3}}{4}, \frac{1+\sqrt{3}}{4} \right\rangle$

18. $\vec{v} = \left\langle \frac{\sqrt{2}}{2}, \frac{\sqrt{2}}{2} \right\rangle$ and $\vec{w} = \left\langle \frac{1}{2}, -\frac{\sqrt{3}}{2} \right\rangle$
$\vec{v} \cdot \vec{w} = \frac{\sqrt{2}-\sqrt{6}}{4}$
$\theta = 105^\circ$
$\text{proj}_{\vec{w}}(\vec{v}) = \left\langle \frac{\sqrt{2}-\sqrt{6}}{8}, \frac{3\sqrt{2}-\sqrt{6}}{8} \right\rangle$
$\vec{q} = \left\langle \frac{3\sqrt{2}+\sqrt{6}}{8}, \frac{\sqrt{2}+\sqrt{6}}{8} \right\rangle$

19. $\vec{v} = \left\langle \frac{\sqrt{3}}{2}, \frac{1}{2} \right\rangle$ and $\vec{w} = \left\langle -\frac{\sqrt{2}}{2}, -\frac{\sqrt{2}}{2} \right\rangle$
$\vec{v} \cdot \vec{w} = -\frac{\sqrt{6}+\sqrt{2}}{4}$
$\theta = 165^\circ$
$\text{proj}_{\vec{w}}(\vec{v}) = \left\langle \frac{\sqrt{3}+1}{4}, \frac{\sqrt{3}+1}{4} \right\rangle$
$\vec{q} = \left\langle \frac{\sqrt{3}-1}{4}, \frac{1-\sqrt{3}}{4} \right\rangle$

20. $\vec{v} = \left\langle \frac{1}{2}, -\frac{\sqrt{3}}{2} \right\rangle$ and $\vec{w} = \left\langle \frac{\sqrt{2}}{2}, -\frac{\sqrt{2}}{2} \right\rangle$
$\vec{v} \cdot \vec{w} = \frac{\sqrt{6}+\sqrt{2}}{4}$
$\theta = 15^\circ$
$\text{proj}_{\vec{w}}(\vec{v}) = \left\langle \frac{\sqrt{3}+1}{4}, -\frac{\sqrt{3}+1}{4} \right\rangle$
$\vec{q} = \left\langle \frac{1-\sqrt{3}}{4}, \frac{1-\sqrt{3}}{4} \right\rangle$

21. $(1500\,\text{pounds})(300\,\text{feet})\cos(0^\circ) = 450,000$ foot-pounds

22. $(10\,\text{pounds})(3\,\text{feet})\cos(0^\circ) = 30$ foot-pounds

23. (13 pounds)(25 feet) $\cos(15^\circ) \approx 313.92$ foot-pounds

24. (100 pounds)(42 feet) $\cos(13^\circ) \approx 4092.35$ foot-pounds

25. (200 pounds)(10 feet) $\cos(77.5^\circ) \approx 432.88$ foot-pounds

11.10 PARAMETRIC EQUATIONS

As we have seen in Exercises 53 - 56 in Section 1.2, Chapter 7 and most recently in Section 11.5, there are scores of interesting curves which, when plotted in the xy-plane, neither represent y as a function of x nor x as a function of y. In this section, we present a new concept which allows us to use functions to study these kinds of curves. To motivate the idea, we imagine a bug crawling across a table top starting at the point O and tracing out a curve C in the plane, as shown below.

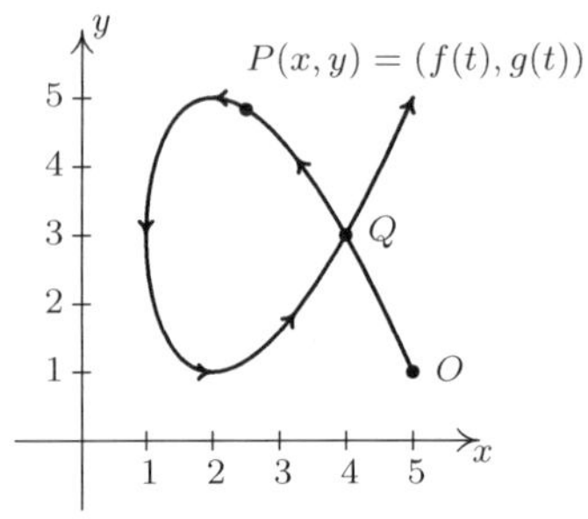

The curve C does not represent y as a function of x because it fails the Vertical Line Test and it does not represent x as a function of y because it fails the Horizontal Line Test. However, since the bug can be in only one *place* $P(x, y)$ at any given time t, we can define the x-coordinate of P as a function of t and the y-coordinate of P as a (usually, but not necessarily) different function of t. (Traditionally, $f(t)$ is used for x and $g(t)$ is used for y.) The independent variable t in this case is called a **parameter** and the system of equations

$$\left\{ \begin{array}{rcl} x & = & f(t) \\ y & = & g(t) \end{array} \right.$$

is called a **system of parametric equations** or a **parametrization** of the curve C.[1] The parametrization of C endows it with an *orientation* and the arrows on C indicate motion in the direction of increasing values of t. In this case, our bug starts at the point O, travels upwards to the left, then loops back around to cross its path[2] at the point Q and finally heads off into the first quadrant. It is important to note that the curve itself is a set of points and as such is devoid of any orientation. The parametrization determines the orientation and as we shall see, different parametrizations can determine different orientations. If all of this seems hauntingly familiar, it should. By definition, the system of equations $\{x = \cos(t),\ y = \sin(t)$ parametrizes the Unit Circle, giving it a counter-clockwise orientation. More generally, the equations of circular motion $\{x = r\cos(\omega t),\ y = r\sin(\omega t)$ developed on page 732 in Section 10.2.1 are parametric equations which trace out a circle of radius r centered at the origin. If $\omega > 0$, the orientation is counter-clockwise; if $\omega < 0$, the orientation is clockwise. The angular frequency ω determines 'how fast' the

[1] Note the use of the indefinite article 'a'. As we shall see, there are infinitely many different parametric representations for any given curve.

[2] Here, the bug reaches the point Q at two different times. While this does not contradict our claim that $f(t)$ and $g(t)$ are *functions* of t, it shows that neither f nor g can be one-to-one. (Think about this before reading on.)

object moves around the circle. In particular, the equations $\left\{x = 2960\cos\left(\frac{\pi}{12}t\right),\ y = 2960\sin\left(\frac{\pi}{12}t\right)\right.$ that model the motion of Lakeland Community College as the earth rotates (see Example 10.2.7 in Section 10.2) parameterize a circle of radius 2960 with a counter-clockwise rotation which completes one revolution as t runs through the interval $[0, 24)$. It is time for another example.

Example 11.10.1. Sketch the curve described by $\left\{\begin{array}{rcl} x & = & t^2 - 3 \\ y & = & 2t - 1 \end{array}\right.$ for $t \geq -2$.

Solution. We follow the same procedure here as we have time and time again when asked to graph anything new – choose friendly values of t, plot the corresponding points and connect the results in a pleasing fashion. Since we are told $t \geq -2$, we start there and as we plot successive points, we draw an arrow to indicate the direction of the path for increasing values of t.

t	$x(t)$	$y(t)$	$(x(t), y(t))$
-2	1	-5	$(1, -5)$
-1	-2	-3	$(-2, -3)$
0	-3	-1	$(-3, -1)$
1	-2	1	$(-2, 1)$
2	1	3	$(1, 3)$
3	6	5	$(6, 5)$

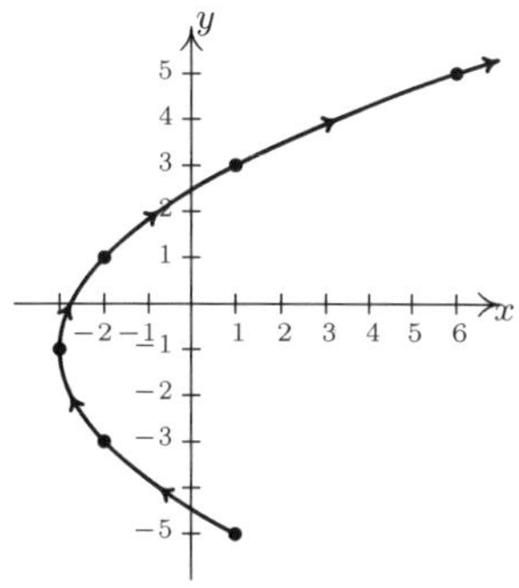

□

The curve sketched out in Example 11.10.1 certainly looks like a parabola, and the presence of the t^2 term in the equation $x = t^2 - 3$ reinforces this hunch. Since the parametric equations $\left\{x = t^2 - 3,\ y = 2t - 1\right.$ given to describe this curve are a *system* of equations, we can use the technique of substitution as described in Section 8.7 to eliminate the parameter t and get an equation involving just x and y. To do so, we choose to solve the equation $y = 2t - 1$ for t to get $t = \frac{y+1}{2}$. Substituting this into the equation $x = t^2 - 3$ yields $x = \left(\frac{y+1}{2}\right)^2 - 3$ or, after some rearrangement, $(y + 1)^2 = 4(x + 3)$. Thinking back to Section 7.3, we see that the graph of this equation is a parabola with vertex $(-3, -1)$ which opens to the right, as required. Technically speaking, the equation $(y + 1)^2 = 4(x + 3)$ describes the *entire* parabola, while the parametric equations $\left\{x = t^2 - 3,\ y = 2t - 1\right.$ for $t \geq -2$ describe only a *portion* of the parabola. In this case,[3] we can remedy this situation by restricting the bounds on y. Since the portion of the parabola we want is exactly the part where $y \geq -5$, the equation $(y+1)^2 = 4(x+3)$ coupled with the restriction $y \geq -5$ describes the same curve as the given parametric equations. The one piece of information we can never recover after eliminating the parameter is the orientation of the curve.

Eliminating the parameter and obtaining an equation in terms of x and y, whenever possible, can be a great help in graphing curves determined by parametric equations. If the system of parametric equations contains algebraic functions, as was the case in Example 11.10.1, then the usual techniques of substitution and elimination as learned in Section 8.7 can be applied to the

[3] We will have an example shortly where no matter how we restrict x and y, we can never accurately describe the curve once we've eliminated the parameter.

system $\{x = f(t),\, y = g(t)$ to eliminate the parameter. If, on the other hand, the parametrization involves the trigonometric functions, the strategy changes slightly. In this case, it is often best to solve for the trigonometric functions and relate them using an identity. We demonstrate these techniques in the following example.

Example 11.10.2. Sketch the curves described by the following parametric equations.

1. $\left\{\begin{array}{rcl} x & = & t^3 \\ y & = & 2t^2 \end{array}\right.$ for $-1 \leq t \leq 1$

2. $\left\{\begin{array}{rcl} x & = & e^{-t} \\ y & = & e^{-2t} \end{array}\right.$ for $t \geq 0$

3. $\left\{\begin{array}{rcl} x & = & \sin(t) \\ y & = & \csc(t) \end{array}\right.$ for $0 < t < \pi$

4. $\left\{\begin{array}{rcl} x & = & 1 + 3\cos(t) \\ y & = & 2\sin(t) \end{array}\right.$ for $0 \leq t \leq \frac{3\pi}{2}$

Solution.

1. To get a feel for the curve described by the system $\{x = t^3,\, y = 2t^2$ we first sketch the graphs of $x = t^3$ and $y = 2t^2$ over the interval $[-1, 1]$. We note that as t takes on values in the interval $[-1, 1]$, $x = t^3$ ranges between -1 and 1, and $y = 2t^2$ ranges between 0 and 2. This means that all of the action is happening on a portion of the plane, namely $\{(x, y) \mid -1 \leq x \leq 1,\, 0 \leq y \leq 2\}$. Next, we plot a few points to get a sense of the position and orientation of the curve. Certainly, $t = -1$ and $t = 1$ are good values to pick since these are the extreme values of t. We also choose $t = 0$, since that corresponds to a relative minimum[4] on the graph of $y = 2t^2$. Plugging in $t = -1$ gives the point $(-1, 2)$, $t = 0$ gives $(0, 0)$ and $t = 1$ gives $(1, 2)$. More generally, we see that $x = t^3$ is *increasing* over the entire interval $[-1, 1]$ whereas $y = 2t^2$ is *decreasing* over the interval $[-1, 0]$ and then *increasing* over $[0, 1]$. Geometrically, this means that in order to trace out the path described by the parametric equations, we start at $(-1, 2)$ (where $t = -1$), then move to the right (since x is increasing) and down (since y is decreasing) to $(0, 0)$ (where $t = 0$). We continue to move to the right (since x is still increasing) but now move upwards (since y is now increasing) until we reach $(1, 2)$ (where $t = 1$). Finally, to get a good sense of the shape of the curve, we eliminate the parameter. Solving $x = t^3$ for t, we get $t = \sqrt[3]{x}$. Substituting this into $y = 2t^2$ gives $y = 2(\sqrt[3]{x})^2 = 2x^{2/3}$. Our experience in Section 5.3 yields the graph of our final answer below.

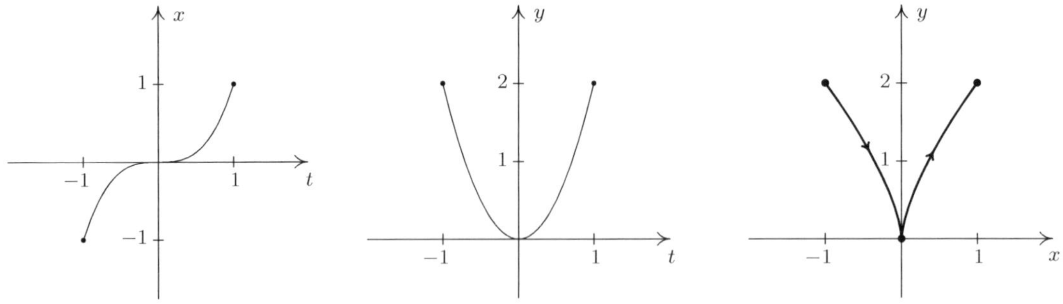

$x = t^3$, $-1 \leq t \leq 1$ $y = 2t^2$, $-1 \leq t \leq 1$ $\{x = t^3,\, y = 2t^2,\, -1 \leq t \leq 1$

[4]You should review Section 1.6.1 if you've forgotten what 'increasing', 'decreasing' and 'relative minimum' mean.

2. For the system $\{x = 2e^{-t},\ y = e^{-2t}$ for $t \geq 0$, we proceed as in the previous example and graph $x = 2e^{-t}$ and $y = e^{-2t}$ over the interval $[0, \infty)$. We find that the range of x in this case is $(0, 2]$ and the range of y is $(0, 1]$. Next, we plug in some friendly values of t to get a sense of the orientation of the curve. Since t lies in the exponent here, 'friendly' values of t involve natural logarithms. Starting with $t = \ln(1) = 0$ we get[5] $(2, 1)$, for $t = \ln(2)$ we get $\left(1, \frac{1}{4}\right)$ and for $t = \ln(3)$ we get $\left(\frac{2}{3}, \frac{1}{9}\right)$. Since t is ranging over the unbounded interval $[0, \infty)$, we take the time to analyze the end behavior of both x and y. As $t \to \infty$, $x = 2e^{-t} \to 0^+$ and $y = e^{-2t} \to 0^+$ as well. This means the graph of $\{x = 2e^{-t},\ y = e^{-2t}$ approaches the point $(0, 0)$. Since both $x = 2e^{-t}$ and $y = e^{-2t}$ are always decreasing for $t \geq 0$, we know that our final graph will start at $(2, 1)$ (where $t = 0$), and move consistently to the left (since x is decreasing) and down (since y is decreasing) to approach the origin. To eliminate the parameter, one way to proceed is to solve $x = 2e^{-t}$ for t to get $t = -\ln\left(\frac{x}{2}\right)$. Substituting this for t in $y = e^{-2t}$ gives $y = e^{-2(-\ln(x/2))} = e^{2\ln(x/2)} = e^{\ln(x/2)^2} = \left(\frac{x}{2}\right)^2 = \frac{x^2}{4}$. Or, we could recognize that $y = c^{-2t} = \left(e^{-t}\right)^2$, and since $x = 2e^{-t}$ means $e^{-t} = \frac{x}{2}$, we get $y = \left(\frac{x}{2}\right)^2 = \frac{x^2}{4}$ this way as well. Either way, the graph of $\{x = 2e^{-t},\ y = e^{-2t}$ for $t \geq 0$ is a portion of the parabola $y = \frac{x^2}{4}$ which starts at the point $(2, 1)$ and heads towards, but never reaches,[6] $(0, 0)$.

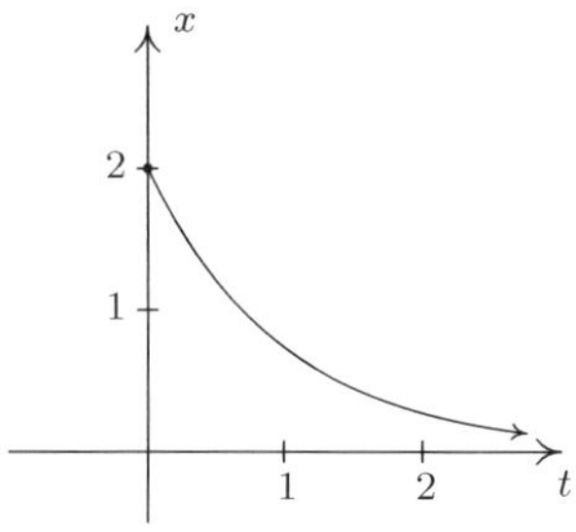

$x = 2e^{-t},\ t \geq 0$

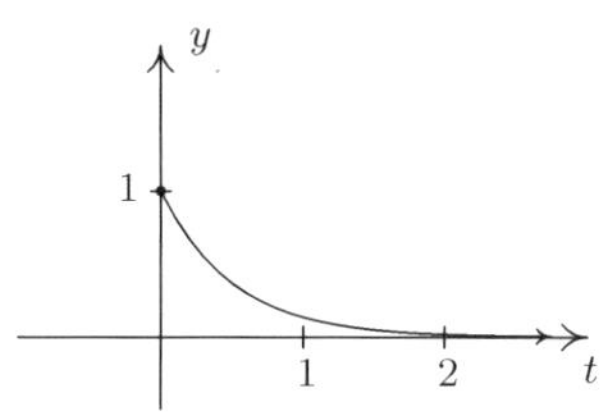

$y = e^{-2t},\ t \geq 0$

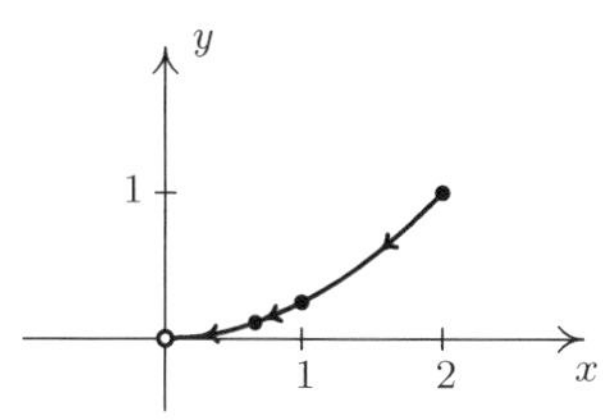

$\{x = 2e^{-t},\ y = e^{-2t},\ t \geq 0$

3. For the system $\{x = \sin(t),\ y = \csc(t)$ for $0 < t < \pi$, we start by graphing $x = \sin(t)$ and $y = \csc(t)$ over the interval $(0, \pi)$. We find that the range of x is $(0, 1]$ while the range of y is $[1, \infty)$. Plotting a few friendly points, we see that $t = \frac{\pi}{6}$ gives the point $\left(\frac{1}{2}, 2\right)$, $t = \frac{\pi}{2}$ gives $(1, 1)$ and $t = \frac{5\pi}{6}$ returns us to $\left(\frac{1}{2}, 2\right)$. Since $t = 0$ and $t = \pi$ aren't included in the domain for t, (because $y = \csc(t)$ is undefined at these t-values), we analyze the behavior of the system as t approaches 0 and π. We find that as $t \to 0^+$ as well as when $t \to \pi^-$, we get $x = \sin(t) \to 0^+$ and $y = \csc(t) \to \infty$. Piecing all of this information together, we get that for t near 0, we have points with very small positive x-values, but very large positive y-values. As t ranges through the interval $\left(0, \frac{\pi}{2}\right]$, $x = \sin(t)$ is increasing and $y = \csc(t)$ is decreasing. This means that we are moving to the right and downwards, through $\left(\frac{1}{2}, 2\right)$ when $t = \frac{\pi}{6}$ to $(1, 1)$ when $t = \frac{\pi}{2}$. Once $t = \frac{\pi}{2}$, the orientation reverses, and we start to head to the left, since $x = \sin(t)$ is now decreasing, and up, since $y = \csc(t)$ is now increasing. We pass back through $\left(\frac{1}{2}, 2\right)$ when $t = \frac{5\pi}{6}$ back to the points with small positive x-coordinates and large

[5] The reader is encouraged to review Sections 6.1 and 6.2 as needed.

[6] Note the open circle at the origin. See the solution to part 3 in Example 1.2.1 on page 22 and Theorem 4.1 in Section 4.1 for a review of this concept.

positive y-coordinates. To better explain this behavior, we eliminate the parameter. Using a reciprocal identity, we write $y = \csc(t) = \frac{1}{\sin(t)}$. Since $x = \sin(t)$, the curve traced out by this parametrization is a portion of the graph of $y = \frac{1}{x}$. We now can explain the unusual behavior as $t \to 0^+$ and $t \to \pi^-$ – for these values of t, we are hugging the vertical asymptote $x = 0$ of the graph of $y = \frac{1}{x}$. We see that the parametrization given above traces out the portion of $y = \frac{1}{x}$ for $0 < x \leq 1$ *twice* as t runs through the interval $(0, \pi)$.

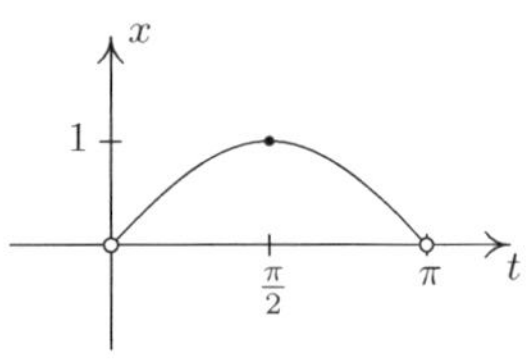

$x = \sin(t)$, $0 < t < \pi$

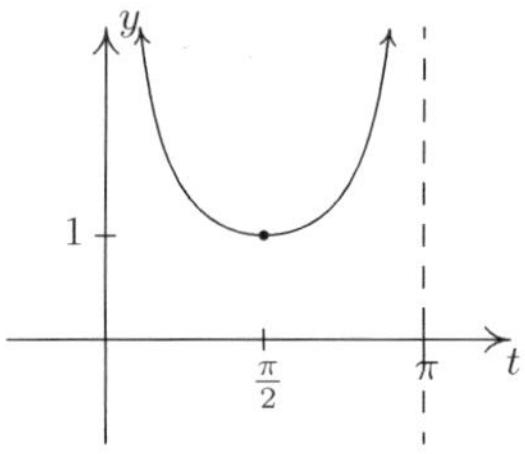

$y = \csc(t)$, $0 < t < \pi$

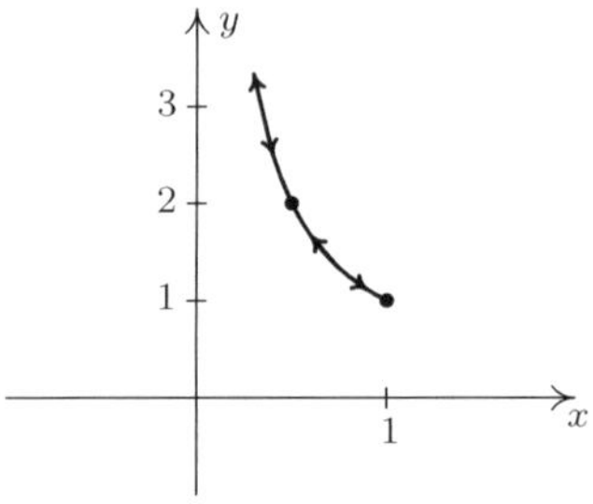

$\{x = \sin(t),\ y = \csc(t),\ 0 < t < \pi$

4. Proceeding as above, we set about graphing $\{x = 1 + 3\cos(t),\ y = 2\sin(t)$ for $0 \leq t \leq \frac{3\pi}{2}$ by first graphing $x = 1 + 3\cos(t)$ and $y = 2\sin(t)$ on the interval $\left[0, \frac{3\pi}{2}\right]$. We see that x ranges from -2 to 4 and y ranges from -2 to 2. Plugging in $t = 0$, $\frac{\pi}{2}$, π and $\frac{3\pi}{2}$ gives the points $(4, 0)$, $(1, 2)$, $(-2, 0)$ and $(1, -2)$, respectively. As t ranges from 0 to $\frac{\pi}{2}$, $x = 1+3\cos(t)$ is decreasing, while $y = 2\sin(t)$ is increasing. This means that we start tracing out our answer at $(4, 0)$ and continue moving to the left and upwards towards $(1, 2)$. For $\frac{\pi}{2} \leq t \leq \pi$, x is decreasing, as is y, so the motion is still right to left, but now is downwards from $(1, 2)$ to $(-2, 0)$. On the interval $\left[\pi, \frac{3\pi}{2}\right]$, x begins to increase, while y continues to decrease. Hence, the motion becomes left to right but continues downwards, connecting $(-2, 0)$ to $(1, -2)$. To eliminate the parameter here, we note that the trigonometric functions involved, namely $\cos(t)$ and $\sin(t)$, are related by the Pythagorean Identity $\cos^2(t) + \sin^2(t) = 1$. Hence, we solve $x = 1 + 3\cos(t)$ for $\cos(t)$ to get $\cos(t) = \frac{x-1}{3}$, and we solve $y = 2\sin(t)$ for $\sin(t)$ to get $\sin(t) = \frac{y}{2}$. Substituting these expressions into $\cos^2(t)+\sin^2(t) = 1$ gives $\left(\frac{x-1}{3}\right)^2 + \left(\frac{y}{2}\right)^2 = 1$, or $\frac{(x-1)^2}{9} + \frac{y^2}{4} = 1$. From Section 7.4, we know that the graph of this equation is an ellipse centered at $(1, 0)$ with vertices at $(-2, 0)$ and $(4, 0)$ with a minor axis of length 4. Our parametric equations here are tracing out three-quarters of this ellipse, in a counter-clockwise direction.

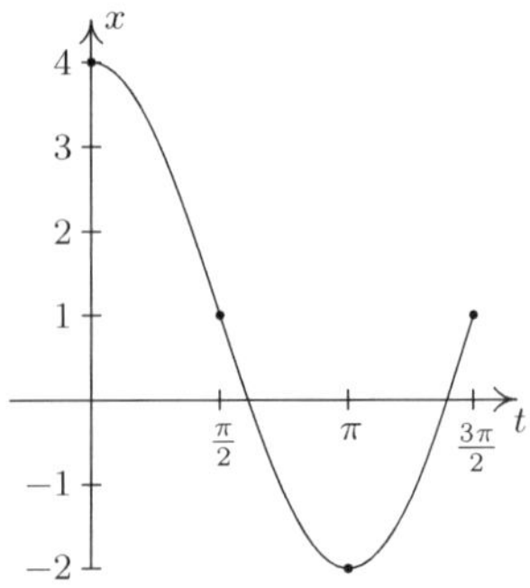

$x = 1 + 3\cos(t)$, $0 \leq t \leq \frac{3\pi}{2}$

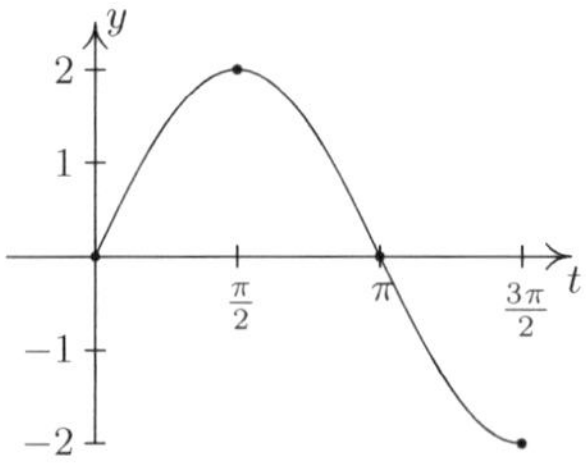

$y = 2\sin(t)$, $0 \leq t \leq \frac{3\pi}{2}$

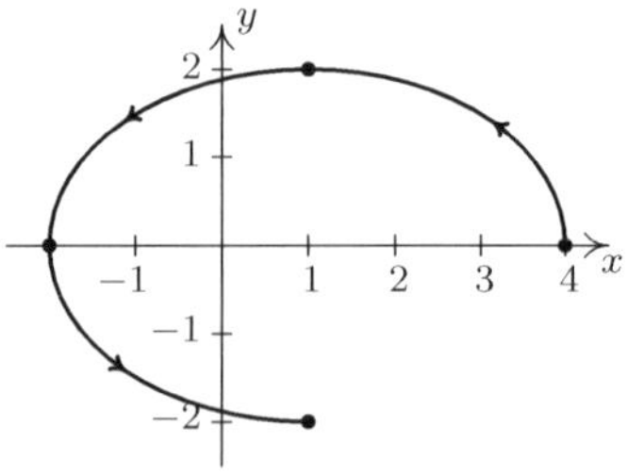

$\{x = 1 + 3\cos(t),\ y = 2\sin(t),\ 0 \leq t \leq \frac{3\pi}{2}$

□

Now that we have had some good practice sketching the graphs of parametric equations, we turn to the problem of finding parametric representations of curves. We start with the following.

Parametrizations of Common Curves

- To parametrize $y = f(x)$ as x runs through some interval I, let $x = t$ and $y = f(t)$ and let t run through I.
- To parametrize $x = g(y)$ as y runs through some interval I, let $x = g(t)$ and $y = t$ and let t run through I.
- To parametrize a directed line segment with initial point (x_0, y_0) and terminal point (x_1, y_1), let $x = x_0 + (x_1 - x_0)t$ and $y = y_0 + (y_1 - y_0)t$ for $0 \leq t \leq 1$.
- To parametrize $\frac{(x-h)^2}{a^2} + \frac{(y-k)^2}{b^2} = 1$ where $a, b > 0$, let $x = h + a\cos(t)$ and $y = k + b\sin(t)$ for $0 \leq t < 2\pi$. (This will impart a *counter-clockwise* orientation.)

The reader is encouraged to verify the above formulas by eliminating the parameter and, when indicated, checking the orientation. We put these formulas to good use in the following example.

Example 11.10.3. Find a parametrization for each of the following curves and check your answers.

1. $y = x^2$ from $x = -3$ to $x = 2$
2. $y = f^{-1}(x)$ where $f(x) = x^5 + 2x + 1$
3. The line segment which starts at $(2, -3)$ and ends at $(1, 5)$
4. The circle $x^2 + 2x + y^2 - 4y = 4$
5. The left half of the ellipse $\frac{x^2}{4} + \frac{y^2}{9} = 1$

Solution.

1. Since $y = x^2$ is written in the form $y = f(x)$, we let $x = t$ and $y = f(t) = t^2$. Since $x = t$, the bounds on t match precisely the bounds on x so we get $\{x = t,\ y = t^2$ for $-3 \leq t \leq 2$. The check is almost trivial; with $x = t$ we have $y = t^2 = x^2$ as $t = x$ runs from -3 to 2.
2. We are told to parametrize $y = f^{-1}(x)$ for $f(x) = x^5 + 2x + 1$ so it is safe to assume that f is one-to-one. (Otherwise, f^{-1} would not exist.) To find a formula $y = f^{-1}(x)$, we follow the procedure outlined on page 384 – we start with the equation $y = f(x)$, interchange x and y and solve for y. Doing so gives us the equation $x = y^5 + 2y + 1$. While we could attempt to solve this equation for y, we don't need to. We can parametrize $x = f(y) = y^5 + 2y + 1$ by setting $y = t$ so that $x = t^5 + 2t + 1$. We know from our work in Section 3.1 that since $f(x) = x^5 + 2x + 1$ is an odd-degree polynomial, the range of $y = f(x) = x^5 + 2x + 1$ is $(-\infty, \infty)$. Hence, in order to trace out the entire graph of $x = f(y) = y^5 + 2y + 1$, we need to let y run through all real numbers. Our final answer to this problem is $\{x = t^5 + 2t + 1,\ y = t$ for $-\infty < t < \infty$. As in the previous problem, our solution is trivial to check.[7]

[7]Provided you followed the inverse function theory, of course.

3. To parametrize line segment which starts at $(2,-3)$ and ends at $(1,5)$, we make use of the formulas $x = x_0+(x_1-x_0)t$ and $y = y_0+(y_1-y_0)t$ for $0 \leq t \leq 1$. While these equations at first glance are quite a handful,[8] they can be summarized as 'starting point + (displacement)t'. To find the equation for x, we have that the line segment *starts* at $x = 2$ and *ends* at $x = 1$. This means the *displacement* in the x-direction is $(1-2) = -1$. Hence, the equation for x is $x = 2+(-1)t = 2-t$. For y, we note that the line segment starts at $y = -3$ and ends at $y = 5$. Hence, the displacement in the y-direction is $(5-(-3)) = 8$, so we get $y = -3+8t$. Our final answer is $\{x = 2-t,\ y = -3+8t$ for $0 \leq t \leq 1$. To check, we can solve $x = 2-t$ for t to get $t = 2-x$. Substituting this into $y = -3+8t$ gives $y = -3+8t = -3+8(2-x)$, or $y = -8x+13$. We know this is the graph of a line, so all we need to check is that it starts and stops at the correct points. When $t = 0$, $x = 2-t = 2$, and when $t = 1$, $x = 2-t = 1$. Plugging in $x = 2$ gives $y = -8(2)+13 = -3$, for an initial point of $(2,-3)$. Plugging in $x = 1$ gives $y = -8(1)+13 = 5$ for an ending point of $(1,5)$, as required.

4. In order to use the formulas above to parametrize the circle $x^2+2x+y^2-4y = 4$, we first need to put it into the correct form. After completing the squares, we get $(x+1)^2+(y-2)^2 = 9$, or $\frac{(x+1)^2}{9}+\frac{(y-2)^2}{9} = 1$. Once again, the formulas $x = h+a\cos(t)$ and $y = k+b\sin(t)$ can be a challenge to memorize, but they come from the Pythagorean Identity $\cos^2(t)+\sin^2(t) = 1$. In the equation $\frac{(x+1)^2}{9}+\frac{(y-2)^2}{9} = 1$, we identify $\cos(t) = \frac{x+1}{3}$ and $\sin(t) = \frac{y-2}{3}$. Rearranging these last two equations, we get $x = -1+3\cos(t)$ and $y = 2+3\sin(t)$. In order to complete one revolution around the circle, we let t range through the interval $[0,2\pi)$. We get as our final answer $\{x = -1+3\cos(t),\ y = 2+3\sin(t)$ for $0 \leq t < 2\pi$. To check our answer, we could eliminate the parameter by solving $x = -1+3\cos(t)$ for $\cos(t)$ and $y = 2+3\sin(t)$ for $\sin(t)$, invoking a Pythagorean Identity, and then manipulating the resulting equation in x and y into the original equation $x^2+2x+y^2-4y = 4$. Instead, we opt for a more direct approach. We substitute $x = -1+3\cos(t)$ and $y = 2+3\sin(t)$ into the equation $x^2+2x+y^2-4y = 4$ and show that the latter is satisfied for all t such that $0 \leq t < 2\pi$.

$$\begin{array}{rcl}
x^2+2x+y^2-4y & = & 4 \\
(-1+3\cos(t))^2+2(-1+3\cos(t))+(2+3\sin(t))^2-4(2+3\sin(t)) & \stackrel{?}{=} & 4 \\
1-6\cos(t)+9\cos^2(t)-2+6\cos(t)+4+12\sin(t)+9\sin^2(t)-8-12\sin(t) & \stackrel{?}{=} & 4 \\
9\cos^2(t)+9\sin^2(t)-5 & \stackrel{?}{=} & 4 \\
9\left(\cos^2(t)+\sin^2(t)\right)-5 & \stackrel{?}{=} & 4 \\
9\,(1)-5 & \stackrel{?}{=} & 4 \\
4 & \stackrel{\checkmark}{=} & 4
\end{array}$$

Now that we know the parametric equations give us points on the circle, we can go through the usual analysis as demonstrated in Example 11.10.2 to show that the entire circle is covered as t ranges through the interval $[0,2\pi)$.

[8] Compare and contrast this with Exercise 65 in Section 11.8.

5. In the equation $\frac{x^2}{4} + \frac{y^2}{9} = 1$, we can either use the formulas above or think back to the Pythagorean Identity to get $x = 2\cos(t)$ and $y = 3\sin(t)$. The normal range on the parameter in this case is $0 \leq t < 2\pi$, but since we are interested in only the left half of the ellipse, we restrict t to the values which correspond to Quadrant II and Quadrant III angles, namely $\frac{\pi}{2} \leq t \leq \frac{3\pi}{2}$. Our final answer is $\{x = 2\cos(t),\ y = 3\sin(t)$ for $\frac{\pi}{2} \leq t \leq \frac{3\pi}{2}$. Substituting $x = 2\cos(t)$ and $y = 3\sin(t)$ into $\frac{x^2}{4} + \frac{y^2}{9} = 1$ gives $\frac{4\cos^2(t)}{4} + \frac{9\sin^2(t)}{9} = 1$, which reduces to the Pythagorean Identity $\cos^2(t) + \sin^2(t) = 1$. This proves that the points generated by the parametric equations $\{x = 2\cos(t),\ y = 3\sin(t)$ lie on the ellipse $\frac{x^2}{4} + \frac{y^2}{9} = 1$. Employing the techniques demonstrated in Example 11.10.2, we find that the restriction $\frac{\pi}{2} \leq t \leq \frac{3\pi}{2}$ generates the left half of the ellipse, as required. □

We note that the formulas given on page 1053 offer only *one* of literally *infinitely* many ways to parametrize the common curves listed there. At times, the formulas offered there need to be altered to suit the situation. Two easy ways to alter parametrizations are given below.

Adjusting Parametric Equations

- **Reversing Orientation:** Replacing every occurrence of t with $-t$ in a parametric description for a curve (including any inequalities which describe the bounds on t) reverses the orientation of the curve.

- **Shift of Parameter:** Replacing every occurrence of t with $(t - c)$ in a parametric description for a curve (including any inequalities which describe the bounds on t) shifts the start of the parameter t ahead by c units.

We demonstrate these techniques in the following example.

Example 11.10.4. Find a parametrization for the following curves.

1. The curve which starts at $(2, 4)$ and follows the parabola $y = x^2$ to end at $(-1, 1)$. Shift the parameter so that the path starts at $t = 0$.

2. The two part path which starts at $(0, 0)$, travels along a line to $(3, 4)$, then travels along a line to $(5, 0)$.

3. The Unit Circle, oriented clockwise, with $t = 0$ corresponding to $(0, -1)$.

Solution.

1. We can parametrize $y = x^2$ from $x = -1$ to $x = 2$ using the formula given on Page 1053 as $\{x = t,\ y = t^2$ for $-1 \leq t \leq 2$. This parametrization, however, starts at $(-1, 1)$ and ends at $(2, 4)$. Hence, we need to reverse the orientation. To do so, we replace every occurrence of t with $-t$ to get $\{x = -t,\ y = (-t)^2$ for $-1 \leq -t \leq 2$. After simplifying, we get $\{x = -t,\ y = t^2$ for $-2 \leq t \leq 1$. We would like t to begin at $t = 0$ instead of $t = -2$. The problem here is that the parametrization we have starts 2 units 'too soon', so we need to introduce a 'time delay' of 2. Replacing every occurrence of t with $(t - 2)$ gives $\{x = -(t - 2),\ y = (t - 2)^2$ for $-2 \leq t - 2 \leq 1$. Simplifying yields $\{x = 2 - t,\ y = t^2 - 4t + 4$ for $0 \leq t \leq 3$.

2. When parameterizing line segments, we think: 'starting point + (displacement)t'. For the first part of the path, we get $\{x = 3t, \ y = 4t$ for $0 \leq t \leq 1$, and for the second part we get $\{x = 3 + 2t, \ y = 4 - 4t$ for $0 \leq t \leq 1$. Since the first parametrization leaves off at $t = 1$, we shift the parameter in the second part so it starts at $t = 1$. Our current description of the second part starts at $t = 0$, so we introduce a 'time delay' of 1 unit to the second set of parametric equations. Replacing t with $(t - 1)$ in the second set of parametric equations gives $\{x = 3 + 2(t - 1), \ y = 4 - 4(t - 1)$ for $0 \leq t - 1 \leq 1$. Simplifying yields $\{x = 1 + 2t, \ y = 8 - 4t$ for $1 \leq t \leq 2$. Hence, we may parametrize the path as $\{x = f(t), \ y = g(t)$ for $0 \leq t \leq 2$ where

$$f(t) = \begin{cases} 3t, & \text{for } 0 \leq t \leq 1 \\ 1 + 2t, & \text{for } 1 \leq t \leq 2 \end{cases} \quad \text{and} \quad g(t) = \begin{cases} 4t, & \text{for } 0 \leq t \leq 1 \\ 8 - 4t, & \text{for } 1 \leq t \leq 2 \end{cases}$$

3. We know that $\{x = \cos(t), \ y = \sin(t)$ for $0 \leq t < 2\pi$ gives a *counter-clockwise* parametrization of the Unit Circle with $t = 0$ corresponding to $(1, 0)$, so the first order of business is to reverse the orientation. Replacing t with $-t$ gives $\{x = \cos(-t), \ y = \sin(-t)$ for $0 \leq -t < 2\pi$, which simplifies[9] to $\{x = \cos(t), \ y = -\sin(t)$ for $-2\pi < t \leq 0$. This parametrization gives a clockwise orientation, but $t = 0$ still corresponds to the point $(1, 0)$; the point $(0, -1)$ is reached when $t = -\frac{3\pi}{2}$. Our strategy is to first get the parametrization to 'start' at the point $(0, -1)$ and then shift the parameter accordingly so the 'start' coincides with $t = 0$. We know that any interval of length 2π will parametrize the entire circle, so we keep the equations $\{x = \cos(t), \ y = -\sin(t)$, but start the parameter t at $-\frac{3\pi}{2}$, and find the upper bound by adding 2π so $-\frac{3\pi}{2} \leq t < \frac{\pi}{2}$. The reader can verify that $\{x = \cos(t), \ y = -\sin(t)$ for $-\frac{3\pi}{2} \leq t < \frac{\pi}{2}$ traces out the Unit Circle clockwise starting at the point $(0, -1)$. We now shift the parameter by introducing a 'time delay' of $\frac{3\pi}{2}$ units by replacing every occurrence of t with $\left(t - \frac{3\pi}{2}\right)$. We get $\left\{x = \cos\left(t - \frac{3\pi}{2}\right), \ y = -\sin\left(t - \frac{3\pi}{2}\right)\right.$ for $-\frac{3\pi}{2} \leq t - \frac{3\pi}{2} < \frac{\pi}{2}$. This simplifies[10] to $\{x = -\sin(t), \ y = -\cos(t)$ for $0 \leq t < 2\pi$, as required. □

We put our answer to Example 11.10.4 number 3 to good use to derive the equation of a cycloid. Suppose a circle of radius r rolls along the positive x-axis at a constant velocity v as pictured below. Let θ be the angle in radians which measures the amount of clockwise rotation experienced by the radius highlighted in the figure.

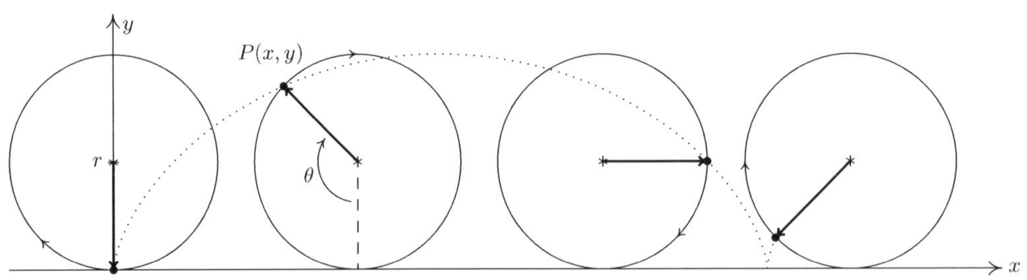

[9]courtesy of the Even/Odd Identities
[10]courtesy of the Sum/Difference Formulas

Our goal is to find parametric equations for the coordinates of the point $P(x, y)$ in terms of θ. From our work in Example 11.10.4 number 3, we know that clockwise motion along the Unit Circle starting at the point $(0, -1)$ can be modeled by the equations $\{x = -\sin(\theta),\ y = -\cos(\theta)$ for $0 \leq \theta < 2\pi$. (We have renamed the parameter 'θ' to match the context of this problem.) To model this motion on a circle of radius r, all we need to do[11] is multiply both x and y by the factor r which yields $\{x = -r\sin(\theta),\ y = -r\cos(\theta)$. We now need to adjust for the fact that the circle isn't stationary with center $(0, 0)$, but rather, is rolling along the positive x-axis. Since the velocity v is constant, we know that at time t, the center of the circle has traveled a distance vt down the positive x-axis. Furthermore, since the radius of the circle is r and the circle isn't moving vertically, we know that the center of the circle is always r units above the x-axis. Putting these two facts together, we have that at time t, the center of the circle is at the point (vt, r). From Section 10.1.1, we know $v = \frac{r\theta}{t}$, or $vt = r\theta$. Hence, the center of the circle, in terms of the parameter θ, is $(r\theta, r)$. As a result, we need to modify the equations $\{x = -r\sin(\theta),\ y = -r\cos(\theta)$ by shifting the x-coordinate to the right $r\theta$ units (by adding $r\theta$ to the expression for x) and the y-coordinate up r units[12] (by adding r to the expression for y). We get $\{x = -r\sin(\theta) + r\theta,\ y = -r\cos(\theta) + r$, which can be written as $\{x = r(\theta - \sin(\theta)),\ y = r(1 - \cos(\theta))$. Since the motion starts at $\theta = 0$ and proceeds indefinitely, we set $\theta \geq 0$.

We end the section with a demonstration of the graphing calculator.

Example 11.10.5. Find the parametric equations of a cycloid which results from a circle of radius 3 rolling down the positive x-axis as described above. Graph your answer using a calculator.

Solution. We have $r = 3$ which gives the equations $\{x = 3(t - \sin(t)),\ y = 3(1 - \cos(t))$ for $t \geq 0$. (Here we have returned to the convention of using t as the parameter.) Sketching the cycloid by hand is a wonderful exercise in Calculus, but for the purposes of this book, we use a graphing utility. Using a calculator to graph parametric equations is very similar to graphing polar equations on a calculator.[13] Ensuring that the calculator is in 'Parametric Mode' and 'radian mode' we enter the equations and advance to the 'Window' screen.

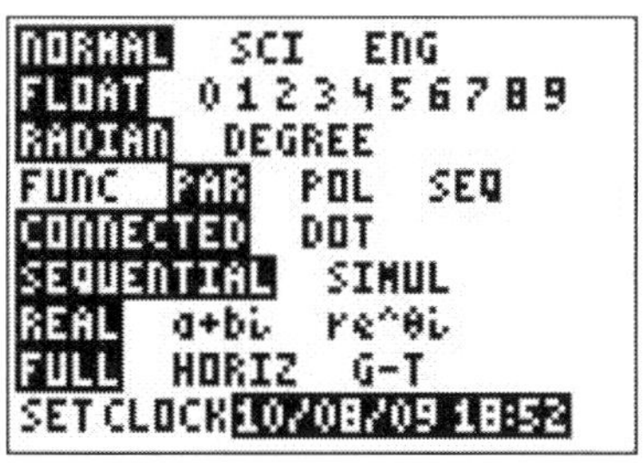

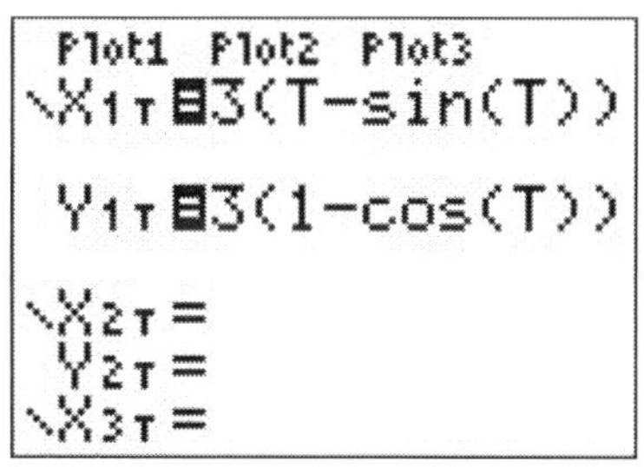

As always, the challenge is to determine appropriate bounds on the parameter, t, as well as for x and y. We know that one full revolution of the circle occurs over the interval $0 \leq t < 2\pi$, so

[11]If we replace x with $\frac{x}{r}$ and y with $\frac{y}{r}$ in the equation for the Unit Circle $x^2 + y^2 = 1$, we obtain $\left(\frac{x}{r}\right)^2 + \left(\frac{y}{r}\right)^2 = 1$ which reduces to $x^2 + y^2 = r^2$. In the language of Section 1.7, we are stretching the graph by a factor of r in both the x- and y-directions. Hence, we multiply both the x- and y-coordinates of points on the graph by r.

[12]Does this seem familiar? See Example 11.1.1 in Section 11.1.

[13]See page 959 in Section 11.5.

it seems reasonable to keep these as our bounds on t. The 'Tstep' seems reasonably small – too large a value here can lead to incorrect graphs.[14] We know from our derivation of the equations of the cycloid that the center of the generating circle has coordinates $(r\theta, r)$, or in this case, $(3t, 3)$. Since t ranges between 0 and 2π, we set x to range between 0 and 6π. The values of y go from the bottom of the circle to the top, so y ranges between 0 and 6.

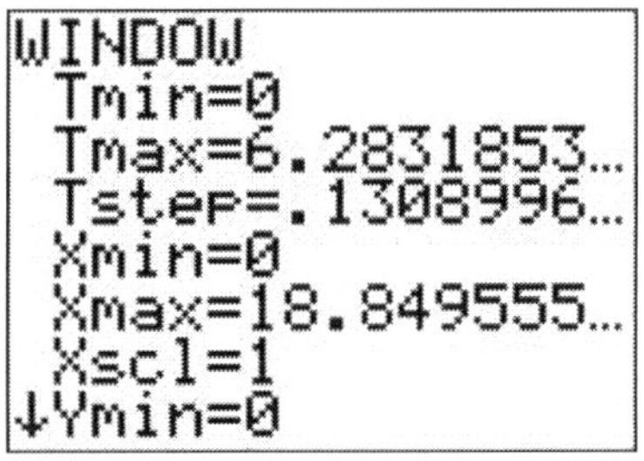

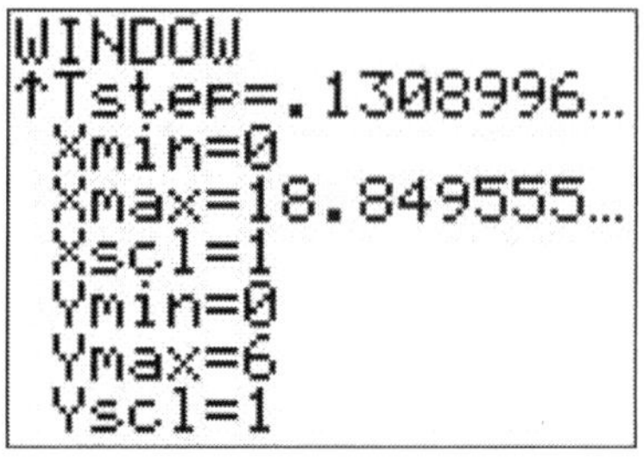

Below we graph the cycloid with these settings, and then extend t to range from 0 to 6π which forces x to range from 0 to 18π yielding three arches of the cycloid. (It is instructive to note that keeping the y settings between 0 and 6 messes up the geometry of the cycloid. The reader is invited to use the Zoom Square feature on the graphing calculator to see what window gives a true geometric perspective of the three arches.)

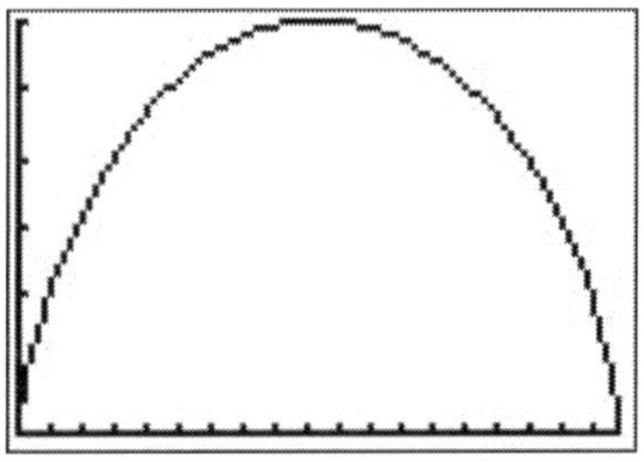

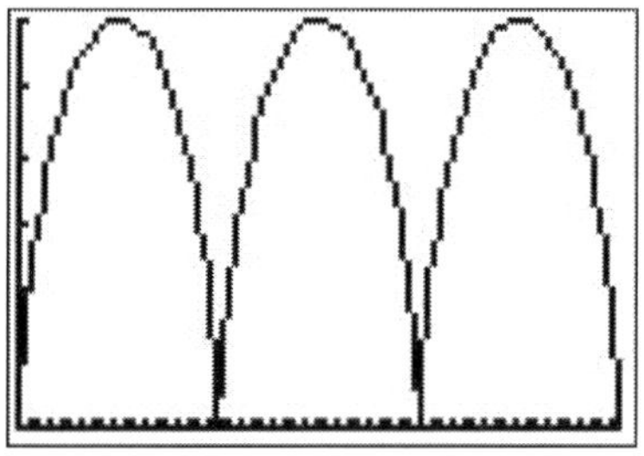

□

[14]Again, see page 959 in Section 11.5.

11.10.1 Exercises

In Exercises 1 - 20, plot the set of parametric equations by hand. Be sure to indicate the orientation imparted on the curve by the parametrization.

1. $\begin{cases} x = 4t - 3 \\ y = 6t - 2 \end{cases}$ for $0 \le t \le 1$

2. $\begin{cases} x = 4t - 1 \\ y = 3 - 4t \end{cases}$ for $0 \le t \le 1$

3. $\begin{cases} x = 2t \\ y = t^2 \end{cases}$ for $-1 \le t \le 2$

4. $\begin{cases} x = t - 1 \\ y = 3 + 2t - t^2 \end{cases}$ for $0 \le t \le 3$

5. $\begin{cases} x = t^2 + 2t + 1 \\ y = t + 1 \end{cases}$ for $t \le 1$

6. $\begin{cases} x = \frac{1}{9}\left(18 - t^2\right) \\ y = \frac{1}{3}t \end{cases}$ for $t \ge -3$

7. $\begin{cases} x = t \\ y = t^3 \end{cases}$ for $-\infty < t < \infty$

8. $\begin{cases} x = t^3 \\ y = t \end{cases}$ for $-\infty < t < \infty$

9. $\begin{cases} x = \cos(t) \\ y = \sin(t) \end{cases}$ for $-\frac{\pi}{2} \le t \le \frac{\pi}{2}$

10. $\begin{cases} x = 3\cos(t) \\ y = 3\sin(t) \end{cases}$ for $0 \le t \le \pi$

11. $\begin{cases} x = -1 + 3\cos(t) \\ y = 4\sin(t) \end{cases}$ for $0 \le t \le 2\pi$

12. $\begin{cases} x = 3\cos(t) \\ y = 2\sin(t) + 1 \end{cases}$ for $\frac{\pi}{2} \le t \le 2\pi$

13. $\begin{cases} x = 2\cos(t) \\ y = \sec(t) \end{cases}$ for $0 \le t < \frac{\pi}{2}$

14. $\begin{cases} x = 2\tan(t) \\ y = \cot(t) \end{cases}$ for $0 < t < \frac{\pi}{2}$

15. $\begin{cases} x = \sec(t) \\ y = \tan(t) \end{cases}$ for $-\frac{\pi}{2} < t < \frac{\pi}{2}$

16. $\begin{cases} x = \sec(t) \\ y = \tan(t) \end{cases}$ for $\frac{\pi}{2} < t < \frac{3\pi}{2}$

17. $\begin{cases} x = \tan(t) \\ y = 2\sec(t) \end{cases}$ for $-\frac{\pi}{2} < t < \frac{\pi}{2}$

18. $\begin{cases} x = \tan(t) \\ y = 2\sec(t) \end{cases}$ for $\frac{\pi}{2} < t < \frac{3\pi}{2}$

19. $\begin{cases} x = \cos(t) \\ y = t \end{cases}$ for $0 \le t \le \pi$

20. $\begin{cases} x = \sin(t) \\ y = t \end{cases}$ for $-\frac{\pi}{2} \le t \le \frac{\pi}{2}$

In Exercises 21 - 24, plot the set of parametric equations with the help of a graphing utility. Be sure to indicate the orientation imparted on the curve by the parametrization.

21. $\begin{cases} x = t^3 - 3t \\ y = t^2 - 4 \end{cases}$ for $-2 \le t \le 2$

22. $\begin{cases} x = 4\cos^3(t) \\ y = 4\sin^3(t) \end{cases}$ for $0 \le t \le 2\pi$

23. $\begin{cases} x = e^t + e^{-t} \\ y = e^t - e^{-t} \end{cases}$ for $-2 \le t \le 2$

24. $\begin{cases} x = \cos(3t) \\ y = \sin(4t) \end{cases}$ for $0 \le t \le 2\pi$

In Exercises 25 - 39, find a parametric description for the given oriented curve.

25. the directed line segment from $(3, -5)$ to $(-2, 2)$

26. the directed line segment from $(-2, -1)$ to $(3, -4)$

27. the curve $y = 4 - x^2$ from $(-2, 0)$ to $(2, 0)$.

28. the curve $y = 4 - x^2$ from $(-2, 0)$ to $(2, 0)$
(Shift the parameter so $t = 0$ corresponds to $(-2, 0)$.)

29. the curve $x = y^2 - 9$ from $(-5, -2)$ to $(0, 3)$.

30. the curve $x = y^2 - 9$ from $(0, 3)$ to $(-5, -2)$.
(Shift the parameter so $t = 0$ corresponds to $(0, 3)$.)

31. the circle $x^2 + y^2 = 25$, oriented counter-clockwise

32. the circle $(x - 1)^2 + y^2 = 4$, oriented counter-clockwise

33. the circle $x^2 + y^2 - 6y = 0$, oriented counter-clockwise

34. the circle $x^2 + y^2 - 6y = 0$, oriented *clockwise*
(Shift the parameter so t begins at 0.)

35. the circle $(x - 3)^2 + (y + 1)^2 = 117$, oriented counter-clockwise

36. the ellipse $(x - 1)^2 + 9y^2 = 9$, oriented counter-clockwise

37. the ellipse $9x^2 + 4y^2 + 24y = 0$, oriented counter-clockwise

38. the ellipse $9x^2 + 4y^2 + 24y = 0$, oriented clockwise
(Shift the parameter so $t = 0$ corresponds to $(0, 0)$.)

39. the triangle with vertices $(0, 0)$, $(3, 0)$, $(0, 4)$, oriented counter-clockwise
(Shift the parameter so $t = 0$ corresponds to $(0, 0)$.)

40. Use parametric equations and a graphing utility to graph the inverse of $f(x) = x^3 + 3x - 4$.

41. Every polar curve $r = f(\theta)$ can be translated to a system of parametric equations with parameter θ by $\{x = r\cos(\theta) = f(\theta)\cos(\theta),\ y = r\sin(\theta) = f(\theta)\sin(\theta)$. Convert $r = 6\cos(2\theta)$ to a system of parametric equations. Check your answer by graphing $r = 6\cos(2\theta)$ by hand using the techniques presented in Section 11.5 and then graphing the parametric equations you found using a graphing utility.

42. Use your results from Exercises 3 and 4 in Section 11.1 to find the parametric equations which model a passenger's position as they ride the London Eye.

Suppose an object, called a projectile, is launched into the air. Ignoring everything except the force gravity, the path of the projectile is given by[15]

$$\begin{cases} x = v_0 \cos(\theta)\, t \\ y = -\dfrac{1}{2}gt^2 + v_0 \sin(\theta)\, t + s_0 \end{cases} \quad \text{for } 0 \leq t \leq T$$

where v_0 is the initial speed of the object, θ is the angle from the horizontal at which the projectile is launched,[16] g is the acceleration due to gravity, s_0 is the initial height of the projectile above the ground and T is the time when the object returns to the ground. (See the figure below.)

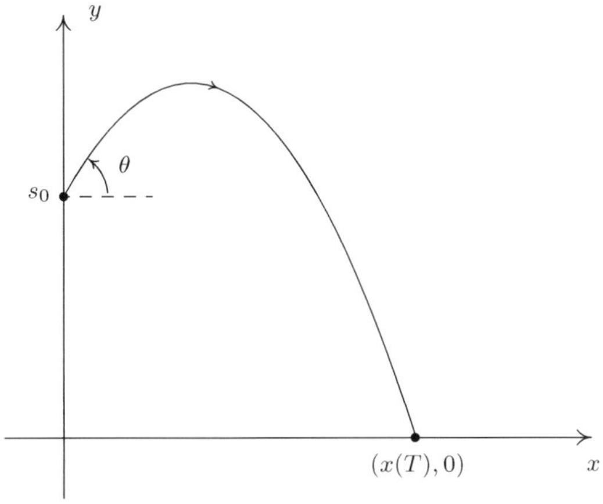

43. Carl's friend Jason competes in Highland Games Competitions across the country. In one event, the 'hammer throw', he throws a 56 pound weight for distance. If the weight is released 6 feet above the ground at an angle of 42° with respect to the horizontal with an initial speed of 33 feet per second, find the parametric equations for the flight of the hammer. (Here, use $g = 32\frac{\text{ft.}}{s^2}$.) When will the hammer hit the ground? How far away will it hit the ground? Check your answer using a graphing utility.

44. Eliminate the parameter in the equations for projectile motion to show that the path of the projectile follows the curve

$$y = -\frac{g\sec^2(\theta)}{2v_0^2}x^2 + \tan(\theta)x + s_0$$

Use the vertex formula (Equation 2.4) to show the maximum height of the projectile is

$$y = \frac{v_0^2 \sin^2(\theta)}{2g} + s_0 \quad \text{when} \quad x = \frac{v_0^2 \sin(2\theta)}{2g}$$

[15] A nice mix of vectors and Calculus are needed to derive this.

[16] We've seen this before. It's the angle of elevation which was defined on page 753.

45. In another event, the 'sheaf toss', Jason throws a 20 pound weight for height. If the weight is released 5 feet above the ground at an angle of 85° with respect to the horizontal and the sheaf reaches a maximum height of 31.5 feet, use your results from part 44 to determine how fast the sheaf was launched into the air. (Once again, use $g = 32 \frac{\text{ft.}}{s^2}$.)

46. Suppose $\theta = \frac{\pi}{2}$. (The projectile was launched vertically.) Simplify the general parametric formula given for $y(t)$ above using $g = 9.8 \, \frac{m}{s^2}$ and compare that to the formula for $s(t)$ given in Exercise 25 in Section 2.3. What is $x(t)$ in this case?

In Exercises 47 - 52, we explore the **hyperbolic cosine** function, denoted $\cosh(t)$, and the **hyperbolic sine** function, denoted $\sinh(t)$, defined below:

$$\cosh(t) = \frac{e^t + e^{-t}}{2} \quad \text{and} \quad \sinh(t) = \frac{e^t - e^{-t}}{2}$$

47. Using a graphing utility as needed, verify that the domain of $\cosh(t)$ is $(-\infty, \infty)$ and the range of $\cosh(t)$ is $[1, \infty)$.

48. Using a graphing utility as needed, verify that the domain and range of $\sinh(t)$ are both $(-\infty, \infty)$.

49. Show that $\{x(t) = \cosh(t), \, y(t) = \sinh(t)$ parametrize the right half of the 'unit' hyperbola $x^2 - y^2 = 1$. (Hence the use of the adjective 'hyperbolic.')

50. Compare the definitions of $\cosh(t)$ and $\sinh(t)$ to the formulas for $\cos(t)$ and $\sin(t)$ given in Exercise 83f in Section 11.7.

51. Four other hyperbolic functions are waiting to be defined: the hyperbolic secant $\operatorname{sech}(t)$, the hyperbolic cosecant $\operatorname{csch}(t)$, the hyperbolic tangent $\tanh(t)$ and the hyperbolic cotangent $\coth(t)$. Define these functions in terms of $\cosh(t)$ and $\sinh(t)$, then convert them to formulas involving e^t and e^{-t}. Consult a suitable reference (a Calculus book, or this entry on the hyperbolic functions) and spend some time reliving the thrills of trigonometry with these 'hyperbolic' functions.

52. If these functions look familiar, they should. Enjoy some nostalgia and revisit Exercise 35 in Section 6.5, Exercise 47 in Section 6.3 and the answer to Exercise 38 in Section 6.4.

11.10.2 Answers

1. $\begin{cases} x = 4t - 3 \\ y = 6t - 2 \end{cases}$ for $0 \leq t \leq 1$

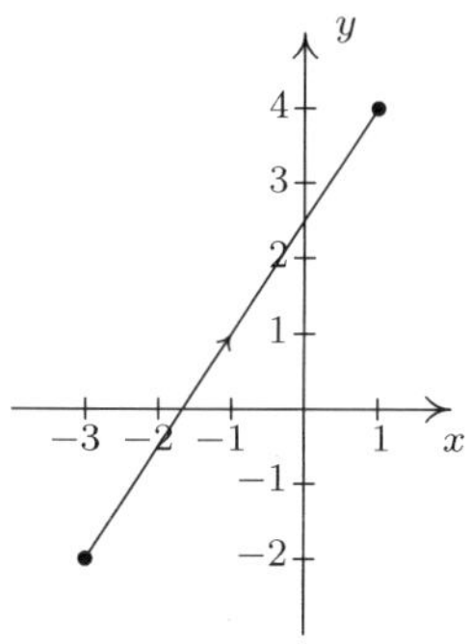

2. $\begin{cases} x = 4t - 1 \\ y = 3 - 4t \end{cases}$ for $0 \leq t \leq 1$

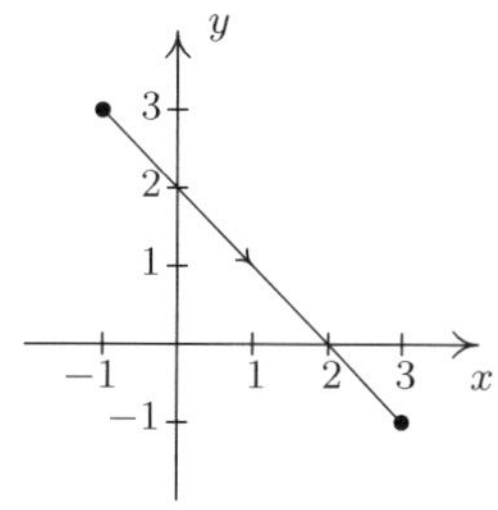

3. $\begin{cases} x = 2t \\ y = t^2 \end{cases}$ for $-1 \leq t \leq 2$

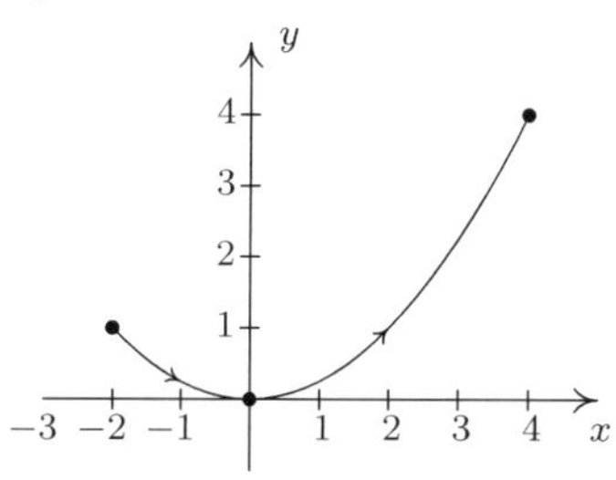

4. $\begin{cases} x = t - 1 \\ y = 3 + 2t - t^2 \end{cases}$ for $0 \leq t \leq 3$

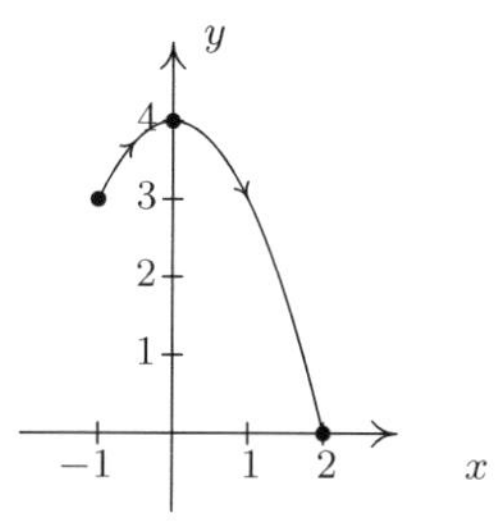

5. $\begin{cases} x = t^2 + 2t + 1 \\ y = t + 1 \end{cases}$ for $t \leq 1$

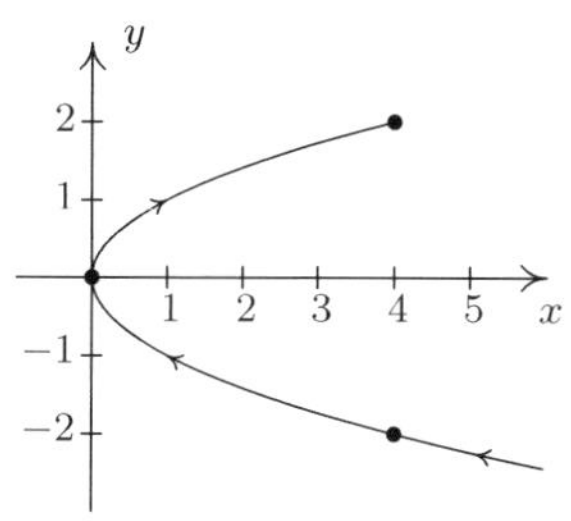

6. $\begin{cases} x = \frac{1}{9}\left(18 - t^2\right) \\ y = \frac{1}{3}t \end{cases}$ for $t \geq -3$

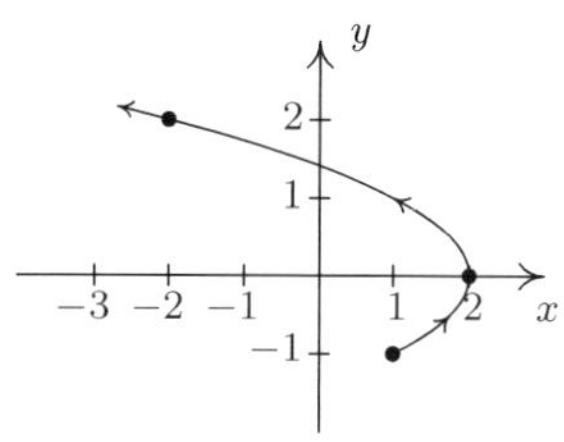

7. $\begin{cases} x = t \\ y = t^3 \end{cases}$ for $-\infty < t < \infty$

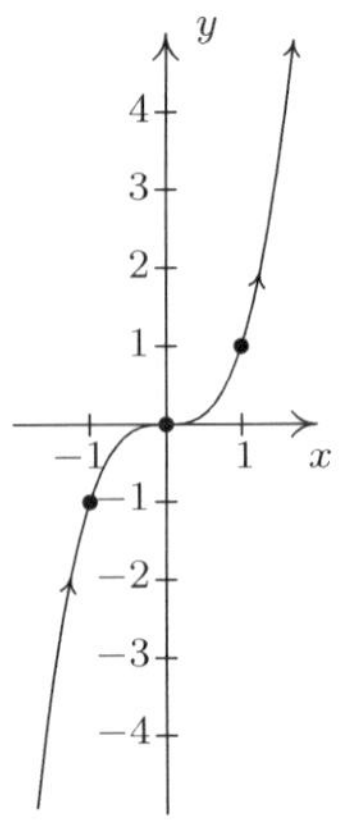

8. $\begin{cases} x = t^3 \\ y = t \end{cases}$ for $-\infty < t < \infty$

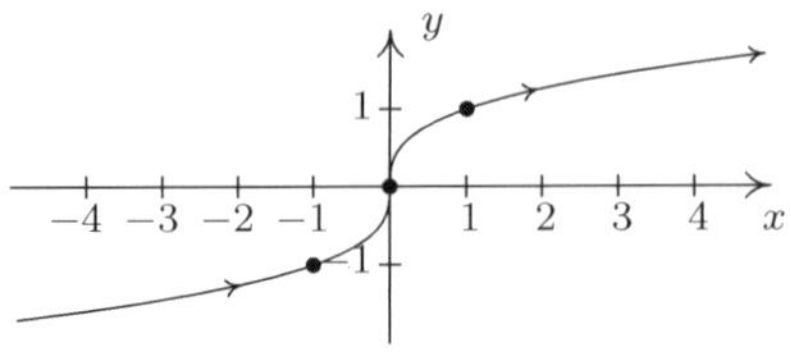

9. $\begin{cases} x = \cos(t) \\ y = \sin(t) \end{cases}$ for $-\dfrac{\pi}{2} \leq t \leq \dfrac{\pi}{2}$

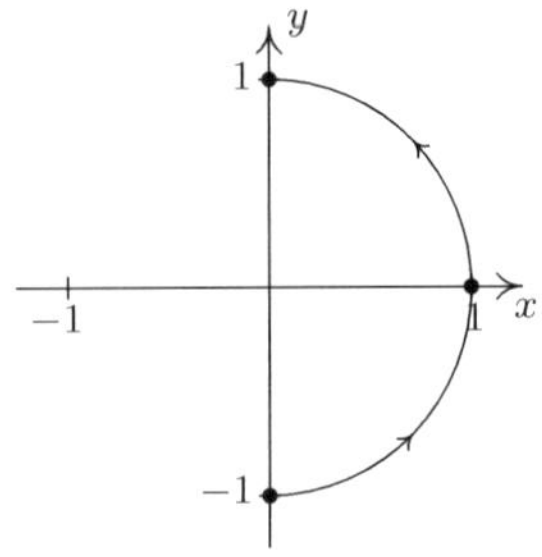

10. $\begin{cases} x = 3\cos(t) \\ y = 3\sin(t) \end{cases}$ for $0 \leq t \leq \pi$

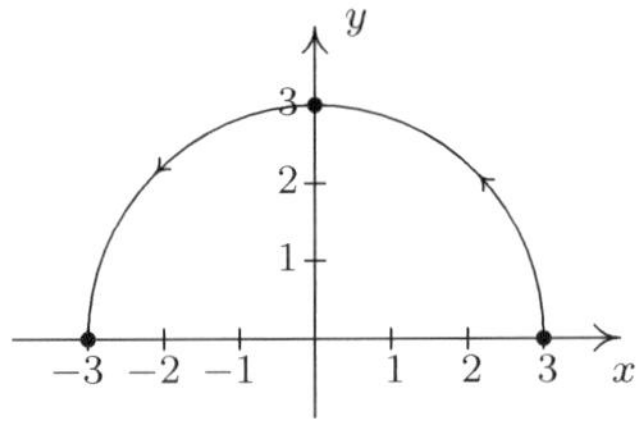

11. $\begin{cases} x = -1 + 3\cos(t) \\ y = 4\sin(t) \end{cases}$ for $0 \leq t \leq 2\pi$

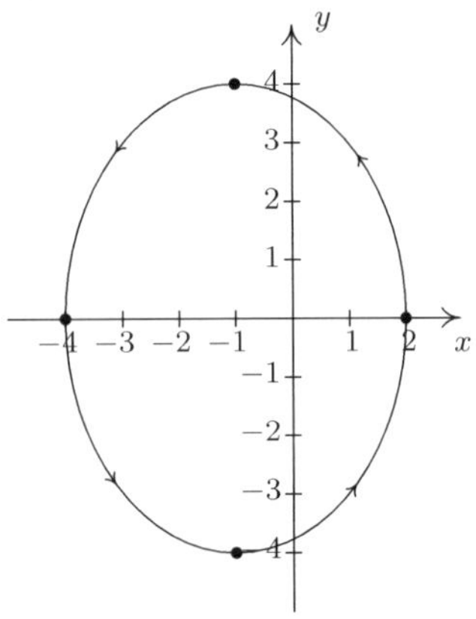

12. $\begin{cases} x = 3\cos(t) \\ y = 2\sin(t) + 1 \end{cases}$ for $\dfrac{\pi}{2} \leq t \leq 2\pi$

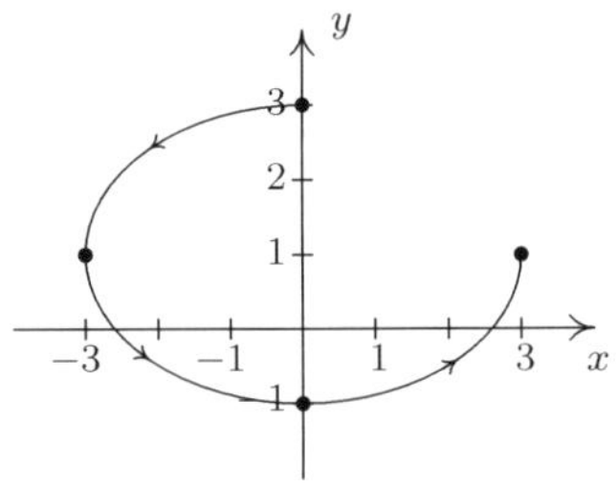

13. $\begin{cases} x = 2\cos(t) \\ y = \sec(t) \end{cases}$ for $0 \leq t < \dfrac{\pi}{2}$

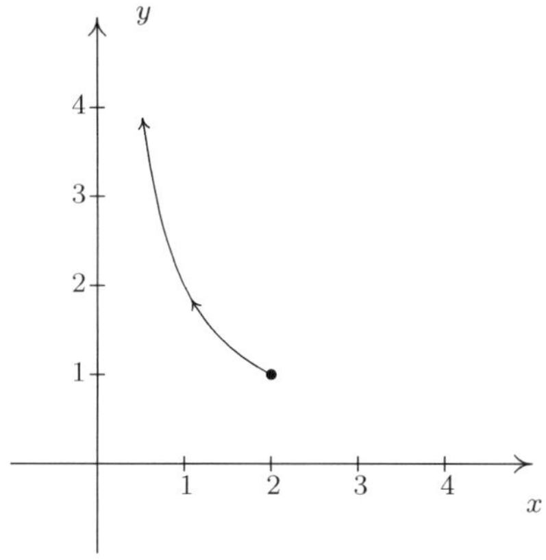

14. $\begin{cases} x = 2\tan(t) \\ y = \cot(t) \end{cases}$ for $0 < t < \dfrac{\pi}{2}$

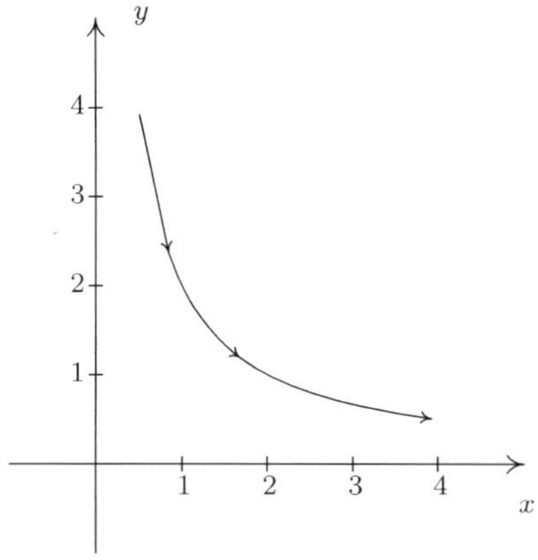

15. $\begin{cases} x = \sec(t) \\ y = \tan(t) \end{cases}$ for $-\dfrac{\pi}{2} < t < \dfrac{\pi}{2}$

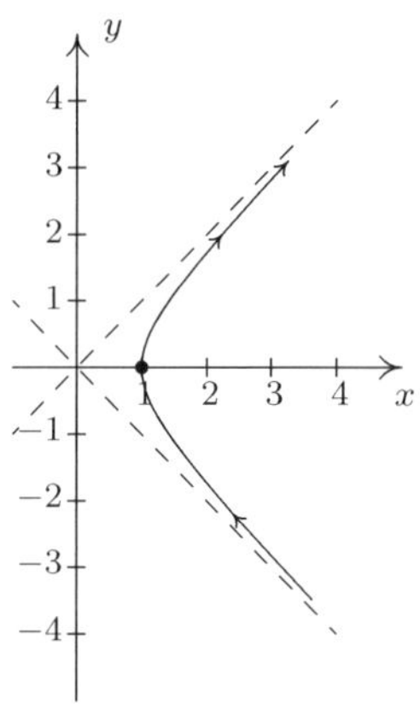

16. $\begin{cases} x = \sec(t) \\ y = \tan(t) \end{cases}$ for $\dfrac{\pi}{2} < t < \dfrac{3\pi}{2}$

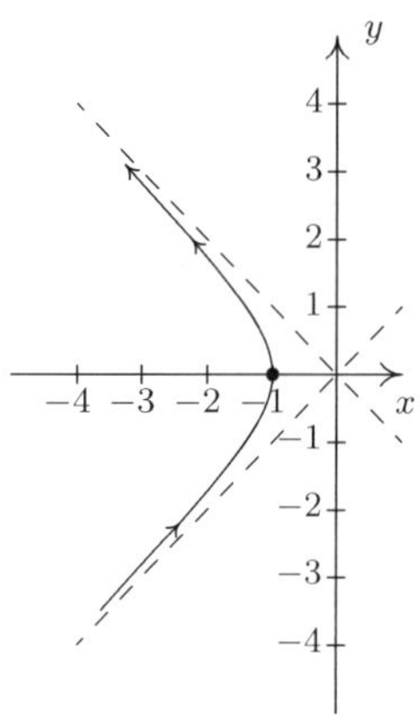

17. $\begin{cases} x = \tan(t) \\ y = 2\sec(t) \end{cases}$ for $-\dfrac{\pi}{2} < t < \dfrac{\pi}{2}$

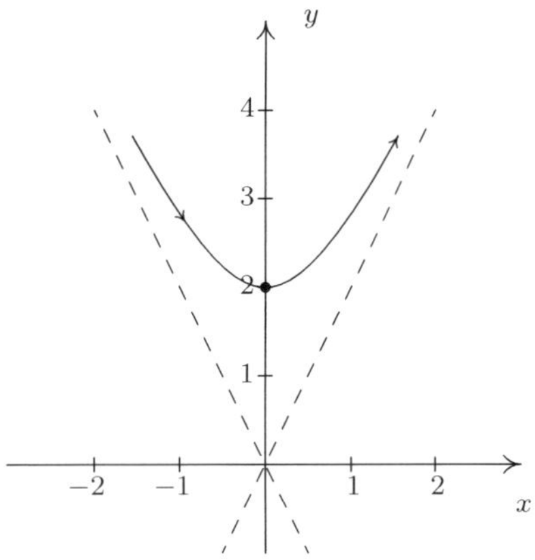

18. $\begin{cases} x = \tan(t) \\ y = 2\sec(t) \end{cases}$ for $\dfrac{\pi}{2} < t < \dfrac{3\pi}{2}$

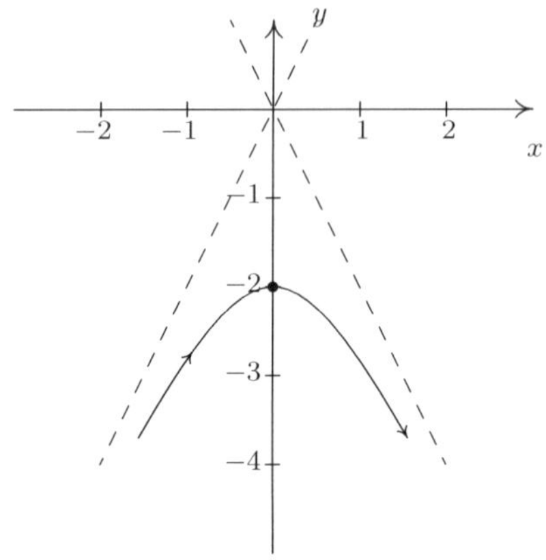

19. $\begin{cases} x = \cos(t) \\ y = t \end{cases}$ for $0 < t < \pi$

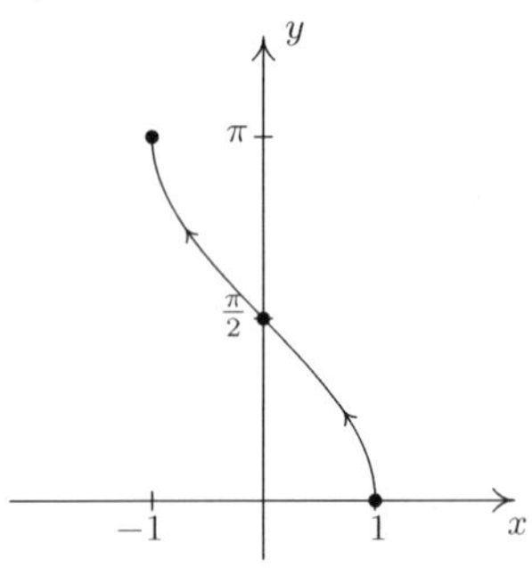

20. $\begin{cases} x = \sin(t) \\ y = t \end{cases}$ for $-\dfrac{\pi}{2} < t < \dfrac{\pi}{2}$

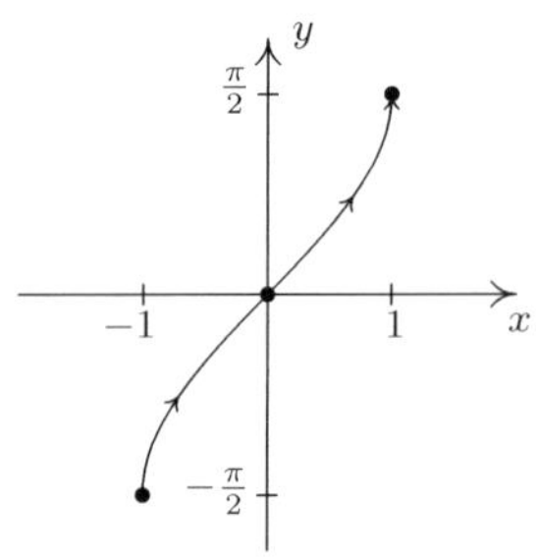

21. $\begin{cases} x = t^3 - 3t \\ y = t^2 - 4 \end{cases}$ for $-2 \le t \le 2$

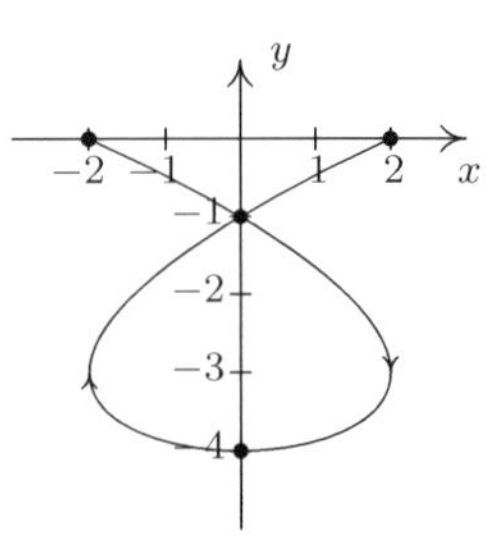

22. $\begin{cases} x = 4\cos^3(t) \\ y = 4\sin^3(t) \end{cases}$ for $0 \le t \le 2\pi$

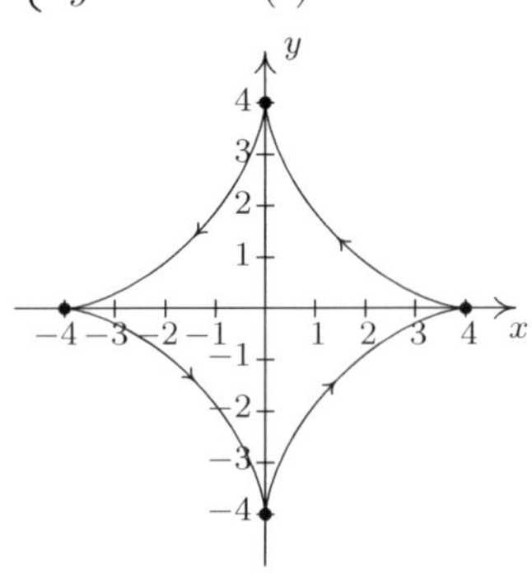

23. $\begin{cases} x = e^t + e^{-t} \\ y = e^t - e^{-t} \end{cases}$ for $-2 \leq t \leq 2$

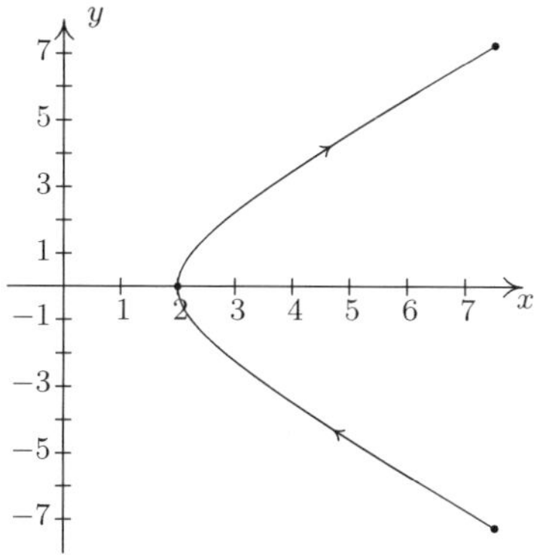

24. $\begin{cases} x = \cos(3t) \\ y = \sin(4t) \end{cases}$ for $0 \leq t \leq 2\pi$

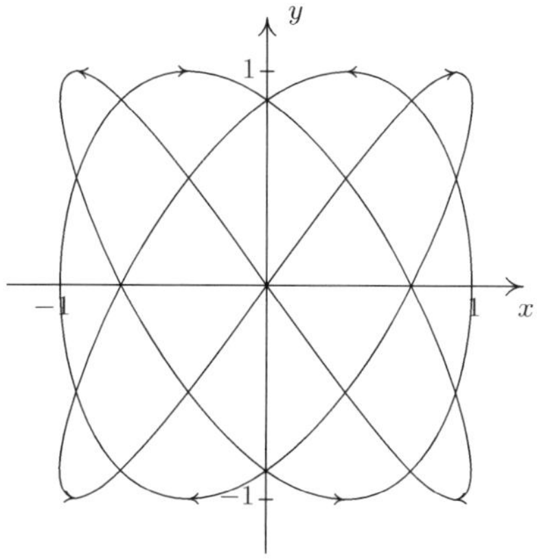

25. $\begin{cases} x = 3 - 5t \\ y = -5 + 7t \end{cases}$ for $0 \leq t \leq 1$

26. $\begin{cases} x = 5t - 2 \\ y = -1 - 3t \end{cases}$ for $0 \leq t \leq 1$

27. $\begin{cases} x = t \\ y = 4 - t^2 \end{cases}$ for $-2 \leq t \leq 2$

28. $\begin{cases} x = t - 2 \\ y = 4t - t^2 \end{cases}$ for $0 \leq t \leq 4$

29. $\begin{cases} x = t^2 - 9 \\ y = t \end{cases}$ for $-2 \leq t \leq 3$

30. $\begin{cases} x = t^2 - 6t \\ y = 3 - t \end{cases}$ for $0 \leq t \leq 5$

31. $\begin{cases} x = 5\cos(t) \\ y = 5\sin(t) \end{cases}$ for $0 \leq t < 2\pi$

32. $\begin{cases} x = 1 + 2\cos(t) \\ y = 2\sin(t) \end{cases}$ for $0 \leq t < 2\pi$

33. $\begin{cases} x = 3\cos(t) \\ y = 3 + 3\sin(t) \end{cases}$ for $0 \leq t < 2\pi$

34. $\begin{cases} x = 3\cos(t) \\ y = 3 - 3\sin(t) \end{cases}$ for $0 \leq t < 2\pi$

35. $\begin{cases} x = 3 + \sqrt{117}\cos(t) \\ y = -1 + \sqrt{117}\sin(t) \end{cases}$ for $0 \leq t < 2\pi$

36. $\begin{cases} x = 1 + 3\cos(t) \\ y = \sin(t) \end{cases}$ for $0 \leq t < 2\pi$

37. $\begin{cases} x = 2\cos(t) \\ y = 3\sin(t) - 3 \end{cases}$ for $0 \leq t < 2\pi$

38. $\begin{cases} x = 2\cos\left(t - \frac{\pi}{2}\right) = 2\sin(t) \\ y = -3 - 3\sin\left(t - \frac{\pi}{2}\right) = -3 + 3\cos(t) \end{cases}$ for $0 \leq t < 2\pi$

39. $\{x(t), y(t)$ where:

$$x(t) = \begin{cases} 3t, & 0 \leq t \leq 1 \\ 6 - 3t, & 1 \leq t \leq 2 \\ 0, & 2 \leq t \leq 3 \end{cases} \qquad y(t) = \begin{cases} 0, & 0 \leq t \leq 1 \\ 4t - 4, & 1 \leq t \leq 2 \\ 12 - 4t, & 2 \leq t \leq 3 \end{cases}$$

40. The parametric equations for the inverse are $\begin{cases} x = t^3 + 3t - 4 \\ y = t \end{cases}$ for $-\infty < t < \infty$

41. $r = 6\cos(2\theta)$ translates to $\begin{cases} x = 6\cos(2\theta)\cos(\theta) \\ y = 6\cos(2\theta)\sin(\theta) \end{cases}$ for $0 \leq \theta < 2\pi$.

42. The parametric equations which describe the locations of passengers on the London Eye are $\begin{cases} x = 67.5\cos\left(\frac{\pi}{15}t - \frac{\pi}{2}\right) = 67.5\sin\left(\frac{\pi}{15}t\right) \\ y = 67.5\sin\left(\frac{\pi}{15}t - \frac{\pi}{2}\right) + 67.5 = 67.5 - 67.5\cos\left(\frac{\pi}{15}t\right) \end{cases}$ for $-\infty < t < \infty$

43. The parametric equations for the hammer throw are $\begin{cases} x = 33\cos(42^\circ)t \\ y = -16t^2 + 33\sin(42^\circ)t + 6 \end{cases}$ for $t \geq 0$. To find when the hammer hits the ground, we solve $y(t) = 0$ and get $t \approx -0.23$ or 1.61. Since $t \geq 0$, the hammer hits the ground after approximately $t = 1.61$ seconds after it was launched into the air. To find how far away the hammer hits the ground, we find $x(1.61) \approx 39.48$ feet from where it was thrown into the air.

45. We solve $y = \dfrac{v_0^2\sin^2(\theta)}{2g} + s_0 = \dfrac{v_0^2\sin^2(85^\circ)}{2(32)} + 5 = 31.5$ to get $v_0 = \pm 41.34$. The initial speed of the sheaf was approximately 41.34 feet per second.

Index

n^{th} root
  of a complex number, 1000, 1001
  principal, 397
n^{th} Roots of Unity, 1006
u-substitution, 273
x-axis, 6
x-coordinate, 6
x-intercept, 25
y-axis, 6
y-coordinate, 6
y-intercept, 25

abscissa, 6
absolute value
  definition of, 173
  inequality, 211
  properties of, 173
acidity of a solution
  pH, 432
acute angle, 694
adjoint of a matrix, 622
alkalinity of a solution
  pH, 432
amplitude, 794, 881
angle
  acute, 694
  between two vectors, 1035, 1036
  central angle, 701
  complementary, 696
  coterminal, 698
  decimal degrees, 695
  definition, 693
  degree, 694
  DMS, 695
  initial side, 698
  measurement, 693
  negative, 698
  obtuse, 694
  of declination, 761
  of depression, 761
  of elevation, 753
  of inclination, 753
  oriented, 697
  positive, 698
  quadrantal, 698
  radian measure, 701
  reference, 721
  right, 694
  standard position, 698
  straight, 693
  supplementary, 696
  terminal side, 698
  vertex, 693
angle side opposite pairs, 896
angular frequency, 708
annuity
  annuity-due, 667
  ordinary
    definition of, 666
    future value, 667
applied domain of a function, 60
arccosecant
  calculus friendly
    definition of, 831
    graph of, 830
    properties of, 831

trigonometry friendly
definition of, 828
graph of, 827
properties of, 828
arccosine
definition of, 820
graph of, 819
properties of, 820
arccotangent
definition of, 824
graph of, 824
properties of, 824
arcsecant
calculus friendly
definition of, 831
graph of, 830
properties of, 831
trigonometry friendly
definition of, 828
graph of, 827
properties of, 828
arcsine
definition of, 820
graph of, 820
properties of, 820
arctangent
definition of, 824
graph of, 823
properties of, 824
argument
of a complex number
definition of, 991
properties of, 995
of a function, 55
of a logarithm, 425
of a trigonometric function, 793
arithmetic sequence, 654
associative property
for function composition, 366
matrix
addition, 579
matrix multiplication, 585
scalar multiplication, 581
vector
addition, 1015
scalar multiplication, 1018
asymptote
horizontal
formal definition of, 304
intuitive definition of, 304
location of, 308
of a hyperbola, 531
slant
determination of, 312
formal definition of, 311
slant (oblique), 311
vertical
formal definition of, 304
intuitive definition of, 304
location of, 306
augmented matrix, 568
average angular velocity, 707
average cost, 346
average cost function, 82
average rate of change, 160
average velocity, 706
axis of symmetry, 191

back substitution, 560
bearings, 905
binomial coefficient, 683
Binomial Theorem, 684
Bisection Method, 277
BMI, body mass index, 355
Boyle's Law, 350
buffer solution, 478

cardioid, 951
Cartesian coordinate plane, 6
Cartesian coordinates, 6
Cauchy's Bound, 269
center
of a circle, 498
of a hyperbola, 531
of an ellipse, 516

central angle, 701
change of base formulas, 442
characteristic polynomial, 626
Charles's Law, 355
circle
  center of, 498
  definition of, 498
  from slicing a cone, 495
  radius of, 498
  standard equation, 498
  standard equation, alternate, 519
circular function, 744
cis(θ), 995
coefficient of determination, 226
cofactor, 616
Cofunction Identities, 773
common base, 420
common logarithm, 422
commutative property
  function composition does not have, 366
  matrix
    addition, 579
  vector
    addition, 1015
    dot product, 1034
complementary angles, 696
Complex Factorization Theorem, 290
complex number
  n^{th} root, 1000, 1001
  n^{th} Roots of Unity, 1006
  argument
    definition of, 991
    properties of, 995
  conjugate
    definition of, 288
    properties of, 289
  definition of, 2, 287, 991
  imaginary part, 991
  imaginary unit, i, 287
  modulus
    definition of, 991
    properties of, 993
  polar form
    cis-notation, 995
  principal argument, 991
  real part, 991
  rectangular form, 991
  set of, 2
complex plane, 991
component form of a vector, 1013
composite function
  definition of, 360
  properties of, 367
compound interest, 470
conic sections
  definition, 495
conjugate axis of a hyperbola, 532
conjugate of a complex number
  definition of, 288
  properties of, 289
Conjugate Pairs Theorem, 291
consistent system, 553
constant function
  as a horizontal line, 156
  formal definition of, 101
  intuitive definition of, 100
constant of proportionality, 350
constant term of a polynomial, 236
continuous, 241
continuously compounded interest, 472
contradiction, 549
coordinates
  Cartesian, 6
  polar, 919
  rectangular, 919
correlation coefficient, 226
cosecant
  graph of, 801
  of an angle, 744, 752
  properties of, 802
cosine
  graph of, 791
  of an angle, 717, 730, 744
  properties of, 791

cost
average, 82, 346
fixed, start-up, 82
variable, 159
cost function, 82
cotangent
graph of, 805
of an angle, 744, 752
properties of, 806
coterminal angle, 698
Coulomb's Law, 355
Cramer's Rule, 619
curve
orientated, 1048
cycloid, 1056

decibel, 431
decimal degrees, 695
decreasing function
formal definition of, 101
intuitive definition of, 100
degree measure, 694
degree of a polynomial, 236
DeMoivre's Theorem, 997
dependent system, 554
dependent variable, 55
depreciation, 420
Descartes' Rule of Signs, 273
determinant of a matrix
definition of, 614
properties of, 616
Difference Identity
for cosine, 771, 775
for sine, 773, 775
for tangent, 775
difference quotient, 79
dimension
of a matrix, 567
direct variation, 350
directrix
of a conic section in polar form, 981
of a parabola, 505
discriminant
of a conic, 979
of a quadratic equation, 195
trichotomy, 195
distance
definition, 10
distance formula, 11
distributive property
matrix
matrix multiplication, 585
scalar multiplication, 581
vector
dot product, 1034
scalar multiplication, 1018
DMS, 695
domain
applied, 60
definition of, 45
implied, 58
dot product
commutative property of, 1034
definition of, 1034
distributive property of, 1034
geometric interpretation, 1035
properties of, 1034
relation to orthogonality, 1037
relation to vector magnitude, 1034
work, 1042
Double Angle Identities, 776

earthquake
Richter Scale, 431
eccentricity, 522, 981
eigenvalue, 626
eigenvector, 626
ellipse
center, 516
definition of, 516
eccentricity, 522
foci, 516
from slicing a cone, 496
guide rectangle, 519
major axis, 516
minor axis, 516

reflective property, 523
standard equation, 519
vertices, 516
ellipsis (...), 31, 651
empty set, 2
end behavior
of $f(x) = ax^n, n$ even, 240
of $f(x) = ax^n, n$ odd, 240
of a function graph, 239
polynomial, 243
entry
in a matrix, 567
equation
contradiction, 549
graph of, 23
identity, 549
linear of n variables, 554
linear of two variables, 549
even function, 95
Even/Odd Identities, 770
exponential function
algebraic properties of, 437
change of base formula, 442
common base, 420
definition of, 418
graphical properties of, 419
inverse properties of, 437
natural base, 420
one-to-one properties of, 437
solving equations with, 448
extended interval notation, 756

Factor Theorem, 258
factorial, 654, 681
fixed cost, 82
focal diameter of a parabola, 507
focal length of a parabola, 506
focus
of a conic section in polar form, 981
focus (foci)
of a hyperbola, 531
of a parabola, 505
of an ellipse, 516
free variable, 552
frequency
angular, 708, 881
of a sinusoid, 795
ordinary, 708, 881
function
(absolute) maximum, 101
(absolute, global) minimum, 101
absolute value, 173
algebraic, 399
argument, 55
arithmetic, 76
as a process, 55, 378
average cost, 82
circular, 744
composite
definition of, 360
properties of, 367
constant, 100, 156
continuous, 241
cost, 82
decreasing, 100
definition as a relation, 43
dependent variable of, 55
difference, 76
difference quotient, 79
domain, 45
even, 95
exponential, 418
Fundamental Graphing Principle, 93
identity, 168
increasing, 100
independent variable of, 55
inverse
definition of, 379
properties of, 379
solving for, 384
uniqueness of, 380
linear, 156
local (relative) maximum, 101
local (relative) minimum, 101
logarithmic, 422

notation, 55
odd, 95
one-to-one, 381
periodic, 790
piecewise-defined, 62
polynomial, 235
price-demand, 82
product, 76
profit, 82
quadratic, 188
quotient, 76
range, 45
rational, 301
revenue, 82
smooth, 241
sum, 76
transformation of graphs, 120, 135
zero, 95
fundamental cycle
of $y = \cos(x)$, 791
Fundamental Graphing Principle
for equations, 23
for functions, 93
for polar equations, 938
Fundamental Theorem of Algebra, 290

Gauss-Jordan Elimination, 571
Gaussian Elimination, 557
geometric sequence, 654
geometric series, 669
graph
hole in, 305
horizontal scaling, 132
horizontal shift, 123
of a function, 93
of a relation, 20
of an equation, 23
rational function, 321
reflection about an axis, 126
transformations, 135
vertical scaling, 130
vertical shift, 121
greatest integer function, 67
growth model
limited, 475
logistic, 475
uninhibited, 472
guide rectangle
for a hyperbola, 532
for an ellipse, 519

Half-Angle Formulas, 779
harmonic motion, 885
Henderson-Hasselbalch Equation, 446
Heron's Formula, 914
hole
in a graph, 305
location of, 306
Hooke's Law, 350
horizontal asymptote
formal definition of, 304
intuitive definition of, 304
location of, 308
horizontal line, 23
Horizontal Line Test (HLT), 381
hyperbola
asymptotes, 531
branch, 531
center, 531
conjugate axis, 532
definition of, 531
foci, 531
from slicing a cone, 496
guide rectangle, 532
standard equation
horizontal, 534
vertical, 534
transverse axis, 531
vertices, 531
hyperbolic cosine, 1062
hyperbolic sine, 1062
hyperboloid, 542

identity
function, 367
matrix, additive, 579

matrix, multiplicative, 585
statement which is always true, 549
imaginary axis, 991
imaginary part of a complex number, 991
imaginary unit, i, 287
implied domain of a function, 58
inconsistent system, 553
increasing function
formal definition of, 101
intuitive definition of, 100
independent system, 554
independent variable, 55
index of a root, 397
induction
base step, 673
induction hypothesis, 673
inductive step, 673
inequality
absolute value, 211
graphical interpretation, 209
non-linear, 643
quadratic, 215
sign diagram, 214
inflection point, 477
information entropy, 477
initial side of an angle, 698
instantaneous rate of change, 161, 472, 707
integer
definition of, 2
greatest integer function, 67
set of, 2
intercept
definition of, 25
location of, 25
interest
compound, 470
compounded continuously, 472
simple, 469
Intermediate Value Theorem
polynomial zero version, 241
interrobang, 321
intersection of two sets, 4
interval
definition of, 3
notation for, 3
notation, extended, 756
inverse
matrix, additive, 579, 581
matrix, multiplicative, 602
of a function
definition of, 379
properties of, 379
solving for, 384
uniqueness of, 380
inverse variation, 350
invertibility
function, 382
invertible
function, 379
matrix, 602
irrational number
definition of, 2
set of, 2
irreducible quadratic, 291

joint variation, 350

Kepler's Third Law of Planetary Motion, 355
Kirchhoff's Voltage Law, 605

latus rectum of a parabola, 507
Law of Cosines, 910
Law of Sines, 897
leading coefficient of a polynomial, 236
leading term of a polynomial, 236
Learning Curve Equation, 315
least squares regression line, 225
lemniscate, 950
limaçon, 950
line
horizontal, 23
least squares regression, 225
linear function, 156
of best fit, 225
parallel, 166

perpendicular, 167
point-slope form, 155
slope of, 151
slope-intercept form, 155
vertical, 23
linear equation
n variables, 554
two variables, 549
linear function, 156
local maximum
formal definition of, 102
intuitive definition of, 101
local minimum
formal definition of, 102
intuitive definition of, 101
logarithm
algebraic properties of, 438
change of base formula, 442
common, 422
general, "base b", 422
graphical properties of, 423
inverse properties of, 437
natural, 422
one-to-one properties of, 437
solving equations with, 459
logarithmic scales, 431
logistic growth, 475
LORAN, 538
lower triangular matrix, 593

main diagonal, 585
major axis of an ellipse, 516
Markov Chain, 592
mathematical model, 60
matrix
addition
associative property, 579
commutative property, 579
definition of, 578
properties of, 579
additive identity, 579
additive inverse, 579
adjoint, 622
augmented, 568
characteristic polynomial, 626
cofactor, 616
definition, 567
determinant
definition of, 614
properties of, 616
dimension, 567
entry, 567
equality, 578
invertible, 602
leading entry, 569
lower triangular, 593
main diagonal, 585
matrix multiplication
associative property of, 585
definition of, 584
distributive property, 585
identity for, 585
properties of, 585
minor, 616
multiplicative inverse, 602
product of row and column, 584
reduced row echelon form, 570
rotation, 986
row echelon form, 569
row operations, 568
scalar multiplication
associative property of, 581
definition of, 580
distributive properties, 581
identity for, 581
properties of, 581
zero product property, 581
size, 567
square matrix, 586
sum, 578
upper triangular, 593
maximum
formal definition of, 102
intuitive definition of, 101
measure of an angle, 693

midpoint
  definition of, 12
  midpoint formula, 13
minimum
  formal definition of, 102
  intuitive definition of, 101
minor, 616
minor axis of an ellipse, 516
model
  mathematical, 60
modulus of a complex number
  definition of, 991
  properties of, 993
multiplicity
  effect on the graph of a polynomial, 245, 249
  of a zero, 244

natural base, 420
natural logarithm, 422
natural number
  definition of, 2
  set of, 2
negative angle, 698
Newton's Law of Cooling, 421, 474
Newton's Law of Universal Gravitation, 351

oblique asymptote, 311
obtuse angle, 694
odd function, 95
Ohm's Law, 350, 605
one-to-one function, 381
ordered pair, 6
ordinary frequency, 708
ordinate, 6
orientation, 1048
oriented angle, 697
oriented arc, 704
origin, 7
orthogonal projection, 1038
orthogonal vectors, 1037
overdetermined system, 554

parabola
  axis of symmetry, 191
  definition of, 505
  directrix, 505
  focal diameter, 507
  focal length, 506
  focus, 505
  from slicing a cone, 496
  graph of a quadratic function, 188
  latus rectum, 507
  reflective property, 510
  standard equation
    horizontal, 508
    vertical, 506
  vertex, 188, 505
  vertex formulas, 194
paraboloid, 510
parallel vectors, 1030
parameter, 1048
parametric equations, 1048
parametric solution, 552
parametrization, 1048
partial fractions, 628
Pascal's Triangle, 688
password strength, 477
period
  circular motion, 708
  of a function, 790
  of a sinusoid, 881
periodic function, 790
pH, 432
phase, 795, 881
phase shift, 795, 881
pi, π, 700
piecewise-defined function, 62
point of diminishing returns, 477
point-slope form of a line, 155
polar coordinates
  conversion into rectangular, 924
  definition of, 919
  equivalent representations of, 923
  polar axis, 919
  pole, 919

polar form of a complex number, 995
polar rose, 950
polynomial division
  dividend, 258
  divisor, 258
  factor, 258
  quotient, 258
  remainder, 258
  synthetic division, 260
polynomial function
  completely factored
    over the complex numbers, 291
    over the real numbers, 291
  constant term, 236
  definition of, 235
  degree, 236
  end behavior, 239
  leading coefficient, 236
  leading term, 236
  variations in sign, 273
  zero
    lower bound, 274
    multiplicity, 244
    upper bound, 274
positive angle, 698
Power Reduction Formulas, 778
power rule
  for absolute value, 173
  for complex numbers, 997
  for exponential functions, 437
  for logarithms, 438
  for radicals, 398
  for the modulus of a complex number, 993
price-demand function, 82
principal, 469
principal n^{th} root, 397
principal argument of a complex number, 991
principal unit vectors, $\hat{\imath}$, $\hat{\jmath}$, 1024
Principle of Mathematical Induction, 673
product rule
  for absolute value, 173
  for complex numbers, 997
  for exponential functions, 437
  for logarithms, 438
  for radicals, 398
  for the modulus of a complex number, 993
Product to Sum Formulas, 780
profit function, 82
projection
  $x-$axis, 45
  $y-$axis, 46
  orthogonal, 1038
Pythagorean Conjugates, 751
Pythagorean Identities, 749

quadrantal angle, 698
quadrants, 8
quadratic formula, 194
quadratic function
  definition of, 188
  general form, 190
  inequality, 215
  irreducible quadratic, 291
  standard form, 190
quadratic regression, 228
Quotient Identities, 745
quotient rule
  for absolute value, 173
  for complex numbers, 997
  for exponential functions, 437
  for logarithms, 438
  for radicals, 398
  for the modulus of a complex number, 993

radian measure, 701
radical
  properties of, 398
radicand, 397
radioactive decay, 473
radius
  of a circle, 498
range
  definition of, 45
rate of change
  average, 160

instantaneous, 161, 472
slope of a line, 154
rational exponent, 398
rational functions, 301
rational number
definition of, 2
set of, 2
Rational Zeros Theorem, 269
ray
definition of, 693
initial point, 693
real axis, 991
Real Factorization Theorem, 292
real number
definition of, 2
set of, 2
real part of a complex number, 991
Reciprocal Identities, 745
rectangular coordinates
also known as Cartesian coordinates, 919
conversion into polar, 924
rectangular form of a complex number, 991
recursion equation, 654
reduced row echelon form, 570
reference angle, 721
Reference Angle Theorem
for cosine and sine, 722
for the circular functions, 747
reflection
of a function graph, 126
of a point, 10
regression
coefficient of determination, 226
correlation coefficient, 226
least squares line, 225
quadratic, 228
total squared error, 225
relation
algebraic description, 23
definition, 20
Fundamental Graphing Principle, 23
Remainder Theorem, 258
revenue function, 82
Richter Scale, 431
right angle, 694
root
index, 397
radicand, 397
Roots of Unity, 1006
rotation matrix, 986
rotation of axes, 974
row echelon form, 569
row operations for a matrix, 568

scalar multiplication
matrix
associative property of, 581
definition of, 580
distributive properties of, 581
properties of, 581
vector
associative property of, 1018
definition of, 1017
distributive properties of, 1018
properties of, 1018
scalar projection, 1039
secant
graph of, 800
of an angle, 744, 752
properties of, 802
secant line, 160
sequence
n^{th} term, 652
alternating, 652
arithmetic
common difference, 654
definition of, 654
formula for n^{th} term, 656
sum of first n terms, 666
definition of, 652
geometric
common ratio, 654
definition of, 654
formula for n^{th} term, 656
sum of first n terms, 666

recursive, 654
series, 668
set
definition of, 1
empty, 2
intersection, 4
roster method, 1
set-builder notation, 1
sets of numbers, 2
union, 4
verbal description, 1
set-builder notation, 1
Side-Angle-Side triangle, 910
Side-Side-Side triangle, 910
sign diagram
algebraic function, 399
for quadratic inequality, 214
polynomial function, 242
rational function, 321
simple interest, 469
sine
graph of, 792
of an angle, 717, 730, 744
properties of, 791
sinusoid
amplitude, 794, 881
baseline, 881
frequency
angular, 881
ordinary, 881
graph of, 795, 882
period, 881
phase, 881
phase shift, 795, 881
properties of, 881
vertical shift, 881
slant asymptote, 311
slant asymptote
determination of, 312
formal definition of, 311
slope
definition, 151
of a line, 151
rate of change, 154
slope-intercept form of a line, 155
smooth, 241
sound intensity level
decibel, 431
square matrix, 586
standard position of a vector, 1019
standard position of an angle, 698
start-up cost, 82
steady state, 592
stochastic process, 592
straight angle, 693
Sum Identity
for cosine, 771, 775
for sine, 773, 775
for tangent, 775
Sum to Product Formulas, 781
summation notation
definition of, 661
index of summation, 661
lower limit of summation, 661
properties of, 664
upper limit of summation, 661
supplementary angles, 696
symmetry
about the x-axis, 9
about the y-axis, 9
about the origin, 9
testing a function graph for, 95
testing an equation for, 26
synthetic division tableau, 260
system of equations
back-substitution, 560
coefficient matrix, 590
consistent, 553
constant matrix, 590
definition, 549
dependent, 554
free variable, 552
Gauss-Jordan Elimination, 571
Gaussian Elimination, 557

inconsistent, 553
independent, 554
leading variable, 556
linear
n variables, 554
two variables, 550
linear in form, 646
non-linear, 637
overdetermined, 554
parametric solution, 552
triangular form, 556
underdetermined, 554
unknowns matrix, 590

tangent
graph of, 804
of an angle, 744, 752
properties of, 806
terminal side of an angle, 698
Thurstone, Louis Leon, 315
total squared error, 225
transformation
non-rigid, 129
rigid, 129
transformations of function graphs, 120, 135
transverse axis of a hyperbola, 531
Triangle Inequality, 183
triangular form, 556

underdetermined system, 554
uninhibited growth, 472
union of two sets, 4
Unit Circle
definition of, 501
important points, 724
unit vector, 1023
Upper and Lower Bounds Theorem, 274
upper triangular matrix, 593

variable
dependent, 55
independent, 55
variable cost, 159
variation
constant of proportionality, 350
direct, 350
inverse, 350
joint, 350
variations in sign, 273
vector
x-component, 1012
y-component, 1012
addition
associative property, 1015
commutative property, 1015
definition of, 1014
properties of, 1015
additive identity, 1015
additive inverse, 1015, 1018
angle between two, 1035, 1036
component form, 1012
Decomposition Theorem
Generalized, 1040
Principal, 1024
definition of, 1012
direction
definition of, 1020
properties of, 1020
dot product
commutative property of, 1034
definition of, 1034
distributive property of, 1034
geometric interpretation, 1035
properties of, 1034
relation to magnitude, 1034
relation to orthogonality, 1037
work, 1042
head, 1012
initial point, 1012
magnitude
definition of, 1020
properties of, 1020
relation to dot product, 1034
normalization, 1024
orthogonal projection, 1038

orthogonal vectors, 1037
parallel, 1030
principal unit vectors, $\hat{\imath}$, $\hat{\jmath}$, 1024
resultant, 1013
scalar multiplication
associative property of, 1018
definition of, 1017
distributive properties, 1018
identity for, 1018
properties of, 1018
zero product property, 1018
scalar product
definition of, 1034
properties of, 1034
scalar projection, 1039
standard position, 1019
tail, 1012
terminal point, 1012
triangle inequality, 1044
unit vector, 1023
velocity
average angular, 707
instantaneous, 707
instantaneous angular, 707
vertex
of a hyperbola, 531
of a parabola, 188, 505
of an angle, 693
of an ellipse, 516
vertical asymptote
formal definition of, 304
intuitive definition of, 304
location of, 306
vertical line, 23
Vertical Line Test (VLT), 43

whole number
definition of, 2
set of, 2
work, 1041
wrapping function, 704

zero
multiplicity of, 244
of a function, 95
upper and lower bounds, 274

Made in the USA
Monee, IL
14 September 2026

62974531R00219